Energy Performance of Buildi

Sofia-Natalia Boemi · Olatz Irulegi
Mattheos Santamouris
Editors

Energy Performance of Buildings

Energy Efficiency and Built Environment in Temperate Climates

Springer

Editors
Sofia-Natalia Boemi
Department of Mechanical Engineering, Process Equipment Design Laboratory
Aristotle University Thessaloniki
Thessaloniki
Greece

Olatz Irulegi
University of the Basque Country
San Sebastián
Spain

Mattheos Santamouris
Physics Department, Group Building Environmental Research Athens
National and Kapodistrian University of Athens
Athens
Greece

ISBN 978-3-319-79319-1 ISBN 978-3-319-20831-2 (eBook)
DOI 10.1007/978-3-319-20831-2

Springer Cham Heidelberg New York Dordrecht London

Softcover re-print of the Hardcover 1st edition 2016

Printed on acid-free paper

Springer International Publishing AG Switzerland is part of Springer Science+Business Media (www.springer.com)

Preface

Ever since the oil crisis in 1973, it has been understood that a large amount of the energy consumed in buildings for their heating, cooling, and lighting is directly linked to the way in which the buildings are designed. Different categories of buildings have different energy needs. But at the same time, energy consumption of buildings for cooling and heating needs is dictated by the climate, the type of building, and the equipment that has been installed. In addition, with recent climactic changes, especially the constant temperature increases, which affect the built environment, the need to record the actual situation and promote good practices becomes imperative.

At the same time, it was realized that a densely built urban environment creates a microclimate on its own, affecting energy balance. It is obvious that without a correct interpretation of climatic, geographic, and location parameters, meeting the goals in a project *a posteriori* would be very difficult. An improved architectural approach improves both energy efficiency and indoor environmental quality and, consequently, the quality of life of the inhabitants. Also, a series of technologies has been incorporated in building design, utilizing solar energy in order to achieve lighting, heating, and cooling, with minimal conventional energy consumption.

In that sense, this book gathers all available information on energy efficiency in the built environment in areas with similar climatic conditions, to southern European countries. Notably, it tries to cover a gap by presenting concentrated information of the most important building sectors: residential, commercial, healthcare, and educational. Emphasis is on the existing building stock because improving its energy performance and using renewables is crucial not only for achieving the EU's 2020 targets, but also for meeting long-term objectives of climate and energy strategies. Also, it presents an overview of the development of energy technology, analyzing the trends and of systems used to decrease demand, as well as examining strategies for energy-saving, evolving, and renewable energies.

The goals of this book are to clarify the present trade-offs inherent in defining sustainability, to study technology and technology-intensive options, and to provide a framework for assessing decision-making. These goals are of key importance, as through the development and implementation of effective energy conservation policies, the success and the need for further energy conservation in the building sector, as well as the importance of the implementation of renewables, are addressed. Special attention is given to the growing debate about the impact of climate change and internal temperatures in buildings—one of the major reasons for fuel poverty and therefore an increase in mortality.

The book is, in fact, a handbook on energy efficiency for readers wanting to understand the performance of different types of buildings. Current regulations in each country and expected mid-term trends provide important context. It includes four different parts with 25 chapters—all contributions from highly recognized experts—covering most of the countries with temperate climates, such as Spain, France, Italy, Greece, and Cyprus. But experts from other countries, such as Ireland and Austria, also contributed to the book, presenting the most promising developments in their fields of expertise.

This volume is organized as a handbook that provides information not only for the scientific world but also for the wider public, which needs to know about energy efficiency in order to discuss and brainstorm among themselves about how the transition to a sustainable energy future can be achieved.

Therefore, Part I focuses on the challenges and priorities for a sustainable built environment. It presents seven chapters on energy and the built environment, its policies, and how building performance is affected by climate change. In addition, users' behavior is analyzed and evaluated in order to understand how it affects building performance and, in reverse, how building performance affects employment. Part II describes the actual performance of building stock and provides data on construction market activity, performance levels of recently built or renovated buildings, and steps beyond sustainability for various buildings. Part III analyzes the maturity, reliability, efficiency, cost, and market availability of technologies that decrease the demand for energy for building supply. Last but not least, Part IV describes tools and strategies for microclimatic analysis of the built environment.

In general, energy-intensive services and "luxuries" are ignored and factors that impact the wider public, such as the use of solar energy, are introduced. From that point of view, the wider objective is to provide readers with the background and methodologies to develop their own conceptualization of energy efficiency and a possible roadmap towards Net Zero Energy Buildings (nZEB).

Contents

Chapter 1
The Built Environment and Its Policies

Eduardo de Oliveira Fernandes

Abstract The built environment with its vast scale and innumerable opportunities for local action is a key element in the pursuit of sustainability, and more so in the advancement of the quintessentially environment-friendly strategies of sufficiency. If pressure on the global environment from energy use is due more to the nature of the source than to the amount of primary energy used, the built environment as an energy system can rely largely on benign passive energy means well adapted to local climate and resources. An architecture of complexity that eschews universal solutions must thus emerge in the future. The institutional framework that regulates buildings and urbanism must be updated and refined to promote rather than hinder this progression of the built environment towards environmental sufficiency.

1.1 Buildings Throughout Time

The first "buildings" made by man were nothing more than temporary, makeshift structures affording some degree of protection against the extremes of the climate, wild animals, and other humans. Primitive man was no different from other animals in exploiting whatever protection natural vegetation and the orography could provide; however, as intelligence allowed humanity to gradually extend its territorial range, domesticating plants and animals and increasing in numbers, it was no longer possible to depend on the occasional cave or tree, and people learned to erect artificial shelters using plant fibre, the animal hides, mud, stone, or any other material obtainable from the surrounding environment.

Man, as a gregarious species, often builds shelters in close proximity. These agglomerations constituted the first human settlements, the organization of which

E. de Oliveira Fernandes (✉)
University of Porto, Porto, Portugal
e-mail: eof@fe.up.pt

S.-N. Boemi et al. (eds.), *Energy Performance of Buildings*,
DOI 10.1007/978-3-319-20831-2_1

was never arbitrary. Kinships, division of labor, control of resources, religious or other institutions, defense against outside attack, and access to the surrounding land that provided the means of subsistence through fishing, hunting and cultivation, all affected the shape and structure of early human settlements. Furthermore, early shelters and settlements in most cases reveal some use of form and disposition to achieve a degree of control of exposure to the sun and to the winds, leading to more pleasant living conditions and to better preservation of foodstuffs among other desirable ends.

This constant "dialogue" between the environments built by man and the natural environment, in time came to be infused with deep cultural values where form, technology, materials, climate, geography, geology and living styles are all intimately interlinked. A large spectrum of building and construction typologies, well adapted to each climate, place and culture, arose around the world, forming a rich patrimony commonly referred to as "vernacular architecture."

The continuous growth of modern urbanization, especially after the beginning of industrialization, introduced new environmental, landscape, and physical parameters that changed the typical urban microclimates, due to new physical forms and to new land uses as well as novel demands for space, mobility, privacy, and socialization.

As a consequence, the modern city tends to challenge and somehow subvert the values from the past. Most buildings erected today lack strong roots in place; they could just as well be located anywhere else on the planet, they use materials sourced worldwide, and even their architecture fails to express an awareness of local climate or solar orientation. An acritical mimicry of extraneous styles, an erroneous belief that any shortcomings of the building envelope can be made up for with artificial means of indoor environmental control, a willful indifference to the concomitant environmental costs and a craven expediency in design, construction, and contracting, lead to this bland kind of building.

The issue, however, should not be reduced to denunciation of such buildings as a class. It is necessary to understand the forces that lead to their appearance. In particular, we must understand how modern urbanization has substituted location for locality. All the rich, "contextual" relationships of the past are deprecated and replaced by locational value, where centrality has a precise commercial translation in terms of economic opportunity and ease of access, with distance as a surrogate for time. This denatured modern understanding of location creates a self-reinforcing demand for huge cities and for high-rise buildings in downtown areas, stimulated as well by other motivations not strictly related to economics. These business districts, along with large-scale airport and seaport complexes and vast industrial districts, have their reason to exist in today's society, driven as it is by economic growth in a globalized economy where personal fulfillment is measured by the consumption of goods, including travel and novelties coming from all over the world.

Yet most families around the world live in homes and neighborhoods, the bleakness of which seems to forbid even dreams of happiness. Should these populations aspire to nothing beyond a dystopia of energy-intensive heating, ventilating and air conditioning (HVAC) systems blowing conditioned air at pre-set temperature and moisture levels?

Countless millions are denied adequate shelter too, lacking space, security, privacy, and even dignity, as a consequence of broader social and economic failures in employment, organization, and public safety policies. Such failings are structural and cannot be answered by quick fixes, just as the problem of sustainable and affordable housing for the twenty-first century cannot be solved with lightweight "drop-in"-type concepts.

1.2 Energy in Buildings: From Sufficiency to Efficiency

Buildings constitute one of the larger economic sectors both in terms of value-added[1] and overall energy use—even without taking into account the "free" solar energy passively captured by buildings. Given the immense scale of the building sector, relatively small interventions in the existing building stock around the world do lead to cumulatively large consequences in terms of CO_2 emissions, impacting the overall climate, and inducing rebound effects that further reinforce the need to tackle buildings as true energy systems.[2]

Life in the twenty-first century is increasingly driven by consumerism and by economic short-termism. The consequent dearth of collective will and vision to tackle societies' gravest and most long-term problems is therefore sad—but not at all surprising. One of the most glaring examples of this is the failure to promote and enforce appropriate criteria for the use of natural resources in general, and of natural renewable energy sources in particular. Regarding the latter, the consumption of energy of fossil origin has reached such levels worldwide that according to the Commission of the United Nations for Climate Change, it is already the major driver of global warming.[3]

It is relevant to recall that the pressure on the global environment from energy use is related more to the nature of the energy sources than to the amount of the primary energy used. The dominant sources of energy on which our modern lifestyles have been based are the fossil fuels (oil, coal and natural gas), plus two other sources for electricity production that fit the top-down centralized paradigm favored by the hegemons that have been dominating the electricity business: large hydro and nuclear.

Electricity is arguably the most significant supply-side energy vector in buildings. It is a convenient final energy form for many services, including artificial lighting, entertainment, and many uses in the kitchen due to its high exergy level. It is easily converted into heat, either with inefficient resistance-based appliances (the "Joule effect") or with the more sophisticated technology of heat pumps and air conditioners.

---

[1]Construction and real estate constituted the 3rd EU sector per value added in 2012; construction per se occupied the 5th place (Eurostat 2012).

[2]e.g., Fig. 1, Modernising Building Energy Codes. IEA, Paris, 2013, p. 14.

[3]Section 1.2 and Fig. SPM.3—Climate Change 2014: Synthesis Report. IPCC, Geneva.

Electricity today is mostly generated from heat generated by burning fossil fuels or by nuclear reactions. The efficiency of this conversion is low—only about 30 %—with the remaining 70 % released into the environment as thermal pollution. To this must be added the chemical and physical pollution from the combustion effluents (gases and particulate matter) and the stockpiles of nuclear waste.

Recent advances in technology promise to change this centralized and wasteful paradigm. Photovoltaic and other solar active technologies will soon be able to satisfy a significant share of the electricity needs in a clean and decentralized fashion, while the modern resources of building simulation and control make it easier to explore the full potential of building design for daylighting, passive cooling, and passive heating (thermal management) through solar shading, collection, distribution, diffusion, and storage.

All this potential inherent in building is still in abeyance, pending wide-scale, systematic implementation of building-related solar technologies for light, heat, and electricity. The best way forward is to explore the concept of buildings as energy systems in themselves, which necessarily has the passive behavior of buildings as its starting point. By playing with buildings as energy systems per se, the time has arrived to leverage all the progress in building science in general, and in the energy technologies applied to buildings in particular, to raise the profile of buildings as relevant actors in the energy arena.

This *desideratum* is still far from fulfilled, yet we can already perceive an excessive and misguided emphasis on the idea of efficiency as the holy grail of environmentally friendly buildings. Efficiency is key in all stages of energy chains originating in conventional sources (fossil fuels, nuclear, natural gas, etc.). The concept of efficiency is, however, only strictly applicable to situations that require obligate "technological" energy systems (whether already in use or planned). Beyond these situations, the concept of efficiency is misleading. Whenever there is an elective aspect, energy efficiency is superseded by a higher-order concept that is very relevant to buildings and cities: energy sufficiency.

Sufficiency means that, for instance, the design of a building can pre-empt part or all of the energy needs for comfort. If the pre-empted fraction is less than 100 %, "add-on" systems are required for the missing fraction, and energy-efficiency concerns are applicable to the latter.

If buildings are reckoned as energy systems in themselves i.e., able to capture, store, and diffuse the energy from the environment, namely from the sun, they are also capable of fulfilling users' comfort requirements with little need for "add-on" energy systems. Then energy sufficiency is enough.

Furthermore, it is a tautology to speak of energy efficiency in this situation, because with renewable energy (especially in the direct satisfaction of end uses), it is meaningless to speak of energy losses. Ambient energy not captured by the application is not "lost", nor does it become a factor of environmental pressure—it simply devolves to the environment that it came from.

If we recall the history of human settlements, one of the first criteria in choosing a location was the availability of resources related to the actual needs of the settlers: water for drinking, cooking, hygiene and cultivation; watercourses for navigation and for transport; wood for cooking and heating; and access to the sun for a more comfortable household. There are many expressions of this wisdom throughout the ages. where the exploration of resources led to effective, although very simple technologies.

The essence of these technologies is their integrated nature. The heavy masonry walls that characterize the vernacular architectures of the Mediterranean basin, for example, provide high thermal inertia and some thermal insulation, but in most cases cannot ensure complete thermal comfort by themselves. They need to be coupled with good building orientation, small windows, sensible urban configurations, such as courtyards and arcades, and judicious use of gardens, fountains and shading; if these are matched with patterns of use, such as opening the windows and spending time outdoors when the outdoor temperatures are pleasant, the end result is living environments that remain unmatched in their delightfulness.

In contrast, the modernization of our societies and the free circulation of all kinds of marketable products, including information, have produced a homogenized built environment where this deep adaptation to time, place, and territory has been lost. We have reached the regrettable situation where the circumambient resources go unused while buildings devour energy supplied by fuels picked up thousands of miles away and costing millions of tons of pollutants.

The calls for action on climate change should make it clear that today—even more than in the past—the design of new dwellings and refurbishment of existing ones, can no longer ignore the different climates, the interdependence of the criteria for comfort, health, wellbeing and the values of sustainability, the most important of which is the containment of CO_2 emissions.

1.3 Requirements for Future Buildings

Buildings as energy systems must respond to the functions and purposes assigned to them, bearing in mind local climatic conditions and ensuring healthy and comfortable conditions for all users, avoiding undue dependency on "external" energy (Fig. 1.1).

Here emerges the need to promote, in particular for the case of the Mediterranean basin, an exercise leading to a rebirth of new building criteria set in the twenty-first century for residential buildings and their attributes, starting from the basic question: up to where must the design, i.e., the architecture and construction of buildings, be required and expected to identify and address relevant societal and environmental values while being likely to fulfill all the typically required attributes of security and constraints of urbanism.

Future buildings, in continuity with the past, will have to provide for a wide spectrum of uses and typologies, from single family housing to offices, malls, shops, museums, and other public buildings. Besides the specific prerequisites, such as for a hospital operating room, a museum, or a school, buildings have to respect basic structural criteria, i.e., the stability of the structure, the durability of

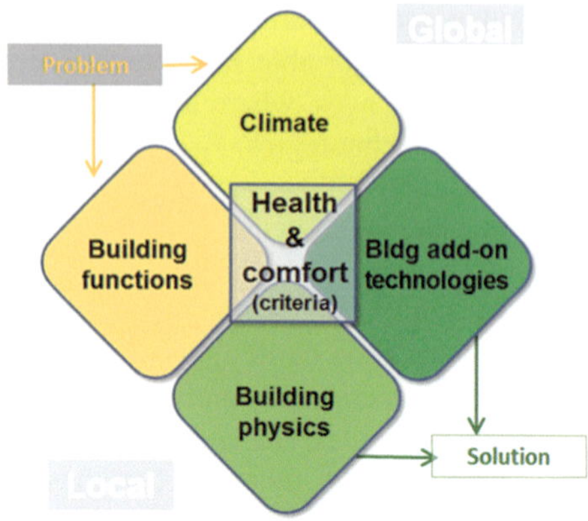

Fig. 1.1 A building exists to fulfill various functions. Within a given climate, the potential of building physics can be explored to reach a local solution that contributes to global sustainability. Only when the potential of building physics has been exhausted will "add-on" technologies be called for

the construction, the impact upon the security of the whole neighborhood in terms of spatial organization and of the activities housed, etc. The building must also ensure urban and architectural coherence and give due consideration to the specific climatic context; it must comply with local, regional and national environmental impact rules in terms of landscape, accessibility, and resources used, among other issues—with solar rights becoming more highly valued.

The planning of future buildings must start with the identification of functional priorities (Fig. 1.1) that follow from past experience and ongoing research, expressed in objective indicators—such as occupational density or its equivalent—regarding products or indoor equipment.

The design process, whatever the level of ambition regarding the architectural expression and cultural impact, will then be organized, bearing in mind the environment *latu* sensu, including the local climate, the urban surroundings (noise and urban/traffic air pollution) and other boundary conditions. This concern with the local environment must be complemented with an equal concern with the global environment, namely, as the requirements associated with the impact on the global climate, primarily from the emissions of CO_2 but also from the use, for instance, of products extracted from tropical forests.

This scrutiny of all factors and impacts of the future building must not, however, supersede the ultimate building criterion: the fulfillment of the predefined functions that will guarantee comfort and health conditions for all users. It is well known that chronic illnesses, such as asthma and other respiratory illnesses are related to indoor air pollution sources, including paints and glues, furniture and cleaning products, and the air-handling components of HVAC systems, in addition to outdoor pollutants that enter the building with the ventilation air. Comfort and health must thus be tackled together, prioritizing the principles and strategies of source control. Architecture and building physics will be the first line of attack in achieving these objectives. The role of architects in environmental design is stressed in some countries more than in others but, generally speaking, the contribution of the relevant engineering specialities is required, hopefully integrated in multidisciplinary teams.

Here enters the concept of sufficiency in giving the particular building the ability to catch and manage ambient energy for the purposes of comfort (heating/cooling/ventilation), resorting to external insulation of walls, thermal inertia of the internal walls, orientation and sizing of openings, shading of the glazed surfaces, variable apertures for natural ventilation, free cooling, etc. The extent to which the building is energy-sufficient will define the environmental impact associated with its energy needs for comfort and health. A separate issue is the set of functions that will take place within the building and that require energy systems. Appropriate, efficient, and environmentally sound systems, respecting global sustainability criteria, must in any case be specified.[4]

What is stated above is applicable to all building types. Residential buildings in particular have requirements that are more flexible in purely functional terms, but that can be more demanding and specific in terms of adaptation to each climate.

A word of caution may be added here regarding the present fashion in EU circles to overvalue the consequences of climate change in the next half-century in terms of the urban thermal environment. The expected rise in global temperatures will be disastrous in some places, due to issues such as flooding, but the temperature changes will have little or no discernible meaning from the point of view of the average costs for heating and cooling buildings in Europe. The expected changes are dwarfed by the climatic differences that today exist between the various EU regions and by microclimatic factors, such as "urban heat islands." It would be much better for the EU to re-evaluate its policies to give due respect to climatic differences. In particular, there is a need to reassess some aspects of the methodology for quantification of energy performance as applied to southern Europe, as the current system incentivizes the inclusion of active energy systems (air conditioning) in residential buildings in those parts of Europe where the climate, the building traditions, and the lifestyle habits favor passive comfort. While this is not the intention of the European policies for energy in buildings, it is too often its unexpected practical outcome. Heating, air conditioning, or airtight envelopes with heat exchangers meting out precisely controlled airflows are technologies with a place in Europe, but the manner in which European policy has been implemented in practice is extending their application in contexts where they are unnecessary and even pernicious "novelties."

1.4 Sustainable Buildings

Future buildings must be affordable in terms of initial investment as well as recurring costs of operation, maintenance, etc. Cutting corners at the expense of comfort, health, functionality, or durability is a false economy, as will be readily

[4]For a schematic of strategies and policies for energy sufficiency relevant to building design see e.g., Fig. 10, *Modernising Building Energy Codes*. IEA, Paris, 2013, p. 28.

revealed by adopting life-cycle frames of reference. This is even truer for sustainability, with its uniquely comprehensive frame of reference.

Sustainability is a concept that has been much abused and needs some clarification. Claims that a particular building is sustainable are easily made but hard to justify; such claims are meaningless in the absence of accepted frames of reference.

There are nowadays several sustainability certification schemes that propose "packs" of criteria to assess and categorize building performance under various perspectives, such as environment, energy, and sustainability. While the pioneering efforts must be praised, the fact remains that all these systems have significant flaws and shortcomings. It is very difficult—if not impossible—to develop truly holistic and universal criteria. The state of the art is such that criteria suited to a specific building type or to a specific country do not work well for other typologies or locations. Furthermore, all schemes that add up disparate criteria to obtain a global assessment are essentially arbitrary, given our current ignorance of the true relative weight of the parameters. Environmental assessment, in a word, is still in its infancy.

Given the current state of knowledge, at this point it is necessary to identify and concentrate efforts on the truly critical aspects of sustainability and create a system where the selection of targets is not bound by any pretense of achieving a supposedly balanced or comprehensive distribution by thematic categories, but is guided exclusively by the critical importance of the selected parameters. To further this discussion, let us refer to the schematic of Fig. 1.2. A pyramid cut in horizontal slices may be used to illustrate a hierarchy by which the interventions in a building should be ordered. From the base to the apex, the sequence is: first

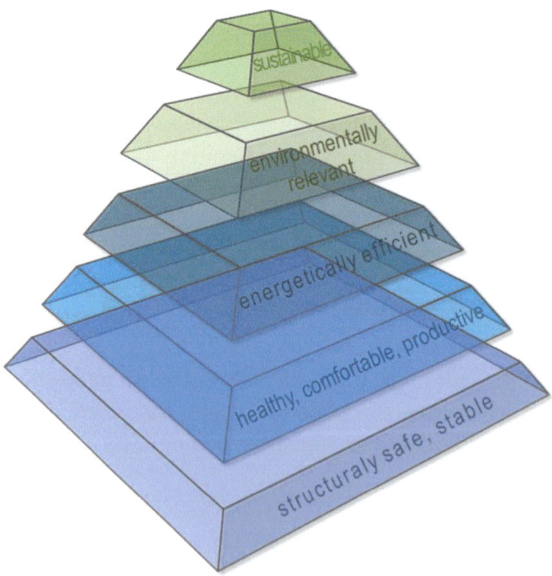

Fig. 1.2 Hierarchy of requirements for sustainable buildings. The pyramid expresses the relative difficulty in terms of rigor of execution and obstacles to implementation

structure (*the firmitas of Vitruvius*); then health and comfort (*utilitas*); environment; energy; and, finally, the crowning criterion—sustainability.

The lower levels of the pyramid express objectives that are more easily defined, and that must be satisfied before the next level can be meaningfully attempted. As we progress upwards, the more difficult it becomes to describe, implement, and evaluate the objectives, while the universe of possible situations fulfilling those requirements becomes smaller and smaller.

The first level addresses the basic and most conventional requirement for any building: it must be structurally safe and must keep its shape over time, thereby ensuring that it can house and support more specific functional requirements (as for a hospital operating room, a museum, or a school). This basic level is also fundamental in terms of the wider effect upon the neighborhood regarding spatial organization, security, and the impact of the supported activities upon the community.

At the second level, the building must respect all indoor environmental values with direct impact on the health of its users. Throughout their useful lives, buildings must ensure the fulfillment of a set of comfort and health requirements. Of course these requirements ought to be properly researched before being adopted by law, a fact that is starkly highlighted by the inconsistencies in the standards and regulations currently in force in different regions.

At the third level, the building must strive to respect all environmental criteria and environmental systems transcending the indoor environment. At this level, there is still a very serious lack of knowledge and of credible methods, in part because the nature and the boundaries of broader environmental systems are not perfectly known, and in part because the diverging interests of stakeholders lead to intentional manipulation and misrepresentation to cut costs, evade responsibilities, or to fit ideological positions. There are several attempts under the form of systems to characterize the environmental quality of buildings. All have their limitations, sometimes balancing realities of a different nature and values, yet a convention must be assumed where the criteria must be explained and justified, including a clear statement for each criterion, its limitations and potential shortcomings if forced beyond what is justified under the current societal values.

At the fourth level, the building must be energy efficient. This concept needs further reflection regarding the extent and limits of the concept of efficiency, which may inspire and provide new avenues of development of the final object that will be built. There is a need to distinguish between efficiency and sufficiency applied, as in the current case, to the case of the energy. A building cannot be considered energy efficient if at its design stage it did not consider the role of sufficiency. With the respect of that priority, the intervention at this stage opens up opportunities to respond to the upper level.

Let's take the case of a new building. In a simplistic way, one can think of building prerequisites as a toolbox where all services, including energy services, are reduced to connections between the building and external energy sources. All internal functions that require energy are in this model fulfillled by energy supplied externally by a network supplying, for instance, gas or electricity. In such a case, the way the systems are selected, sized, installed, and operated will affect the amounts of energy used. In this context, there is a rationale to state levels of efficiency, the more efficient systems (which may include the grid and/or gas network) being those that use less energy overall. At this level, the efficiency is related only to the physical energy "good," and not to its market value, which will often vary widely for a given energy vector.

Otherwise, the building may be conceived having in mind the climatic conditions and the fact that a building is in itself an energy system that can catch energy from the sun, directly as light and heat, being able to store some of the heat in its walls, floors, ceilings, etc. Every building "manages" heat through natural collection or dissipation through the envelope, thermal storage, and re-radiation exchanges among the various surfaces of the internal blackbody represented by each closed space. Properly designed buildings can manage these exchanges in a purposeful and effective way, leading to drastically reduced final energy needs for comfort. This result is not achieved by any "efficiency" type of action, but because the building was conceived and designed to make it as "sufficient" as possible in terms of energy for health and comfort.

With such a building, the "external" energy supply system will only have to deliver the energy to reach the stated conditions for that space—also, of course, with the same prerequisite to make it so that the energy services ensure an efficient supply at the user level. The energy "provided" by the building as an energy system is a result of careful implementation of sufficiency strategies. Sufficiency is of course applicable to other issues beyond energy and, even within the energy field, is not exclusive to buildings. Let's take the case of mobility in a city. There is a wide spectrum of possible options, from soft mobility (walking, biking, etc.) to collective means of transport (heavy and light rail, buses, etc.), where the energy demand, in comparison with the generalized reliance on cars, is much lower. Electrification of vehicles can make the end result even better. Comparing the two situations, it is clear that the reduction of energy use for mobility is due primarily to measures of sufficiency, and secondarily to efficiency measures.

Both sufficiency and efficiency contribute to the "happy ending": less energy use. But they would not necessarily lead to the same outcome. The above needs to be made clear, as nowadays when one speaks about energy efficiency in buildings, for example, the default professional practice and, unfortunately, the regulations, operate under the assumption that the building will definitely have a heating/cooling system, and the discussion moves on to the efficiency of the "add-on" systems. The higher-order sufficiency options are penalized under this system.

Arriving at the highest level of the pyramid, the building must be sustainable. Sustainability is a difficult and misleading word, so a brief review is in order. We could start from Go Brundtland's report which, by the way, has become a historical landmark, but let us return instead to what we learned in primary school about the cycles of nature. There are two interlinked cycles: the water cycle, and the carbon cycle; the latter bears a curious interdependence between the vegetal and the animal worlds, the first processing CO_2 and the second favoring O_2 while releasing CO_2. One may say that these two cycles are the hallmark of sustainability and control the basic climatic conditions of the earth. Before the industrial revolution, the actions of man were too feeble to disturb the global cycles of nature.

Pre-industrial mankind did cause quite substantial impacts at the local level, with land being deforested, or local fisheries being overexploited, for example. Regarding the global cycles, the water cycle, which has more local aspects, was the first where people started to realize that they were overtaxing the resources available in many locations. Furthermore, overexploitation of water resources was often combined with and aggravated by pollution of the sources, due to the concentration of demand in cities generating vast amounts of sewage that was not able to re-enter the cycle at the same pace as potable water was used up.

With significant financial resources, it was possible to create water treatment stations that accelerate the process of recovering the original water condition. But these energy-intensive solutions to the water problem, together with the myriad other energy-hungry processes taking place in modern cities, led to an explosion in energy consumption, mostly supported on fossil fuels. This trend was already taxing the global cycles when done on behalf of only one-quarter of the world's population. When China, India, and many other countries started to catch up to the way of life of the developed world, this further accelerated the use of fossil energy resources and the growth in the concentration of CO_2 in the atmosphere, causing higher planetary heat retention, and therefore global warming and potential for climate change.

Human development is a moral imperative and is also thought to be a necessary step before individuals and societies can aspire to longer term values including environmental responsibility. Thus it is ethically and politically necessary that all mankind achieve a higher standard of living. It will be very challenging to reconcile this development with lower global CO_2 emissions. It will be necessary to act at all levels, by substituting energy resources—namely, by using more renewable resources, but also by increasing the efficiency of all systems still relying on fossil fuel while promoting sufficiency in systems using energy, such as mobility and buildings.

A sustainable building is the ultimate and most difficult step (Fig. 1.2). Further issues must be solved. For example, buildings currently use materials that are sourced from everywhere in the world, travelling great distances and thus amassing very large CO_2 "bills" aggravated by the waste and energy use at extraction sites and at building sites. When extraction takes place in sensitive environments, e.g., in tropical forests, there is also the destruction of sensitive habitats and the loss of biodiversity as ecosystems that took millennia to form—all, say, for the temporary luxury of an exotic wood veneer finish.

Sustainability is a word that has been used with a wide spectrum of meanings, but sustainability could probably be better understood if the environmental impact of a building could integrate three basic components: (1) impacts associated with the nearby environment, i.e., local "pollution"; (2) the so-called "footprint" that encompasses the use of natural resources that cannot be "recreated" or "regenerated" within a reasonable amount of time, or not at all; and (3) everything related to the global warming and consequent climate change. Building functions and the conditions for performing those functions have been set, modified, adapted and refined throughout the ages since the specific needs of society at each time have changed. Integration seemed to be a basic criterion for a building design starting with its location and the overall architecture, bearing in mind the functions expected from the building and the performance of the building as a user of environmental resources, and specifically as an energy system in itself—namely, by the daily use of environmental resources such as climate, water, wood, energy, and fresh air.

Traditionally, almost all those "services" were satisfied by the buildings in themselves, according to the available technologies and the standards of living throughout time. In northern Portugal, the rural houses counted on the fireplace to warm the rooms and to produce hot water. Also, the heat of domesticated animals, such as cows, was very often used to create warm buffer zones, as in the animal stalls below the inhabited floors of many rural houses in this region.

Modern answers to these challenges can range from "add-on" solar systems to collecting solar energy that can be reverted to the benefit of the urban needs to district heating and cooling systems at the scale of the neighborhood or at even larger scales. Nowadays there are more resources to support the design of buildings, and it is possible to verify in great detail what the performance of a building will be, so from this point of view it can be said that the work of the designers has become easier. However, this is offset by much more complex boundary conditions.

Regarding the building in itself, knowledge about physical phenomena, such as heat conduction, heat storage, air circulation, and their relative potential contribution to indoor comfort, and the existence of means to simulate all associated physical phenomena, allows for optimized designs and for enhanced performance concerning the indoor thermal conditions in view of human comfort in both the winter and summer seasons.

The external boundary conditions are, nevertheless, ever more challenging. The general quality of urban air in many locations, and the rigidity of urban planning regulations—eventually for good reasons—severely restrict the opportunities for optimization of each building as an intrinsic energy system.

There are truly novel challenges that need to be wisely identified, characterized, and tackled. Industrialization and globalization led to some overriding of the climatological differences leading to the generalization of the architectural stereotypes and the adoption of "add-on" equipment as a fatality—in nature as well as in size. Even the regulatory environment can stimulate or detract in this context, and this is not just a matter of technology, but rather a matter of people, culture, and means. Nobody will impose energy systems, but rather thermal requirements or indoor air quality requirements. It is already scientifically proven, although not yet widely understood by the public and the professions, that many buildings with sophisticated technical environmental systems do not provide comfortable and good-quality indoor environments everywhere, as seen in many elementary schools in Europe.

In addition, there is another challenge that is associated with the design approach: large architectural offices that produce designs for buildings all over the world. Perhaps there is less justification for regional variations in skyscrapers designed for Hong Kong, São Paulo or Chicago—although this is far from having been proven—but in the case of most other building types, including, obviously, housing—universal designs are simply unacceptable. In particular, we emphasize the fact that the design of buildings for the Mediterranean requires a very specific approach because the potential for passive design is so high in this type of climate.

Many excellent tools for detailed building simulation are now available, such as the very powerful ESP-r simulation software developed in the 1990s by our colleagues at the University of Glasgow. The bottleneck nowadays is not the capacity of the tools but the difficulties of modelling. Simulation is not useful for guiding architectural concepts if the overhead required is too great, or if the models can only be run when most details have been specified. There is an urgent need to have simple, but sturdy tools more oriented to use by architects in the earlier stages of design; this is particularly true for the buildings of the Mediterranean basin, which is the land of the "architecture of complexity"—to use the memorable words of Professor Rafael Serra of the University of Barcelona.

Simulation models tend to stereotype and simplify, and cannot substitute for careful building design. Unfortunately, equipment is often sized on the basis of hasty simulations, not even pausing to ponder which type of solution is best, but simply diving headlong into the specification of equipment. The "add-on" systems are also very often designed and sized based on questionable indicators in what regards their need and their success, such as ventilation rates and operating and management strategies.

Going back to housing, the same trends are visible. For instance, air conditioning technology appeared due to the need to remove heat and moisture in tropical climates and from spaces with high densities of occupation. But nowadays one finds houses that need air conditioning in temperate climates, where high temperatures and high relative humidity never coincide, just because the sun crosses the badly oriented and oversized glazed bays and no shading was foreseen.

In a recent doctoral thesis in Porto, the question was asked: "Which are the critical parameters that architects must consider from the earliest design stages for temperate climates?" Two parameters turned out to be very clearly critical: thermal insulation, and thermal mass. But what is the use of increasing thermal insulation if the internal surfaces are clad with gypsum board, negating thermal inertia? Both properties are essential, and more of one cannot substitute for a lack of the other.

1.5 The Built Environment and Its Policies: the Case of the Mediterranean Basin

The previous paragraphs develop a position regarding future actions on built environment that is based on an interpretation of the progress of knowledge in the last forty years. New evidence has confirmed the centrality of energy within the values of sustainability, and the need for new energy perspectives and technologies that address both local conditions-as the climate-and the global environmental constraints-mainly expressed as risks of climate change. A focus was put on the holistic role of building physics, giving precedence to comfort and health as the main criteria of environmental quality.

A statistical concept of comfort has dominated the field of thermal comfort since the early seventies, following the publication of Professor Ole Fanger's doctoral thesis in 1970 in Copenhagen. While the limitations of this model have become apparent, we have not yet been able to make a clear-cut distinction between what can be done within the more recent "adaptive comfort" concept and what effectively requires fully controlled climatization though technological equipment, in particular for cooling and ventilating buildings in summer. Current building simulation and evaluation models often assume that buildings are already provided with artificial heating and air conditioning technologies, thereby missing or misrepresenting the role of building physics which, in the context of the Mediterranean basin, is much more significant than in northern Europe or even in the USA..

Within this time period, new pathways of theory and action were explored regarding passive solar technologies, with the PLEA (Passive and Low Energy Architecture) Conferences playing an important role. By the early eighties a deep disconnect had formed between the proponents of passive solar approaches, claiming air temperature swings of up to 3–4 °C indoors to be acceptable, and practitioners within the umbrella of professional air conditioning industry bodies such as ASHRAE, who pointed to complaints from customers and building users arising from temperature variations as small as 0.5 °C! The latter was a result of a statistical comfort concept, while the former heralded the emerging adaptive comfort concept.

The Mediterranean basin is characterized by a very rich history, as the amenity of its natural climate has, for millennia, favored human settlements on its margins. Yet there are many other areas in the world with similar climates. Without attempting exhaustive cataloguing, it can be said that very broad areas in America, Asia, and even in Africa can be likened to the climate of the Mediterranean basin.

Therefore, the built environment policies in the Mediterranean basin must liberate themselves from the "perversities" of the technological "contamination" from other latitudes, and starting from a basic model such as that in Fig. 1.1, giving particular consideration to the following points:

- Buildings to be seen by themselves or by clusters in their regional and urban context
- Special care must be paid to building orientation and the location and sizing of the windows
- Local climate and overall comfort and health conditions associated with the built environment have to be updated and readjusted, given the advances regarding comfort and indoor environment quality
- Adaptive comfort must be the first target for indoor comfort
- Source control of indoor and outdoor pollution must be the first priority for health

- Ventilation will be used for people's comfort, and neither to control undue pollutant loads nor to carry on heat
- Priority must go to buildings as energy systems in themselves, before the eventually necessary "add-on" technologies/equipment
- Energy sufficiency must be explored before energy efficiency
- Design models and tools and critical parameters must be called upon at an earlier stage of the design phase.

In this context, the question is to know whether everything has been done in the technological era to respond to the comfort and health challenges in the Mediterranean climate, giving first priority to energy sufficiency.

Fanger's well-known comfort concept was born in Denmark and had a strong impact in the USA and Japan as those countries associated "dated" responses to the climate for comfort together with economic growth. It has been necessary to reach 2004 to recognize the "adaptive comfort" as a concept of comfort that is more flexible and adapted to climates like the Mediterranean's, with mild winters and summers, and where high air temperatures and high relative air humidity do not coincide, as they do in the tropical zones of the planet. Those given climatic conditions permit the design of buildings with a set of rules that play with an acceptable range of variable temperatures by managing the storage and caption of heat gains through wise shading, indoor thermal inertia, and a building envelope insulated from the outside.

After the adaptive comfort, there is still time to adapt minds, approaches, and tool thinking, designing and organizing buildings for people in their particular places—while bearing in mind the global responsibility, i.e., the sustainability that is made up of everyone's molecular contribution, that those better than us will one day be able to assess and judge.

Part I
Challenges and Priorities for a Sustainable Built Environment

Chapter 2
Climatic Change in the Built Environment in Temperate Climates with Emphasis on the Mediterranean Area

Constantinos Cartalis

Abstract Climate change in the built environment is described by means of a multi-fold relationship: cities are major contributors to CO_2 emissions; climate change poses key threats to urban infrastructure and quality of life; climate change interacts with urbanization, population aging and socio-economic issues; and how cities grow and operate matters for energy demand and thus for greenhouse gas emissions. This relationship is particularly important since nearly 73 % of the European population lives in cities, and this is projected to reach 82 % in 2050. At the same time, climate change is occurring in Europe, as mean temperature has increased and precipitation patterns have changed. In terms of the Mediterranean area, increasing droughts and extreme climate phenomena (heat spells/heat waves) are key future characteristics that will put the whole region under serious pressure until 2100. They will also directly influence the built environment, including its resilience, and are expected to impact more cities with higher concentrations of vulnerable people, including the elderly, who are considered to be more sensitive to various climatic stress factors as compared to people of working age.

2.1 Introduction

Climate change is taking place and affecting cities, including their built environment as well as their residents—especially the most vulnerable ones. Impacts have already been reported and highly reliable estimates show the relationship between the increased frequency of climate extremes and the severity of the impacts. A growing number of cities in Europe have taken initiatives to modify their energy production, consumption patterns and greenhouse gas emissions and to develop adaptation policies to climate change,

C. Cartalis (✉)
Department of Environmental Physics, University of Athens, Athens, Greece
e-mail: ckartali@phys.uoa.gr

S.-N. Boemi et al. (eds.), *Energy Performance of Buildings*,
DOI 10.1007/978-3-319-20831-2_2

including policies for the built environment. Furthermore, the role of cities is critical in achieving the objectives of European Union for a low-carbon, resource-efficient, ecosystems-resilient society. In this chapter, climate change in the built environment is examined, with the examination concentrating on temperate climates (range of latitude between 23.5° and 66.5°), with emphasis on the Mediterranean area, and addressing changes in climatic characteristics that need to be considered in climate change adaptation and mitigation plans.

2.2 The Multi-Fold Relationship Between Cities and Climate Change

Several publications have explained the state of the environment in cities, especially in light of new challenges such as climate change (Cartalis 2014; European Environment Agency 2012, 2013; International Council for Local Environmental Initiatives 2013; Intergovernmental Panel for Climate Change—IPCC, Climate Change 2013 and European Commission 2013). In particular, the International Council for Local Environmental Initiatives (2013) speaks about the urgent need to build cities that are resilient to climate change. IPCC (2013) also refers to the impacts of climate change on cities, mostly in terms of the increase in temperatures as well as of the increased occurrence of tropical days and heat waves. The European Environment Agency (2013), in its publication on the impacts of climate change, also refers to the impact of climate change on the European continent and correlates the increase of heat waves in Europe and the spatial and temporal enhancement of thermal heat islands in several European cities to the intensified urbanization trends in Europe.

Climate change in the built environment (hereinafter referred to as "cities") is pragmatically described by means of a multifold relationship (EEA 2010a, b; Kamal and Alexis 2009; Stone et al. 2012):

1. Cities are major contributors to CO_2 emissions. Roughly half of the world's population lives in urban areas; this share is increasing over time and is projected to reach 60 % by 2030. Cities consume most—between 60 and 80 %—of the energy production worldwide and account for a roughly equal share of the global CO_2 emissions. GHG emissions in cities are increasingly driven less by industrial activities and more by the energy services required for lighting, heating and cooling, appliance use, electronics use, and mobility. Growing urbanization will lead to a significant increase in energy use and CO_2 emissions, particularly in countries where urban energy use is likely to shift from CO_2-neutral energy sources (biomass and waste) to CO_2-intensive energy sources.

2. Climate change poses key threats to urban infrastructure and quality of life due to rising sea levels, more extreme storms and flooding, and extreme heat events. Heat events in particular can compromise urban infrastructure since they are expected to be felt more strongly in urban areas due to urban heat island (UHI) effects, which are strongly related to changes in land use/land cover (LULC). In particular, due to the large amount of concrete and asphalt in cities, the difference in average annual temperatures compared to rural areas ranges from 3.5 to 4.5 °C (up to a difference of 10 °C in large cities), and is expected to increase by 1 °C per decade (Santamouris 2007; Stone et al. 2010). In addition, due to urban sprawl, hotspots are developed even at distances from the city centers.
3. Climate change interacts with urbanization, population aging and socio-economic issues; it also interacts with the economic development of cities. As cities influence to considerable extent Europe's economy, any climate-related problems may place Europe's economy and quality of life under threat. At the same time, demographic trends result in an aging population, a fact that increases the number of people vulnerable to heat waves. Urbanization also reduces the area available for natural flood management, thus exposing cities to the adverse effects of extreme weather events associated with heavy storms or rainfall. The close interaction of climate change to socio-economic changes on an urban scale may increase the vulnerability of people, property and ecosystems; for the vulnerability to be faced, adaptation measures are urgently needed.
4. How cities grow and operate matters for energy demand and thus for GHG emissions. Urban density and spatial planning and organization are key factors that influence energy consumption, especially in the transportation and building sectors. The acceleration of urbanization since 1950 has been accompanied by urban sprawl, with urban land area doubling in developed countries and increasing five-fold in the rest of the world. The expansion of built-up areas through suburbanization has been particularly prominent among OECD metropolitan areas (66 of the 78 largest OECD cities experienced a faster growth of their suburban belt than their urban core between 1995 and 2005), whereas increasing density could significantly reduce energy consumption in urban areas (Kamal and Alexis 2009). It should be mentioned that the role of land in climate change and GHG mitigation has received less attention than energy systems—despite the fact that evidence from urbanized regions, where land use activities have resulted in significant changes to land cover, suggest land use to be a significant and measurable driver of climate change as well as one that operates through a physical mechanism independent of GHG emissions. In light of this evidence, the rate of warming at regional to local scales can be slowed through a reversal of land cover change activities that serve to reduce surface albedo or produce a shift in the surface energy balance.

2.3 Urbanization in Europe

Nearly 73 % of the European population lives in cities, and this is projected to reach 82 % by 2050 (United Nations 2014). Urbanization is associated with changes in materials and energy flow in the cities (Chrysoulakis et al. 2013), can increase pressures on the environment, and may support climate change mechanisms through enhanced emissions of greenhouse gases. It may also reduce the resilience of cities to extreme events associated with climate change due to urban land take and the limited capacity to regenerate the building stock of the cities. Table 2.1 provides the urban population as a percentage of the total population (for 2011 and the projection for 2050) for selected countries in the Mediterranean area (UN 2014). It also provides the urbanization rate, which describes the average rate of change per year of the urban population, as deduced for 2010 until 2015 (UN 2014).

The interplay between the urbanization process, local environmental change, and accelerating climate change needs to be carefully assessed. According to several researchers (Stone et al. 2010; Kolokotroni et al. 2010; Grimmond 2011; Santamouris 2014), the long-term trend in surface air temperature in urban centers is also associated with the intensity of urbanization.

Urbanization is also linked to urban sprawl; emissions of greenhouse gases are higher in commuter towns not only because of car dependency but also due to the characteristics of the buildings (EEA 2012; IPCC 2014). Typically, European cities are densely populated, whereas large differences are observed in the level of sprawl, both between and within European countries (EEA 2015).

Table 2.1 Urban population as percentage of the total population (for 2011 and projection for 2050) for selected countries in the Mediterranean area (UN 2014)

Country	Urban population (% of the total population 2014)	Urban population (% of total population 2050)	Average annual rate of change 2010–2015 (%)
Albania	56	76	1.9
Croatia	59	72	0.5
Cyprus	67	72	−0.2
France	79	86	0.3
Greece	78	86	0.4
Italy	69	78	0.2
Malta	95	97	0.2
Portugal	63	77	0.9
Serbia	55	67	0.1
Spain	79	86	0.3
Slovenia	50	61	−0.2
Turkey	73	84	0.7

It also provides the urbanization rate, which describes the average rate of change per year of the urban population, as deduced for 2010–2015 (UN 2014)

Table 2.2 Urban sprawl from 2000 until 2006, using the urban sprawl index, which measures the growth in built-up areas over time, adjusted for the growth in population (OECD 2013a, b)

Country	Urban sprawl index (%)
France	−2.1
Greece	1.8
Italy	0.9
Spain	3.8
Portugal	6.2
Slovenia	−4.2

Table 2.3 Population (in million inhabitants) in major cities in the Mediterranean area (2005–2010–2015)

Heading	Athens	Rome	Madrid	Lisbon	Porto	Naples	Marseille	Valencia
2005	3.251	3.345	5.619	2.747	1.303	2.255	1.413	0.804
2010	3.381	3.306	5.487	2.824	1.355	2.348	1.472	0.798
2015	3.550	3.302	5.891	2.944	1.426	2.463	1.570	0.800
Change from 2005 to 2015	9.2 %	−1.1 %	4.8 %	7.1 %	9.4 %	9.2 %	11.1 %	−0.1 %

Table 2.2 provides a view of urban sprawl between 2000 and 2006, using the urban sprawl index (OECD 2013a, b), which measures the growth in built-up areas over time, adjusted for the growth in population. When the population changes, the index measures the increase in the built-up area over time relative to a benchmark where the built-up area would have increased in line with population growth. The index equals zero when both population and the built-up area are stable over time. It is larger (smaller) than zero when the growth of the built-up area is greater (smaller) than the growth of the population, i.e., the density of the metropolitan area has decreased (increased). Urban sprawl index is higher in Portugal and Spain, at medium values in Italy and Greece, whereas it reflects negative values for France and Slovenia.

A close look at major cities in the Mediterranean area is provided in Table 2.3 for 2005, 2010, and 2015. An increasing trend is observed in most of the cities, with the most significant ones for Athens, Porto, Naples, and Istanbul.

2.4 Climate Change in Europe

Climate change is occurring in Europe as mean temperatures have increased and precipitation patterns have changed (Alcamo et al. 2007). Climate change is a stress factor for cities and ecosystems (EEA 2012), whereas climate-related extreme weather events, such as cold spells and heat waves, result in health and social impacts in Europe and may impact considerably urban infrastructure and the built environment (EEA 2010a, 2012).

High temperature extremes (hot days, tropical nights, and heat waves) have become more frequent (Vautard et al. 2013), while low temperature extremes (cold spells, frost days) have become less frequent in Europe (EEA 2012).

Climate models show significant agreement in warming, with the strongest warming in southern Europe in summer (Giorgi and Bi 2005; Hertig and Jacobeit 2008; Goodess et al. 2009). The likely increase in the frequency and intensity of heat waves, particularly in southern Europe, is projected to increase heat-attributable deaths unless adaptation measures are undertaken (Baccini et al. 2011; WHO 2011a, b; IPCC 2014). Without adaptation, between 60,000 and 165,000 additional heat-related deaths per year in the EU are projected by the 2080s, depending on the scenario (Ciscar et al. 2011).

Finally, even under a climate warming limited to 2 °C compared to pre-industrial times, the climate of Europe is projected to change significantly from today's climate over the next few decades (Van der Linden and Mitchell 2009).

2.5 Climate in the Mediterranean Area

The climate in the Mediterranean area is affected by interactions between mid-latitude and tropical processes (Giorgi 2008; Xoplaki et al. 2003), whereas it has shown large climate shifts in the past (Luterbacher et al.2007). Overall, the climate is mild and wet during the winter and hot and dry during the summer (Gao et al. 2006); however, a main characteristic of the Mediterranean area is the high spatial variability in seasonal mean temperature and total precipitation (Table 2.4). Such variability is enhanced by the orography of the area as well as land–sea interactions.

The winter climate is mostly dominated by the westward movement of storms originating over the Atlantic and impinging upon the western European coasts. The winter Mediterranean climate, and, most importantly, precipitation, are thus affected by the North Atlantic oscillation over its western areas (e.g., Hurrell et al. 1995), the East Atlantic, and other patterns over its northern and eastern areas (Xoplaki et al. 2004). In addition to Atlantic storms, Mediterranean storms can be produced internally to the region in correspondence to cyclogenetic areas, such as the the Gulf of Lyon and the Gulf of Genoa (Lionello et al. 2006b), or depressions originating from the southwestern Mediterranean Sea. Table 2.4 summarizes the Mediterranean climate in winter and summer (Ulbrich et al. 2006; IPCC 2013).

In addition to planetary scale processes and teleconnections, the climate of the Mediterranean is affected by local processes induced by the complex physiography of the region and the presence of a large body of water. For example, the Mediterranean Sea is an important source of moisture and energy for storms (Lionello et al. 2006a, b; Ulbrich et al. 2006) as thermodynamic instability is developed and sustained due to synoptic systems developed in the southern Mediterranean and spreading to the northeast.

Table 2.4 Summary of climate patterns and characteristics in the Mediterranean area in winter and summer

Period	Overall pattern	Climate characteristic
Mediterranean winter (DJF)	From very cold mountainous (Alps, Dinaric Alps, Pyrenees) to mild areas along the coastlines of to the south of the peninsulas, the eastern basin and North Africa	The total winter precipitation amounts range from 50 to100 mm (North Africa) to over 500 mm (along western coasts of the peninsulas enhanced due to orographic forcing and land sea interactions)
Mediterranean summer (JJA)	Mean summer (JJA) temperatures show a gradient from north (cool) to south (warm) and exceed 30 °C in the southeast. High pressure and descending motions dominate the region, l eading to dry conditions, particularly over the southern Mediterranean	Large areas receive no rain during summer, while in mountainous areas, precipitation totals can reach 400 mm

2.6 Climate Change in the Mediterranean Area

Increasing droughts and extreme climate phenomena (heat spells/heat waves) are key future characteristics that will put the whole region under serious pressure until 2100 (IPCC 2013). In particular, the Mediterranean area has been identified as one of the most prominent "hot-spots" (a region whose climate is especially responsive to global change) in future climate change projections (Giorgi et al. 2006).

An analysis of near-surface temperature observations over the Mediterranean land areas during the recent past shows increasing trends (EEA 2012; Ulbrich et al. 2012; IPCC 2013, 2014). In particular, the region surrounding the Mediterranean Sea has been warming during most of the twentieth century, with the warming trend being, during the period 1951–2000, about 0.1 °C/decade. The warming is found to have a remarkable spatial and seasonal modulation. During the summer in the north, the trend over western and central Europe appears to be greater than in the winter season. The largest trend (up to 0.2 °C/decade) is found in summer (JJA) over the Iberian Peninsula and western part of North Africa.

In accordance with the wide-spread increase of hot extremes, the first decade of the twenty-first century has been characterized by frequent heat waves within the Mediterranean-European region (EEA 2012).

The heat wave of summer 2003 struck Western Europe, including France, Spain and Portugal. Moreover, according to state-of-the-art regional multi-model experiments, the probability of a summer experiencing "mega-heatwaves" will increase 5–10 times within the next 40 years (Barriopedro et al. 2011). Based on the above findings, a synopsis of the changes per climatic parameter for the Mediterranean area is provided in Table 2.5.

Table 2.5 Synopsis of change per climatic parameter for the Mediterranean area

Climatic parameter	Observed change
Temperature	A general decreasing trend of cold extremes is found. Since 1970, an increase in air temperature of almost 2 °C has been recorded in southwestern Europe (Iberian Peninsula and south of France). In summer, a prominent increase is found almost everywhere and especially over the sea. Increasing trends have been particularly strong over the last 20 years (1989–2008) over the Central and Eastern Mediterranean
Precipitation	Decreasing precipitation throughout the region (mainly western Mediterranean, southeast Europe and Middle East). Southern countries are among the most water-stressed ones
Sea level	The global mean of sea-level rise was around 3 mm per year over the last two decades. The Intergovernmental Panel on Climate change predicts a sea-level rise of 0.1–0.3 m by 2050 and of 0.1–0.9 m by 2100, with significant (and possibly higher) impacts in the southern Mediterranean area
Cyclones and wind storms	General decrease in the density of cyclone tracks; general decrease in the frequency of cyclones associated with extreme winds
Overall	The Mediterranean climate would become progressively warmer, drier, and less windy in the twenty-first century

Table 2.6 Climate extremes occurring in the Mediterranean area

Extreme temperatures	General increase in the number of very hot days and nights as well as longer warm spells and heat waves, with the largest increases over the Iberian Peninsula in summer. Hot extremes in winter also increase in the western Mediterranean, whereas in the eastern Mediterranean, some decrease is observed. General decrease in the number of very cold days and nights and shorter cold spells
Extreme precipitation	Increase in heavy daily precipitation in winter over the Iberian Peninsula except for the south, southern Italy and the Aegean, and a weak general increase in the percentage of winter precipitation in association with strong daily precipitation events; decrease in the number of days with heavy precipitation over the western and central part of the Mediterranean region and an increase over the northeastern part; general increase in the intensity of heavy precipitation events over the entire Mediterranean region in all seasons except for the southwestern part with a reduction in the warm season, mainly caused by the increased atmospheric moisture content
Cyclones and wind storms	No significant change in the intensity of the most extreme wind storms

Specific reference is made to the climate change impacts with respect to climate extremes occurring in the Mediterranean area (Table 2.6).

In Fig. 2.1, a combined view of "Aggregate potential impact of climate change," "Overall capacity to adapt to climate change," and "Potential vulnerability to climate change" is provided (EEA 2012). The Mediterranean area, including

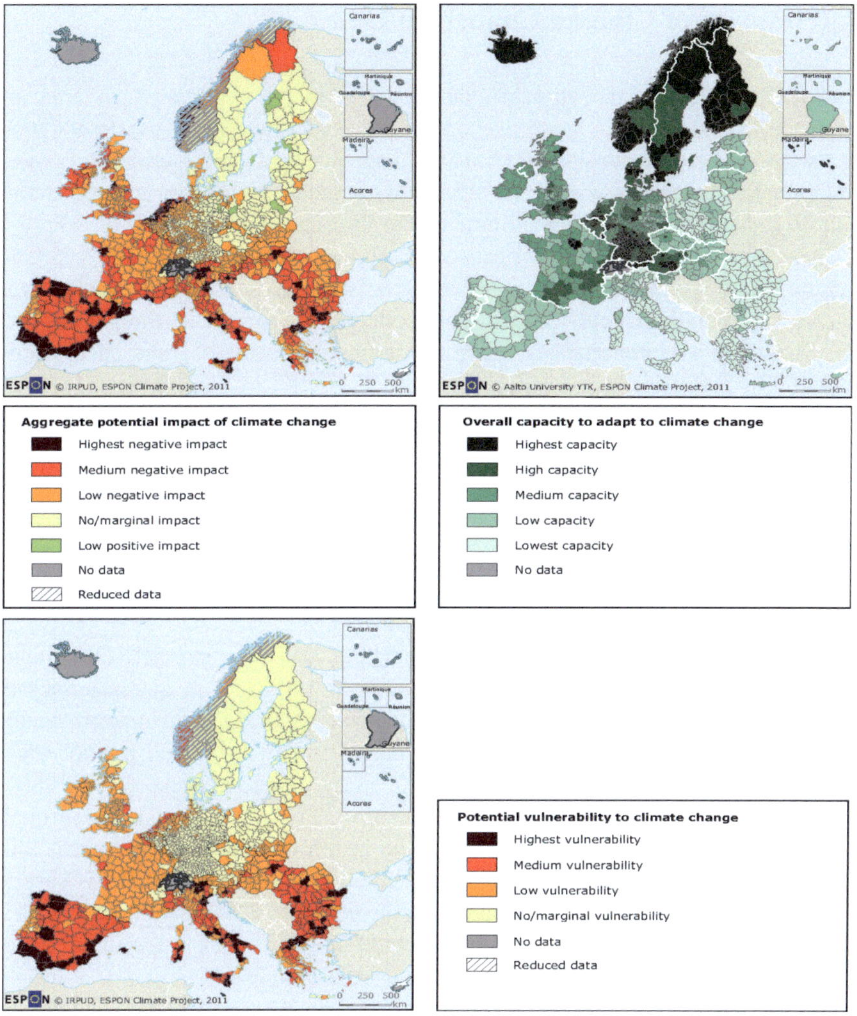

Fig. 2.1 A combined view of "Aggregate potential impact of climate change," "Overall capacity to adapt to climate change," and "Potential vulnerability to climate change" for the Mediterranean area (http://www.eea.europa.eu/legal/copyright; © European Environment Agency)

urban areas, shows, overall, medium negative potential impact; coastal areas in Spain and coastal/inland areas in Italy show the highest negative potential impact of climate change. Furthermore, the Mediterranean area shows medium to highest vulnerability to climate change (in the same overall areas as before), a fact that if linked to the limited capacity to adapt to climate change, reflects the sensitivity of the area in climate change and the urgent need to adapt measures in full temporal and spatial scales and a wide sectoral range.

2.7 Impacts of Climate Change on Cities

Potential changes in the climate characteristics in the Mediterranean area are expected to impact cities, buildings and infrastructure, and the ecosystem. Cities, in particular, may be considered climate change hot spots in the sense that potential climate change impacts on the urban environment or different activity sectors can be particularly pronounced. Some of the impacts are determined by complex relationships between various aspects of the climate changes, i.e., changes in temperature and in precipitation, urban characteristics, such as urban density, type and age of buildings, and various socio-economic aspects, i.e., the possibilities for adaptation to these changes, share of vulnerable population, and energy poverty (Wilby 2007).

1. Detection of climate-driven changes and trends at the scale of individual cities is problematic due to the complexity of a number of coupled processes and the high inter-annual variability of local weather and factors, such as land-use change or urbanization effects. Furthermore, climate change can influence the dynamics of a city, and a city may modify the climate at the local level. Table 2.7 provides a number of coupled processes that may be influenced by the interaction of urbanization and climate change.
2. The physical constituents of built areas within cities also interact with climate drivers. For example, runoff from impervious surfaces can increase risks of flooding and erosion, and will reduce evapotranspiration with impact on the energy budget of the respective area, whereas it may also result in poor water quality due to uncontrolled discharging of storm water.
3. Atmospheric circulation patterns are a major factor affecting ambient air quality and pollution episodes, and hence the health of urban populations as witnessed during the 2003 heat wave (EEA 2003; Gryparis et al. 2004). Several studies have indicated that weather patterns favoring air stagnation, heat waves, lower rainfall and ventilation could become more frequent in the future, leading to deteriorating air quality (Leung and Gustafson 2005).
4. Built areas have UHIs that may be up to 5–6 °C warmer than in the surrounding countryside (Oke 1982; Santamouris 2007; Mihalakakou et al. 2004 Santamouris et al. 2015). Building materials retain more solar energy during the day and have lower rates of radiant cooling during the night. Urban areas

Table 2.7 Coupled processes that may be influenced by the interaction of urbanization and climate change

Coupled process	Example
Altering small-scale local processes	Land–sea breeze
Modifying synoptic meteorology	Changes in the position of high pressure systems in relation to UHI events
Enhancing radiative forcing	Changes in the energy budget due to increase of GHGs
Enhancing urban heat islands (UHI)	Changes in land use/land cover

also have lower wind speeds, less convective heat losses and evapotranspiration (due to the limited share of such land cover as bare soil or green areas), yielding more energy for surface warming. Artificial heating and cooling of buildings, transportation and industrial processes introduce additional sources of heat into the urban environment causing distinct weekly cycles in UHI intensity (Wilby 2003) and its overall increase with years. Using statistical methods, it has been found (Santamouris 2012) that the nocturnal UHI in Athens could further intensify during the summer by the 2050s. This translates into a significant increase (more than 30 %) in the number of nights with intense UHI episodes.

5. With increasing surface and near-surface air temperatures, heat waves are expected to increase in frequency and severity in a warmer world (Meehl and Tebaldi 2004; Santamouris et al. 2015). UHIs will enhance the effects of regional warming by increasing summer temperatures relative to peri-urban and rural areas.

In Fig. 2.2, the modeled number of heat wave days [termed as combined tropical nights (days in which the lowest temperature is higher than 20 °C) and hot days (>35 °C)] are provided in the background. Higher values (in purple) are seen in the Mediterranean area, mostly in Greece, Italy, and southern Spain and Portugal. In the same figure, urban areas are depicted graphically in terms of their population density and the share of green and blue areas; it can be seen that the share of green and blue areas in southern cities is less than 30 % and in many cases even less than 20 %. The higher the population density and the lower the share in green and blue areas, the higher the intensity of the UHI. To this end, a large number of cities in Europe, and in the Mediterranean area in particular, are considered to have strong UHI potential and are thus expected to experience relatively strong increases in heat load in the future.

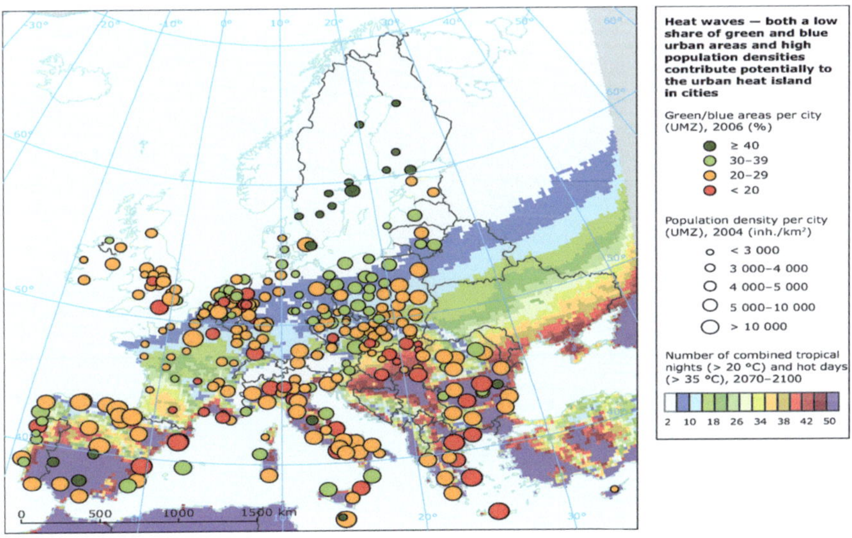

Fig. 2.2 Projection of heat wave days for Europe for the period 2070–2100 (http://www.eea.europa.eu/legal/copyright; © European Environment Agency)

6. Mortality increases in hot weather, especially among the most vulnerable, including the elderly. During the great European heat wave of 2003, urban centers, such as Paris, were particularly affected due to extreme day temperatures (Tobias et al. 2010). Under existing air pollution abatement policies, 311,000 premature deaths are projected for 2030 due to ground-level ozone and fine particles (EEA 2006). Full implementation of measures to achieve the long-term climate objective of the EU in terms of limiting global mean temperature increases to 2 °C, would reduce premature deaths by over 20,000 by 2030.
7. Urban air pollution concentrations may also increase during heat waves, with significant consequences for mortality as in the summer of 2003. This is because high temperatures and solar radiation stimulate the production of photochemical smog as well as ozone precursor volatile organic compounds (VOCs).
8. The impacts of climate change, including increasingly hot weather and heat waves, are expected to rise in areas with higher concentrations of vulnerable people, including the elderly, who are considered to be a group more sensitive to various climatic stress factors than younger people. As seen in Fig. 2.3, a significant number of cities in the Mediterranean area have proportions of vulnerable people (aging 65 and over), in medium and high percentages: in the range of 15–17 % for Athens, Naples, Porto, Valencia, etc., of 17–20 % for

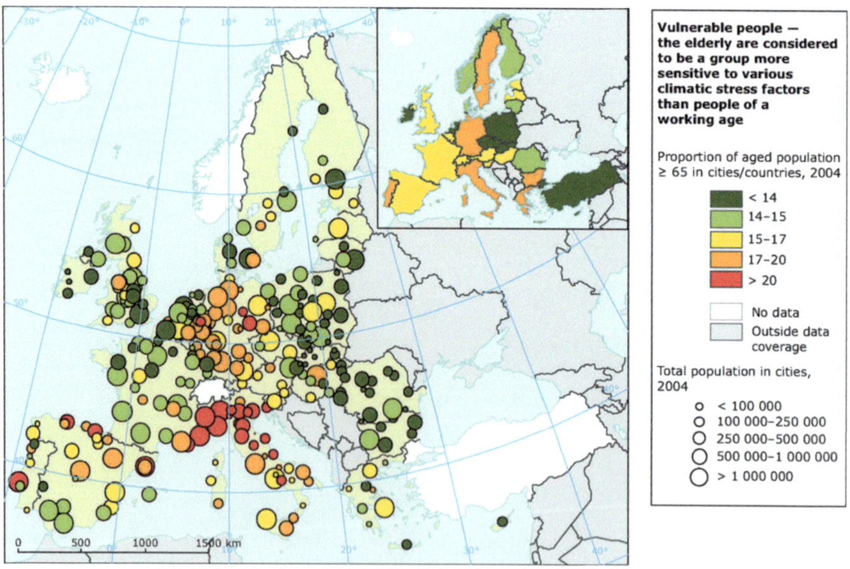

Fig. 2.3 Proportion of population over 65 years old in cities and countries. Elderly people are considered to be more vulnerable to various climatic stress factors (http://www.eea.europa.eu/legal/copyright; © European Environment Agency) the urban

Barcelona, Madrid, Marseille, Naples, etc., and more than 20 % for Lisbon, Nice, and Milan.

9. Thermal comfort is important for the built environment for two reasons: (1) the temperatures that people set in their homes are an important factor for energy consumption, and (2) whether a person feels comfort or discomfort depends on the state of the thermal environment, the quality of the building stock they occupy, and their ability to afford to mechanically heat or cool the building. Since thermal comfort depends directly on air temperature, it is expected to deteriorate in warming cities (Polydoros and Cartalis 2014).
10. Housing infrastructure is vulnerable to extreme weather events that may be caused due to climate change. In practical terms, buildings that were originally designed for certain thermal conditions will need to operate in drier and hotter climates in the future (WHO 2008).
11. Climate change is expected to result in the increased use of cooling energy and reduced use of heating energy (Cartalis et al. 2001; Santamouris 2001, 2012). Asimakopoulos et al. (2012), in his projections for Greece, indicates reductions of energy use for heating of up to 25 %, and highly increasing needs for additional energy for cooling. Future estimates show that 15 more days of heavy cooling will be required over parts of southern Spain and Italy, eastern Greece, and western Turkey and Cyprus (Giannakopoulos et al. 2009a). Elsewhere, increases of fewer than 15 days are evident for the near future (2021–2050). In another study, Giannakopoulos et al. (2009b) states that an additional two to three weeks along the Mediterranean coast will require cooling, whereas up to five more weeks of cooling will be needed inland by the end of the twenty-first century, i.e., by 2071–2100.
12. Regarding both heating and cooling, the total annual energy demand for the Mediterranean area as a whole is estimated to increase in future decades due to climate change (Giannakopoulos et al. 2009a; Santamouris et al. 2015); as a result, increases in the net annual electricity generation costs in most of the Mediterranean countries is estimated (Mirasgedis et al. 2007).
13. The cooling potential of natural ventilation falls with rising outdoor temperatures; as a result the demand for summer cooling could grow as internal temperatures rise during heat waves. For example, a study of energy demand in Athens showed a 30 % increase by the 2080s during July and August (Giannakopoulos and Psiloglou 2006).
14. Climate change will affect the built environment that is culturally valued through extreme events and chronic damage to materials (Brimblecombe 2010). In particular, marble monuments in the Mediterranean area will continue to experience high levels of thermal stress (Bonazza et al. 2009).
15. Green space is regarded by many as a crucial component of urban landscapes: for countering the UHI, reducing flood risk, improving air quality, and promoting habitat availability/connectivity (Julia et al. 2009). Green spaces are also susceptible to climate change, which implies that consideration needs to be given so that vegetation will not lead to local drying of soils.

16. Coastal cities are particularly vulnerable to climate change as this increases their vulnerability to rising sea levels and storm surge, presenting severe risks to the building stock and urban infrastructure (Nicholls 2004). Assessments of future flood risk for coastal cities reflect the added complexity of interactions between sea level rise, tidal surges, and storminess (Nicholls 2010).

2.8 Conclusion

Increasing droughts and extreme climate phenomena (heat spells/heat waves) are key future characteristics that will put the entire region, and especially cities and their built environments, under serious pressure until 2100 (IPCC 2013). In light of this, the Mediterranean area has been identified as one of the most prominent "hot spots" in future climate change projections (Giorgi 2006).

The high density and the poor (from an energy efficiency point of view) building stock of several Mediterranean cities, and the limited share of green areas within the boundaries of the cities, in conjunction with the higher percentage shares of vulnerable people (including elderly and urban poor), are key factors raising their vulnerability to climate change, including climate extremes (Grimmond 2011; IPCC 2014).

Building cities that are resilient and sustainable should be the basis of any local, regional, or national adaptation plan to climate change. This is highly important as if urbanization trends in the Mediterranean area are sustained, nearly 85 % of the population in the northern countries of the area will live in cities. Plans need to reflect multidisciplinary cooperation and sector policy reforms, such as urban transport policies, low carbon, and energy efficiency measures with emphasis on the building stock, and sustainable and resilient city planning. Decisions taken today define the future of many Mediterranean cities, in which the building and urban infrastructure for 2050 is being built today, yet 2050 will be very different from today—at least from a climatic point of view.

References

Alcamo J, Moreno JM, Nováky B, Bindi M, Corobov R, Devoy RJN, Giannakopoulos C, Martin E, Olesen JE, Shvidenko A (2007) Europe and climate change 2007: impacts, adaptation and vulnerability. In: Parry ML, Canziani OF, Palutikof JP, van der Linden PJ, Hanson CE (eds) Contribution of working group II to the fourth assessment report of the intergovernmental panel on climate change (IPCC). Cambridge University, Cambridge, pp 541–580

Asimakopoulos DN, Santamouris M, Farrou I, Laskari M, Saliari M, Zanis G, Giannakidis G, Tigas K, Kapsomenakis J, Douvis C, Zerefos SC, Antonakaki T, Giannakopoulos C (2012) Modelling the energy demand projection of the building sector in Greece in the 21st century. Energy Build 49:488–498

Baccini M, Kosatsky T, Analitis HR, Anderson MD, Ovidio M, Menne B, Michelozzi P, Biggeri A (2011) Impact of heat on mortality in 15 European cities: attributable deaths under different weather scenarios. J Epidemiol Community Heatlh 65:64–70

Barriopedro D, Fischer EM, Luterbacher J, Trigo RM, García-Herrera R (2011) The hot summer of 2010: redrawing the temperature record map of Europe, Science, 332, 6026, doi:10.1126/science.1201224

Bonazza A, Messina P, Sabbioni C, Grossi CM, Brimblecombe P (2009) Mapping the impact of climate change on surface recession of carbonate buildings in Europe. Sci Total Environ 407:2039–2050

Brimblecombe P (2010) Climate change and cultural heritage. In: Lefevre R, Sabbioni C (eds) Heritage Climatology. Edipuglia, Bari, pp 18–30

Cartalis C (2014) Towards resilient cities—a review of definitions, challenges and prospects. Adv Build Energy Res 8:259–266

Cartalis C, Synodinou A, Proedrou M, Tsangrasoulis A, Santamouris M (2001) Modifications in energy demand in urban areas as a result of climate changes: an assessment for the south east Mediterranean region. Energy Convers Manag 42:1647–1656

Chrysoulakis N, Lopes M, San José R, Grimmond CSB, Jones MB, Magliulo V, Klostermann JEM, Synnefa A, Mitraka Z, Castro E, González A, Vogt R, Vesal T, Spano D, Pigeon G, Freer-Smith P, Staszewski T, Hodges N, Mills G, Cartalis C (2013) Sustainable urban metabolism as a link between bio-physical sciences and urban planning: the BRIDGE project. Landscape and Urban Planning 112:100–117

Ciscar JC, Iglesias A, Feyen L, Szabo L, Van Regemorter D, Amelung B, Nicholls R, Watkiss P, Christensen O, Dankers R, Garrote L, Goodess C, Hunt A, Moreno A, Richards J, Soria A (2011) Physical and economic consequences of climate change in Europe. Proc Natl Acad Sci USA 108:2678–2683. doi:10.1073/pnas.1011612108

EEA (2003) Air Pollution by Ozone in Europe in Summer 2003: Overview of Exceedances of EC Ozone Threshold Values during the Summer Season April-August 2003 and Comparisons with Previous Years. Available via DIALOG http://reports.eea.europa.eu/topic_report_2003_3/en. Cited 30 Mar 2015

EEA (2006) Environment and human health, EEA Report No 5/2013, Joint EEA-JRC report

EEA (2010a) European Environment State and Outlook 2010: Urban Environment. SOER 2010 Report, European Environment Agency, Copenhagen (doi: 10.2800/57739)

EEA (2010b) The GMES Urban Atlas. European Environment Agency, Copenhagen

EEA (2012) Climate change, impacts and vulnerability in Europe, An indicator-based report, Report No 12/2012, pp. 300

EEA (2013) Adaptation in Europe—Addressing risks and opportunities from climate change in the context of socio-economic developments, Report No 3/2013, pp. 136

EEA (2015) SOER 2015—The European environment—state and outlook 2015. A comprehensive assessment of the European environment's state, trends and prospects, in a global context, European Environment Agency. http://www.eea.europa.eu/soer

European Commission (2013) An EU strategy on adaptation to climate change, 216 Final, 2013, pp. 11

Gao X, Pal JS, Giorgi F (2006) Projected changes in mean and extreme precipitation over the Mediterranean region from high resolution double nested RCMsimulation. Geophys Res Lett 33:L03706. doi:10.1029/2005GL024954

Giannakopoulos C, Psiloglou B (2006) Trends in energy load demand for Athens, Greece: weather and non-weather related factors. Clim Res 31:97–108

Giannakopoulos C, Hadjinicolaou P, Zerefos SC, Demosthenous G (2009a) Changing Energy Requirements in the Mediterranean Under Changing Climatic Conditions. Energies 2:805–815. doi:10.3390/en20400805

Giannakopoulos C, Le Sager P, Bindi M, Moriondo M, Kostopoulou E, Goodess CM (2009b) Climatic changes and associated impacts in the Mediterranean resulting from a 2 °C global warming. Global Planet Change 68:209–224

Giorgi F (2006) Climate change hot-spots. Geophys Res Lett 33:L08707. doi:10.1029/2006GL025734

Giorgi F, Bi X (2005) Updated regional precipitation and temperature changes for the 21st century from ensembles of recent AOGCM simulations. Geophys Res Lett 32:L21715

Goodess CM, Jacob D, Deque M, Guttierrez JM, Huth R, Kendon E, Leckebusch GC, Lorenz P, Pavan V (2009) Downscaling methods, data and tools for input to impact assessments. In: van der Linden P, Mitchell JFB (eds) ENSEMBLES: Climate Change and its Impacts: Summary of research and results from the ENSEMBLES project. UK, Met Office Hadley Centre, pp 59–78

Grimmond S (2011) Climate in Cities. In: Douglas I, Goode D, Houck M, Wang R (eds) The Routledge Handbook of Urban Ecology. Routledge, London

Gryparis A, Forsberg B, Katsouyanni K, Analitis A, Touloumi G, Schwartz J (2004) Acute effects of ozone on mortality from the "air pollution and health: a European approach" project. Am J Respir Crit Care Med 170:1080–1087

Hertig E, Jacobeit J (2008) Downscaling future climate change: Temperature scenarios for the Mediterranean area. Global Planet Change 63:127–131

Hurrell JW (1995) Decadal trends in the North Atlantic Oscillation: regional temperature and precipitation. Science 269:676–679

International Council for Local Environmental Initiatives (2013) Available via DIALOG http://resilientcities.iclei.org/resilient-cities-hub-site/about-the-global-forum/resilient-cities-2013/. Cited 30 Mar 2015

IPCC (2013) Climate change 2013: the physical science Basis'. contribution of working group i to the fifth assessment report of the intergovernmental panel on climate change. In: Stocker TF, Qin D, Plattner G-K, Tignor M, Allen SK, Boschung J, Nauels A, Xia Y, Bex V, Midgley PM (eds) Intergovernmental panel on climate change. Cambridge University Press, Cambridge

IPCC (2014) Summary for Policymakers. In: climate change 2014: impacts, adaptation, and vulnerability. Contribution of working group ii to the fifth assessment report of the intergovernmental panel on climate change. Intergovernmental panel on climate change. Cambridge University Press, Cambridge

Julia E, Santamouris M, Dimoudi A (2009) Monitoring the effect of urban green areas on the heat island in Athens. Environ Monit Assess 156:275–292

Kamal-Chaoui L, Alexis R (2009) competitive cities and climate change, regional development working paper no 2, pp 172

Kolokotroni M, Davies M, Croxford B, Bhuyan S, Mavrogianni A (2010) A validated methodology for the prediction of heating and cooling energy demand for buildings within the urban heat island: case study of London. Solar Energy 84:2246–2255

Leung LR, Gustafson WI Jr. (2005) Potential regional climate change and implications to U.S. air quality. Geophys Res Lett 32. doi:10.1029/2005GL022911

Lionello P, Malanotte P, Boscolo R (2006a) Mediterranean climate variability. In: Malanotte P, Boscolo R (eds) Lionello P. Elsevier, Amsterdam

Lionello P, Malanotte-Rizzoli P, Boscolo R, Alpert P, Artale V, Li L, Luterbacher J, May W, Trigo R, Tsimplis M, Ulbrich U, Xoplaki E (2006b) The Mediterranean climate: an overview of the main characteristics and issues. In: Lionello P, Malanotte-Rizzoli P, Boscolo R (eds) Mediterranean Climate var, pp1–18

Luterbacher J, Liniger M, Menzel A, Estrella N, Della-Marta PM, Pfister C, Rutishauser T, Xoplaki E (2007) Exceptional European warmth of autumn 2006 and winter 2007: Historical context, the underlying dynamics, and its phenological impacts. Geophys Res Lett. doi:10.1029/2007GL029951, vol. 34

Meehl G, Tebaldi C (2004) More Intense, More Frequent, and Longer Lasting Heat Waves in the 21st Century. Science 305:994–997. doi:10.1126/science.1098704

Mihalakakou G, Santamouris M, Papanikolaou N, Cartalis C, Tsagrassoulis A (2004) Simulation of the urban heat island phenomenon in Mediterranean climates. Pure appl Geophys 161:429–451

Mirasgedis S, Sarafidis Y, Georgopoulou E, Kotroni V, Lagouvardos K, Lalas DP (2007) Modelling framework for estimating impacts of climate change on electricity demand at regional level: case of Greece. Energy Convers Manag 48:1737–1750

Nicholls RJ (2004) Coastal flooding and wetland loss in the 21st century: changes under the SRES Climate and socio-economic scenarios. Glob Environ Change 14:69–86
Nicholls RJ, Cazenave A (2010) Sea level rise and its impact on coastal zones. Science 329:1517–1520
OECD (2013a) Regions at a Glance. http://dx.doi.org/10.1787/reg_glance-2013-en
OECD (2013b) Metropolitan areas. OECD Regional Statistics (database). doi:10.1787/data-00531-en
Polydoros A, Cartalis C (2014) Assessing thermal risk in urban areas—an application for the urban agglomeration of Athens. Adv Build Energy Res 8:74–83. doi:10.1080/17512549.2014.890536
Santamouris M (2001) Energy and climate in the urban built environment. James and James Science Publishers, London
Santamouris M (2007) Heat Island Research in Europe—The State of the Art. J Adv Build Energy Res 1:123–150 Review Paper
Santamouris M (2012) Cooling the cities—a review of reflective and green roof mitigation technologies to fight heat island and improve comfort in urban environments. Solar Energy 103:682–703
Santamouris M, Cartalis C, Synnefa A, Kolokotsa D (2015) On the impact of urban heat island and global warming on the power demand and electricity consumption of buildings—a review. Energy Build. doi:10.1016/j.enbuild.2014.09.052
Stone B, Hess JJ, Frumkin H (2010) Urban form and extreme heat events: are sprawling cities more vulnerable to climate change than compact cities? Environ Health Perspect 118:1425–1428
Stone B, Vargo J, Habeeb D (2012) Managing climate change in cities: will climate action plans work? Landscape and Urban Planning 107(3):263–270
Tobias A, Garcia de Olalla P, Linares C, Bleda MJ, Cayla JA, Diaz J (2010) Short term effects of extreme hot summer temperatures on total daily mortality in Barcelona, Spain. Int J Biometel 54:115–117
Ulbrich U, May W, Li L, Lionello P, Pinto JG, Somot S (2006) The Mediterranean climate change under global warming. In: Lionello P, Malanotte-Rizzoli P, Boscolo R (eds) Mediterranean climate variability. Elsevier, Amsterdam, pp 398–415
Ulbrich U, Lionello P, Belusic D, Jacobeit J, Knippertz P, Kuglitsch FG, Leckebusch GC, Luterbacher J, Mauden M, Maheras P, Nissen KM, Pavan V, Pinto JG, Saaroni H, Seubert S, Toreti A, Xoplaki E, Ziv B (2012) Climate of the mediterranean: synoptic patterns, temperature, winds, and their extremes, chapter 5 in: the climate of the mediterranean region. Elsevier, Amsterdam
United Nations, Department of Economic and Social Affairs, Population Division (2014) World urbanization prospects: the 2014 revision, highlights (ST/ESA/SER.A/352)
Van der Linden P, Mitchell JFB (2009) ENSEMBLES: climate change and its impacts: summary of research and results from the ENSEMBLES project (160 pp). Met Office Hadley Centre, UK
Vautard R, Gobiet A, Jacob D, Belda M, Colette A, Deque M, Fernandez J, Garcia-Diez M, Georgen K, Guttler I, Halkenka T, Karacostas T, Katragkou E, Keuler K, Kotlarski S, Mayer S, van Meijgaard E, Nikulin G, Patarcic M, Scinocca J, Sobolowski S, Suklitsch M, Teichmann C, Warrach-Sagi K, Wulfemeyer V, Yiou P (2013) The simulation of European heat waves from an ensemble of regional climate models within the EURO-CORDEX project. Clim Dyn 41:2555–2575
WHO (2008) Protecting health in Europe from climate change. World Health Organization, Geneva
WHO (2011a) Urban outdoor air pollution database. World Health Organization, Geneva
WHO (2011b) Air quality and health, Fact sheet No 313. World Health Organization, Geneva
Wilby RL (2003) Weekly warming. Weather 58:446–447

Wilby RL (2007) A review of climate change impacts on the built environment. Built Environ 33:31–45. doi:10.2148/benv.33.1.31

Xoplaki E, Gonzalez-Rouco FJ, Luterbacher J, Wanner H (2003) Mediterranean summer air temperature variability and its connection to the large-scale atmospheric circulation and SSTs. Clim Dyn 20:723–739. doi:10.1007/s00382-003-0304-x

Xoplaki E, Gonzalez-Rouco FJ, Luterbacher J, Wanner H (2004) Wet season Mediterranean precipitation variability: influence of large-scale dynamics. Clim Dyn 23:63–78. doi:10.1007/s00382-004-0422-0

Chapter 3
The Role of Buildings in Energy Systems

Argiro Dimoudi and Stamatis Zoras

Abstract The construction sector consumes significant amounts of natural resources (raw materials, water, energy) during the various phases of its activity that covers construction, operation, and demolition of structures. A large amount of energy is required for the operation of buildings during their overall lives in order to meet their habitats needs—about 40 % of the final energy consumption in the EU in 2012. The EU has set a target for all new buildings to be "nearly zero-energy" by 2020. Considering that construction of new buildings has declined over the last few years and that there is a large stock of old buildings that were constructed without any thermal or energy regulations in several European countries, energy renovation in existing buildings has a high potential for energy efficiency. As environmental issues become increasingly significant, buildings become more energy efficient and the energy needs during their operation decreases, aimed at "nearly zero energy" buildings. Thus, the energy required for construction and, consequently, for the material production, is becoming more and more important. A significant contribution in the efforts to reduce the environmental impacts from construction activities is evaluating their environmental consequences in each stage of their life-cycle. This has led to the development of different "environmental life-cycle" assessment approaches.

A. Dimoudi (✉) · S. Zoras
Laboratory of Environmental and Energy Design of Buildings and Settlements,
Department of Environmental Engineering, School of Engineering,
Democritus University of Thrace, Vass. Sofias 12, 67 100 Xanthi, Greece
e-mail: adimoudi@env.duth.gr

S.-N. Boemi et al. (eds.), *Energy Performance of Buildings*,
DOI 10.1007/978-3-319-20831-2_3

3.1 Sustainability and Construction Activity

The construction sector consumes significant amounts of natural resources during the various phases of its activity that covers construction, operation, and demolition of structures. The natural resources that are consumed during these phases are the raw materials and water used for the manufacture of the structural materials and structural components and the energy required for the extraction of the raw materials, production of structural materials and components, for the structures' construction and demolition. A large amount of energy is also required for the operation of buildings during their overall lives in order to meet their habitats' needs.

The construction and use of buildings in the EU accounts for about half of all extracted materials and about a third of water consumption in the EU, with amounts varying from country to country (COM 445 2014). Energy used in the manufacture of construction products and the construction process also plays a major role in the overall environmental impact of a building. Studies show that between 5 and 10 % of total energy consumption across the EU is related to the production of construction products with the embodied greenhouse gas emissions of a building increasing—which can comprise a significant share of total greenhouse gas emissions (COM 445 2014). The construction sector also generates about one-third of all waste and is associated with environmental pressures that arise at different stages of a building's life-cycle, including the manufacturing of construction products, building construction, use, renovation, and the management of building waste (COM 445 2014).

The construction sector is a very important economic sector. According to data from the European Construction Industry Federation (FIEC), in 2013 the construction sector contributed 8.8 % of the GDP of the EU-28, accounting for a €1.162 billion output. Buildings account for the majority of the construction activities, as civil works (e.g., roads. bridges, wastewater treatment, etc.) account for only by 21.0 % (Fig. 3.1) (FIEC 2014). Compared to 2006, a decrease was seen in the construction of residential buildings, while there was an increase in the maintenance and renovation works in existing buildings (Fig. 3.2) (Campogrande 2007).

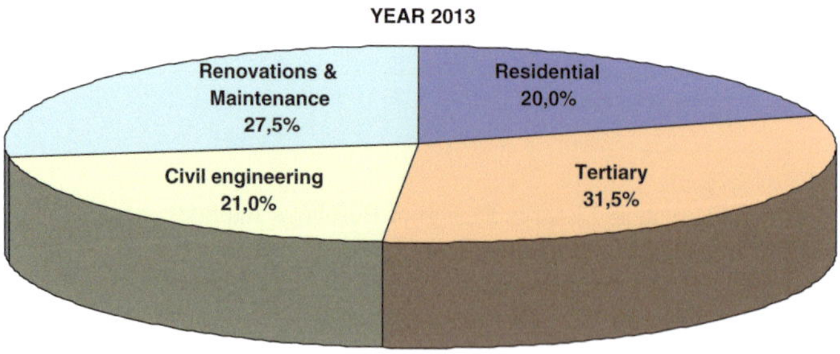

Fig. 3.1 Construction activity in Europe (based on data by FIEC 2014)

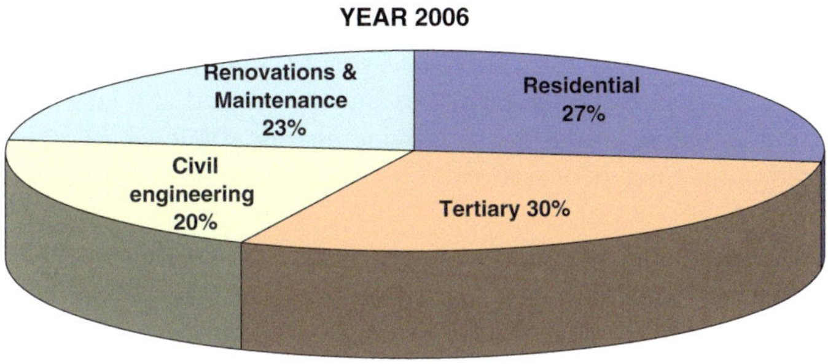

Fig. 3.2 Construction activity in the E.E.27 (based on data by Campogrande 2007)

For many years, the greatest importance from the environmental point of view was given at the operation phase of buildings—which is actually the longest period in a buildings life. After the first oil crisis in the 1980s, priority was given to a reduction of the energy that is consumed by buildings during their operation. However, since the beginning of the 1990s, when the environmental consequences from the other three construction phases began to increase (e.g., reduction of natural resource reserves, management of construction waste, difficulties in finding safe sites for construction waste deposits), a holistic approach was initiated that considered the environmental performance of structures and the consequences of the construction works on the environment. This approach resulted in implementing the "sustainability" as a new concept in structures consideration, aiming at the least possible burden on the environment during all phases of construction.

It is evident that even the construction of low-energy buildings, which do not depend very much on conventional fuel, consume, mainly through the use of construction materials and components, their manufacturing procedures and their construction, considerable amounts of non-renewable energy thus resulting in the release of pollutant emissions to the environment.

Since 2010, the EU2020 strategy has been setting the framework for the European economy for the following decade and beyond by focusing on three main priorities: smart, sustainable, and inclusive growth. As a follow-up, the "Resource Efficiency Roadmap" was adopted by the European Commission in September 2011. It concluded that existing policies on buildings, mainly linked to energy efficiency, need to be complemented with policies for resource efficiency looking at a wider range of resource use and environmental impacts, across the life-cycle of buildings. Such policies would "contribute to a competitive construction sector and to the development of a resource efficient building stock". As energy efficiency during the buildings' operation is already addressed by existing policies, the focus should be on resources such as materials (including waste), water, and embedded energy, addressing resource use and related environmental impacts all along the life cycle of buildings, from the extraction of building materials to demolition and recycling of materials (end of life) (EC–Sustainable Buildings 2015).

Energy efficiency measures should be considered in conjunction with their environmental impacts over the entire life cycle of a building; otherwise, impacts may be overlooked or additional problems may be created at a later stage of their life cycles, e.g., some solutions for improving energy efficiency of a building during its operation may make later recycling more difficult and expensive (COM 445 2014).

A question usually arises among designers, engineers, and users about the additional cost that new practices and requirements put at a building's cost. A study conducted by QUALITEL in France concluded that the extra cost for constructing sustainable residential buildings—as opposed to standard ones—has gone from 10 % in 2003 to below 1 % in 2014. A similar trend has also been noted in the UK (COM 445 2014).

3.2 Energy Consumption in Buildings

3.2.1 Overall Energy Consumption in the Building Sector

Buildings (both households and the tertiary sector) are the larger energy consumers in Europe, accounting for approximately 40 % of the overall energy consumption in 2012 (Fig. 3.3) (Saheb et al. 2015). While several countries exceed 40 %, such as Croatia and Greece sharing 43.1 and 42.1 %, while other countries account for lower percentages, such as Spain and Portugal, covering 30.7 and 28 % of the overall national energy consumption (Table 3.1) (EU 2014).

Residential buildings are the biggest energy consumers among building categories, in 2012 reaching up to 66 % of a building's total final energy consumption in the EU (Saheb et al. 2015). In 2012, the specific energy consumption was around 220 kWh/m^2, with a large difference between residential (200 kWh/m^2) and tertiary sector buildings (300 kWh/m^2) (Odyssee-mure.eu 2015).

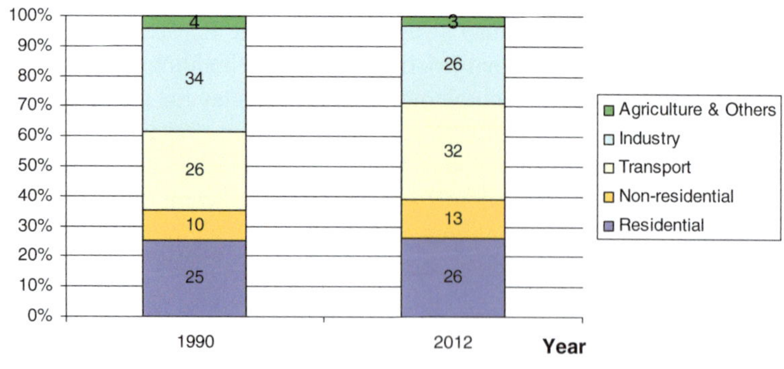

Fig. 3.3 Final energy consumption by sector in EU (based on data by Saheb et al. 2015)

Table 3.1 Final energy consumption by sector, in selected countries, in Mtoe for 1995–2012 (based on EU data 2014)

	1995	2000	2005	2010	2011	2012
GREECE						
Industry	4	4.5	4.2	3.5	3.3	3
Transport	6.5	7.3	8.2	8.2	7.4	6.4
Households	3.3	4.5	5.5	4.6	5.5	5
Services	0.9	1.3	1.9	2	1.9	2.2
Agriculture and fishing	1	1.1	1.1	0.8	0.7	0.3
Other	–	–	–	–	0.1	0.2
SPAIN						
Industry	20.5	25.4	31	21.4	21.4	20.8
Transport	26.4	33.2	39.9	37.2	36	33.3
Households	10	12	15.1	16.9	15.6	15.5
Services	4.3	6.7	8.4	9.8	10.2	10
Agriculture and fishing	2.2	2.6	3.1	2.2	2.4	2.7
Other	0.5	0	0.2	1.5	1	0.7
CROATIA						
Industry	1.3	1.4	1.6	1.4	1.3	1.1
Transport	1.2	1.5	1.9	2.1	2	2
Households	1.4	1.7	1.9	1.9	1.9	1.8
Services	0.4	0.5	0.7	0.8	0.8	0.7
Agriculture and fishing	0.2	0.3	0.2	0.2	0.2	0.2
Other	–	–	–	–	–	–
PORTUGAL						
Industry	4.9	6.3	5.8	5.5	5.3	4.7
Transport	4.9	6.6	7.2	7.3	6.9	6.5
Households	2.6	2.8	3.2	3	2.8	2.7
Services	0.9	1.4	2.2	1.9	1.8	1.8
Agriculture and fishing	0.5	0.7	0.6	0.5	0.4	0.4
Other	–	–	–	–	–	–

The average energy consumption per dwelling was reduced in about two-thirds of EU countries after 2000—a reduction of 1.5 % /year at the EU level between 2000 and 2010; this was mainly attributed to space heating, where a reduction of 3 %, amounting to 25 Mtoe, was reported. Space heating is the end-use that consumes the highest amounts of energy (67 % of total buildings' energy consumption), followed by water heating (13 %), electrical appliances (11 %), cooking (6 %), lighting only with 2 %, and a very small amount for space cooling (about 0.5 %) (Fig. 3.4). A relative growth in the energy consumption of electrical appliances was reported over the last years (from 9 to 11 % between 2000 and 2012), which resulted, looking at the overall energy balance, in a reduction of the relative fraction of energy attributed to space (Lapillone et al. 2014a, b).

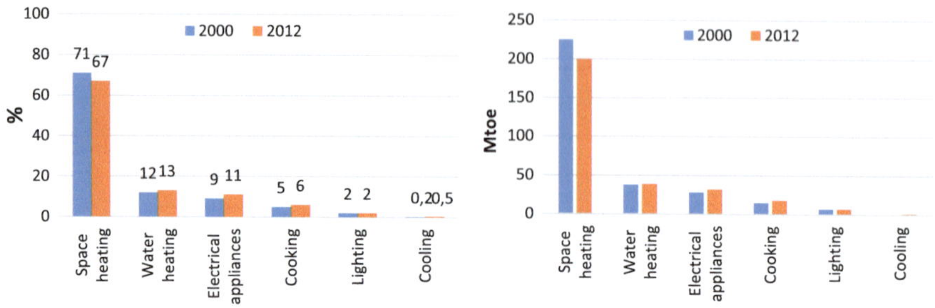

Fig. 3.4 Energy consumption breakdown in households in the EU countries (based on data by Lapillone et al. 2014b)

Regarding air conditioning energy consumption, although it can reach up to only 10 % of the total electricity consumption in countries using much air-conditioning (Cyprus, Malta, and Bulgaria), the average consumption per dwelling for this end-use is increasing as the installation of air conditioning units is increasing (Lapillone et al. 2014b).

3.2.2 Energy Consumption Per Fuel Type and Renewable Energy Sources (RES)

The main fuel consumed in 2012 in Europe (EU-28) was petroleum, with a share of 39 %, followed by gas, with a 22.9 % share (Fig. 3.5). Electricity was consumed by 21.8 % and renewable energy sources (RES) covered about 7.2 %. A decrease was reported in the use of conventional fuels, especially of petroleum, following the recession in Europe, while RES consumption is following an increasing trend as imposed by the 20-20-20 target in the EU (20 % energy reduction—20 % RES-produced energy—20 % CO_2 emissions reduction) (EU 2014).

In dwellings (Fig. 3.6), natural gas is the dominant energy source (37 %), followed by electricity (25 %) and wood (14 %), both of them presenting an increasing trend, as opposed to oil (with a 13 % share) that is gradually being phased out in the EU but remains significant in island countries (Lapillone et al. 2014b). In the tertiary sector (Fig. 3.6), the main energy carriers are electricity and gas, covering about 78 % of total energy consumption. A remarkable increase is observed in electricity consumption, from 33 % in 1990 to 43 % in 2011, while gas remains stable at 35 % since 2000 (Lapillone et al. 2014a).

Specific area electricity consumption varies significantly by building type and country, being higher in Nordic countries (Norway (166 kWh/m^2), Sweden, and Finland (128 kWh/m^2)), due the use of electricity for space heating, while the average value in the EU is 67 kWh/m^2.

Although energy efficiency measures have been introduced in several countries over the last years, about 25 % of the progress in energy efficiency for space

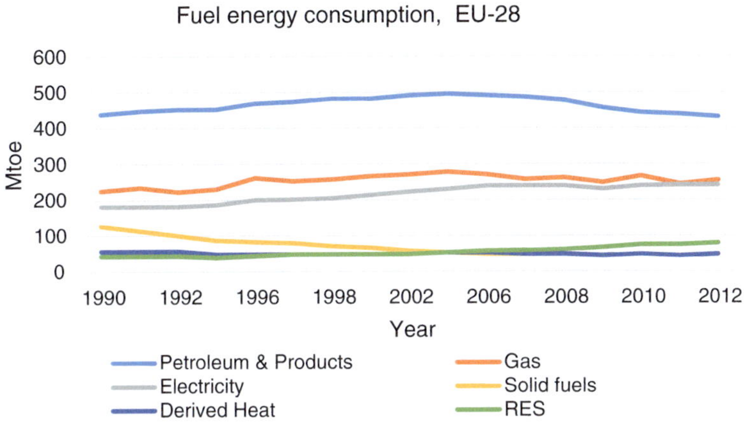

Fig. 3.5 Energy consumption trend in the EU-28 by fuel type (based on data by EU 2014)

heating has been offset by the larger area of new dwellings and by increased heating comfort standards (wider use of central heating systems). An increased use of RES (Fig. 3.7), especially in buildings, has been reported over the last years (Fig. 3.7 and Table 3.2); this is attributed to incentives introduced in all Member Countries for mandatory use of RES—e.g., the mandatory use of solar collectors to partially cover hot water needs in order to reach the renewable energy targets by 2020. The reduction in the cost of RES systems also contributed to this outcome.

The dominant renewable energy source in the building sector is solar energy, using collectors for heating production—mainly for domestic hot water—and photovoltaics for electricity production. Cyprus has the highest installed area of solar collectors in Europe, with 78 % of dwellings having installed solar collectors in 2011, followed by Greece with about 30 % (Fig. 3.8). Austria is in third place, with about 18 % of the dwellings having installed solar collectors in 2011 as compared to 3 % in 1990. Austria is considered the benchmark for solar collectors for countries with medium solar radiation, and Cyprus for countries with high solar radiation (Lapilloneet al. 2014b).

3.3 Means of Reducing Energy Consumption

3.3.1 Energy Efficiency

As buildings are responsible for 38 % of the EU's total CO_2 emissions (Saheb et al. 2015), the reduction of the energy demand of buildings and decarbonization of the energy supply for residential and tertiary buildings is of great importance in

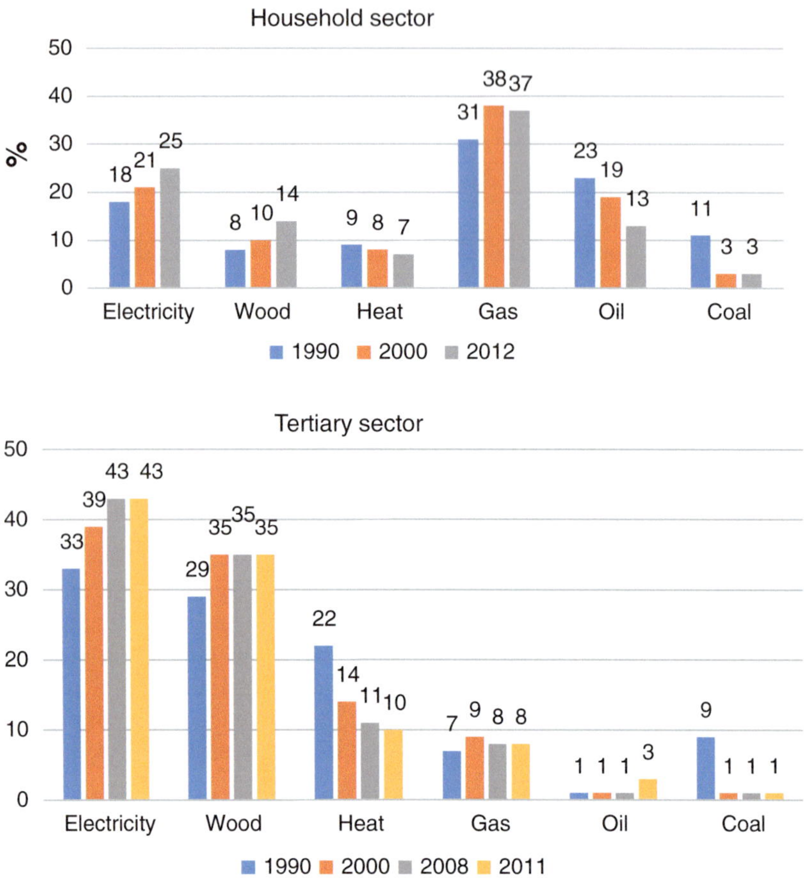

Fig. 3.6 Energy consumption by fuel type in households and in the tertiary sector (based on data by Lapillone et al. 2014a, b)

order to achieve the EU climate and energy goals for 2020. The EU has set a target for all new buildings to be nearly zero-energy: 2020 (DIRECTIVE 2010). Energy efficiency is the first priority for reaching this target. "Nearly zero energy" buildings are characterized by very high energy performance, while the small amount of energy required should be covered by RES.

Considering that construction of new buildings has been declining over the last years (Fig. 3.1), and that many older buildings were constructed without any thermal or energy regulation in several European countries, the energy renovation of existing buildings has a high potential for achieving energy efficiency.

About 35 % of the EU's buildings are over 50 years old. Residential buildings comprise the largest segment of the European building stock; these buildings are responsible for most of the building sector's energy consumption and CO_2 emissions (Chadiarakou and Santamouris 2015). In Greece, there are approximately

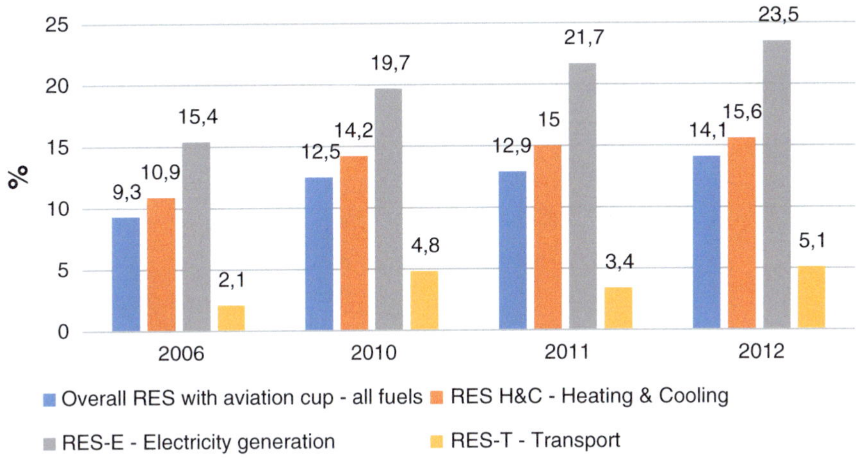

Fig. 3.7 RES share in the gross final energy in EU-28 (based on data from the EU 2014)

Table 3.2 RES for heating and cooling share in the gross final energy in selected countries (based on data from the EU 2014)

	2010	2011	2012
Country		%	
Greece	17.8	19.4	24.4
Croatia	13.0	15.6	18.3
Spain	12.6	13.6	14.0
Portugal	33.7	35.0	33.0

4,000,000 buildings; around 70 % of them were constructed before the implementation of the Thermal Insulation Regulation (TIR) in 1989 and so they do not have any kind of thermal insulation (Chadiarakou and Santamouris 2015). In Portugal, 50 % of dwellings were constructed before the enactment of the TIR in 1990, indicating that there is a great potential for energy efficiency renewal in these buildings.

New buildings in general consume less than 3–5 l of heating oil/m^2/ year, while older buildings on average consume about 25 l, even as many as 60 l/m^2/year. It is estimated that by improving the energy efficiency of buildings, the total EU energy consumption can be reduced by 5–6 %, resulting in a reduction of CO_2 emissions by about 5 % (EU—Buildings 2015).

The quality of the buildings determines the energy requirements, which in turn affects energy consumption and energy bills. The usual way to reduce energy expenses is to compromise on comfort for heating and cooling. In 2012, 11 % of the EU population was unable to keep their homes warm in the winter, and 19 % living in dwellings not comfortably cool in the summer. There is growing concern regarding the number of EU citizens facing fuel poverty, as over 30 % of people in EU member states with per capita GDPs below the EU average faced fuel

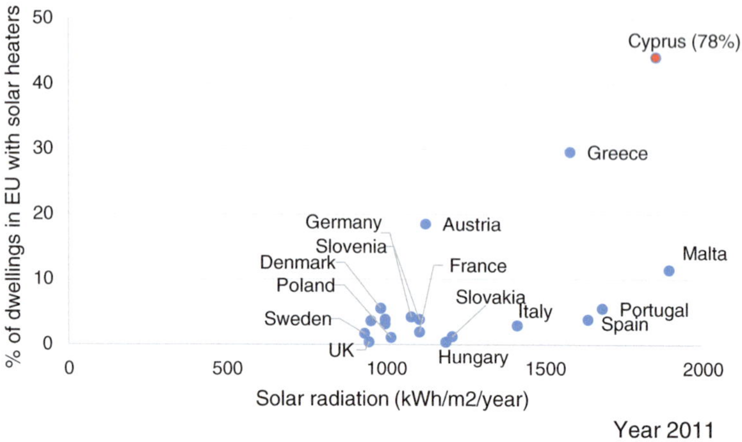

Fig. 3.8 Illustration of % installation of solar water heaters in dwellings and solar radiation (Year 2011) (based on data by Lapillone et al. 2014b)

poverty in Europe (Saheb et al. 2015). In a survey carried out in a sample of about 600 households in Greece, it was found that 2 % of the higher income households and 14 % of the lower-income households were below the fuel poverty threshold (Santamouris et al. 2013).

The key to reducing energy needs and consequently household energy bills is the improvement of a building's quality by insulating the building's envelope (walls, roof, and floor), replacing windows with new, high thermal performance ones, and maintaining or replacing heating and cooling systems by the best available technologies. At the EU level, energy renovation is given high priority in the "Strategy for a Resilient Energy Union" in order to meet the targets of its climate change policy (COM 2015, 80). Energy renovation costs should be less than 25 % of the value of the home; otherwise it is considered more sensible to construct a completely new building than to renovate the existing one (Saheb et al. 2015).

In order to achieve high energy efficiency and even buildings that produce energy, new technologies are required. Energy renovation will stimulate technological innovation in the chain of the building sector (Table 3.3). In the future, buildings need to become "smart," connected to storage systems, smart grids, and vehicle/transport systems.

3.4 Embodied Energy of Structural Materials and Components

The fossil fuel consumed by the construction sector is not only restricted by the amounts of energy required for the operation of buildings (operation of electrical and mechanical installations for heating, cooling, lighting, etc.), but it also

Table 3.3 Technological innovation needs for the building sector (based on Saheb et al. 2015)

Economic activity	Building component and/or systems	Technology
Floor and wall covering	Building envelope	Advanced insulation material such as VIPs for nZEBs and space constrained applications
		Air sealing testing methodologies
		Reflective surfaces for roofing materials for southern Europe and dense urban areas
Painting and glazing	Windows	Double low-e conductive frames
		Triple glazing for northern Europe with low-e and low-conductive frames
		Energy plus windows with dynamic solar control and glass that optimize daylight
		Automatic solar controls and exterior solar shades and blinds with low e-film and high insulation
Plastering, joinery installation	Joinery	Air sealing
Plumbing, heat, electrical and air conditioning installation	Heating systems	Active solar thermal systems fully integrated to buildings
		Thermal energy storage (TES) materials and full systems with integrated ICT
	Heating cooling and hot water systems	Cost-efficient heat pumps
	Heating cooling and electricity production	Efficient and smart CHP
	Cooling systems	Sorption cooling systems driven by hot water
Building completion and finishing	Energy management	Building information and management (BIM)
		ICT for grid integration and consumers information

covers the fossil fuel energy consumed during the manufacture of structural materials and components, the energy required during the construction, maintenance, replacement and demolition works of structures. The amounts of energy hidden in building materials and components and associated construction work are called "embodied" or "embedded energy."

The amounts of energy needed for the production of structural materials and components include the energy needed for the extraction of the raw materials, transportation from the extraction site to the manufacturing installation, and for the manufacturing process. The energy used for the transportation of the raw materials depends on the availability of the raw materials in the region and on the means of transportation (e.g., truck, train, ship) and the distance. For local materials, which are usually simple materials, such as wood, stone, or adobe, transportation distances are relatively short. For high technology materials, such as metals or plastics, transportation distances may be long, and in some cases even several thousand kilometers. The energy consumed for the manufacturing process depends on the different machinery types used in the process. Ideally, secondary energy consumption during the manufacturing process should be considered that includes the energy consumed for the operation of the manufacturing installation e.g., for heating, cooling, lighting systems, the energy to manufacture major machinery and for machinery maintenance. The amount of energy required for the manufacture of building materials and components is known as "cradle-to-gate" embodied energy, which includes all the energy (in primary form) required to produce a finished product until it leaves the factory gate, without any further considerations. It is the most common specified boundary condition.

The energy needed for the transportation of finished products to a construction site also depends on the distance of the site from the factory and means of freight transportation. When all energy required for all phases, including raw materials extraction to delivery of the products to the construction site is considered, then the "cradle-to-site" embodied energy is implied. This definition is useful when looking at the comparative scale of building components and neglects any maintenance or end-of-life costs (Anderson 2015).

The amounts of energy required for the construction, maintenance, replacement and demolition of structures include the energy consumed by the mechanical equipment for the construction (referred to as "initial energy" accounting for all phases since raw material extraction), for maintenance and replacement (recurring energy) and for the demolition of structures, transportation of structural waste to the waste deposit site, and for waste management (demolition waste). The whole life cycle energy of a structure component is known as "cradle-to-grav": embodied energy, which defines energy spent by a building component or the whole building throughout its life. This embodied energy does not imply that waste is landfilled. If products are recycled at the end of life (Anderson 2015), assessment would account for this, ensuring that the benefits of recycling for a material are not doubly counted for both the use of recycled content and its recycling at the end of life. The definition of "cradle-to-grave" embodied energy is a far more useful one when looking at a building or project holistically, although it is much more complicated to estimate it (Anderson 2015). In order to determine the magnitude of the embodied energy, an accounting methodology is required that sums up the energy inputs over the major part of the materials supply chain or the whole life-cycle (Hammond and Jones 2008).

The emissions related to energy consumption that are responsible for global warming and climate change, e.g., CO_2 and other greenhouse gases, are also considered and give rise to the concept of "embodied carbon." Thus,the embodied carbon footprint is the amount of carbon (CO_2 or greenhouse gases that cause climate change added up as CO_2 equivalents (CO_2 eq)) released to the atmosphere for the manufacturing of a product and/or construction work.

The embodied energy is commonly expressed in terms of MJ per unit weight (kg or tonne) or area (m^2). It is significant to split the required energy into conventional energy and renewable energy. The environmental implications of the energy needed to make a kilogram of product are expressed in terms of the quantity of gas emissions of "CO_2"and "SO_2" that are emitted at the atmosphere during the construction activity. The amounts of produced CO_2 and SO_2 emissions depend on the fuel type (oil, gas, etc.) that is used during the various stages of the construction activities but also at the equivalence that exists on the national level between the quantities of gas emissions per produced kWh of fuel.

In order to determine the embodied energy, it is necessary to trace the flow of procedures and energy-related actions in the various stages of the life cycle of construction—depending on the boundaries set for the analysis, i.e., whether the factory, site, or whole life cycle. Any change in the production process will have an impact on the embodied energy and the associated environmental parameters of the produced structural materials. Knowledge of the quantities of raw materials and the amounts of energy used throughout the various stages of the production procedure (e.g., extraction and manufacturing) are needed to estimate the embodied energy of a product.

Values of embodied energy and carbon (dioxide) emissions for common structural materials are depicted in Table 3.4, as drawn from the *Inventory of Carbon and Energy* (ICE) (Hammond and Jones 2011), an open-access database listing almost 200 different materials, developed at Bath University (UK).

3.5 Assessment Methods

3.5.1 Introduction

As environmental issues continue to become increasingly significant, buildings become more energy efficient and the energy needs during their operation decreases, aimed at "nearly zero energy" buildings. Thus, the energy required for construction, and consequently for the material production, is becoming more and more important.

A significant contribution to the efforts for reducing the environmental impacts from construction activities is to evaluate the environmental consequences of a

Table 3.4 Embodied energy and equivalent emissions of CO_2 of structural products

Material	Embodied energy (MJ/kg)	Embodied carbon ($kg/CO_2/kg$)	Embodied carbon ($kg/CO_{2\,e}/kg$)
Stone	(Difficult to select data, high standard deviations)		
General	1.26 (?)	0.073	0.079
Granite	11.00	0.64	0.70
Limestone	1.50	0.087	0.09
Marble	2.00	0.116	0.13
Marble tile	3.33	0.192	0.21
Sandstone	1.00 (?)	0.05	–
Slate	0.1–1.0	0.006–0.058	0.007–0.063
Steel			
Steel (General UK (EU)—Average recycled content of 59 %)	20.10	1.37	1.46
Virgin	35.40	2.71	2.89
Recycled	9.40	0.44	0.47
Stainless steel	56.70	6.15	–
Aluminium			
Aluminium general, incl. 33 % recycled	155	8.24	9.16
Virgin	218	11.46	12.79
Recycled	29	1.69	1.81
Copper			
Copper (average incl. 37 % recycled)	42	2.60	2.71
Bricks			
Common clay brick (simple baked, incl. Terracotta and bricks)	3.0	0.23	0.24
Mortar			
Mortar (1:3 cement: sand mix)	1.33	0.208	0.221
Mortar (1:6)	0.85	0.127	0.136
Mortar (1:1:6 cement: lime: sand mix)	1.11	0.163	0.174
Mortar (1:2:9)	1.03	0.145	0.155
Tiles – Flooring			
Clay tile	6.5	0.45	0.48
Ceramic tiles and cladding panels	12.00	0.74	0.78
Terrazzo tiles	1.40	0.12	–
Vinyl flooring (General)	68.60	2.61	3.19
Insulation			
General insulation	45.00	1.86	0.24

(continued)

Table 3.4 (continued)

Material	Embodied energy (MJ/kg)	Embodied carbon ($kg/CO_2/kg$)	Embodied carbon ($kg/CO_{2e}/kg$)
Rockwool (Slab) (gradle to grave)	16.80	1.05	1.12
Fiberglass (Glasswool)	28.00	1.35	–
Mineral wool	16.60	1.20	1.28
Cellular glass insulation	27.00	–	–
Cellulose insulation	0.94–3.3	–	–
Expanded polystyrene	88.60	2.55	3.29
Polyurethane insulation (rigid foam)	101.50	3.48	4.26
Woodwool (board)	20.00	0.98	–
Wool (recycled)	20.90	–	–
Cork	4.0	0.19	–
Straw	0.24	0.01	–
Perlite expanded	10.00	0.52	–
Perlite natural	0.66	0.03	–
Windows			
Glass (primary)	15.00	0.86	0.91
1.2*1.2 double glazed (air or argon filled):			
Aluminium framed	5470	279	–
PVC framed	2150–2470	110–126	–
Timber framed	230–490	12–25	–

construction activity or product manufacture during each stage of its life cycle. This has led to the development of various environmental life cycle assessment approaches, in which not only is the embodied energy used, but the raw materials and water (input), together with the pollutants and wastes (output) released into the environment (air, soil, water) during the construction activity or manufacturing procedure, are traced and quantified.

The environmental assessment methods can be classified thus:

Environmental assessment of structural materials and processes:

- assessment of the manufacturing procedure where all phases of the production procedure, up to the delivery of the final construction product, are assessed
- assessment and classification of structural materials and components through certification schemes (e.g., eco-label).

Pieces of information obtained from both approaches can be used for environmental assessment of alternative structural materials that can be used in construction.

Environmental assessment of the overall construction project, considering the environmental assessment of the individual construction materials and components.

The *Life Cycle Analysis* (LCA) was introduced in 1997 with the International Standard ISO 14040, which described the principles and framework of LCA (ISO 14040: Environmental Management—Life cycle Assessment—Principles and Framework). In 2006, two new standards were released, covering all relevant standards issued up until this year, the ISO 14040 2006 and ISO 14044 2006, with ISO 14040 referred to the *Principles and Framework* and 14044 to the *Requirements and Guidelines of LCA*. These standards stimulated the development of several national methodologies of structural products. One of the first ones was the "Environmental Profile Methodology" developed by Building Research Establishment (BRE) (UK) in 1999 (*BRE Methodology for Environmental Profiles of Construction Materials, Components and Buildings*), for assessing the cradle-to-grave environmental impact of construction materials (BRE 2007).

A set of international standards was issued dealing with environmental labelling (ISO 14020 to 14025) with ISO 14025, published in 2006 (ISO 14025:2006: *Environmental Labels and Declarations-Type III Environmental Declarations—Principles and Procedure*) that established the principles and specified the procedures for developing *Environmental Product Declarations* (EPD). An ISO specialized-for-construction materials was published in 2007, the ISO 21930 (ISO 21930:2007: *Sustainability in Buildings and Civil Engineering Works—Environmental Declaration of Building Products*) that provides the principles and requirements for Type III environmental declarations (EPD) of building products and encompassed the different national approaches while ISO 21931 (ISO 21931-1:2010. Sustainability in building construction—Framework for methods of assessment of the environmental performance of construction works—Part 1: Buildings—identifies and describes issues to be taken into account in the use and development of methods of assessment of the environmental performance for new or existing buildings in their design, construction, operation, maintenance and refurbishment, and in the deconstruction stages. The suite of standards covering issues of sustainability of buildings is synoptically described in Fig. 3.9.

The need for harmonization into a common methodology across Europe of the different approaches for assessment of the environmental impacts associated with construction products and buildings, was covered with the issue of European Standards:

- the EN 15804 (EN 15804+A1: 2012. Sustainability of construction works—environmental product declarations—core rules for the product category of construction products) for development of EPDs of construction products and services and
- the EN 15978 (EN 15978: 2011. *Sustainability of Construction Works—Assessment of Environmental Performance of Buildings—Calculation Method*) that specifies the calculation method, based on Life Cycle Assessment (LCA) and other quantified environmental information, to assess the environmental performance of a building, where a set of environmental indicators are assessed over the full life cycle of a building. The approach to the assessment covers all stages of the building's life cycle and is based on data obtained from EPD (EN 15804) and other information necessary and relevant for carrying out the assessment (Fig. 3.10).

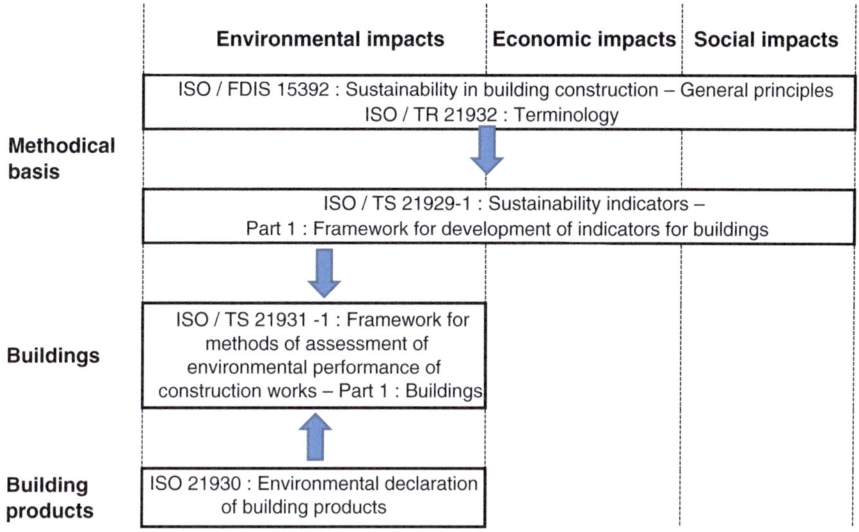

Fig. 3.9 Schematic representation of international standards dealing with sustainability in building construction (based on ISO 21930:2007(en): ISO-Online Browsing Platform)

Several countries, such as the UK, France, Germany, the Netherlands, Sweden, Norway, Spain, Portugal, Italy, and the United States have adopted the EN 15804 for structural materials.

The European Commission recognized the need to complement existing energy policies for buildings with policies for resource efficiency, looking at a wider range of resource use and environmental impacts during the life cycles of buildings, and thus the *Resource Efficiency Roadmap* was adopted in September 2011 (EU Roadmap 2015). It also explained in the publication on *Strategy for the Sustainable Competitiveness of the Construction Sector and Its Enterprises* on July 31, 2012 (COM 2012, 433), that among the main challenges that the construction sector faces up until 2020—in order to grow stronger and more viable in the future—is improvement in resource efficiency and environmental performance. It thus highlighted the need for the development of methods to assess the "environmental performance" (EC-Sustainable Buildings 2015).

As stated in *Resource Efficiency Opportunities in the Building Sector*' (COM 2014, 445), the European Commission will establish a common framework of core indicators to be used to assess the environmental performance of buildings throughout their life cycles, and thus allow comparability and provide users and policy makers with easier access to reliable and consistent information. The framework will focus on the most essential aspects of environmental impacts and be flexible so that it can be used on its own or incorporated as a module in existing and new assessment schemes next to their larger sets of indicators. It will be based partly on existing work, such as the EN15978 technical standard, as well as existing voluntary commercial certification schemes for buildings.

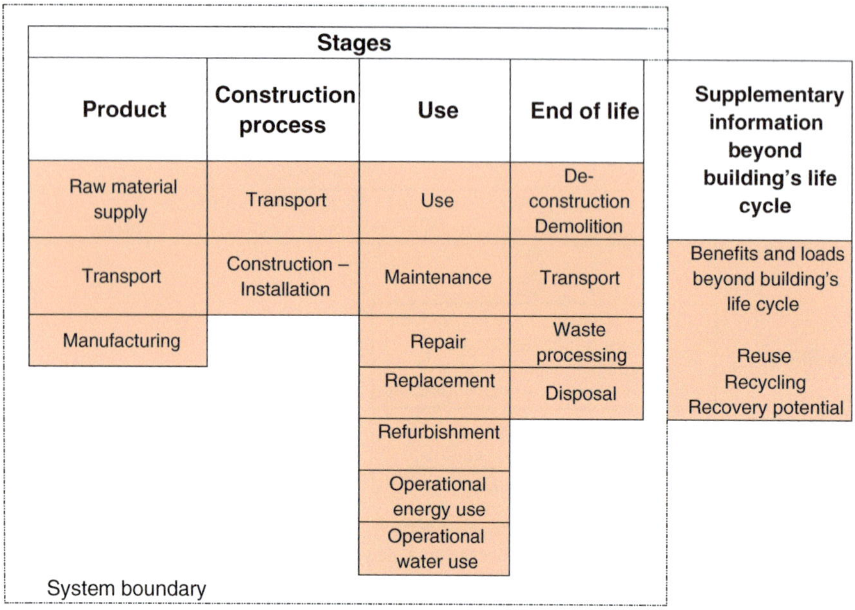

Fig. 3.10 Illustration of the modules considered in analysis of EN 15804/15978

3.5.2 Environmental Assessment of Structural Products and Processes

The LCA is one of the most popular techniques used as an analytical tool for the assessment of the environmental impacts caused by a material, a process, or an application during its whole life cycle. It identifies and quantifies great amounts of input about the raw materials, ingredients, products, manufacturing process, and energy use, as well as input for the environmental aggregation during all stages of the life cycle of the products and processes. This means:

- the environmental destruction from the raw materials extraction
- the pollutant emissions produced during the production process
- the availability of raw materials
- the recovery of materials
- the energy consumed for the manufacture of the final products
- the amounts of waste produced during the manufacturing process and the degree of pollution caused, and
- the implications for air quality and human health.

A wide range of the environmental aspects of building materials could be aggregated and quantified with an inventory analysis into a limited set of the recognizable impact indicators (e.g., global warming, ozone depletion, acidification) that are used to quantify and aggregate all environmental aspects of building materials and processes. LCA models differ, depending on the boundaries of the system they examine (linked to the goal and scope of the environmental assessment and the boundaries set e.g., cradle-to-grave), the input/output environmental flows examined (linked to the life cycle inventory (LCI) model), and the type and number of indicators assessed (associated with the evaluation model). The differentiation among the available assessment tools depends on their user friendliness, the magnitude of their data bases, and their subject specification as well as purchase cost.

The tools can be classified into:

1. Those assessing construction products such as the detailed SimaPro (UK) and GaBi (Germany) tools, the Ide-Mat (Netherlands), EQUER (France), KCL-ECO (Finland), BEES-Building for Environmental and Economic Sustainability (USA), ATHENA Impact Estimator for Buildings (Canada), and LISA-LCA in Sustainable Architecture (Australia) tools, and
2. the tools for environmental assessment of production processes, such as the GEMIS (Germany), UBERTO (Switzerland), TEAM-Tool for Environmental Analysis and Management (France), and BOUSTEAD (UK).

The LCA requires a great amount of input, and in several cases the collection and process of the required pieces of information was difficult and time-consuming, with a high overall cost because many companies consider these data to be confidential, or else they do not have detailed records.

3.5.3 Environmental Assessment Methods for Buildings and Construction Works

There has been a growing movement towards sustainable construction since the second half of the 1980s, leading to the development of various methods for evaluating the environmental performance of buildings. This category covers methods and rating schemes for environmental assessment of the entire construction project. Most existing building assessment systems attempt to serve two functions at once (Larsson 2014): to guide developers and designers in their attempts to design for high performance, and to measure and assess building performance in as objective a way as possible.

The first rating tool was launched in 1990 with the introduction of the British BREEAM rating tool, and five years later was followed by the French HQE system, and the GBTool launched by the Natural Resources Canada in 1996, and by LEED in 2000 in the USA. A year later, the GASBEE followed in 2001 (Japan), in 2002 the Green Globe (BREEAM Canada) and the Green Star in Australia, and in 2006 the DGNB from the German Green Building Council (GBC).

3.5.3.1 BREEAM (BRE Environmental Assessment Method)

The BREEAM method was developed by the Building Research Establishment (BRE) in the United Kingdom and was first launched in 1990 (www.breeam.org). It is one of the first methods developed for the environmental assessment of new and existing buildings and one of the world's foremost environmental assessment methods and rating systems for buildings. It assesses the environmental performance of buildings in the following areas (BREEAM 2015):

- Management: overall management policy, commissioning site management and procedural issues, waste recycling, pollution minimization
- Energy: energy consumption and CO_2 emissions
- Health and well-being: indoor and outdoor environment issues affecting health and well-being, such as adequate ventilation, lighting, thermal comfort
- Air and water pollution issues: leakage detection systems, in situ processes, RES, pollution prevention plan
- Transport: transport-related pollutants and location-related factors
- Land use: greenfield and brownfield sites, reuse of site soil, polluted soil use
- Ecology: ecological value conservation and enhancement of the site, land with low or the lowest ecological value, preservation of significant ecological systems in the land's vicinity, minimization of the impact in the biodiversity of the area
- Materials: environmental implications of structural materials throughout their life cycle
- Water: water consumption and efficiency, reduction of consumption, leakage detection.

Developers and designers are encouraged to consider these issues from the earliest phases of the project in order to achieve a high BREEAM rating. A credit is awarded for each area and a set of environmental weightings enables the credits to be added together to produce the final overall score of the project. The certificate awarded to a building has one of the following ratings: Pass, Good, Very good, Excellent, Outstanding, or a five-star rating system is provided.

The BREEAM method covers not only various building types in the tertiary sector, residential buildings and communities at the design stage, but also existing buildings. It covers offices, schools, healthcare buildings, retail outlets and shopping malls, industrial buildings, etc., while for the residential buildings there are various certification schemes: (1) The British Government's Code for Sustainable Homes (CSH) replaced EcoHomes for the assessment of new housing in England, Wales and Northern Ireland; (2) BREEAM EcoHomes for new homes in Scotland; and (3) BREEAM Multi-residential covering buildings housing many individuals and offering shared facilities. BREEAM Communities is a certification scheme to certify development proposals at the planning stage. BREEAM In-Use is a scheme to help building managers reduce the running costs and improve the environmental performance of existing buildings. BREEAM is used not only in the UK but in several other countries under National Scheme Operators (NSOs) as well.

3.5.3.2 SBTOOL (Sustainable Buildings Tool)

The SBTOOL is an international system of environmental and sustainability assessment of buildings. It is a generic framework to support the sustainability performance assessment of buildings that can be easily adapted to regional and building type variations, and to the use of different languages. It has been under development since 1996, and various versions of the system have been tested since 1998. It is the successor of the GBTool software tool (green building tool) that was launched by Natural Resources Canada for the needs of the International Initiative "Green Building Challenge" (GBC), an international program of more than 25 countries that aimed at setting common criteria for naming a building a "green building." Since 2002, the responsibility of the tool has been handed over to the International Initiative for a Sustainable Built Environment (iiSBE) (www.iisbe.org). The tool is in spreadsheet form and includes criteria for various parameters, such as site selection, energy program and development, indoor environmental quality, long-life efficiency, and social and financial criteria. Weighting factors are defined for each criterion, based on national or local assessment conditions and construction practices. The scoring process in SBTool relies on a series of comparisons between the characteristics of object building and national or regional references for minimally acceptable practice, "good" practice, and "best" practice. Buildings are rated in the range of -1 for non-acceptable performance, and from $+1$ up to $+5$ for "minimum accepted" performance, to "best practice" accordingly, with $+3$ corresponding to "good practice."

The SBTool system consists of two distinct assessment modules that are linked to phases of the life-cycle; one for Site Assessment, carried out in the Pre-Design phase, and another for Building Assessment, carried out in the Design, Construction or Operations phases (Larsson 2014). The SBTool has been improved and updated over the years; the last version, which was updated in 2014, is calibrated from the countries participating at iiSBE and its results have been presented at the SB-Sustainable Buildings Conferences.

3.5.3.3 Green Globes

Green Globes (Greenglobe 2015) is an online green building rating and certification tool that is used for the assessment of existing and new buildings, primarily in Canada and the USA. There are three Green Globe modules, for:

- New construction/significant renovations (for commercial, institutional and multi-residential building categories, including offices, school, hospitals, hotels, academic and industrial facilities, warehouses, laboratories, sports facilities, and multi-residential buildings)
- Interiors of commercial buildings (i.e., office fit set-ups)
- Existing buildings (offices, multi-residential, retail, health care, light industrial).

The Green Globes for Existing Buildings initially developed in 2000, is based on the BREEAM Canada edition and shortly afterward the New Buildings Canada module was launched. The system was adopted in the US in 2004 as an alternative to the LEED building rating system with the US rights acquired by the Green Building Initiative. It is an online assessment protocol, rating system, and guidance for design, operation, and management of sustainable buildings. It is questionnaire-based with pop-up tips that show the applicable technical tables that are needed to reply to the questions, and is structured as a self-assessment to be done in-house, by a design team and the project manager. Users can see how points are being awarded and how they are scoring.

Green Globes for Existing Buildings utilizes weighted criteria where projects earn points relative to their impact on (or benefit to) the sustainability of the building on a 1000-point scale, in seven categories. A minimum number of points in each category is required. The three key performance indicators with the highest points are energy efficiency, materials choices and resource consumption, and indoor environmental quality. Additional environmental assessment areas include project management, water, and emissions and other impacts (e.g., site for new constructions).

3.5.3.4 LEED® (Leadership in Energy and Environmental Design)

The LEED® is one of the dominant assessment methods in the USA and worldwide. It was initially developed in the US in 1998; its development was supervised by the USGBC (USGBC 2015), a non-profit coalition of building industry leaders.

It encompasses five rating systems: (1) building design and construction (new construction, core and shell, schools, retail, hospitality, data centers, warehouses and distribution centers, and healthcare); (2) interior design and construction (commercial interiors, retail and hospitality); (3) operation and maintenance of existing buildings (schools, retail, hospitality, data centers, and warehouses and distribution centers); (4) neighborhood development, i.e., new land development projects or redevelopment projects containing residential uses, nonresidential uses, or a mix; and (5) homes (single family, low-rise multi-family, i.e., one to three stories, or four to six story mid-rise multi-family, including homes and multi-family low-rise and mid-rise).

Each rating system comprises different credit categories, such as materials and resources, energy and atmosphere, water efficiency, location and transportation, sustainable sites, indoor environmental quality, innovation, and regional priority credits. Additional credit categories are examined for neighborhood development: smart location and linkage, neighborhood pattern and design, and green infrastructure and buildings.

A project receives LEED certification if it satisfies prerequisites and earns points in each credit category. Prerequisites and credits differ for each rating system; there are four levels of certification according to the points awarded to a project: certified (40–49 points); silver (50–59 points); gold (60–79 points); and platinum (>80 points).

3.5.3.5 CASBEE (Comprehensive Assessment System for Building Environmental Efficiency)

The CASBEE method has been in existence since 2001 from the "Japan Sustainable Building Consortium" for the needs of the Japanese construction market CASBBEE (2015). An English version is also available (www.ibec.or.jp/CASBBEE/english).

CASBEE considers the whole architectural design process, starting from the pre-design stage and continuing through design and post-design stages, covering the various building categories, e.g., offices, schools, and apartments. It is composed of four basic assessment tools: (1) CASBEE for Pre-design; (2) CASBEE for New Construction; (3) CASBEE for Existing Building; and (4) CASBEE for Renovation. Tools for specific applications were also developed—for example CASBEE for Market Promotion, CSABBE for Heat Island, CASBEE for Urban Development, CASBEE for Cities, and CASBEE for Home (detached houses).

It recognizes four assessment fields: (1) energy efficiency; (2) resource efficiency; (3) local environment; and (4) indoor environment. An indicator is derived, and the "built environment efficiency" (BEE), which is the core concept of CASBEE, is calculated as the ratio of parameters describing the "built environmental quality" (Q) to parameters describing the "built environment load" (L).

$$\text{Built Environment Efficiency (BEE)} = \frac{Q\,(\text{Built environment quality})}{L\,(\text{Built environment load})}$$

The "built environmental quality" (Q), is defined in terms of $Q1$ indoor environment, $Q2$ quality of services, and $Q3$ outdoor environment on site and the "built environment load" (L), in terms of $L1$ energy, $L2$ resources and materials, and $L3$ off-site environment.

Some local governments in Japan have introduced as mandatory for building permits to include assessment with the CASBEE method, in the same way as the Energy Saving Plan. Since December 2011, 24 local governments have been introduced the reporting system as their environmental policies.

3.6 Discussion

Buildings are the largest energy consumers in Europe, accounting for approximately 40 % of the final energy consumption in 2012, while the construction and use of buildings accounts for about half of all extracted materials and about a third of water consumption in the EU. The construction sector also generates about one-third of all waste.

For many years, the greatest importance from the environmental point of view was given at the operation phase of buildings—which is actually the longest period in a construction's life. After the first oil crisis in the 1980s, priority was given to reducing the energy that is consumed from a building during its operation. However, since the beginning of the 1990s, when the environmental consequences from the other three construction phases began to increase (e.g., reduction of natural resources reserves, management of construction waste, and difficulties in finding safe sites for construction waste deposits), a holistic approach was initiated that considered the environmental performance of structures and the consequences of the construction works on the environment. This approach resulted in implementing "sustainability" as a new concept in structures' consideration, aiming for the least possible burden on the environment during all phases of construction.

Environmental concerns resulted in an increased use of RES and especially in buildings over the last few years, attributed to incentives introduced in all EU member countries for the mandatory use of RES and reduction in the cost of RES systems. The dominant renewable energy source in the building sector is solar energy, with the use of solar collectors, mainly for domestic hot water and photovoltaics for electricity production.

But a growing concern in the EU is the increasing number of people facing "fuel poverty," as more than 30 % of people in EU member states with per capita GDPs below the EU average face fuel poverty. The key to reducing energy needs, and consequently household energy bills, is the thermal improvement of buildings (envelope insulation, window replacement, and maintaining or replacing heating and cooling systems by the best available technologies). Considering also that construction of new buildings has declined lately and that there are many old buildings in several European countries that were constructed without any thermal or energy regulations, the energy renovation of existing buildings has great potential for energy efficiency.

At the EU level, energy renovation is given high priority in the *Strategy for a Resilient Energy Union* in order to meet the targets of its climate change policy. It is also expected that energy renovation will stimulate technological innovation in the chain of the building sector. In the future, buildings need to become smart, connected to storage systems, smart grids, and vehicles/transport systems. It is estimated that by improving the energy efficiency of buildings, the total EU energy consumption can be reduced by 5–6 %, resulting in a reduction of CO_2 emissions by about 5 %.

The European Commission recognized the need to complement existing energy policies for buildings with policies for resource efficiency, looking at a wider range of resource use and environmental impacts, across the life cycle of buildings (*Resource Efficiency Roadmap,* 2011). Such policies would "contribute to a competitive construction sector and to the development of a resource efficient building stock." But energy efficiency measures should be examined, taking into consideration their environmental impacts over the entire life cycle of a building; otherwise, impacts may be overlooked or additional problems may be created at a later stage of their life cycle—for example, some steps tp improve a building's energy efficiency during its operation may result in expensive and difficult recycling in the future.

The fossil fuel consumed in the construction sector is not only restricted by the amounts of energy required for the operation of buildings (operation of electrical and mechanical installations for heating, cooling, lighting, etc.), it also covers the fossil fuel energy consumed during the manufacture of structural materials and components, the energy required during the construction, maintenance, replacement, and demolition of structures. These amounts of energy hidden in building materials and components and associated construction work are called "embodied or embedded energy," while the corresponding environmental implications are expressed in terms of the quantity of emissions of CO_2 and SO_2 that are emitted into the atmosphere during the construction and give rise to the notion of "embodied carbon."

As environmental issues continue to become increasingly important, buildings become more energy efficient and the energy needs during their operation decrease, aiming at "nearly zero energy" buildings. Thus the energy required for construction and, consequently, for the material production, is becoming more and more important.

A significant contribution in the efforts for reduction of the environmental impacts from construction activities is to evaluate the environmental consequences of a construction activity or product manufacture in each stage of its life cycle. This has led to the development of various environmental life cycle assessment approaches. In these approaches, not only the embodied energy, but the raw materials and water used (input), together with the pollutants and wastes (output) released into environment (air, soil, water) during the construction activity or manufacture procedure are traced and quantified. Across these lines, a set of International and European Standards was issued on environmental labelling of products (ISO 14020 to 14025), specialized for construction materials (ISO 21930, EN 15804), or assessment methods of the environmental performance for the whole building (ISO 21931, EN 15978) was developed.

As buildings are responsible for about 38 % of the EU's total CO_2 emissions, reduction of the energy demand of buildings and decarbonization of the energy supply in buildings is of great importance in order to achieve the EU's climate and energy strategy for 2020. Decarbonization of the construction sector is the challenge for the future in order to develop environmental friendly construction.

References

Anderson J (2015) Embodied carbon and EPDs. Available via DIALOG. http://www.greenspec.co.uk/building-design/embodied-energy/. Cited 1 Mar 2015

BRE (2007) Methodology for environmental profiles of construction products. Watford, BRE. Available via DIALOG. http://www.bre.co.uk/filelibrary/greenguide/PDF/Environmental_Profiles_Methodology_2007_-_Draft.pdf). Cited 10 Mar 2015

BREEAM (2015). Available via DIALOG. http://www.breeam.org. Cited 10 Mar 2015

Campogrande D (2007) The European construction industry—facts and trends. European Construction Industry Federation (FIEC), ERA Convention, Berlin, 5–6 June

CASBBEE (2015). Available www.ibec.or.jp/CASBBEE/english. Cited 10 Mar 2015

Chadiarakou S, Santamouris M (2015) Field survey on multi-family buildings in order to depict their energy characteristics. Int J Sust Energy 34:271–281

COM (2012) 433 (2012) Communication from the commission on 'strategy for the sustainable competitiveness of the construction sector and its enterprises'. Brussels 31 July 2012
COM (2014) 445 (2014) Communication from the commission on 'resource efficiency opportunities in the building sector'. Brussels 01 July 2014
COM (2015) 80 (2015) Communication from the commission on 'a framework strategy for a resilient energy union with a forward-looking climate change policy'. Brussels 25 Feb 2015
DIRECTIVE 2010/31/EU of the European Parliament and of the Council of 19 May 2010 on the energy performance of buildings (recast), Official Journal of the European Union, 18 06 2010
EC-Sustainable buildings (2015) In: EC/Environment/Sustainable consumption and production/Sustainable buildings (last updated 04 Mar 2015)
EU (2014) EU energy in figures—pocketbook 2014. Luxembourg. ISBN 978-92-79-29317-7
EU—Buildings (2015) Available via DIALOG. http://ec.europa.eu/energy/en/topics/energy-efficiency/buildings. Cited 10 Mar 2015
FIEC (2014) Construction in Europe—key figures; Activity 2013. European Construction Industry Federation (FIEC) Available via DIALOG. http://www.fiec.eu. Cited 5 Mar 2015
Greenglobe (2015). Available via DIALOG. http://www.greenglobes.com. Cited 10 Mar 2015
Hammond G, Jones C (2011) Inventory of carbon and energy, 2nd edn. University of Bath, Bath
Hammond G, Jones C (2008) Embodied energy and carbon in construction materials. In: Proceedings of the Institution of Civil Engineers, Energy 161(Issue EN2): 87–98. doi:10.1680/ener.2008.161.2.87
ISO 21930:2007(en): Sustainability in building construction – Environmental declaration of building components. ISO-Online Browsing Platform. Available via https://www.iso.org/obp/ui/#iso:std:iso:21930:ed-1:v1:en. Cited in 20 April 2015
Lapillone B, Pollier K, Mairet N (2014a) Energy efficiency trends in tertiary in the EU. EU—Odyssey_Enerdata
Lapillone B, Pollier K, Samci N (2014b) Energy efficiency trends for households in the EU. EU—Odyssey_Enerdata
Larsson N (2014) Overview of the SBTool assessment framework. http://iisbe.org/system/files/SBTool%202014%20description%2016Jul14.pdf
Odyssee-mure.eu (2015) Available via DIALOG. http://www.odyssee-mure.eu/publications/efficiency-by-sector/buildings/buildings10.pdf. Cited 10 Mar 2015
Santamouris M et al (2013) Financial crisis and energy consumption: a household survey in Greece. Energy Build 65:477–487
Saheb Y, Bodis K, Szabo S, Ossenbrink H, Panev S (2015) Energy renovation: the Trupm card for the new start for Europe. JRC Science and Policy Reports, Luxembourg: Report EUR 26888, ISBN 978-92-79-43603-1 (PDF)
USGBC (2015). Available via DIALOG. http://www.usgbc.org. Cited 10 Mar 2015

Chapter 4
Challenges and Priorities for a Sustainable Built Environment in Southern Europe—The Impact of Energy Efficiency Measures and Renewable Energies on Employment

Mattheos Santamouris

Abstract The present chapter tries to identify the main advantages of energy conservation measures as well as of the renewable energies on economy and labour. Most of the existing studies foccusing in Southern European climates are presented discussed and analysed.

4.1 Introduction

This article discusses the main problems and challenges regarding the energy and environmental quality of the building sector in southern Europe—specifically, Cyprus, Italy, Malta, Greece, Portugal, and Spain. Construction presents one of the most dynamic sectors of the national economies in Europe. According to the European Construction Industry Federation (2013), the European construction sector in 2013 had a total construction output of €1162 trillion, which represents 51.5 % of the gross fixed capital formation in Europe and 8.8 % of the GDP. The magnitude in the Southern European countries was close to €245.5 billion, which represents almost 21 % of all European activities in construction. In parallel, the construction sector in Europe consists of 2.9 million enterprises, while almost 38 % of them are in the above-mentioned countries. According to the same source, 20.9 % of the economic activities had to do with civil engineering projects; 31.5 % were related to the construction of non-residential buildings, 27.5 % was oriented to the rehabilitation and maintenance of the buildings, and finally, 20.1 % was spent on the construction of new residences.

The construction sector is responsible for a very high percentage of the employment in Europe; the total number of jobs provided by construction in Europe is close to 13,870,000, of which 22.4 % are in the southern European countries.

M. Santamouris (✉)
Physics Department, Group Building Environmental Research Athens,
National and Kapodistrian University of Athens, Athens, Greece
e-mail: msantam@phys.uoa.gr

S.-N. Boemi et al. (eds.), *Energy Performance of Buildings*,
DOI 10.1007/978-3-319-20831-2_4

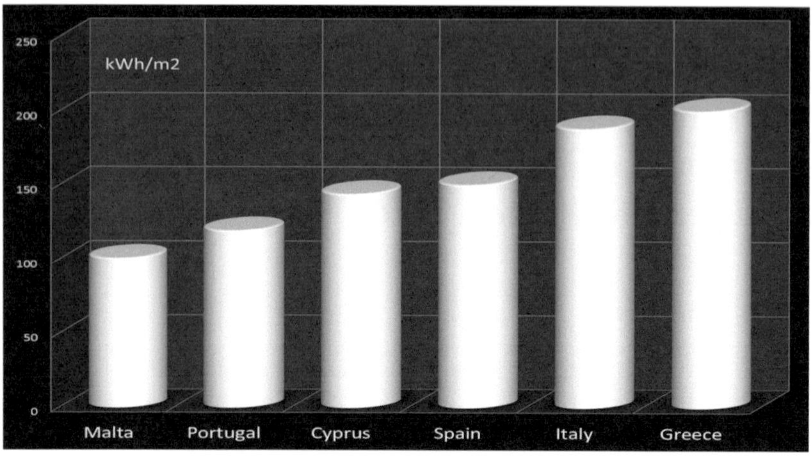

Fig. 4.1 Total unit consumption per m^2 in buildings (Entranze 2015)

The absolute energy consumption of the buildings in the European south is not among the highest in the European Union. Data provided by various databases (Entrance 2015) show that the average energy consumption of buildings in these countries varies between 101 and 201 kWh/m^2/year (Fig. 4.1). In comparison, the corresponding energy consumption in Germany, the UK, Latvia, and Luxemburg are 238, 265, 332, and 522 kWh/m^2/year respectively. However, a direct comparison of the energy consumption without a climatic homogenization does not provide a clear picture of the existing situation. When specific energy consumption data are put together regarding the climatic conditions, and reported under the same European average climate, one sees that the building's energy consumption in southern European is the highest on the continent (Fig. 4.2). Thus there is a

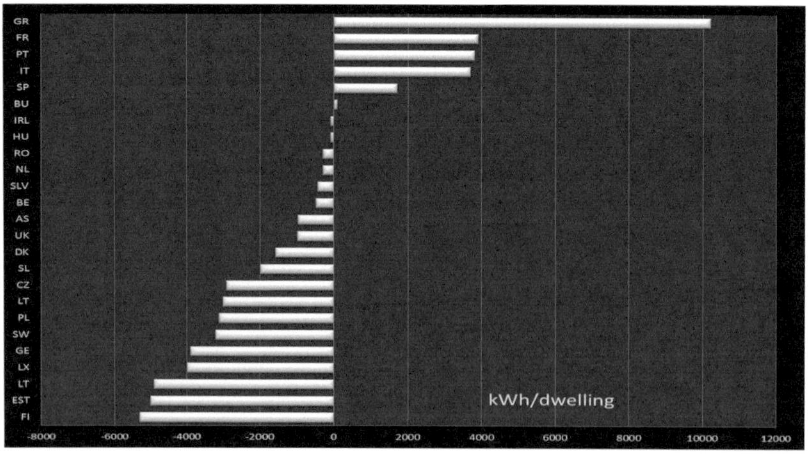

Fig. 4.2 Difference in household energy consumption per dwelling between the local climate and average EU climate by country, for space heating (Nikiel and Oxley 2011)

very high potential for energy conservation and substitution while the necessary activities to improve the energy and environmental quality of the building sector could be a first-order promoter for the local economies.

4.2 The Built Environment—Defining the Challenges and Priorities in Southern Europe

The relatively high energy consumption of the building sector of the southern European countries, in combination with the obligation of the European Union to fight climate change and decrease CO_2 emissions, offers important economic opportunities. Retrofitting the existing building stock can boost local economies, increase the standards of living, enhance employment, improve health conditions in the built environment, and protect the low-income households from the main economic and environmental challenges.

In particular, the main challenges and priorities regarding the possible policies related to the energy and environmental quality of the built environment should address the following areas of concern:

1. Fighting "energy poverty" and alleviating the possible impacts of the economic and social differentiation and stratification. Actually, and according to official European statistics, almost 80,000,000 people in the 28 countries of the EU lived below the poverty line in 2010 (Population and Social Conditions 2014); this corresponds to about 16.4 % of the total EU population. In parallel, almost 23 % of the European population is at risk of poverty or social exclusion. Energy poverty is defined as a situation where specific households are not able to satisfy the required energy services of the households, both from the social and material points of view (Bouzarovski 2011). Energy poverty results from specific economic problems characterized by low family income, living in non-appropriate and high energy-consuming houses (Lampietti and Meyer 2002). Households living under energy poverty conditions face serious problems of indoor environmental quality, have significant health problems, and may have much higher mortality and morbidity rates, while their quality of life has seriously deteriorated (Bouzarovski 2011).
2. To fight local and global climate change and protect citizens from extreme climatic phenomena using appropriate mitigation and adaptation techniques. Local climate change is related to higher urban ambient temperatures compared to the surrounding rural and suburban areas. The phenomenon is known as an "urban heat island" and is the most documented phenomenon of climate change (Santamouris 2001). Higher urban temperatures have a very serious impact on the energy consumption of buildings, affecting human health and outdoor comfort, while significantly increasing the ecological footprints of the cities (Santamouris 2015). As mentioned in Santamouris (2014a), the Global Energy Penalty per unit of city Surface, GEPS, varies between 1.1

and 5.5 kWh/m^2. In parallel, the Global Energy Penalty per person (GEPP) ranges from 104 to 405 kWh. It is also calculated that the UHI causes an average energy penalty per unit of surface close to 2.4 kWh/m^2, a Global Energy Penalty per unit of surface and per degree of the UHI intensity around 0.74 $kWh/m^2/K$, and a GEPP close to 237 kWh/p. In parallel, many studies investigating the impact of high ambient temperatures on hospital admission and on human mortality rates conclude that during periods of very high ambient temperatures (heat waves), the rate of hospital admissions for heat stroke, heat exhaustion, etc., increases considerably (Pirard et al. 2005). Also, all recent studies have shown that there is a very significant correlation between high ambient temperature and mortality. Most of the studies conclude that heat-related mortality increases rapidly for temperatures above 22–30 °C (Huynen et al. 2001). To compensate for the impact of global and local climate change, specific mitigation and adaptation technologies are proposed and applied. Among the most important of the techniques are those aiming to increase the albedo of the cities and the green spaces in cities (Santamouris 2014b). Results from existing applications of these mitigation techniques have shown that it is possible to decrease the average ambient temperature of cities up to 2.0 K and the maximum ambient temperature up to 4.0 K.

3. To increase the intelligence and smartness in the built and urban environment through the use of advanced ICT technologies able to provide enhanced services to the citizens and improve their quality of life. Recent technological developments in the field of ICT permit improving the indoor environmental quality of buildings and also provide intelligence at the urban level. Hundreds of smart industrial products are available and their applications are found in millions of buildings. The monitoring of many buildings equipped with advanced intelligence systems shows that important benefits of energy consumption of the buildings are achieved when the global environmental quality is seriously improved. In parallel, intelligent systems applied at the urban level provide additional useful information to the citizens and are an excellent coordination of the urban systems and infrastructures.
4. Minimize the energy consumption in the building sector through the use of clean and renewable energy systems and techniques presenting a significant added value for the local economies ability to contribute to the economic development. In fact, energy efficiency technologies and renewable energy systems, when applied to buildings, can significantly reduce the energy consumption and improve the indoor environmental quality with a very reasonable cost. The European Directive on the energy efficiency of buildings that asks for the design of near-zero energy buildings after the end of this decade offers a huge challenge to improve the quality of the existing building stock and create the market mechanisms necessary to promote economic development through the retrofitting of the existing buildings and the construction of new ones. The whole process cannot be seen as a technological problem, rather as a political and economic one. Although the necessary energy technology is commercially

available, economic, social, and political obstacles still exist and delay the implementation of energy-efficient technologies in the building sector.

5. Improve the indoor environmental quality of buildings, enhance thermal comfort throughout the year, and decrease problems of indoor air quality. Pure indoor environmental quality is a serious problem not only in southern European buildings. Low levels of thermal comfort and high indoor pollutant concentrations destroy the quality of life of European citizens, and in some cases threaten their lives. Using proper ventilation technologies and systems as well as low pollution materials and systems can enhance the quality of buildings and help citizens to live under proper conditions.
6. Make the advanced systems and technologies, as well as the appropriate energy services, able to improve the quality of life available and affordable to everyone, and educate the citizens in order to improve their knowledge and preferences regarding sustainable procedures and policies. It is extremely important that new developments in the field of building technologies are commercialized at reasonable prices and become available to the whole population. In parallel, it is very important to educate citizens on issues related to sustainability in order to understand and valorize all aspects related to the energy and environmental quality of the building sector.

In the following, the specific contribution of energy efficiency and renewable energy technologies to economic development and increase in employment is analyzed for the specific countries. Most of the existing studies and their results are presented.

4.2.1 Fighting Economic and Social Stratification Discrimination Through Energy Investment

It is well accepted that investment in the energy and environmental sectors have a very positive impact on the global development of a country or region while helping to increase employment and fight social discrimination. Specific energy and environmental investments have a direct impact on employment, creating the additional jobs necessary for the operational needs of the new structures and developments while greatly, although indirectly, contributing to the generation of additional employment and business opportunities. In particular, indirect capital and employment benefits are generated by the increase of the demand in the supply chains associated with the specific investments and also the rise of the general consumption patterns induced by the increased spending capacity of the benefitting population.

Several studies have been carried out to estimate the direct, indirect, and induced economic and employment benefits of the energy and environmental investments in Europe; most of them are based on an input-output approach and use economic and employment multipliers to estimate the impacts of new investments. In particular, Jeeninga et al. (1999) found that that energy efficiency measures have a “positive but relatively small” employment effect, while Wade et al. (2000) estimated that an investment of €1,000,000 on energy efficiency measures can generate 26.6 additional jobs. An average value of €17,000,000 to €19,000,000 invested in energy efficiency measures is proposed by BPIE (2011). In the case where additional private funds are added to the investments of the state, in an analogy close to 5 to 1, the estimated potential for creating additional jobs may reach close to 90 jobs created for every €1,000,000 spent on energy conservation measures (Jansen 2012).

The market opportunities related to the energy efficiency measures and the possible application of renewable energies are quite high. Butson (1998) has estimated that the market opportunities of the energy service market in Europe are between €5–10 billion, while the European Commission has estimated that a 20 % decrease in the energy consumption in Europe by 2020 may create 1,000,000 additional jobs (European Commission 2003).

Turkolias and Mirasgedis (2011) have calculated the employment effects of the possible investments necessary in Greece to satisfy the energy targets met by the EU. Table 4.1 gives the specific direct, indirect, and induced results. The study showed that most of the development and fabrication of the systems under consideration have already been implemented in Greece. As shown, the employment effects of the specific investments vary from 265 man-years/TWh for geothermal energy to 1503 man-years for photovoltaics. It also concluded that the employment effect per installed MW of capacity will be: wind (construction): 17.2 man-years/MW; pv (construction): 33.7 man-years/MW; hydro (construction): 27 man-years/MW; geothermal (construction): 25.3 man-years/MW; and biomass (construction): 46.9 man-years/MW.

Markaki et al. (2013) have estimated the employment effects of additional investments in renewable energies and energy conservation in the building sector. Table 4.2 shows the jobs to be generated in the global economy per €1,000,000 of investment. Data are given in full-time equivalent jobs and include the direct, indirect, and induced impacts. It is concluded that investments in energy conservation of buildings may create up to 26.35 equivalent jobs per €1,000,000 of investment, while the corresponding figure for the installation of renewable energies is lower and close to 20.47.

A similar study was done by Moreno and Lopez (2008) in the Asturias area of Spain. It investigated the employment benefits of additional investments on the renewable energy sources sector and arrived at the following results regarding the labor impact of the various renewable sources: wind: 13.2 jobs/MW; solar thermal: 7.5 jobs/1000 m^2; pv: 37.3 jobs/MW; biofuels: 6.5 jobs/kt/year; hydro: 20 jobs/MW; biomass thermal: 0.13 jobs/tep; biomass electric: 4.14 jobs/MW: and biogas: 31 jobs/MW. The specific figures are quite similar to the corresponding figures calculated for Greece by Turkolias and Mirasgedis (2011).

A study aiming to investigate the impact of various energy conservation measures in the building sector is presented by Oliveira et al. (2014). Table 4.3 shows the results in full-time employment (FTE), created in a direct, indirect, and

Table 4.1 Calculated employment effects due to the utilization of renewable technologies in Greece (in man-years/TWh) (Tourkolias and Mirasgedis 2011)

	Wind		PV		Hydro		Geothermal		Biomass	
	Construction	Operation	Construction	Operation	Construction	Operation	Construction	Operation	Construction	Operation
Direct	160.3		136.9		612.2		146.8		83.3	
Indirect	88.2		61.6		333.7		56.4		39.5	
Induced	66.3		74.7		255.6		98.0		31.5	
Total	314.8		273.2		1201.5		301.3		154.3	

Table 4.2 Employment per €1,000,000 of investments by type of measures (in full-time equivalent jobs) (Markaki et al. 2013)

Type of intervention	Direct impacts	Indirect impacts	Induced impacts	Total
Wind farms	8.96	6.87	3.98	19.80
Off shore wind farms	9.82	7.26	4.20	21.28
Photovoltaic/solar thermal units	9.39	7.11	4.18	20.68
Small hydro	10.83	7.71	4.15	22.69
Pumped storage hydro	9.83	7.57	3.72	21.13
Geothermal units	8.80	6.97	4.24	20.01
Biomass units	12.62	6.71	3.98	23.31
Insulation of buildings	2.16	11.13	8.83	3.66
Replacement of old boilers	1.86	7.65	5.36	3.69
Solar collectors for hot water and space heating	1.95	8.71	5.96	4.12
Use of low energy bulbs and appliances	1.93	27.91	4.73	4.55
Use of smart metering systems	1.93	27.91	4.73	4.66
Co-generation/district heating	1.92	8.35	6.05	3.68
Research in energy conservation	1.27	0.13	2.06	0.83

Table 4.3 Direct, indirect and induced job totals per each type of building and retrofit measures (Oliveira et al. 2014)

Building function building type	No. of façades	Retrofit investment	Direct jobs (FTE)	Indirect jobs (FTE)	Induced jobs (FTE)	Total (FTE)
Residential (<1945)	4	Window frame	214	178	375	768
		Window glaze	127	107	198	432
		Facade insulation	293	274	527	1093
		Roof insulation	77	76	147	300
	2	Window frame	124	104	218	446
		Window glaze	74	62	115	251
		Facade insulation	170	159	306	636
		Roof insulation	77	76	147	300

(continued)

Table 4.3 (continued)

Building function building type	No. of façades	Retrofit investment	Direct jobs (FTE)	Indirect jobs (FTE)	Induced jobs (FTE)	Total (FTE)
Apartment building, 3 floors (1946–1990)	4	Window frame	525	437	921	1883
		Window glaze	312	262	485	1059
		Façade insulation	402	375	722	1499
		Roof insulation	103	101	196	400
	2	Window frame	310	258	544	1111
		Window glaze	184	154	286	625
		Façade insulation	237	221	426	885
		Roof insulation	103	101	196	400
Private services, apartment building, 3 floors (<1945)	4	Window frame	951	792	1669	3412
		Window glaze	565	474	879	1919
		Façade insulation	226	212	407	845
		Roof insulation	61	60	116	237
Apartment building, 4 floors, (1946–1990)	2	Window frame	1327	1105	2329	4761
		Window glaze	789	662	1227	2677
		Façade insulation	113	105	203	421
		Roof insulation	83	81	157	321
Public services, apartment building, 3 floors (<1945)	4	Window frame	188	157	330	675
		Window glaze	112	94	174	380
		Façade insulation	45	42	81	167
		Roof insulation	30	30	58	119
Public services, apartment building, 3 floors (1946–1990)	4	Window frame	188	157	330	675
		Window glaze	112	94	174	380
		Façade insulation	32	30	58	119
		Roof insulation	22	22	42	86

induced way, per building type and retrofit investment. As shown, retrofitting of the existing building stock in Portugal can create thousands of jobs. The retrofitting measure presenting the higher impact on employment is the replacement of single-glazed windows and the insulation of the façades in older single dwellings.

The cost of the additional jobs created in the various sectors of the renewable energy industry in Spain is evaluated by Alvarez et al. (2009). The specific results of the study are given in Table 4.4. As shown, the calculated cost per additional job is found to be very high and outside the limits of the actual economic conditions. In parallel, the same study has evaluated the number of jobs lost in the other sectors of the energy industry due to the replacement of conventional sources. It also found that the employment impact on renewable energy is quite negative; in particular, for each MW of installed wind capacity, almost 4.27 jobs are lost in the global economy; for mini-hydro, 5.05 jobs lost per MW; and finally, for solar systems, almost 12.7 jobs per MW were lost.

However, as mentioned by Kammen et al. (2004), it is almost impossible to directly compare the results of the various studies on employment, as the boundary conditions among them vary significantly. Three specific studies, (Renewable Energy Policy Project, REPP 2001; Greenpeace 2001; Greenpeace and EWEA 2003), have proposed specific figures regarding the employment potential of renewable energy sources. In parallel, estimations about the employment impact of coal and gas are given. The specific results are shown in Table 4.5.

As shown, renewable energy technologies generate many more jobs than conventional energy sources per average megawatt. It is evident that strategic investments in energy conservation technologies for the building sector and renewable energy sources for power and energy production in the southern European countries may have a very positive impact on employment, generate wealth in the corresponding countries, and boost the national economies.

4.3 Conclusions

The building sector in the south of Europe is characterized by significant energy consumption; inappropriate indoor environmental quality for all households; important health problems related to energy-poor citizens; and significant problems' adaptation related to actual and future climatic change. The development of advanced energy efficiency technologies and the possible application of renewable energy technologies offer very significant economic, technical and social opportunities for the countries under discussion. In parallel, the development of efficient climatic mitigation and adaptation techniques to fight local and global climate change can improve the quality in the built environment and protect vulnerable populations from extreme climatic events.

Several studies have been carried out to identify the impact on employment of energy efficiency and renewable energies. It is a common conclusion from all

Table 4.4 Subsidy and investment per worker for the various types of renewable energies in Spain, (Alvarez et al. 2009)

	Number of direct jobs	Number of indirect jobs (difference)	Total jobs	Total subsidy (spent and committed) in M€, NPV @ 4 %	Subsidy M€/job	Total Investment (in M€)	Investment (in M€)/job
Wind	6825	8.175	15,000	16,436.38	1.09	14,723	0.98
Hydro	1475	3225	4700	2551.28	0.54	1067.04	0.22
Photovoltaic	14,500	0	14,500	9683.48	0.66	16,131.5	1.11

Table 4.5 Comparison of jobs/MWp, jobs/MWa and person-years/GWh among various renewable and conventional technologies for 2000 working hours/year (Kammen et al. 2004)

Energy technologies	Source	Capacity factors (%)	Equipment lifetime (years)	Average employment over life of facility					
				Total (jobs/MWp)		Total (jobs/MWa)		Total (person-years/GWh)	
				Construction, manufacturing, installation	O&M and fuel processing	Construction, manufacturing, installation	O&M and fuel processing	Construction, manufacturing, installation	O&M and fuel processing
Part a									
PV 1	REPP (2001)	21	25	1.29	0.25	6.21	1.2	0.71	0.14
PV 2	Greenpeace(2001)	21	25	1.2	1	5.76	4.8	0.66	0.55
Wind 1	REPP (2001)	35	25	0.15	0.1	0.43	0.27	0.05	0.03
Wind 2	EWEA/Greenpeace (Greenpeace 2003)	35	25	0.88	0.1	2.51	0.27	0.29	0.03
Biomass—high estimate	REPP (2001)	85	25	0.34	2.08	0.4	2.44	0.05	0.28
Biomass—low estimate	REPP (2001)	85	25	0.34	0.32	0.4	0.38	0.05	0.04
Coal	REPP (2001); Kammen, from REPP (2001)	80	40	0.21	0.59	0.27	0.74	0.03	0.08
Gas	CALPIRG (2003), BLS (2004)	85	40	0.21	0.6	0.25	0.7	0.03	0.08

Energy technology	Source of numbers	Capacity factor (%)	Equipment lifetime (years)	Employment components		
				Construction, manufacturing and installation (person-years/MWp)	Operation and maintenance (jobs/MWp)	Fuel extraction and processing (person-years/GWh)
Part b						
PV 1	REPP (2001)	21	25	32.33	0.25	0

(continued)

Table 4.5 (continued)

Energy technology	Source of numbers	Capacity factor (%)	Equipment lifetime (years)	Employment components		
				Construction, manufacturing and installation (person-years/MWp)	Operation and maintenance (jobs/MWp)	Fuel extraction and processing (person-years/GWh)
PV 2	Greenpeace (2001)	21	25	30	1	0
Wind 1	REPP (2001)	35	25	3.8	0.1	0
Wind 2	EWEA/Greenpeace (2003)	35	25	22	0.1	0
Biomass—high estimate	REPP (2001)	85	25	8.5	0.44	0.22
Biomass—low estimate	REPP (2001)	85	25	8.5	0.04	0.04
Coal	REPP (2001); Kammen, from REPP (2001)	80	40	8.5	0.18	0.06
Gas	CALPIRG (2003), BLS (2004)	85	40	8.5	0.1	0.07

studies (except for one),that these innovative energy systems can generate a very high number of new jobs and boost the national economies through specific investments related to the rehabilitation of the existing building stock and the replacement of traditional energy-generating plants with new ones based on renewable energy sources.

The energy requirements set by the recent European Directives, asking for the design and application of near-zero energy buildings at the end of the decade, can offer the necessary momentum to enhance the market opportunities and improve the quality of life of the citizens.

References

Alvarez CG, Jara RM, Julian JRR, Bielsa JIG (2009) Study of the effects on employment of public aid to renewable energy sources. Technical report prepared by University Rey Juan Carlos. Available via dialog. http://www.juandemariana.org/pdf/090327-employment-public-aid-renewable.pdf. Cited 20 Mar 2015

Bouzarovski S (2011) Energy poverty in the EU: a review of the evidence. DG Regio workshop on 'Cohesion policy investing in energy efficiency in buildings', Brussels

Buildings performance Institute Europe (BPIE) (2011) Europe's buildings under the microscope, Brussels. Available via dialog. www.bpie.eu/eu_buildings_under_microscope.html. Cited 10 Mar 2015

Butson J (1998) The potential for energy service companies in the European Union. Amsterdam

Entranze (2015) Total building consumption per square meter in Europe. Available via dialog. http://www.entranze.enerdata.eu/total-unit-consumption-per-m2-in-buildings-at-normal-climate.html. Cited 10 Mar 2015

European Commission (2003) Proposal for a Directive of the European Parliament and of the Council on energy end-use efficiency and energy services

FIEC (2013) Construction in Europe, key figures of 2013. Available via dialog. http://www.sefifrance.fr/images/documents/fiec_keyfigures2013_final.pdf. Cited 5 Mar 2015

Greenpeace (2001) 2 million jobs by 2020. Solar generation. Solar electricity for over 1 billion people and 2 million jobs by 2020

Greenpeace (2003) A blueprint to achieve 12 % of the world's electricity from wind power by 2020. European Wind Energy Association Wind Force 12

Huynen MMTE, Martens P, Schram D, Weijenberg MP, Kunst AE (2001) The impact of heat waves and cold spells on mortality rates in the Dutch population. Environ Health Perspect 109:463–470

Jansen R (2012) Creating jobs in the energy efficiency field: not an easy job, but worth it. Available via dialog. http://energyindemand.com/2012/02/14/creating-jobs-in-the-energy-efficiency-field-not-an-easy-job-but-worth-it/. Cited 23 Feb 2012

Jeeninga H, Weber C, Mäenpää I, Rivero García F, Wiltshire V, Wade J (1999) Employment impacts of energy conservation: schemes in the residential sector, a contribution to the save employment project. Report no. ECN-C-99-082, ECN, The Netherlands

Kammen D, Kammal K, Fripp M (2004) Putting renewables to work: how many jobs can the clean energy industry generate? RAEL report, University of California, Berkeley

Lampietti J, Meyer A (2002) When heat is a luxury: helping the urban poor of Europe and Central Asia cope with the cold. World Bank, Washington, DC

Markaki M, Belegri-Roboli A, Michaelides P, Mirasgedis S, Lalas DP (2013) The impact of clean energy investments on the Greek economy: an input–output analysis (2010–2020). Energy Policy 57:263–275

Moreno B, Lopez AJ (2008) The effect of renewable energy on employment. The case of Asturias (Spain). Renew Sustain Energy Rev 12:732–751
Nikiel A, Oxley S (2011) European energy efficiency trends—household energy consumption UK energy statistics
Oliveira C, Coelho D, Pereira da Silva P (2014) A prospective analysis of the employment impacts of energy efficiency retrofit investment in the Portuguese building stock by 2020. Int J Sustain Urban Planning Manage 2:81–92
Pirard P, Vandentorren S, Pascal M, Laaidi K, Le Tertre A, Cassadou S, Ledrans M (2005) Summary of the mortality impact assessment of the 2003 heat wave in France. Eur Surveill 10:153–156
Population and Social Conditions (2014) European Statistical Office
Renewable Energy Policy Project (2001) The work that goes into renewable energy
Santamouris M (2001) Energy and climate in the urban built environment. Earthscan, London, UK
Santamouris M (2014a) On the energy impact of urban heat island and global warming on buildings. Energy Build 82:100–113
Santamouris M (2014b) Cooling the cities—a review of reflective and green roof mitigation technologies to fight heat island and improve comfort in urban environments. Solar Energy 103:682–703
Santamouris M (2015) Regulating the damaged thermostat of the cities—status, impacts and mitigation challenges. Energy Build 91:43–56
Tourkolias C, Mirasgedis S (2011) Quantification and monetization of employment benefits associated with renewable energy technologies in Greece. Renew Sustain Energy Rev 15:2876–2886
Wade J, Wiltshire V, Crase I (2000) National and local employment impacts of energy efficiency investment programmes. Final report to the European Commission, vol 1, summary report, Association for the Conservation of Energy, UK

Chapter 5
Indicators for Buildings' Energy Performance

Sofia-Natalia Boemi and Charalampos Tziogas

Abstract Over the last few decades, a wide variety of papers have aimed at developing, enhancing, and categorizing the sustainability and energy efficiency indicators. Those efforts assess the economic, social and environmental indicators in isolation to each to other without energy efficiency, material efficiency, and resources sustainability. These studies' main focus is to provide important information to policy-makers to understand the implications and impacts of various energy programs, alternative policies, strategies, and plans in shaping development within the countries. In fact, indicators—when properly analyzed and interpreted—can be useful tools for communicating data relating to energy and sustainable development issues to policy-makers and to the public, and for promoting institutional dialogue. This research is a first-time effort towards mapping the existing indicators of energy efficiency in the built sector by grouping all the aforementioned indicators into five main parameters. The success of the critical taxonomy is based on the association of sustainable performance indicators with energy sustainability. The link between those parameters provide a structure and clarify statistical data to give better insight into the factors that affect energy efficiency, environment, economics, and social well-being, and how they might be influenced and trends improved. Indicators can be used to monitor the progress of past policies, and to provide a "reality check" on strategies for future sustainable development.

5.1 Introduction

The world's population is projected to reach 9.1 billion by 2050, while the level of global urbanization is expected to rise from 52 to 69 % during that time (Voskamp and Van de Vena 2015). Although urbanized land area occupies only 2 % of the

S.-N. Boemi (✉) · C. Tziogas
Process Equipment Design Laboratory, Department of Mechanical Engineering, Aristotle University of Thessaloniki, P.O. Box 487, 54124 Thessaloniki, Greece
e-mail: nboemi@gmail.com

S.-N. Boemi et al. (eds.), *Energy Performance of Buildings*,
DOI 10.1007/978-3-319-20831-2_5

planet's surface, more than half of the world's population inhabits urban areas (World Bank 2012). To this effect, the need to develop the corresponding buildings' infrastructure to support human life and well-being is further highlighted (Poyil and Misra 2015). However, intense urbanization has major ramifications on the functioning of local or even global ecosystems (Wan et al. 2015). Therefore, modern needs and trends necessitate the development of buildings that can foster sustainable living along the three pillars of the environment, the society, and the economy (Zuo and Zhao 2014).

It has been said that "If putting a man on the moon was one of the greatest challenges the twentieth century faced, tackling climate change is a much bigger challenge that we in the twenty-first century are confronted with" (Kamilaris et al. 2014; Smart 2020 2008). Today, the building sector accounts for about one-third of global carbon dioxide (CO_2) emissions, while the overall annual growth rate in the sector's energy consumption is estimated at around 36 % (IEA 2013). Specifically, the energy use in European buildings represents about 40 % of the European Union's (EU) total energy consumption (BPIE 2011). Therefore, the effective design and construction of buildings is necessary for energy conservation, while it assists in minimizing greenhouse gas emissions and ensuring proper comfort in living conditions. Hence, integrated sustainability assessment frameworks are imperative for evaluating the energy performance of buildings (Pulselli et al. 2007), and for supporting the decision-making process for a sustainable urban future (Vučićević et al. 2013).

Nevertheless, the assessment of a building's energy performance requires that sustainability indicators be identified and clearly defined, realistic, understandable, and practical (Todorovic and Kim 2012). However, a building's sustainability assessment framework that elaborates generally approved indicators does not yet exist, while generic frameworks such as the "green building challenge" can only serve for reference purposes (Kajikawa et al. 2011). Furthermore, most of the relevant studies only myopically examine the sustainability performance in the building sector, while mainly focusing on residences (Asdrubali et al. 2015; Vučićević et al. 2013). Buildings in the health, education or accommodation services sectors are rather neglected (Michailidou et al. 2015; Mori and Yamashita 2015); in addition, the limited enforcement of obligatory legislative measures that promote buildings' sustainability hinders the progress in the research field.

This challenge is even more prominent in the building sector in Greece. The national building landscape has been characterized by a complete lack of modern legislation on energy and environmental protection of buildings for about 40 years (IEA 2013). Indicatively, Eurostat and the European Environmental Agency report that Greek households have the highest energy consumption among EU member states. Specifically, Greek households' energy consumption is 30 % greater than that of Spanish households, and nearly twice that of Portugal. The high energy consumption has an important social dimension, because in the contemporary financial crisis era, the low income citizens live in thermally unprotected buildings and have increased expenses for heating purposes per person and unit of surface (Mihalakakou et al. 2002; Santamouris et al. 2007a, b).

Notably, this absence is evident even in buildings of social interest such as schools and hospitals.

In addition, since the energy demand of buildings is directly related to the climate of each region, it is evident that climate will have major consequences on the built environment (Santamouris 2007a; Cartalis et al. 2011). Energy policies have been analyzed with simulations and measurements in buildings located in both cold and warm conditions (Mlakar and Štrancar 2011; Persson and Westermark 2010). However, analyses of the energy-saving potential in buildings (excluding the residential sector), in temperate conditions have not been carried out.

To that end, the purpose of this study is to provide a critical review of key sustainability performance indicators in buildings—excluding residential ones; it focuses on buildings in countries with temperate climates, such as Portugal, Spain, France, Italy, Malta, Croatia, Slovenia, Greece, Cyprus, Israel, Albania, and Turkey. We further propose a methodological framework that can be used as a decision support system allowing for the decision-making upon indicators to monitor and foster the overall sustainability performance in hospitals, schools, and hotels.

This investigation is a first-time effort towards mapping the existing indicators of energy efficiency in the built sector. Initially, background research on the energy-related institutional landscape is presented; following that, a critical taxonomy of relevant indicators has been provided with a view towards mapping the existing research efforts. Then, the key steps of the decision-making process are recognized in order to design and manage building sustainability, and finally, an articulation of the major findings are presented and discussed.

5.1.1 Background

5.1.1.1 Buildings' Energy Analysis

Policy-makers need simple methods for measuring and assessing current and future effects of energy use towards a sustainable future. To that end, several frameworks of energy-related indicators have been developed to assess the trends towards sustainable development (Abdallah et al. 2015; Tsai 2010). However, the reported frameworks are often case-dependent and tailored to the specific characteristics of the decision-making problem to be monitored and tackled. Nevertheless, a comprehensive framework that incorporates energy assessment indicators does not exist; thus, an integrated framework of analytically and politically related indicators is needed (Patlitzianas et al. 2008). Such a framework should also encapsulate the characteristics that each type of building needs to serve (e.g., school, hotel, hospital).

The indicators can be used to measure the influence of changes in the energy demand or of changes in all activities requiring energy. In addition, the indicators help connect energy use to financial and technological parameters, such as energy price, economic growth, and new technologies. Moreover, the indicators and the comparisons that they provide, both for the particular country and internationally, provide analysts with managerial insights towards the effects of relative policies (Schipper and Haas 1997).

Therefore, appropriate energy-related KPIs in buildings can be useful tools for communicating data relating to energy and sustainable development issues to policy-makers and to the public, and for promoting institutional dialogue. On the other hand, upgrading (heating and watering) equipment and building systems are, in fact, potential ways to increase energy efficiency, but there is no guarantee that these practices will have a positive effect in the long run (Abu Bakar et al. 2015).

As required by the OECD (2001), the indicators must be as specific as possible so as to be understood and mutually acceptable to the above-mentioned participants. In particular, they can assist policy-makers in (Patlitzianas et al. 2008):

- Monitoring of targets, which are set by national and international levels,
- Estimating policy impacts that had already been implemented, and
- Planning of future policy actions, establishing priorities, and recognizing basic key parameters that can influence the energy market.

5.1.2 European Landscape

The main EU policy documents and directives that focus on promoting sustainable development highlight the issues of energy efficiency and the use of renewable energy sources as strategic options in the energy sector (Streimikiene and Sivickas 2008). Specifically, the key EU energy policy priorities that stem from the Commission's energy policy agenda aim to (1) reduce energy impact in the environment; (2) promote improvements in energy generation and foster efficient use of energy; (3) increase the reliability and security of energy supply; and (4) promote renewable energy sources and climate change mitigation.

In order to assess the implementation progress of the articulated EU energy directives, several qualitative and quantitative key performance indicators (KPIs) are proposed in the scientific literature. To that end, an integrated methodological framework to support energy policy analysis and monitoring could be a useful tool. Such a tool might be applied by EU member states to harmonize EU energy policies and enhance their implementation on the country level.

Today, the main European legislative instrument for improving the energy efficiency of the European Building Stock is the European Directive 2002/91/EC on the Energy Performance of Buildings (EPBD) and its recast, Directive 2010/31/EC. The aforementioned directives align with the European commitments in the Kyoto Protocol about climate change, environmental protection, and security in energy supply. Furthermore, the EU Green Paper on European Strategy for Sustainable, Competitive and Secure Energy (SEC (2006) 317) sets the main priorities for the EU energy strategy.

Apart from the legislative framework, the Organization for Economic Cooperation and Development (OECD 2001) defines indicators to measure the influence of changes on energy demand or of changes in all activities using energy (Schipper and Grubb 2000). In addition, the indicators assist in exemplifying the way energy use is related to economic and technological parameters, such as energy prices, economic growth, and new technologies (Patlitzianas et al. 2008). Table 5.1 summarizes and elaborates general indicators for energy policy analysis.

The articulation of specific legislative and regulatory frameworks appears to have a positive effect on the accomplishment of the Europe 2020 objectives, referring to the target countries of Portugal, Spain, France, Italy, Malta, Croatia, Slovenia, Greece, Cyprus, Israel, Albania, and Turkey (see Fig. 5.1).

Table 5.1 General indicators for energy policy analysis

Policy	Indicators	Reference
Energy efficiency	End-use energy intensity of GDP	Directive 2006/32/EC on end-use efficiency and energy services
	Energy saved in buildings	2002/91/EC directive on the energy performance of buildings
	Savings of primary energy supply	The Commission's new Green Paper on energy efficiency COM (2005) 265
Use of renewables (RES)	The share of CHP in electricity production	2004/8/EC directive on the promotion of co-generation national energy strategy
	The share of renewables in primary energy supply	The White Paper on renewable sources
	The share of renewables in primary energy supply	The White Paper on renewable sources
	The share of renewables in electricity generation	Directive 2001/77/EC on the promotion of electricity produced from renewable energy sources in the internal electricity market

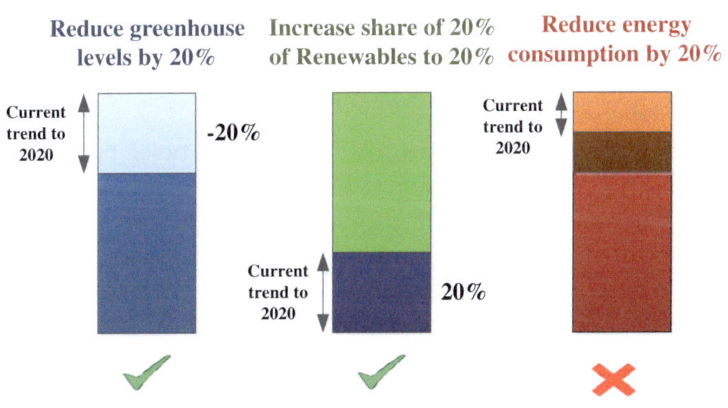

Fig. 5.1 Trends towards the EU energy policy priorities by 2020 (Barroso 2011)

5.2 The Resulting Taxonomy

Following the work of Patlitzianas et al. (2008) and Schipper and Grubb (2000), in this study we performed an extensive and up-to-date literature review to firstly identify the KPIs that are most commonly used. The critical synthesis of the review reveals that energy indicators in the building sector can be categorized as follows:

- Descriptive indicators—for example, the percentages of energy per fuel
- Basic normalized indicators, which indicate the use of energy of every field, divided by population or Gross Domestic Product (GDP)
- Comparative indicators, which try to indicate the similarity in characteristics between different countries, using necessary normalizations
- Structural indicators, which distribute the economic or human activities in several scales or forms of performance
- Intensity indicators, which relate the use of energy to a specific activity and are mostly connected with energy performance
- Decomposition indicators, which indicate how different sections of total use of energy influence the total releases
- Causal indicators, which indicate which kind of fundamental economic, demographic, or geographical parameters most influence the use of energy
- Consequential indicators, e.g., the exhaust gas, which measure the connection between the human activities and the use of energy, in terms of the environmental pollution or other disturbances in the environment
- Physical indicators, which indicate the change of the energy performance and productivity.

Table 2.2 presents a comprehensive synopsis of the matching of the type of building with critical energy-related KPIs and the relevant state-of-the-art research efforts. Our taxonomy could assist greatly in drafting the agenda for future research by first identifying the existing gaps and overlaps in current research. In general, despite the fact that there is a plethora of studies that examine various KPIs and methodological approaches, there is a lack of holistic frameworks and a general roadmap for the design of sustainable buildings. This research gap is even more evident in the case of buildings of social interest. The building types were categorized based on their use.

The taxonomy provided indicates that research on hotel buildings has been quite significant. On the other hand, research efforts should focus more on the design of schools and, in general, on buildings for education purposes. Moreover, up until now, significant attention has been paid to the sustainability of building in the sensitive health sector. Additionally, Table 5.2 reveals that research efforts need to emphasize the design of sustainable office buildings and further scrutinize the adoption of quantitative KPIs. Finally, a significant gap in the existing body of research concerns shopping centers, despite their economic significance, as the lack of elaborated energy performance indicators is obvious.

Table 5.2 Review of recent publications with indicators that affect the energy performance of a building

Type of building	Indicators	Reference
Hotel	Energy performance indicator, based on primary energy use (kgep/m^2/year)	Goncalves et al. (2012)
	Exergy efficiency (the exergy efficiency indicator is applied, comparing exergy levels between supply and demand)	
	Energy use index (EUI or energy use per unit of gross floor area)	Deng and Burnett (2002)
	Water use index (WUI or water use per unit of gross floor area)	
	Total energy per unit area, EUI (kWh/m^2/year)	Wang (2012), Priyadarsini et al. (2009), Santamouris et al. (1996)
	Total energy per guest room (MWh/room/year)	
	Total energy per guest room night (kWh)/(number of rooms × occupancy × 365)	
	Total energy per guest night (kWh)/number of guests	
	Energy conservation	Teng et al. (2012), Pieri et al. (2015)
	Carbon reduction (ECCR)	
	Total energy consumption	Lu et al. (2013)
	Carbon efficiency	Filimonau et al. (2011)
School	Total building size and specific surface indicators (building size, shape and layout)	Katafygiotou and Serghides (2014), Beusker et al. (2012), Dimoudi and Kostarela (2009), Hernandez et al. (2008), Santamouris et al. (2007a, b), Butala and Novak (1999)
	Compactness (the building's geometric shape represents a relevant impact factor on heating energy consumption)	
	Thermal mass	
	Standard of thermal insulation	
	Standard, condition and percentage of exterior glass surfaces	
	Type of energy source	
	Standard and condition of technical installations	
	Occupancy patterns	
Hospital	Bed density (beds/m^2)	Rasouli et al. (2014), Saidur et al. (2010), Lozano et al. (2009), Salem Szklo et al. (2004)
	Total monthly energy intensity—kWh/bed	
	Share held by electricity (%)	

(continued)

Table 5.2 (continued)

Type of building	Indicators	Reference
	Power load factor	
	End use of electricity (%) (lighting, air conditioning, water heating, total)	
	Physical indicators [lighting (W/m^2), air conditioning ($TR/100\ m^2$), hot bath water (m^3/bed/month), participation of central air conditioning in installed cooling capacity (%)]	
	Heat load factor (%)	
	Efficiency rating—central steam generation system (%)	
	Bed density ($beds/m^2$)	
	Total monthly energy intensity—kWh/bed	
	Share held by electricity (%)	
	Number of users per floor area	Sanz-Calcedo et al. (2011), García-Sanz-Calcedo et al. (2014)
	Global heating potential (GHP) as a function of the number of users	
	Annual energy consumption and number of users	
Office buildings	Primary energy	Pikas et al. (2014)
	Total energy cost per year	
	Net present value (NPV)—Calculation of optimal PV panel size to achieve nZEB level	
	Environmental indicators (CO_2, SO_2 emissions) of the construction materials	Dimoudi and Tompa (2008)
	Operational energy of the buildings	Dimoudi and Tompa (2008), Burdová and Vilčeková (2012)
	Energy use (for heating, cooling, lighting, appliances, hot water production)	Svensson et al. (2006), Burdová and Vilčeková (2012)
	Energy use in concept if using primary energy in a life cycle perspective	Svensson et al. (2006)
	Environmental pressure	
	Active solar design	Burdová and Vilčeková (2012)
	System of energy management	
	Use of environmental labels	Nunes et al. (2013)
Shopping centers	There is nothing specific	

(continued)

Table 5.2 (continued)

Type of building	Indicators	Reference
All buildings	Energy efficiency	Iwaro et al. (2014) and all the above
	External benefit	
	Economic efficiency	
	Material efficiency	
	Environmental impact	
	Regulation efficiency	
	U-factor (W/m^2 K) and R-value	Bellamy (2014)

5.3 Decision-Making Framework

Designing, managing and operating sustainable buildings involves a complex and integrated decision-making process. This is even more accentuated when buildings serve social purposes, such as for hospitals and schools. Generally, for assessing and calculating the energy performance of buildings, four (4) methodologies are applied (Caldera et al. 2008).

The first method represents an evaluation of a building's energy demand based on standard assumptions that are described in national or international standards. The second approach includes a calculation of the actual energy consumption and continuous adjustments of documented consumption indicators in terms of designated factors (e.g., climatic data, operating time). The third uses simulation techniques to specify the energy demand in reference to physical and thermal quantities of individual buildings. According to Beusker et al. (2012), such a detailed analysis is unlikely to become a common tool for the evaluation of large building stocks. The fourth approach describes the development of a simplified model to assess the energy performance of existing buildings based on quantitative parameters. A move toward establishing such quantitative parameters through large-scale developments is evident in several empirical studies, e.g., for schools, hotels and sports facilities (Beusker et al. 2012; Corgnati et al. 2007; Santamouris et al. 1996) and, in particular, for residential and office buildings (Caldera et al. 2008; Teng et al. 2012). These investigations underline the desperate need for statistically valid data to provide accurate estimates and reliable benchmarks.

Based on an extensive synthesis of the literature, we provide a first comprehensive methodological draft of all the major parameters that affect the energy performance of a building. The inclusive framework is presented in Fig. 5.2. This framework is by no means a rigid model, nor is it based on an exhaustive list of all relevant research efforts, but rather acts as a roadmap and synthesis of all key steps that we have identified as part of our research.

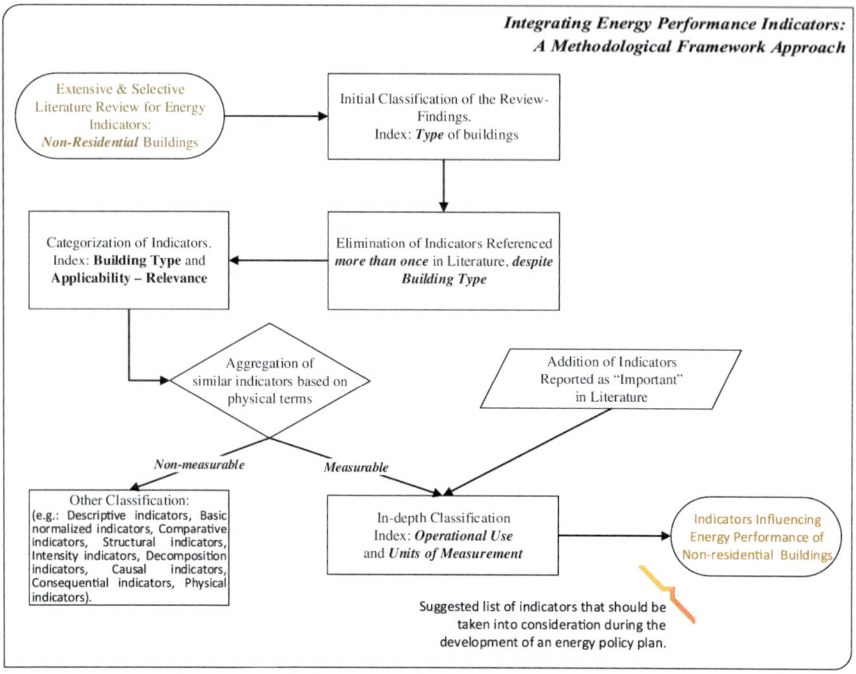

Fig. 5.2 Decision-making framework

5.4 Findings

Our study reveals that many challenges still remain in the domain of energy efficiency indicators, including more accurate KPIs that take into account building profiles, occupant behavior and environmental context. In particular, a few factors that should usually be considered in an energy analysis approach for buildings—both existing and those under construction—are: (1) accuracy; (2) sensitivity; (3) speed; (4) reproducibility; (5) ease of use and level of detail; (6) availability of required data; (7) quality of output; and (8) stage of the project (for structures during the construction phase).

After applying the developed framework, we concluded that for each building type, the following indicators described in Table 5.3 should be taken into consideration. The indicators are further classified into five (5) broad categories: performance, operational, physical, energy consumption, and carbon footprint indicators.

Table 5.3 Proposed indicators

Type of building	Type of indicators	Commonly used
Hotels Office buildings energy policy	Building's performance	• Energy performance indicators are based on: primary energy use (kgep/m^2/year)
		• Primary energy
		• Energy use in concept if using primary energy in a life cycle perspective—Global heating potential (GHP) as a function of the number of users
		• Heat load factor (%)
		• End use of electricity (%) (lighting, air conditioning, water heating, total)
		• Efficiency rating–central steam generation system (%)—end use of electricity (%) (lighting, air conditioning, water heating, total)
Office buildings Schools Hospitals	Operational profile	• Number of users per floor area
		• Energy use index (EUI or energy use per unit of gross floor area)
		• Total monthly energy intensity—kWh/bed
		• Exergy efficiency (the exergy efficiency indicator, is applied, regarding to compare exergy levels between supply and demand)
Hotels		• Bed density (beds/m^2)
Hospitals Office buildings Schools Housing	Physical characteristics	• Lighting (W/m^2), air conditioning (TR/100 m^2), Hot bath water (m^3/bed/month), participation of central air conditioning in in-stalled cooling capacity (%)
		• Energy use (for heating, cooling, lighting, appliances, hot water production)
		• Standard and condition of technical installations
		• R-value and U-factor (W/m^2 K)
Hotels Office buildings Hospitals	Total energy consumption	• Total energy cost per year
		• Annual energy consumption/energy consumption per capita
		• Total energy per unit area, EUI (kWh/m^2/year)
		• Total energy per guest room-night, (kWh)/(number of rooms × occupancy × 365)

(continued)

Table 5.3 (continued)

Type of building	Type of indicators	Commonly used
		• Total energy per guest room (MWh/room/year)
		• Total energy per guest night (kWh)/number of guests
Hotels	Carbon footprint	• Environmental pressure
		• Environmental impact
Office buildings		• Environmental indicators (CO_2, SO_2 emissions) of the construction materials
Buildings general		
		• Material efficiency

5.5 Discussion

It is important for policy-makers to understand the implications and impacts of various energy programs, alternative policies, strategies, and plans in shaping development within the countries. Indicators—when properly analyzed and interpreted—can be useful tools for communicating data relating to energy and sustainable development issues to policy-makers and to the public, and for promoting institutional dialogue.

These policies provide a structure and clarify statistical data to give better insight into the factors that affect energy efficiency, environment, economics, and social well-being, and how these might be influenced and trends improved. Indicators can be used to monitor the progress of past policies, and provide a "reality-check" on strategies for future sustainable development. This cannot be achieved without critical analysis of the underlying causal and driving factors.

Sustainability assessment methods to evaluate energy efficiency, material efficiency and resources sustainability in economic, social, and environmental terms. In this study, an effort to realize sustainability in the built environment establishing sustainable building assessment methods and energy performance sustainable indicators were assessed together.

Also, it succeeded in grouping and categorizing the energy efficiency indicators of five important parameters that a designer of a policy should take into consideration: a building's performance, operational profile, physical characteristics, total energy consumption, and carbon footprint. This information is essential to have available in order to comprehend building energy information and allow suitable analysis, and efficiently plan energy policies for the future. In that respect, the study developed a list of indicators to assess the energy efficiency of non-residential buildings. That list can be used as a valuable tool for energy managers and energy policy makers to measure their effectiveness towards environmental and energy terms and to create energy strategic plans.

References

Abdallah M, El-Rayes K (2015) Optimizing the selection of building upgrade measures to minimize the operational negative environmental impacts of existing buildings. Building Environ 84:32–43

Abu Bakar NN , Yusri Hassan M, Abdullah H, Rahman HA, Abdullah MdP, Hussin F, Bandi M (2015) Energy efficiency index as an indicator for measuring building energy performance: A review. Renew Sustain Energy Rev 44:1–11

Asdrubali F, D'Alessandro F, Schiavoni S (2015) Sustain Mater Technol 4:1–17

Barroso (2011) Energy Priorities for Europe. Available via DIALOG. http://ec.europa.eu/europe2020/pdf/energy_en.pdf

Bellamy L (2014) Towards the development of new energy performance indicators for the external walls of residential buildings. Energy Build 68:696–702

Beusker E, Stoy C, Pollalis SN (2012) Estimation model and benchmarks for heating energy consumption of schools and sport facilities in Germany. Build Environ 49:324–335

BPIE—Buildings Performance Institute Europe (2011) Europe's buildings under the microscope. A country-by-country review of the energy. Available via dialog. http://www.europeanclimate.org/documents/LR_%20CbC_study.pdf. Cited 21 April 2015

Burdová EK, Vilčeková S (2012) Energy performance indicators developing. Energy Procedia 14:1175–1180

Butala V, Novak P (1999) Energy consumption and potential energy savings in old school buildings. Energy Build 29:241–246

Caldera M, Corgnati SP, Filippi M (2008) Energy demand for space heating through a statistical approach: application to residential buildings. Energy Build 40:1972–1983

Cartalis C, Synodinou A, Proedrou M, Tsangrassoulis A, Santamouris M (2011) Modifications in energy demand in urban areas as a result of climate changes: an assessment for the southeast Mediterranean region. Energy Convers Manage 42:1647–1656

COM (2005) The Commission's new Green Paper on energy efficiency COM

Corgnati SP, Filippi M, Viazzo S (2007) Perception of the thermal environment in high school and university classrooms: subjective preferences and thermal comfort. Build Environ 42:951–959

Deng S-M, Burnett J (2002) Water use in hotels in Hong Kong. Int J Hospitality Manage 21:57–66

Dimoudi A, Kostarela P (2009) Energy monitoring and conservation potential in school buildings in the C′ climatic zone of Greece. Renew Energy 34:289–296

Dimoudi A, Tompa C (2008) Energy and environmental indicators related to construction of office buildings. Resour Conserv Recycl 53:86–95

IEA-International Energy Agency (2013) Transition to sustainable buildings—strategies and opportunities to 2050. Available via DIALOG. http://www.iea.org/media/training/presentations/etw2014/publications/Sustainable_Buildings_2013.pdf

Filimonau V, Dickinson J, Robbins D, Huijbregts MAJ (2011) Reviewing the carbon footprint analysis of hotels: Life Cycle Energy Analysis (LCEA) as a holistic method for carbon impact appraisal of tourist accommodation. J Clean Prod 19:1917–1930

García-Sanz-Calcedo J, López-Rodríguez F, Cuadros F (2014) Quantitative analysis on energy efficiency of health centers according to their size. Energy Build 73:7–12

Gonçalves P, Gaspar AR, Gameiro da Silva M (2012) Energy and exergy-based indicators for the energy performance assessment of a hotel building. Energy Build 52:181–188

Hernandez P, Burke K, Lewis JO (2008) Development of energy performance benchmarks and building energy ratings for non-domestic buildings: an example for Irish primary schools. Energy Build 40:249–254

Iwaro J, Mwasha A, Williams RG, Zico R (2014) An integrated criteria weighting framework for the sustainable performance assessment and design of building envelope. Renew Sustain Energy Rev 29:417–434

Kajikawa Y, Inoue T, Goh TN (2011) Analysis of building environment assessment frameworks and their implications for sustainability indicators. Sustain Sci 6:233–246
Kamilaris A, Kalluri B, Kondepudi S, Wai TK (2014) A literature survey on measuring energy usage for miscellaneous electric loads in offices and commercial buildings. Renew Sustain Energy Rev 34:536–550
Katafygiotou MC, Serghides DK (2014) Analysis of structural elements and energy consumption of school building stock in Cyprus: energy simulations and upgrade scenarios of a typical school. Energy Build 72:8–16
Lozano MA, Ramos JC, Carvalho M, Serra LM (2009) Structure optimization of energy supply systems in tertiary sector buildings. Energy Build 41:1063–1075
Lu S, Wei S, Zhang K, Kong X, Wu W (2013) Investigation and analysis on the energy consumption of starred hotel buildings in Hainan Province, the tropical region of China. Energy Convers Manage 75:570–580
Michailidou AV, Vlachokostas C, Moussiopoulos N (2015) A methodology to assess the overall environmental pressure attributed to tourism areas: a combined approach for typical all-sized hotels in Chalkidiki. Greece Ecol Ind 50:108–119
Mihalakakou G, Santamouris M, Tsagrassoulis A (2002) On the energy consumption in residential buildings. Energy Build 34:727–736
Mlakar J, Štrancar J (2011) Overheating in residential passive house: solution strategies revealed and confirmed through data analysis and simulations. Energy Build 43:1443–1451
Mori K, Yamashita T (2015) Methodological framework of sustainability assessment in City Sustainability Index (CSI): a concept of constraint and maximisation indicators. Habitat Int 45:10–14
Nunes P, Lerer MM, Carrilho da Graça G (2013) Energy certification of existing office buildings: analysis of two case studies and qualitative reflection. Sustain Cities Soc 9:81–95
OECD (2001) Environmental indicators: towards sustainable development. Available via dialog. http://www.oecd.org/site/worldforum/33703867.pdf. Cited 21 Mar 2015
Patlitzianas KD, Doukas H, Kagiannas AG, Psarras J (2008) Sustainable energy policy indicators: review and recommendations. Renew Energy 33:966–973
Persson J, Westermark M (2010) Phase change material cool storage for a Swedish passive house. Energy Build 54:490–495
Pieri SP, Tzouvadakis I, Santamouris M (2015) Identifying energy consumption patterns in the Attica hotel sector using cluster analysis techniques with the aim of reducing hotels' CO_2 footprint. Energy Build 94:252–262
Pikas E, Thalfeldt M, Kurnitski J (2014) Cost optimal and nearly zero energy building solutions for office buildings. Energy Build 74:30–42
Poyil RP, Misra AK (2015) Urban agglomeration impact analysis using remote sensing and GIS techniques in Malegaon city, India. Int J Sustain Built Environ (in press)
Priyadarsini R, Xuchao W, Siew Eang L (2009) A study on energy performance of hotel buildings in Singapore. Energy Build 41:1319–1324
Pulselli RM, Simoncini E, Pulselli FM, Bastianoni S (2007) Emergy analysis of building manufacturing, maintenance and use: em-building indices to evaluate housing sustainability. Energy Build 39:620–628
Rasouli M, Akbari S, Simonson CJ, Besant RW (2014) Energetic, economic and environmental analysis of a health-care facility HVAC system equipped with a run-around membrane energy exchanger. Energy Build 69:112–121
Saidur R, Hasanuzzaman M, Yogeswaran S, Mohammed HA, Hossain MS (2010) An end-use energy analysis in a Malaysian public hospital. Energy 35:4780–4785
Salem Szklo A, Borghetti Soares A, Tiomno Tolmasquim M (2004) Energy consumption indicators and CHP technical potential in the Brazilian hospital sector. Energy Convers Manage 45:2075–2091
Santamouris M, Balaras CA, Dascalaki E, Argiriou A, Gaglia A (1996) Energy conservation and retrofitting potential in Hellenic hotels. Energy Build 24:65–75

Santamouris M, Kapsis K, Korres D, Livada I, Pavlou C, Assimakopoulos MN (2007a) On the relation between the energy and social characteristics of the residential sector. Energy Build 39:893–905

Santamouris M, Mihalakakou G, Patargias P, Gaitani N, Sfakianaki K, Papaglastra M, Pavlou C, Doukas P, Primikiri E, Geros C, Assimakopoulos MN, Mitoula R, Zerefos S (2007b) Using intelligent clustering techniques to classify the energy performance of school buildings. Energy Build 39:45–51

Sanz-Calcedo JG, Blázquez FC, Rodríguez FL, Ruiz-Celma A (2011) Influence of the number of users on the energy efficiency of health centres. Energy Build 43:1544–1548

Schipper L, Grubb M (2000) On the rebound? Feedback between energy intensities and energy uses in IEA countries. Energy Policy 28:367–388

Schipper L, Haas R (1997) The political relevance of energy and CO_2 indicators—an introduction. Energy Policy 25:639–649

Smart 2020 (2008) The Climate Group, Smart 2020: enabling the low carbon economy in the information age. A report by The Climate Group on behalf of the Global eSustainability Initiative (GeSI)

Streimikiene D, Šivickas G (2008) The EU sustainable energy policy indicators framework. Environ Int 34:1227–1240

Svensson N, Roth L, Eklund M, Mårtensson A (2006) Environmental relevance and use of energy indicators in environmental management and research. J Clean Prod 14:134–145

Teng C-C, Horng J-S, Hu M-L, Chien L-H, Shen Y-C (2012) Developing energy conservation and carbon reduction indicators for the hotel industry in Taiwan. Int J Hospitality Manage 31:199–208

Todorovic MS, Kim JT (2012) Beyond the science and art of the healthy buildings daylighting dynamic control's performance prediction and validation. Energy Build 46:159–166

Tsai W-T (2010) Energy sustainability from analysis of sustainable development indicators: a case study in Taiwan. Renew Sustain Energy Rev 14:2131–2138

Voskamp IM, Van de Vena FHM (2015) Planning support system for climate adaptation: composing effective sets of blue-green measures to reduce urban vulnerability to extreme weather events. Build Environ 83:159–167

Vučićević B, Stojiljković M, Afgan N, Turanjanin V, Jovanović M, Bakić V (2013) Sustainability assessment of residential buildings by non-linear normalization procedure. Energy Build 58:348–354

Wan L, Ye X, Lee J, Lu X, Zheng L, Wu K (2015) Effects of urbanization on ecosystem service values in a mineral resource-based city. Habitat Int 46:54–63

Wang JC (2012) A study on the energy performance of hotel buildings in Taiwan. Energy Build 49:268–275

World Bank (2012) World development report, gender equality and development. Available via dialog. https://siteresources.worldbank.org/INTWDR2012/Resources/7778105-1299699968583/7786210-1315936222006/Complete-Repo

Zuo J, Zhao Z-Y (2014) Green building research–current status and future agenda: A review. Renew Sustain Energy Rev 30:271–281

Chapter 6
Life Cycle Versus Carbon Footprint Analysis for Construction Materials

Efrosini Giama

Abstract Climatic change, with the increase of ambient temperature, undoubtedly affects the built environment and leads to a significant increase in energy consumption in the building sector—which is already the biggest energy consumer in Europe. Environmental assessment tools provide an effective framework with appropriate indicators for evaluating the environmental performance of buildings and materials for targeting the integration of the sustainable development of a building's entire life cycle, based on ISO standards and environmental legislation, into the design, construction, and marketing. Direct benefits associated with the building certification schemes include the fostering of energy and CO_2 emissions reductions, as well as broader environmental benefits, increased public awareness of energy and environmental issues, and achieving lower operational costs for users—especially now that the ongoing economic recession has affected so many European countries.

6.1 Introduction

Buildings make a considerable environmental impact, with almost 30 % of the global carbon footprint due to buildings, and with a prediction for future growth (Wiedmann 2014). The European Commission has been trying since 2002 to establish a common policy for sustainable buildings (European Commission) and low environmental impact materials promoting energy efficiency and reduction of greenhouse gas (GHG) based on directives and policies. The Construction Products Regulation (CPR) is to ensure reliable information on construction products in relation to their performances (European Commission 2014). In order to provide reliable data concerning construction materials performance there are several standards

E. Giama (✉)
Aristotle University of Thessaloniki, School of Engineering, Thessaloniki 54124, Greece
e-mail: fgiama@auth.gr

S.-N. Boemi et al. (eds.), *Energy Performance of Buildings*,
DOI 10.1007/978-3-319-20831-2_6

and policies. These documents provide a "common technical language" presentig assessment methods for the construction materials performance. The legislative framework for GHG emissions is supported by a number of international standards such as the Greenhouse Gas Protocol Initiative (Greenhouse Gas Protocol), the ISO 14064-65 series of standards (International Organization for Standardization), which are fully compatible with the GHG Protocol, and the Intergovernmental Panel on Climate Change (Intergovernmental Panel on Climate Change 2006, 2010).

The energy and environmental certification processes, as a result of the overall legislative framework for sustainable management of buildings, promoted environmental evaluation aspects relevant to buildings' life cycles. Therefore, and aiming to improve energy efficiency in the overall life cycle of buildings, emphasis must be put on the materials' selection, focusing on low environmental impact during the production process and on the use of natural, recyclable, or recycled materials, ideally of local origin, so as to minimize transportation emissions.

On the other hand, taking into consideration the ongoing economic recession that affects the construction industry in many European countries, but also considering the energy poverty that makes significant sectors of the populations suffer, it is not unexpected that sustainability is no longer among the top priorities for many businesses and organizations (Slini et al. 2015; Santamouris et al. 2013). However, and despite this rather bleak outlook, there is also evidence that environmental management, by means of measuring processes, monitoring indicators in the area of energy and materials consumption, and streamlining these procedures, could lead to cost reduction, sustainable development and improvement of the living standards (Karkanias et al. 2010).

In order to include environmental aspects in planning and policy-making, a number of different tools for analyzing environmental impacts of different systems have been developed and have been widely applied. These include process tools, such as Strategic Environmental Assessment (SEA) and Environmental Management Systems (EMS), as well as more analytical tools, such as Life Cycle Assessment (LCA) and Carbon Footprint Analysis (Ahlroth et al. 2011). Although the environmental assessment tools focus on the evaluation of the environmental criteria assisting decision-makers to include the environmental parameter in the decision-making process, there are some differences in their structure that make them more suitable or less suitable, depending on the type of the system studied. For instance, some tools, such as Life Cycle Analysis and Carbon Footprint Analysis, have been widely used for the products' and production systems' environmental evaluation, compared to Environmental Managements Systems EMS (EMAS, ISO 14001), which are used for companies and organization agencies. Focusing on construction materials and buildings' evaluation, the assessment tools are classified into three levels: Level Three focuses on whole-building assessment and includes methodologies such as BREEAM and LEED (Giama and Papadopoulos 2012). Level Two includes decision support tools, such as Envest (UK), Athena (Canada), and Ecoquantum (NL). Finally, Level One deals with product evaluation and includes supporting tools such as SimaPro (NL) and Gemis (Germany). Several reviews on environmental evaluation of constructions monitored more than 20 LCA applications in buildings, evaluating the whole construction process, and 15 evaluations of construction materials between 2000 and 2009 (Ortiz et al. 2009; Bribián et al. 2011a, b; Fuertes et al. 2013).

The degree of sustainability in a production process can be measured by several criteria, such as total energy content, i.e., the energy required to produce, package, distribute, use, and dispose of a specific product; emissions—greenhouse gases, dust, and other chemical and natural substances; raw materials' use; and waste generation and recyclability (Bribián et al. 2011a, b).

It is generally accepted that the manufacture and transport processes are responsible for a significant amount of the environmental impact on the construction materials' life cycle. The novelty—and at the same time the aim of this chapter—is to verify this perception and quantify it, mainly by using two environmental assessment methods: the Life Cycle Analysis (LCA) and the Carbon Footprint Analysis. At the same time, it is of interest to compare the applicability, the performance, and the user-friendliness of those two methods for the evaluation of materials and their contribution to construction materials' ecolabelling certification. To do this, the two methods are described in detail and compared with respect to the input needed, their core process, the form of the results obtained, and the compatibility with other assessment tools, systems, and standards.

The construction materials evaluated in terms of their environmental impact are bricks, cement, steel, concrete, and cement mortar, as well as insulation materials (stonewool, extruded polystyrene, expanded polystyrene, glasswool, and polyurethane foam). These products have been selected because they are the most common ones in the European building construction process and are therefore used in significant quantities. The initial data were obtained from industrial producers, while a comparative analysis with published data was carried out in order to verify the reliability of the data.

6.2 Methodological Approach

Life Cycle Analysis is a popular environmental tool that has been applied since the early 1980s to a plethora of products and processes, examining the environmental performance of the selected reference systems "from the cradle to the grave." In brief, the concept of LCA is based on (1) the consideration of the entire life cycle, which includes raw material extraction and processing, the production, and the use of the product up to recycling and/or disposal; (2) the coverage of all environmental impacts connected with the products' life cycle, such as emissions to air, water, and soil, waste, raw material consumption, or land use; and (3) the aggregation of the environmental effects in considering possible impacts and their evaluation in order to give guided environmental decision support. LCA therefore offers a comprehensive analysis that links actions with environmental impacts. At the same time, it provides quantitative and qualitative results, and, taking into consideration the link between the system's functions and environmental impacts, it is easy to identify the issues that need improvement.

There are four main stages for the implementation of LCA: (1) planning; (2) inventory analysis; (3) impact assessment; and (4) improvement analysis. The main purpose of planning is to define the scope and goal as well as the boundaries of the system studied. During the planning, the objective should be clear and the data collection sources defined. The inventory analysis is the stage where the inputs and outputs are quantified; energy, raw materials, water consumption, air emissions, and solid waste are quantified, either through measurements or database searches, surveys, or software calculations. During impact assessment, the environmental impact is calculated as a result of the inventory analysis. The input and output data are translated to environmental impacts such as climate change, acidification, eutrophication, and photochemical oxidation. Finally, the stage of improvement analysis involves discussion and improvement suggestions (Bribián et al. 2011a, b).

The system boundaries for each construction material studied consisted of two main subsystems: production processes of the material, including the extraction of raw materials and energy use, and auxiliary activities (for instance, resin production which is required in the manufacturing of stone-wool), and the product's packaging, storage and transportation. In order to implement LCA methodology a reference system is necessary. According to the processes included in the reference system the lca is defined as gradle to gate or gradle to grave approach. If for instance, final disposal process is included at the analysis we have gradle to grave approach if the system's processes end at the production process we have a gradle to gate approach. The reason for choosing a cradle-to-gate approach has to do with the quality and reliability of the initial data used for the inventory analysis (Cabeza et al. 2014; Bribián et al. 2011a, b).

The second environmental assessment tool applied for the materials evaluation is the Carbon Footprint Analysis, which focuses on mapping greenhouse gas emissions throughout the products or processes, aiming for sustainability and economic benefits. A "carbon footprint" is in that sense a measure of the greenhouse gas emissions associated with an activity or group of activities of a product or process (Wiedmann and Minx 2008).

The task of calculating carbon footprints can be approached methodologically from two different directions: bottom-up, based on Process Analysis (PA)—or top-down, based on Environmental Input/Output (EIO) analysis. Both PA and EIO deal with the aforementioned challenges and strive to capture the full life cycle impacts (Pandey et al. 2010).

Process analysis (PA) is a bottom-up method that has been developed to understand the environmental impacts of individual products from the cradle to the grave. The bottom-up nature of PA-LCAs (process-based LCAs) means that they suffer from a system boundary problem—only on-site, most first-order, and some second-order impacts are considered. If PA-LCAs are used for calculating carbon footprint estimates, a strong emphasis therefore needs to be put on the identification of appropriate system boundaries, which minimize this truncation error. PA-based LCAs run into further difficulties once carbon footprints for larger entities, such as government, households, or particular industrial sectors have to be established. Environmental input-output (EIO) analysis provides an alternative top-down approach to carbon footprinting. Input-output tables are economic accounts providing a picture of all economic activities at the meso (sector) level. In combination with consistent environmental account data, they can be used to establish carbon footprint estimates in a comprehensive and robust way, taking into account all higher order impacts and setting the whole economic system as a boundary (Monahan and Powell 2011; Greenhouse Gas Protocol 2010).

Carbon Footprint Analysis compared to LCA is easier to implement, as it focuses on a specific category of emissions and is limited to certain processes; both tools can be applied to products and processes as well (Table 6.1). There are a number of newly developed Carbon Footprint Analysis web-based software tools, such as TerraPass, Carbonica, Carbonfund, and Carbon Counter, which support the carbon footprint calculations for selected products and activities (Giama and Papadopoulos 2012).

The two basic types of indicators are used for environmental evaluation are Environmental Condition Indicators (ECI) and Environmental Performance Indicators (EPI). The ECI indicators describe effects in the environment, for example, they give emphasis on gas emissions' effect in the local atmospheric quality, or the effect of humid waste in water channels near a production area. This indicators' category focus on specified environmental impacts and evaluate impacts such as greenhouse effect, eutrophication, biodiversity reduction, etc. Environmental Performance Indicators (EPI) are subdivided into Management Performance Indicators (MPI) and Operational Performance Indicators (OPI). The EPI refer to environmental indicators that provide information mainly on the administration's efforts, measures, and contribution to the overall organization's environmental management. OPIs are related mainly to the materials' consumption, energy management, waste, and emissions production, and they evaluate the "real" environmental aspects of organizations. They are subdivided into quantitative and energy indicators and are usually set after an inflow-outflow analysis. Examples of such indicators include the electricity or heat per product's production unit consumption, the total waste produced per product's production unit, etc. (Table 6.2). OPIs constitute the base of internal and external communication for environmental data when EMS (especially on EMAS implementation) are implemented within the organization's structure.

Table 6.1 Environmental assessment tools (type and application)

Tools	Type	Approach	Application
LCA	Process tool		Per process Per product
Carbon footprint analysis	Process tool	Cradle-to-grave or cradle-to-gate	Per person, Per household Per process Per product
Ecolabel	Process tool		Per product
Environmental management systems (EMS)	Analytical tool	Evaluation of all environmental aspects	Per organization

Table 6.2 Environmental assessment tools (standards and indicators)

Tools	Emissions/Impacts	Standarization	Quantitative indicators
LCA	Air emissions, solid and liquid waste	ISO 14040-14044 ISO 14031	Operational performance indicator (OPI): Energy (kWh/kg material) Heating (lt/kg material) Raw material (kg/kg material) Packing material (kg/kg material)
	Climate change Acidification Eutrophication Smog Solid wastes Water eutrophication		Environmental condition indicators (ECI): CO_2 equivalent, SO_2 equivalent, PO_4 equivalent, SPM equivalent, C_2H_4 equivalent/kg material produced
Carbon footprint analysis	Greenhouse gases	PAS 2050 ISO 14064-65 ISO 14031	Operational performance indicator (OPI): Energy (kWh/kg material) Heating (lt/kg material) Raw material (kg/kg material) Packing material (kg/kg material)
	Climate change		Environmental condition indicators (ECI): CO_2 equivalent, /kg material produced
Ecolabel	Air emissions, solid and liquid waste	ISO 14040-14044 and PAS 2050 ISO 14064-65 ISO 14031	LCA indicators (OPI and EPI)
Environmental management systems (EMS)	Climate change Acidification Eutrophication Smog solid wastes Solid wastes Water eutrophication	EMAS, ISO 14001 ISO 14031	LCA indicators and management performance indicators (MPI)

6.3 Results and Discussion

For the scope of this study, data on raw materials and energy flows have been acquired by means of an extensive survey, with the participation of major industrial producers. For the output data, namely emissions from mining, production, packaging, storage and transportation at the inventory phase, two software tools were used for the results' reliability control: the SimaPro LCA software, which is a life cycle analysis model with embodied EcoInvent LCA database

(Pré Consultants 2001, 2009). At the environmental impact assessment phase (normalization and weighting), two sets of indicators were used, one derived from the CML 2 baseline 2000 m method (Center of Environmental Studies) and the other from the Eco Indicator 95 method (Eco Indicator). The functional unit selected for the materials' environmental evaluation is kg emission/kg building material and MJ/kg building material for the embodied energy.

The output data indicators provided information about specific air emissions from raw materials extraction, production processes, transportation, and storage procedures such as CO_2 equivalent, SO_2 equivalent, PO_4 equivalent, SPM equivalent, C_2H_4 equivalent, and environmental impacts such as climate change, acidification, and eutrophication.

For clay brick production, the main procedures studied and evaluated were: (1) mining and storage: surface clays, shales, and some fire clays are mined in open pits with power equipment. Then the clay or shale mixtures are transported to plant storage areas; (2) preparation: to break up large clay lumps and stones, the material is processed through size-reduction machines before mixing the raw materials. Usually the material is processed through inclined vibrating screens to control the particles' size; and (3) forming: tempering, the first step in the forming process, which produces a homogeneous, plastic clay mass. Usually, this is achieved by adding water to the clay in a pugmill, which is a mixing chamber with one or more revolving shafts with blade extensions. After pugging, the plastic clay mass is ready for forming. There are three principal processes for forming brick: stiff-mud, soft-mud and dry-press.

The cement production is divided into five major processes: (1) extraction of raw materials such as stone and limestone; (2) transportation processes; (3) "farina" production; (4) "klinker" production; and (5) cement production. The cement production system is an energy-increased process with subsystems associated with each of the five stages mentioned above. The concrete production is based on cement production and input of auxiliary materials, such as sand and gravel. The functional unit for the emissions at the concrete production system is kg of emissions/m3 of concrete.

The steel's production process consists of the following steps: supplying the production line with the initial mixture of the recycled scrap; loading the recycling scrap and infusing the mixture under high-pressure conditions and temperature in electric furnace; and the final process and smelting of the mixture, reheating, and forming into the final product. Finally, cement mortar production is actually a mixture of cement with water, limestone, and sand.

As far as the insulation materials' involves, the most widely used categories of insulating materials are inorganic fibrous (glass-wool and stone-wool) and organic foamy ones (expanded and extruded polystyrene and, to a lesser extent, polyurethane), while all other materials cover the remaining 10 % of the market (including wood-wool, regips, perlite, etc.). The production of stone wool includes the following main processes: creation of the raw material mixture, which consists of bauxite and amphibolite; melting the rock mixture in an electric furnace; producing resin; adding resin to the rock mixture; adding silicone, water, and oil to the new mixture; forming the new mixture and Hardening it in a polymerization oven; compressing and cutting the final product; product packaging; temporary product storage; and, finally, product transport.

The extruded polystyrene's production process consists of the following steps: supplying the production line, which consists of two extruders; supplying the first extruder with styrol and additive substances; mixing and increasing the mixture's viscosity; infusing the mixture under high pressure conditions and temperatures (200 °C); diffusing the mixture; diffusing complete additives in the polymer's mass and controlling progressive refrigeration of the mixture in the second extruder; changing the material's flow from cylindrical to a flat form in the head drawing; the mixture's exit in atmospheric pressure conditions; the mixture's expansion at the appropriate thickness to form the forming plates; cutting and freezing the final product at an ambient temperature; packaging and temporarily storing the product; and transporting the product.

The expanded polystyrene's production process consists of the following steps: supplying the production line with the initial mixture; heating and humidifying of the initial mixture; drying the mixture and then infusing it under high pressure conditions to form blocks; cutting it into plates; and packaging and transporting the final product.

Based on the LCA evaluation results, the production procedures as well as the transportation process contribute mainly to air emissions and more specifically to CO_2 production. Actually, the CO_2 is the most significant emission in quantity of all the construction products studied (Table 6.3) (Anastaselos et al. 2009; Giama and Papadopoulos 2015). Nevertheless, the production processes contribute the most, compared to other processes (mining, transportation, packaging), to air emissions production (Table 6.4).

Table 6.3 Environmental condition indicators (ECI) for construction and insulation materials

Material	kg CO_{2eq}	kg SO_{2eq}	kg PO_{4eq}	Kg C_2H_{4eq}	EmbEn [MJ]
Steel	0.63761	0.00395	0.00018	0.00016	9.76
Reinforced concrete	0.34	0.90673	0.08972	0.03596	0.48
Cement plaster	0.22134	0.00050	0.00005	0.00002	1.42
Cement portland	0.85807	0.00131	0.00018	0.00005	3.33
Brick	0.23595	0.00070	0.00007	0.00005	2.76
Stone	1.01494	0.00671	0.00057	0.00024	16.73
Plaster board	0.39033	0.00167	0.00019	0.00007	6.03
Common plaster	0.26146	0.00036	0.00005	0.00003	1.45
Acrylic plaster	0.20961	0.00087	0.00007	0.00009	4.96
Ceramic tiles	0.95001	0.00418	0.00031	0.00021	15.72
Expanded polystyrene (EPS)	3.24197	0.01268	0.00096	0.00054	76.16
Extuded polystyrene (XPS)	4.04462	0.01646	0.00125	0.00088	92.38
Glasswool (GW)	3.30205	0.01904	0.00158	0.00105	60.10
Polyurethane foam (PUR)	4.42797	0.01934	0.00279	0.00212	92.30
Stonewool (SW)	2.17293	0.01303	0.00132	0.00059	24.90

Table 6.4 Carbon footprint analysis results per process (tCO_{2eq} /Kg of building material)

Material	Mining	Production	Transport	Total
Cement	0.000261	0.0017	0.000163	0.002124
Reinforced concrete	–	0.114	0.317	0.431
Steel	–	0.000561	0.000081	0.000642
Brick	0.00077	0.13	0.0055	0.13627
Plasterboard	–	0.0001		0.0001
Concrete plaster	0.000444	0.00032	0.0000136	0.000778
Stonewool	0.0003	0.002	0.000015	0.002302
Extruded polystyrene	0.0005	0.001	0.000025	0.001525
Expanded polystyrene	0.0005	0.003	0.000025	0.003525

The results for Carbon Footprint Analysis were obtained from the Footprint Reporter, which is a web-supported software. The differences in the environmental performance indicators calculated by the LCA approach are due to the various software and databases used.

Nevertheless, in both environmental evaluation methodologies, the production processes proved to be responsible for most of the energy consumption and the respective CO_2 emissions, compared to other processes, such as mining, transportation, or packaging. The emissions depend to a high degree on the national energy mixture, for which the producer is obviously not responsible. The use of renewable energy sources for the energy used during the production process—biomass, for instance—would improve the carbon footprint of the construction materials and the building's carbon footprint as a result. Moreover, end-of-life management, recycling building materials, upgrading the industrial infrastructure (not necessarily by spending money on refurbishment, but improving the monitoring and control process), and implementing standards such as ISO 50001 that focus on Energy Systems Management, will definitely contribute to the environmental impacts reduction. Finally, measures such as the promotion of locally extracted raw materials can reduce the carbon footprint to a significant extent.

6.4 Conclusions

Carbon Footprint Analysis has been fairly recently introduced as a promising environmental assessment tool. Compared to LCA, which is well established, having been applied to products and processes for more than 25 years, it is clear that LCA is obviously the more mature environmental tool, with abundant literature available, highly developed, verified and validated software tools and extensive data bases with information for a plethora of materials. At the same time, LCA is definitely more complicated compared to Carbon Footprint Analysis. Carbon Footprint Analysis follows the "cradle-to-grave" approach, by focusing on mapping the greenhouse gas emissions throughout the reference systems of products or processes, targeting sustainability and economic benefits. As a new tool, there are still significant gaps in the literature, especially with regard to properties of materials. It is, however, definitely more comprehensive as a methodology to the receivers of information and at the same time less complicated for the auditor; therefore it can be more easily applied, especially to small and medium-size enterprises, where monitoring and recording schemes are not always implemented. It can also improve public awareness on environmental issues, as it can link everyday processes to CO_2 emissions in a way easily understood by the final consumer. Nevertheless, both methodologies are included in the ecolabelling schemes as criteria for product certification.

The production of construction materials accounts for significant quantities of raw materials and large amounts of energy, respectively. The fact that construction materials contribute in a most decisive way to sustainable building management has been proven by many studies; they are rightly, therefore, important elements for the energy-conscious and bioclimatic design and construction of new buildings.

Their exact environmental features, however, have to be determined on a product-specific base, as they depend on the raw material used, the manufacturing process applied, and the energy sources used for the production. Furthermore, the determination of the materials' environmental impact can be carried out in many methodological ways. The energy consumption from the production processes, based on the LCA and Carbon Footprint Analysis, comes to 85–95 % and from the mining process 5–10 %, while transportation contributes up to 6 % to the final energy consumption, and packaging only 0.04 %. Considering the construction materials' environmental impact, it is necessary to promote the best techniques available and introduce innovative solutions in the production processes in order to reduce the depletion of the natural, finite resources.

Another interesting point that emerged from this study is the need to minimize the transport of raw materials. In that sense, promoting the use of resources produced locally is a further important measure to reduce transport emissions and, not to be forgotten, costs.

Finally, an optimistic result of the field study is that despite the ongoing economic recession, there is evidence that environmental management in the area of energy and materials production is perceived as positive by the industry, since it can lead to cost reductions in the manufacturing process and to certified sustainability, which is becoming an important cause in this highly competitive market.

References

Ahlroth S, Nilsson M, Finnveden G, Hjelm O, Hochschorner E (2011) Weighting and valuation in selected environmental systems analysis tools—suggestions for further developments. J Cleaner Prod 19:145–156

Anastaselos D, Giama E, Papadopoulos AM (2009) An assessment tool for the energy, economic and environmental evaluation of thermal insulation solutions. Energy Build 41:1165–1171

Bribián IZ, Capilla AV, Uson AA (2011a) Life cycle assessment of building materials: comparative analysis of energy and environmental impacts evaluation of the eco-efficiency improvement potential. Build Environ 46(5):1133–1140

Bribián IZ, Capilla AV, Uson AA (2011b) Life cycle assessment of buildings impacts and evaluation of the eco-efficiency. Build Environ 46:113–114

Cabeza LF, Rincon L, Vilarino V, Perez G, Castell A (2014) Life Cycle assessment (LCA) and life cycle energy analysis (LCEA) of buildings and the building sector: a review. Renew Sustain Energy Rev 29:394–416

CML 2 baseline method (2000) Centre for environmental studies (CML), University of Leiden, 2001. Available at http://www.leidenuniv.nl/interfac/cml/ssp/LCA2/index.html, Switzerland

Directive 2002/91/EC Energy Performance of Buildings, http://eur-lex.europa.eu/LexUriServ/LexUriServ.do?uri=OJ:L:2003:001:0065:0065:EN:PDF. 04/04/2014

European Regulation 305/2011 Construction Products Regulation, http://eur-lex.europa.eu/LexUriServ/LexUriServ.do?uri=OJ:L:2011:088:0005:0043:EN:PDF. 04/04/2014

European Regulation 89/106 Construction Products Regulation, http://eur-lex.europa.eu/LexUriServ/LexUriServ.do?uri=OJ:L:1989:040:0012:0026:EN:PDF. 06/04/2014

Fuertes M, Casals M, Gangolles N, Forcada M, Macarulla X, Roca A (2013) An environmental impact casual model for improving the environmental performance of construction processes. J Cleaner Pord 52:425–437

Giama E, Papadopoulos AM (2012) Sustainable building management: an overview of certification schemes and standards. Adv Build Energy Res. doi:10.1080/17512549.2012.740905

Giama E, Papadopoulos AM (2015) Assessment tools for the environmental evaluation of concrete, plaster and brick elements production. J Cleaner Prod 1–11. doi:10.1016/j.jclepro.2015.03.006

Greenhouse Gas Protocol (2010) Calculation tools from http://www.ghgprotocol.org/calculation-tools. 4 June 2014

IPCC—Intergovernmental Panel on Climate Change 2006 (2006) IPCC guidelines for national greenhouse gas inventories. Downloaded from http://www.ipcc-nggip.iges.or.jp/public/2006gol/index.html

IPCC—Intergovernmental Panel on Climate Change 2010 (2010) Task on National Greenhouse Gas Inventories. Emission factor database. Downloaded from http://www.ipcc-nggip.iges.or.jp/EFDB/main.php

Karkanias C, Boemi SN, Papadopoulos AM, Tsoutsos TD, Karagiannidis A (2010) Energy efficiency in the hellenic building sector: an assessment of the restrictions and perspectives of the market. Energy Policy 38(6):2776–2784

Monahan J, Powell JC (2011) An embodied carbon and energy analysis of modern methods of construction in housing: a case study using a lifecycle assessment framework. Energy Build 43:179–188

Ortiz O, Castells F, Sonemann G (2009) Sustainability in the construction industry: a review of recent developments based on LCA. Constr Build Mater 23:28–39

Pandey D, Agrawal M, Pandey JS (2010) Carbon Footprint: Current methods of estimation. Environ Monit Assess. doi:10.1007/s10661-010-1678-y

Pré Consultants (2001) The eco indicator 99: a damage oriented method for life cycle impact assessment. Manual for designers, 3rd edn. www.pre.nl, Switzerland

Pré Consultants (2009) SimaPro LCA software, Version 7.1., Product écology Consultants, Netherlands. www.pre.nl/simapro, Switzerland

Santamouris M, Paravantis J, Founda D, Kolokotsa D, Michalakakou P, Papadopoulos AM, Kontoulis N, Tzavali A, Stigka E, Ioannidis Z, Mehilli A, Matthiessen A, Servou E (2013) Fuel/energy poverty and the financial crisis: A household survey in Greece. Energy Build 65:477–487

Slini T, Giama E, Papadopoulos AM (2015) The impact of economic recession on domestic energy consumption. Int J Sustain Energ 34(3–4):259–270. doi:10.1080/14786451.2014.882335

Wiedmann T (2014) Editorial: carbon footprint and input-output analysis—an introduction. Econ Syst Res 21

Wiedmann T, Minx J (2008) A Definition of 'Carbon Footprint'. J Ecol Econ Res Trends 1:1–11

Chapter 7
Economic Experiments Used for the Evaluation of Building Users' Energy-Saving Behavior

Nieves García Martín, Gerardo Sabater-Grande, Aurora García-Gallego, Nikolaos Georgantzis, Iván Barreda-Tarrazona and Enrique Belenguer

Abstract Several treatments that could be implemented in the home environment are evaluated with the objective of reaching a more rational and efficient use of energy. We feel that detailed knowledge of energy-consuming behavior is paramount for the development and implementation of new technologies, services, and even policies that could result in more rational energy use. The proposed evaluation methodology is based on the development of economic experiments implemented in an experimental economics laboratory, where the behavior of individuals when making decisions related to energy use in the domestic environment can be tested.

N.G. Martín (✉) · E. Belenguer (✉)
Department of Industrial Systems Engineering, Universitat Jaume I, Castellón, Spain
e-mail: ngarcia8281@cv.gva.es

E. Belenguer
e-mail: efbeleng@uji.es

G. Sabater-Grande (✉) · A. García-Gallego (✉) · I. Barreda-Tarrazona (✉)
Laboratorio de Economía Experimental and Economics Department, Universitat Jaume I, Castellón, Spain
e-mail: sabater@uji.es

A. García-Gallego
e-mail: mgarcia@eco.uji.es

I. Barreda-Tarrazona
e-mail: ivan.barreda@eco.uji.es

N. Georgantzis (✉)
School of Agriculture Policy and Development, University of Reading, Reading, UK
e-mail: georgant@eco.uji.es

S.-N. Boemi et al. (eds.), *Energy Performance of Buildings*,
DOI 10.1007/978-3-319-20831-2_7

7.1 Introduction

The demand for energy in the residential sector in developed countries has been rising steadily over the last 30 years, although this upward trend has slowed down significantly since 2008 due to the economic crisis. In Spain in 2011, the energy consumed in the residential sector accounted for 25 % of the country's total electricity consumption, as well as 17 % of the total energy consumption. There is a similar proportion of energy consumption in other temperate countries, such as 21 % of the total amount of energy in the USA, 29 % in the UK (in 2005), or 33 % in Denmark.

Given the importance of the residential sector in the consumption of energy in a country, researchers have proposed various methods to reduce household consumption, but test results are not conclusive. On the one hand, informative campaigns have yielded poor results or have not been time-persistent. On the other, the results of more sophisticated methods, such as feedback information and control technologies, are difficult to compare and validate, due to the use of different sample sizes and their implementation in countries with very different climates and consumption habits. In this sense, it may be important to distinguish among developed countries that, belonging to the temperate zone of the earth, may present large climatic differences (basically Mediterranean, continental, and Atlantic climates), implying different energy-consuming requirements and patterns.

In this paper we present the results of an economic experiment that was designed and performed in a laboratory with two main objectives. The first was to analyze various systems or treatments for improving energy efficiency at home in a controlled environment, regardless of geographic and climatic variations, with a similar number of samples under equal conditions. This provides us with data from various treatments that can be compared properly, thereby simplifying the data acquisition process and reducing the cost of the test development.

The second objective was related to the importance of human behavior in the success of each treatment. It is well known that human behavior consists of a combination of rational and irrational (emotional) decisions. A detailed analysis of this behavior is critical, in our opinion, to determine the scope and effects of each method.

We cannot reasonably expect a realistic representation of a household to be obtained just by using a computer simulation, and therefore the results of the experiment about energy savings, as absolute values, will not be exact. Nevertheless, the experiment is useful as a means for analyzing and quantifying the differences among the methods, so the proportional variations among them will be established, that is, the experiment will provide a scale.

7.2 Literature Review

Feedback information varies from advice on energy usage or comparison with peers to more complex feedback technologies providing real-time consumption information through Internet-enabled devices. Estimates of the energy savings from feedback technologies vary widely from none, to as much as 20 % (Ehrhardt-Martinez et al. 2010). Delmas et al. (2013) offer the most comprehensive meta-analysis of information-based energy conservation experiments conducted to date. The study shows that strategies providing individualized audits and consulting are comparatively more effective for conservation behavior than strategies that provide historical, peer-comparison energy feedback. Overall, this meta-analysis suggests that information strategies do induce energy conservation, but it is less clear which strategies work best, in part because many experiments simultaneously use more than one strategy, thus leading to confounding effects. Delmas et al. (2013) claimed that additional experiments are needed to better identify the winning strategies.

Studies that have analyzed the impact of peer comparisons on conservation have yielded mixed results. For example, Goldstein et al. (2008) found that social norms can increase towel reuse by hotel guests. In a review of the literature on the effect of feedback on home energy consumption, Fischer (2008) noted that of the dozen studies testing the impact of comparisons with others that she reviewed, none showed an effect. She attributed the failure to the boomerang problem, where informing individuals of typical peer behavior inadvertently inspires those who have been underestimating the prevalence of an activity to increase the unwanted behavior. Cialdini et al. (1991) argued that combining injunctive norms (those that express social values rather than actual behavior) with descriptive norms can neutralize the boomerang effect. Schultz et al. (2007) conducted a randomized field study in San Marcos, California, about the effectiveness of social norms messaging (alongside energy-saving tips) to reduce home energy consumption. They found that combining descriptive and injunctive messages lowered energy consumption and reduced the undesirable boomerang effect. Ayres et al. (2013) found that by providing customers with feedback on home electricity and natural gas usage with a focus on peer comparisons, utilities can reduce energy consumption at a low cost.

Today it is possible to think of sophisticated demand response systems, where smart meters could allow consumers to adjust consumption in response to price signals that vary in time, leading to a more efficient use of electricity.[1] Brophy et al. (2009) investigated the international experience in liberalized electricity markets and highlighted smart metering as a necessary tool to be able to overcome barriers such as information asymmetry.

[1]Conchado and Linares (2010) is an interesting survey of the state of the art in the quantification of demand response program benefits.

Fisher (2008) offered a theoretical perspective on the matter and introduced a psychological model to explain why feedback works. She maintains that successful feedback has to capture consumers' attention, link specific actions to their effects, and activate various motives. One unanimous finding is that feedback is more effective when it is more detailed and more closely linked to consumption actions.

Several[2] field studies analyze feedback-induced energy savings using various in-home monitors, including aggregate-level and appliance-specific real-time feedback. Other studies are focused on web-based feedback technologies. In this line, Abrahamse et al. (2005) found that European households were capable of generating average savings of 5.1 % using an Internet-based tool. Benders et al. (2006) concluded that a web-based application produced average household savings of 8.5 % in the Netherlands. Additionally, Ueno et al. (2006a, b) found average savings of 18 % in Japan.

Generally speaking, the most directly related experimental literature on electricity markets can be classified in two main categories: field experiments and lab experiments. In the first category, Battalio et al. (1979) ran an interesting field experiment designed to determine the effects of various price and non-price policies on electricity consumption. Five different price policies were tested, and the results show that price policies work much better than informational policies. Faruqui and Sergici (2009) reviewed the most recent experimental evidence on the effectiveness of residential dynamic pricing programs. Their review of 15 different field experiments on pricing reveals that the demand response impacts from various pilot programs vary from modest to substantial, largely depending on the data used in the experiments and the availability of enabling technologies.

In the category of lab experiments, Adilov et al. (2004, 2005) conducted experiments on demand-side and full two-sided electricity markets, respectively. The aim of both papers was to test the efficiency of two alternative forms of active demand-side participation. They evaluated various experimental price structures that represent end-use consumers who can substitute part of their usage between day and night: fixed price, a demand-response program with fixed price and credit for reduced purchase, and a real-time pricing system where prices are forecast for the coming day/night. Their main conclusion is that the real-time program results in the greatest market efficiency.

Barreda et al. (2012) studied consumers' behavior in experimental electricity markets, finding that a dynamic system for prices is more efficient than a fixed one. They showed that a dynamic scheme with sanctions, although less preferred by consumers, is more effective than the one with bonuses in order to reduce peak consumption.

[2]See Carroll and Mark (2009) and Carroll et al. (2009), for example.

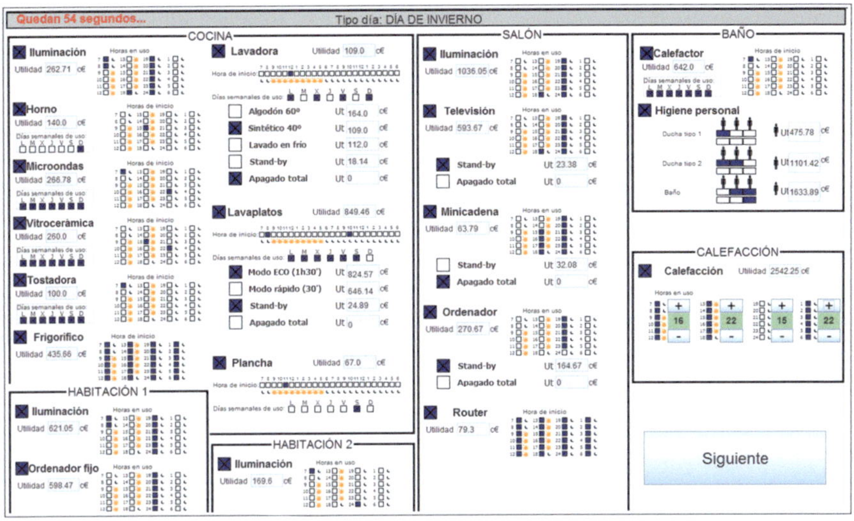

Fig. 7.1 Programming Screen (PS)

7.3 Experimental Design

We conducted an experiment consisting of simulating consumers' energy behavior in a virtual home according to the information they were provided with. Participants were selected among students who volunteered at Universitat Jaume I, Castellón, Spain.

We assumed that the residence was occupied by three members (father, mother and son) and consisted of a living room, kitchen, two bedrooms, and a bathroom. Heating and hot water ran on gas, whereas lighting and all other appliances worked on electric energy. In this framework, energy consumption had to be chosen by subjects for each period representing a typical winter or summer day.[3] The 18-period sequence implemented included 3 winter days, 3 summer days, 3 winter days, and so on.

In order to simulate this environment, we designed a Programming Screen (PS) (see Fig. 7.1), on which subjects decided which appliances to switch on or off in order to maximize their net gains, calculated as the difference between their valuation of the energy consumed and the money spent to buy this energy.

As is common in laboratory experiments, the valuation of purchases was preassigned to buyers through an Induced Utility Function[4] (IUF), which determined

[3]Differences between winter and summer days are focused on the possibility of using a heating system that consumes gas energy.

[4]The utility provided by the use of electricity depends on the amount consumed according to a concave function. Parameterizations of these functions for each appliance are available upon request.

the maximum price the consumer was willing to pay for each unit of energy. On the PS, subjects decided on energy consumption for seven different groups of items:

1. Lighting. We assumed that there were four lighting selection[5] points: kitchen, living room and the two bedrooms.
2. Oven, microwave, toaster, iron and heater. For this group of appliances, subjects had to choose whether to use each of them, the number of days per week they were to be used, and the number of hours per day.
3. Washing machine and dishwasher. In addition to the features considered in the previous group, subjects had to select the type of wash program and, when not in use, whether to leave the appliances in standby mode or switch them off.
4. Refrigerator and router. Neither of them allowed the subjects any kind of choice, as they had to be running 24 h a day. However, they represented an energy consumption shown on the PS.
5. TV, stereo, desktop computer, and laptop. In this group, participants had to select the number of hours they wanted to use them and, in the case of no use, choose between the stand-by or off modes.
6. Personal hygiene. In this case, the subject had to choose the type (quick shower, slow shower, or bath) and the number of daily washes (none, one, or two for all types) for each family member.
7. Heating. Subjects could choose use and temperature for four time slots (7:00–12:00, 13:00–18:00, 19:00–24:00 and 1:00–6:00), but only for winter days. On summer days this item was removed from the PS.

Participants were not given any time restriction when it came to making their choices. However, there was a timer on their computer screen counting down the seconds from 300 to 0. Once the first five minutes were over, the counter turned red, indicating that a reasonable amount of time had already passed. Once decisions had been made for these seven groups of items, subjects received information about their gas and electricity expenditure along with utility and net gains achieved in ExCUs (experimental currency units) on the invoice screen (see Fig. 7.2).

In order to study the effect of information on energy consumption, five different treatments were considered. In the baseline treatment (T0), subjects were provided with a gas and electricity bill for each period after their decisions had been made. In treatment 1 (T1), before energy consumption was chosen, participants were additionally provided with advice[6] about energy savings, which was repeated in

---

[5]Lighting is selected 24 h per period by default, and subjects must turn off the light at undesired hours by clicking on the square to the right of the corresponding hour.

[6]Advice that was provided included sentences such as: (a) Turn off equipment when it's not in use; (b) The microwave uses less energy than the oven and it can cook the same foods; (c) Remember to turn off lights when you leave the room; and (d) A bath costs more than a shower, etc.

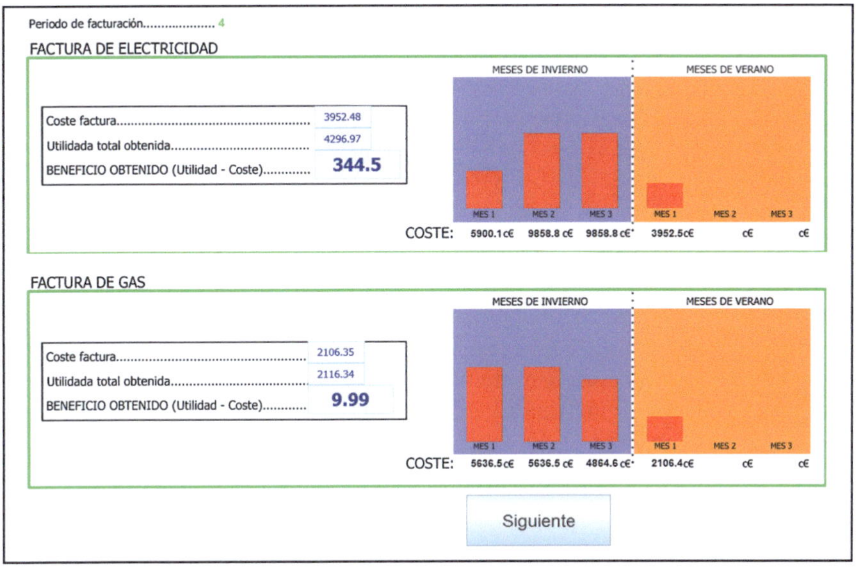

Fig. 7.2 Invoice screen

every period. In treatment 2 (T2), along with the aforementioned bill, individuals also received information about instantaneous gas and electric consumption in kWh and their corresponding cost in ExCUs for the entire household through a smart meter display. In treatment 3 (T3), subjects disposed of a set of five smart meters, showing the instantaneous consumption of each separate group[7] of appliances in kWh or ExCUS. In treatment 4 (T4), after consumption decisions had been made for each period, subjects were provided with comparative information about average and minimum energy consumptions in their market, along with the standard gas and electricity bill. T4 was repeated in two separate sessions in order to increase the number of independent observations.

The differences introduced on the PS were minimal, consisting of the type of information provided depending on the treatment that was implemented. In Table 7.1 we present a summary of the experimental sessions that were held.

A total of 173 subjects (30 per session, with the exception of T2, where there were 23) were recruited among undergraduate students from various economics- or business-related courses, using standard recruitment procedures with an open call for subjects through our lab's Web site. Before the beginning of each session, subjects were given written instructions, which were also read aloud by the organizers, and any remaining issues were resolved on an individual basis.

[7]Independent smart meters are offered for lighting, kitchen, gas consumption, leisure (TV, stereo, and computers) and others (washer, dishwasher, iron, and bathroom heater).

Table 7.1 Summary of experimental treatments

Treatment	Information	Sessions	Subjects
T0	No	1	30
T1	Advice	1	23
T2	Aggregated consumption	1	30
T3	Disaggregated consumption	1	30
T4	Consumption comparison	2	60

At the end of each session, which typically lasted 90–120 min (depending on the experimental treatment), subjects responded to an environmental awareness questionnaire. After that, they were paid privately in cash, the average earnings being around €20. All sessions were computerized and carried out in a specialized computer lab, using original software programmed in JAVA.

7.4 Results

In this section, we analyze the gas and electricity consumption of participants and corresponding consumer surplus obtained in the five implemented treatments. In Table 7.2 we observe that, with the exception of treatment 3, all treatments showed a decrease in electricity consumption relative to the baseline treatment. However this reduction is not associated with an increase in surplus. On the contrary, all treatments exhibit a higher surplus than the baseline treatment.

Regarding gas consumption, with the exception of the advice treatment, we observe in Table 7.3 that all treatments present higher consumption and lower surplus than the baseline treatment.

Figure 7.3 shows the average values of electricity consumption for each treatment as a percentage with respect to the optimum. The optimum point is defined as the use of each piece of equipment at its point of maximum benefit, and it was designed as the most energy-efficient behavior without loss of comfort. The values at the optimum point are shown in Table 7.4.

Table 7.2 Summary of statistical descriptions. Electricity

Treatment	Electricity consumption (c€)		Electricity surplus (c€)	
	Average	Standard deviation	Average	Standard deviation
T0	3363.75	1059.77	501.78	67.03
T1	3154.37	1464.79	507.31	85.10
T2	3179.01	1180.91	539.40	92.12
T3	3856.77	1420.80	567.18	89.56
T4	3091.17	1185.52	508.13	94.20

Table 7.3 Summary of statistical descriptions—gas

Treatment	Gas surplus (winter) (c€)		Gas surplus (summer) (c€)	
	Average	Standard deviation	Average	Standard deviation
T0	1738.77	754.1	91.70	30.29
T1	1646.11	701.27	94.00	33.91
T2	2371.54	1378.61	85.72	36.16
T3	1986.00	1065.06	86.08	27.97
T4	1874.96	711.47	82.27	29.90

As the difference between summer and winter in electricity is minor, the average for the whole year is shown. The statistical significance of all the differences has been tested[8] using Mann-Whitney tests; results can be checked in Table 7.6.

Figure 7.3 shows how all the treatments were thriftier than the optimum. In fact, the subjects displayed very conservative behavior when consuming energy due to a kind of suggestion effect. We consider that this effect is related to different behavioral phenomena, such as the motivation during the introduction to the experiments, the belief that great energy savings would result in a great benefit, and also the presence in most of the subjects of home-grown values related to the importance of energy saving, ecological behavior, etc.

When comparing the various methods, it can be concluded that the treatments that achieve the highest savings are T1 (advice) and T4 (information sharing). These are the treatments with the highest pressure regarding saving and also with the poorest level of feedback about one's own energy consumption. We call them "saving maximizers," and they are basically characterized by a compulsion to save.

The behavior of the subjects is not wholly explained just by studying the energy consumption. It is also necessary to look at the benefits in order to fully understand the results of the experiment. Results regarding the average values of electrical benefits are shown in Fig. 7.4; they are shown as a percentage with respect to the optimum, just as in the case of electricity consumption.

In this case, the most successful treatments are T2 (aggregated consumption feedback) and T3 (disaggregated consumption feedback); this means that the ones with the best feedback give the highest benefits. We can see how the treatments with the least feedback are the ones that sacrifice benefits, i.e., comfort.

Because of the way in which the utility function is constructed, maximizing benefits means maximizing satisfaction obtained by the use of appliances and at the same time controlling the amount of consumption. Given that the experiment

[8]Due to the fact that only a part of the studied variables are normal, we have done all the statistical research using Mann-Whitney tests (the results of the Shapiro-Wilk test over those variables are shown in Table 7.5). Only in the particular case when comparing two variables with a normal distribution, the Mann-Whitney and t-test have been done. We have not detected any difference between the results of both tests in any case.

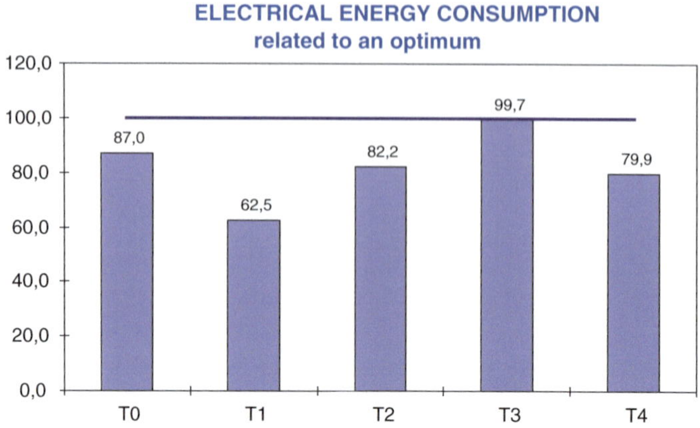

Fig. 7.3 Average values of the consumption

Table 7.4 Electricity and gas. Optimum values

Variable	Season	Optimum value (c€)
Electricity consumption	Winter	4150.1
	Summer	3586.4
	Average	3868.3
Electricity surplus	Winter	941.3
	Summer	816.4
	Average	878.9
Gas consumption	Winter	3306.7
	Summer	567.0
	Average	Not applicable
Gas surplus	Winter	207.5
	Summer	75
	Average	Not applicable

does not seek to reduce comfort but to use the energy needed to maintain it in an efficient way, we can conclude that T2 and T3 are the most successful treatments of all. The treatments with enough information to reach maximum benefits have been called "benefit maximizers."

On the other hand, the evolution over time of results regarding gas consumption from water heating and house heating is shown in Figs. 7.5 and 7.6. House heating is shown only during winter months (months 1, 2, 3, 7, 8, 9, 13, 14 and 15), while water heating is shown throughout the duration of the experiment. Each treatment is shown in a different color, and the optimum treatment is also shown as a horizontal line.

Figure 7.6 shows how the subjects have under-used house heating. In fact, from the individual results of the majority of the subjects, it can be seen that most of

Table 7.5 Shapiro-Wilk tests (significantly non-normal in italics)

	Electricity consumption *p*-value			Electricity surplus *p*-value		
Treatment	Winter	Summer	All	Winter	Summer	All
T0	*0.001*	*0.001*	*0.001*	0.545	0.781	0.773
T1	*0.001*	*0.005*	*0.006*	0.379	0.827	0.815
T2	0.059	0.238	0.093	0.905	0.662	0.878
T3	*0.019*	*0.032*	*0.016*	*0.030*	0.632	0.548
T4	*0.039*	*0.002*	*0.015*	0.692	*0.024*	0.232

Table 7.6 Mann-Whitney test of the variable electrical consumption item[a]

Treatment	Optimum	T0	T1	T2	T3
T0	<				
T1	<	<			
T2	<	<	>		
T3	<	>	<	<	
T4	<	<	<	<	<

aNote "<" means "row < column"

	p-value > 0.1
	0.05 < p-value < 0.1
	p-value < 0.05

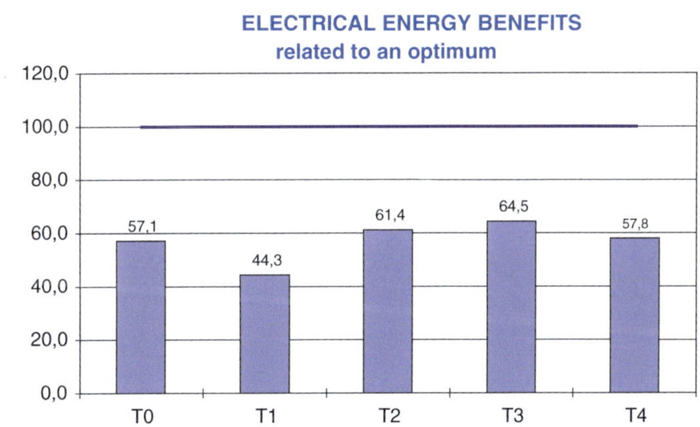

Fig. 7.4 Average values of the benefit

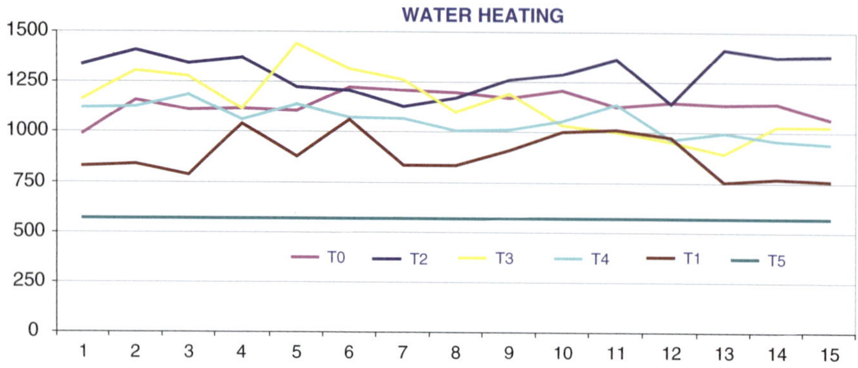

Fig. 7.5 Water heating

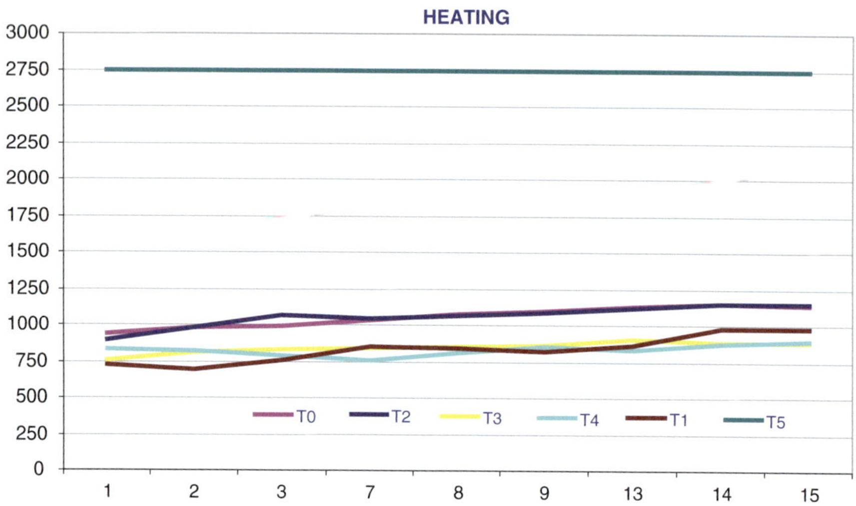

Fig. 7.6 House heating

them did not turn on the heat. Figure 7.5 shows erratic behavior—the only thrifty one—and with no apparent improvement during the experiment. No learning process can be appreciated in the figures, so we can conclude that the subjects had problems understanding the tool.

The results can be divided into those for electricity and gas, due to the fact that they were treated in separate fields throughout the whole experiment, both on the programming screen and also on the invoice screen. The gas results can therefore be rejected until the tool has been improved.

7.5 Conclusions

Regarding the results obtained from the experiments conducted in this study, first the numeric values of the modification of the behavior obtained with various treatments are going to be discussed. It is important to emphasize that the variations have been obtained in a relatively simple computer simulation, so they are not directly applicable to reality. Nevertheless, the existence of significant modifications in the behavior, as well as the proportion of the measured variations among the treatments, can be extrapolated to reality.

There are three main specific conclusions from this paper. The first and most significant is the importance of the citizens' environmental awareness when designing any energy-saving campaign. Reductions of up to 13 % points are achieved just through suggestion, without any other indication regarding their behavior. The reduction reaches 31.4 % points if the suggestion is supported by feedback consisting of repeating messages related to energy savings.

This awareness, without any further information, achieves good results in terms of savings at the expense of the sacrifices on the consumer's part, as can be concluded from the "saving-maximizing" treatments. In order to maintain the savings in the long run, it is necessary to guarantee the occupant's satisfaction; otherwise, the tool of suggestion is likely to lose effectiveness with time.

The way to turn the new behavior into a habit is to strengthen environmental awareness with methods that provide citizens with the tools needed to reach the same levels of savings without loss of comfort. The results about aggregated and disaggregated consumption feedback clearly show how people can rationalize their energy behavior when they are provided with enough real-time information about their behavior's results. Combining suggestions and smart meters appears to be the most successful method tested—awareness as the motor promoting changes, and feedback as the tool allowing the citizen to apply them efficiently.

The second remarkable result is related to consumer comparison. With this, considerable reductions in consumption without loss of benefits have been achieved. This method is not only successful, but also easy and cheap to apply, at least in neighborhoods where supply companies have installed telematic reading meters. The value used as a reference can be the neighborhood average consumption or a residential energy efficiency index, which can motivate citizens to reduce their bills even more than by comparing themselves with their neighbors.

The third conclusion is about the need to develop more sophisticated methods to support reductions in the consumption of gas, which is the energy source for heating and hot water in the simulation. The subjects have demonstrated a limited capability to manage complexity in the program, which is seen quite clearly through the meaningless results in gas consumption.

In temperate countries of the developed world, heating is the main energy-consuming system of a household; the second is hot water. The proportion of the total is variable, but in most cases heating exceeds 50 % of the total consumption of the energy at home, the 70 % adding the hot water. The development of methods to reduce its consumption is essential, and the fact that it is not easily understood by the subjects is something to take into account. A simpler interface for heating and hot water needs to be developed for the simulation, but it is important to note that the same problem can appear in reality: a brochure with advice regarding the use of gas, which is easy to understand and simple to apply, has to precede every other feedback method.

7.6 Further Investigations

This kind of economic experiment is very innovative from two points of view: its methodology, and its object of study. With regard to the methodology, the virtual environment designed for conducting the experiments is considerably more complex than any developed to date. The number of decisions that the subject has to make throughout the sessions, the amount of information he or she is given, and the complexity of changes in the behavior needed to obtain measurable results, have never before been attempted in a laboratory—at least not successfully. Having obtained significant results in such a complicated environment opens up a new field of research in experimental economics based on the use of simulations that are more complex and therefore more similar to reality.

With regard to the goal of this particular study, the variations in the energy behavior at home depending on the quality of the information provided have never before been tested as an economic experiment. This opens another new path of investigation in the field of engineering: every method of energy saving can now be checked simply, in the short term and, with a minimal budget, in the laboratory. More significant results should be tested in reality, but a lot of important previous observations can be carried out at low cost.

We plan to develop new projects on the same subject in order to test new methods, combinations of methods, and variations in the population from which the sample is extracted. We believe that it can be especially interesting to examine various dynamic pricing methods in order to determine, before implementing them in reality, which kinds of combinations are simple enough to be understood and processed by a non-technical population.

On the other hand, as a second part of this research, plans are under way to install smart meters in a group of households in Castellón, Spain. By knowing the scale of efficiency of each method and the actual efficiency of one of them, the real efficiency of each one can be established.

Acknowledgments Financial support by BP Chair of Energy Efficiency—UJI, the Spanish Ministry of Science and Innovation (project ECO2011-23634), the Spanish Ministry of Economics and Competitiveness (project ECO2013-44409-P) and the Bank of Spain Excellence Chair in Computational Economics (project 11I229.01/1) is gratefully acknowledged.

References

Abrahamse W, Steg L, Vlek C, Rothengatter T (2005) A review of intervention studies aimed at household energy conservation. J Environ Psychol 25:273–291

Adilov N, Schuler RE, Schulze W, Toomey DE (2004) The effect of customer participation in electricity markets: an experimental analysis of alternative market structures. In: Proceedings of the 37th Annual Hawaii international conference on system sciences

Adilov N, Schuler RE, Schulze W, Toomey DE, Zimmerman RZ (2005) Market structure and the predictability of electricity system line flows: An experimental analysis. In: Presented at 38th Annual Hawaii international conference on systems science, Waikoloa, HI

Ayres I, Raseman S, Shih A (2013) Evidence from two large field experiments that peer comparison feedback can reduce residential energy usage. J Law Econ Organ 29:992–1022

Barreda-Tarrazona I, García-Gallego MA, Pavan M, Sabater-Grande G (2012) Demand response in experimental electricity markets. Revista Internacional de Sociología 70:27–65

Battalio RC, Kagel J, Winkler RC, Winett RA (1979) Residential electricity demand: An experimental analysis. Rev Econ Stat 61(2):180–189. doi:10.2307/1924585

Benders RMJ, Kok R, Moll HC, Wiersma G, Noorman KJ (2006) New approaches for household energy conservation—in search of personal household energy budgets and energy reduction options. Energy Policy 34(18):3612–3622

Brophy Haney A, Jamasb T, Pollitt MG (2009) Smart metering and electricity demand: Technology, economics and international experience. Cambridge Working Papers in Economics, Number 0905

Carroll E, Mark B (2009) Research to inform design of residential energy use behavior change pilot. Conservation Improvement Program Presentation for the Minnesota Office of Energy Security. (July) Franklin Energy

Carroll E, Eric H, Mark B (2009) Residential energy use behavior change pilot. CMFS project code B21383. (April) Franklin Energy

Cialdini, RB., Kallgren CA, Reno RR (1991) A Focus theory of normative conduct: a theoretical refinement and reevaluation of the role of norms in human behavior. In: Zanna ZP (ed) Advances in experimental social psychology, vol 24. Academic Press, New York, pp. 201–234

Conchado A, Linares P (2010) Estimación de los beneficios de la gestión activa de la demanda. Revisión del estado del arte y propuestas. Cuadernos Económicos de ICE 79:187–212

Delmas MA, Fischlein M, Asensio OI (2013) Information strategies and energy conservation behavior: a meta-analysis of experimental studies from 1975 to 2012. Energy Policy 61:729–739

Ehrhardt-Martinez K, Donnelly K, Laitner J (2010) Advanced metering initiatives and residential feedback programs: a meta-review for household electricity-saving opportunities. Technical Report E105, American Council for an Energy Efficient Economy

Faruqui A, Sergici S (2009) Household response to dynamic pricing of electricity: a survey of the experimental evidence. SSRN Report no. 1134132, Social Science Research Network, 10 Jan 2009. http://papers.ssrn.com/sol3/papers.cfm?abstract_id=1134132

Fisher C (2008) Feedback on household electricity consumption: a tool for saving energy? Energ Effi 1:79–104. doi:10.1007/s12053-008-9009-7

Goldstein NJ, Cialdini RB, Griskevicius V (2008) A room with a viewpoint: using social norms to motivate environmental conservation in hotels. J Consum Res 35(3):472–482

Schultz, PW, Nolan JM, Cialdini RB, Goldstein NJ, Vladas G (2007) The constructive, destructive, and reconstructive power of social norms. Psychol Sci 18:429–434

Ueno T, Sano F, Saeki O, Tsuji K (2006a) Effectiveness of an energy consumption information system on energy savings in residential houses based on monitored data. Appl Energy 83(8):166–183

Ueno T, Tsuji K, Nakano Y (2006b) Effectiveness of displaying energy consumption data in residential buildings: to know is to change. In: Proceedings of the ACEEE 2006 summer study on energy efficiency in buildings, vol 7. American Council for an Energy-Efficient Economy, Washington, D.C., pp 264–277

Chapter 8
Technologies and Socio-economic Strategies to nZEB in the Building Stock of the Mediterranean Area

Annarita Ferrante

Abstract The greatest potential for energy savings in the EU is in its existing buildings. The concept of "Nearly Zero Energy Building" (nZEB), which represents the main future target for the design of new buildings, is now gaining increasing attention in relation to the renovation of existing buildings as well. This is exceptionally challenging, considering the economic crisis in the EU and, in particular, in the Mediterranean areas, which are currently experiencing high levels of unemployment, poverty, and social exclusion. This chapter presents and discusses some progress in low- and zero-energy research and practice. It contains a brief review of the current policy background and case studies, and a set of demonstration projects containing evaluation and demonstration procedures that consider the technical, economic, and social feasibility of nearly zero energy buildings in the Athens metropolitan area (AMA).

A. Ferrante (✉)
Department of Architecture, School of Engineering and Architecture,
UNIBO, Viale Risorgimento 2, 40126 Bologna, Italy
e-mail: annarita.ferrante@unibo.it

A. Ferrante
Physics Department, Group Building Environmental Research,
UOA, National and Kapodistrian University of Athens,
Panepistimioupolis, 15784 Athens, Greece

S.-N. Boemi et al. (eds.), *Energy Performance of Buildings*,
DOI 10.1007/978-3-319-20831-2_8

8.1 Towards Nearly Zero Energy Urban Settings in the Mediterranean Climate

8.1.1 State of the Art and Crucial Issues in the Urban Environment of the Mediterranean Areas. A Case Study of the Athens Metropolitan Area (AMA)

Today, more than two-thirds of the European population lives in cities and their surrounding urban areas (EU Report, Cities of Tomorrow 2011), which conversely consume 80 % of the final energy in Europe (EU Report, World and European Sustainable Cities, EU Report 2010; Energy Cities 2012).[1] Increasing urbanization and deficiencies in development control in the urban environment have important consequences on the thermal degradation of urban climate and the environmental efficiency of buildings. As a result of heat balance, air temperatures in densely built urban areas are higher than the temperatures of the surrounding rural zones. This phenomenon, known as "heat island" (HI), is due to many factors (Santamouris 2001a, b; Yamashita 1996): the canyon geometry, the thermal properties of materials increasing storage of sensible heat in the fabric of the city, the anthropogenic heat, and the urban greenhouse all contribute to an increase in the urban HI effect. Research on this subject usually refer to the "urban HI intensity," which is the maximum temperature difference between the city and the surrounding area (Santamouris 2001a, b).

In this context, the city of Athens is a significant pilot study: many studies have confirmed the existence of a strong HI phenomenon (Mihalakakou et al. 2004; Santamouris et al. 1999a, b; Akbari et al. 2009). The association of the HI with synoptic climatic conditions has been identified (Niachou et al. 2001), while the influence of surface temperature and wind conditions have been analyzed (Hassid et al. 2000; Papanikolaou et al. 2008; Stathopoulou et al. 2009). In parallel, the impact of various mitigation techniques involving cooling and reflective materials has been identified (Doulos et al. 2001; Karlessi et al. 2009; Synnefa et al. 2007; Santamouris 2001a, b, Santamouris et al. 2007, 2010; Giannopoulou et al. 2010; Livada et al. 2002, 2007). As already stated, all did

[1]Some significant and alarming figures: the world population has grown from 2 to 6 billion, and soon will reach 7 billion, while the percentage of human beings living in cities has increased from 3 % in 1800 to 14 % in 1900 and is estimated to rise from the current 50 % to 75 % in 2050. The figure for Europe is still higher: 83 % of the population are expected to live in cities by 2050 (EU Report, Brussels, 2010). The average temperature on the Earth's surface has suffered an increase of +0.6 % and is estimated to reach 1.5 % by 2030. The progressive increase of global warming will specifically raise urban temperatures and heat island effect. After the Messina earthquake of 1908 (which caused about 83,000 deaths) the hot summer of 2003 with ~70,000 deaths, mostly in the cities, was the second heaviest natural disaster of the last 100 years in Europe.

research on this subject, referring to the "urban HI intensity" as the maximum temperature difference between the city and the surrounding area, demonstrating that the AMA represents a highly significant pilot study: During the hot summers (corresponding to the HI upper limit during them), urban stations present temperatures significantly higher than those recorded in the comparable suburban stations (the gap varies from 5–15 °C). Furthermore, a detailed statistical analysis of the HI characteristics and distribution in the greater Athens area was carried out (Giannopoulou et al. 2011) using temperature data of 25 stations around the city. Measurements' results show that HI intensity presents its maximum concentration in the center and the western part of the Athens area, with up to a 5 °C higher temperature regime, due to the lack of green, the densely built areas, and anthropogenic heat; lower values were found at the northern and the eastern sectors of the AMA. As a result, the cooling load of reference buildings in the city center is about twice the value of similar buildings in rural areas.

Previous research done in Athens (Ferrante et al. 1998, Sfakianaki et al. 2009), has shown some appropriate procedures for redesigning urban areas using natural components, such as green roofs and permeable surfaces. In fact, the design of outdoor spaces—even if reduced to the envelope of the buildings because of existing urban constraints within thickly-built urban areas as well as the use of natural components—are key means to improving urban conditions in relation to both microclimate and reduction of pollutants. The results clearly indicated that outer surfaces' alternative designs act as a prior microclimate modifier and greatly improve outdoor air climate and quality (up to 2/3 °C reduction in ambient temperature) (Santamouris 2001a, b). Other significant physical factors in the thermal performance of urban environments are wind flow and air circulation (Santamouris et al. 1999a, b, Ricciardelli and Polimeno 2006) as well as air stratification and distribution in urban canyons and open areas. It is clear that in particular, the HI effect and the microclimatic conditions typical of open urban canyons (Bitan 1992) appear to be greatly influenced by the thermal properties of the materials and components used in the buildings and on the streets (Buttstädt et al. 2010). The comparative research carried out by the Host Institution demonstrated that the use of light-colored materials (Synnefa et al. 2007) and thermo-chromic building coatings can contribute to energy savings in buildings, providing a thermally comfortable indoor environment and improved urban microclimatic conditions (Karlessi et al. 2009a, b).

Since the morphological and spatial geometry of the urban "textures," and the thermal properties of surface coatings and green surfaces have a strong potential for the energy performance and cooling demand reduction of existing buildings and urban settings, it is therefore evident that they have to be conceived and investigated as a whole, consisting of the buildings blocks and the related open areas (streets, squares, courtyards).

8.1.2 Policy Background and Zero Energy Case Studies

Over the last few decades, energy-oriented innovations in building technology have emerged in many areas of the building construction sector[2] (Brown and Vergragt 2008), until the most recent experiences aiming at setting to zero the carbon emission of new developments[3] and even of a whole city[4]: a pilot city plan to set to zero the carbon emissions of Copenhagen has been developed.[5] The increasing interest in nZEBs, recent European (EU Parliament 2009) and national directives on Energy Performance of Buildings (EPB),[6] more easily accessible Best Available Techniques (BATs) and Renewable Energy Sources (RES), all seem to point to further exploitation of BAT and better penetration of RES into new building construction (EU Communication 2011a, b) .

Most of the existing pilot nZEBs refer to newly conceived buildings and development plans. Furthermore, in spite of growing investments in RES technology (Bürer and Wüstenhagen 2009), feed-in tariffs and, in general, policy incentives (Bulkeley 2010), additional investments are needed to reduce carbon emissions and fossil fuel consumption.[7] "Needless to say, this is particularly challenging in a context of global economic slowdown such as the one the world is currently experiencing" (Masini et al. 2010).

Thus, the challenge is now to expand technical ZEB knowledge in existing built environments. We need to shift our technical understanding of EE from new developments and buildings to existing buildings, within the real urban environments of active

---

[2]Green buildings now belong to the "history of architecture": the first prototype buildings and their attempts to achieve zero-heating in the form of solar houses date back to 1950s (Hernandez and Kenny 2010). Among the recent experiences is the well known urban village BedZED (Beddington Zero Emission Development), winner of the prestigious Energy Awards in Linz, Housing and Building category, Austria, 2002 (Marsh 2002).

[3]Bill Dunster, Craig Simmons, Bobby Gilbert, The ZEDbook, solutions for a Shrinking World, Taylor and Francis, 2008 Zed Factory Ltd.

[4]A first zero waste-zero carbon emission City is to be constructed in Abu Dhabi, Masdar City, designed by N. Foster. Despite its location (the oil rich and hot part of the world) the development is designed as a huge, positive energy building, resulting in a self-sustaining, car-free city.

[5]The climate Plan (City of Copenhagen 2009) demonstrates how to make Copenhagen the world's first carbon neutral capital by 2025 by means of using biomass in power stations, erecting windmill parks, increasing reliance on geothermal power and renovating the district heating network.

[6]In the frame of the legislative plane, recently the European Parliament (Directive 2010/31/EU on the EPB), amending the previous 2002 EPB Directive, has approved a recast, proposing that by 31 December 2020 all new buildings will be nearly zero-energy consumption and will have to produce as much energy as they consume on-site. See also Task 40/Towards Zero Energy Solar Buildings, IEA SHC /ECBCS Project, Annex 52.

[7]The treaty issued by several NGOs calls for a doubling of market investments by 2012 and quadrupling by 2020 to attain the proposed carbon emission reduction targets (Meyer et al. 2009). As reported by Guy (2006), according to the United Nations Environmental Program (UNEP) there is still an "*urgent need for the incorporation of EE issues to be included in urban planning and construction*".

cities, since the large amount of existing stock represents the wider potential in terms of carbon. In this context, recent studies and design proposals on building energy retrofitting (Ferrante and Semprini 2011) has proved that huge energy savings can be achieved by adding different coatings to existing buildings. In particular, the combination of new building coatings, such as sunspaces, buffer zones, or RES, drastically reduce the energy performance indexes of the buildings up to the target of nZEB.

8.1.3 Low Carbon Communities and Grass-Roots Initiatives in the Urban Environment

Despite excellent studies and pilot zero-energy cities, an increasing societal need for human recognition in a depersonalized urban environment is emerging. Furthermore, as a combined outcome of the labor market crisis on the one hand, and the functional or "Ford" city of the twentieth century on the other, much of the urban space is devalued: the "transparency" of the façades at street level are destroyed either by closing all potential social/public attractiveness at street level, or by the infill of garages.

Redesigning energy technologies in the urban areas is certainly a major scientific challenge. However, succeeding in this endeavor requires more than getting the engineering right (Webler and Tuler 2010) because energy efficiency in urban settings is more than a technical problem.

Recent studies have suggested that more focus should be put on the social aspects at the community level, and that energy users should be engaged in their role of citizens. In fact, developing more sustainable consumption and production systems depends upon consumers' willingness to engage in "greener" and more collective behaviors (Peattie 2010). In this framework, local urban communities have inimitable advantages in providing infrastructure for more sustainable consumption environments; in fact, different types of low-carbon communities as social contexts gathered to reduce carbon intensity are emerging at different scales (Heiskanen et al.). Existing literature (Mulugetta et al. 2010; Guy 2006; Masini and Menichetti 2010) stresses the need for a clear transfer from a technically/economically based urban theory to a human- based and socio-technical urban vision to achieve greener behavior in an urban environment. Furthermore, some of Europe's leading nations in innovation have included user-driven or user-centerd innovation as a way of providing innovative products and services that correspond better to user needs and are therefore more competitive. User-driven innovation (EU 2009) is closely associated with design, and involves tools and methodologies developed and used by designers.[8] Practices of (re)design of existing buildings by engaging final users are also taking place in various contexts in the EU and worldwide.

---

[8]EU (Commission of) Communities 2009, "Design as a driver of user-centerd innovation", Brussels, 7.4.2009, SEC(2009)501.

In brief, there is a special need for efficient schemes where zero-energy requirements, energy construction quality, and technological flexibility can fit with different users' needs and expectations, taking into account the various economic and social possibilities within the large building stock throughout European cities. Existing buildings in urban environments represent the biggest challenge, both in terms of carbon—because of the large amount of existing stock—and for the social impact they may generate on the relationship between human behavior and urban sites.

Finally, Real Urban Environments (RUEs, i.e.,streets, squares, courtyards, and connected residential buildings) are the core of the search for new intersections between urban dwellers and energy-related issues. An effective research action should explore the socio-technical mechanisms that can promote the concrete synergies between economic constraints, users' expectations, EE systems, and production, in renewed forms of urban self-expression.

8.2 Towards "Nearly Zero Energy" and Socio-oriented Urban Settings in the Mediterranean Climate

It is well known that a bioclimatic approach to the planning phase is of strategic importance to pursue the goal of sustainable architecture (Ferrante et al. 1999; Ferrante and Mihalakakou 2001); the proper selection of a site, the building orientation, and the organization of the blocks in the urban sector play a very important role in defining the future quality of life in the interior spaces and the future consumptions (Ferrante and Cascella 2011; Ferrante 2012, 2014). When we deal with existing buildings, however, all these passive components are not variables to be defined, but have already been defined as critical features or strengths and have been used as input for starting the design stage of refurbishment intervention.

To reach the ambitious target of nZEBs—EPBD recast 2010/31/EU—the technical feasibility in general is not sufficient to widely spread nZEB into current building practice, especially since the quality of the large amount of existing building stock represents a huge potential in carbon terms. As a matter of fact, the existing building stock in the EU consists of the great majority (about 60 %) of buildings having been built after the Second World War (1960 s and 1980 s). This percentage even increases (about 70–75 %) if we confine the analysis within the boundaries of the Mediterranean European countries (Greece, Cyprus, Spain, Portugal, etc.), and slightly increases for Italy as well (about 65 %). Furthermore, housing in the EU represents a huge percentage of the building stock. The EU dwelling stock accounts for about 200,000,000 units, representing around 27 % of energy consumption in the EU; the potential reduction in CO_2 emissions that energy-efficient housing would provide cannot be underestimated. But of Europe's existing buildings, only about 1.2 % are renovated per year.[9] Since the energy

[9]JWG: Towards assisting EU Member States on developing long-term strategies for mobilising investment in building energy renovation (per EU Energy Efficiency Directive Article 4), Composite Document of the Joint Working Group of CA EED, CA EPBD and CA RES, November 2013 (http://www.ca-eed.eu/reports/art-4-guidance-document/eed-article-4-assistance-document).

requirements applicable to housing before 1995 were absent or very limited, the majority of existing buildings in the EU were built well before the application of regulatory measures on energy consumption reduction; as a result, the existing building stock across the EU presents very low-standard energy performance (EU Report, Housing Statistics in the European Union 2010).

A survey of social providers across the EU (Cecodhas, POWER HOUSE, Nearly ZERO Energy CHALLENGE, Progress Report March 2013) has identified five key categories of barriers—economic, technical, credibility, social, legislative—in delivering new construction and retrofitting to nZEB standards; in particular, among others, the lack of access to available and affordable finance to retrofit existing stock towards nZEBs is one of the major barriers.

Nonetheless, studies and analyses done in various EU contexts as well have shown wide margins of Energy Efficiency (EE) in the retrofitting through standard technical interventions such as HVAC plant systems's retrofitting or U-value reduction of building components (Ferrante 2011, 2012, 2014). High up-front costs of investment for energy retrofitting have driven some experiences at EU level (i.e., SuRE-Fit, Reshape, Solar Decathlon, etc.) as well as pilot cases in Austria and The Netherlands, to focus on the strategy of rooftop extensions as additional volumes whose economic value in the housing real estate market counterbalances the energy retrofitting costs. Within the framework of the social housing sector, many experiences have highlighted the coincidence between low-quality and high-energy management costs, often associated with the deterioration of both buildings and urban contexts (Santamouris et al. 2007). Thus, promoting energy improvement measures for social housing is an urgent and challenging goal.

The opportunity for interventions to reduce energy costs must also be considered in terms of the renovation and valorization of buildings, which are corresponding key benefits. In fact, operating a mere energy retrofitting on individual residential units or building blocks in urban contexts in severe degradation conditions may imply a devaluation of the same retrofitting actions; in other words, energy retrofitting in the context of existing housing stock has a strong influence on the quality of the built environment that cannot be ignored since they have the potential of a key added value in terms of social, technical, and economical feasibility. This added value can be the trigger for activating a multiple-oriented strategy in a comprehensive integrated solution responding to various sectorial needs. In fact, the functionally oriented approach and the mono-perspective use of architectural building types have to be considered a past attitude, since it is now necessary to shift the analysis and the research from one singular point of view to multiple approaches in order to expand research in architecture to incorporate the voices of many in a user-driven, bottom-up design approach to stimulate the process of renewal according to a socio-oriented use of energy solutions and technologies in the urban environment. In a complexly structured context where the number of decision-makers and cultural scenarios overlap, where the temporal dimensions and social background of the citizens are dissimilar, and where local and global, physical and virtual dimensions co-exist and where, finally, it is no longer possible to ignore the direct and indirect relations with the context, it is necessary to identify design procedures that can quickly adapt to environmental variations and new requirements.

(Re)design processes for existing urban buildings are called for to respond coherently to social and end-users' requirements, to current energy and safety regulations/requirements and more permanent components (structural and functional invariants) as well. In this process, a certain degree of adaptability/flexibility and the adoption of processes directly engaging the inhabitants could offer a real solution—both to the anonymity and standardization of these housing complexes.

8.3 Energy Retrofitting Scenarios of Existing Buildings to Achieve nZEBs: The Case Study of the Peristeri Workers' Houses' Urban Compound

Thus, in consonance with the physical aspects highlighted in Sect. 8.1, the research was conducted in a case study specifically selected for its representativeness, both in terms of geographical location (the western part of the AMA) and for its constructive type. In fact, the building types—a series of block buildings with a structure made of reinforced concrete and infill walls—is massively present throughout AMA suburbs and typically connected with similar building blocks all over Europe. They form, as a whole, the greater majority (about 60 %) of the existing building stock in the EU.

In this typical urban setting, starting from simulations resulting in extremely low energy performance at the initial state of the existing compound's buildings, alternative retrofitting design scenarios have been studied and validated to prove the technical feasibility of nearly ZEBs. Further reflection on the estimated costs of proposed energy retrofitting options clearly indicate the need for additional tools and measures to be developed at the public, legislative, and market levels to counterbalance the large pay-back times of energy retrofitting measures.

Simulations have been done using DesignBuilder, a fully featured interface using the EnergyPlus platform to detect energy, carbon, lighting, and comfort performance of buildings. EnergyPlus is the U.S. Department of Energy's third-generation dynamic building energy simulation engine for modelling building, heating, cooling, lighting, ventilation, and other energy flows. It has been validated under the comparative Standard Method of Testing for the Evaluation of Building Energy Analysis Computer Programs BESTEST/ASHARE STD 140.

Thus, the research study has hypothesized the introduction of both dense and mass consistent coating materials to achieve improved thermal performance in hot summers and cold winters. To evaluate the effectiveness of the proposed solution, a simulation of the building before and after the intervention was carried out on an annual scale, and indexes of primary energy for heating in winters (Epi) and for summer air conditioning (Epe) were calculated and used as a benchmark for assessing the effectiveness of the intervention, along with the amortization time of the investment. For these evaluations, it was decided not to make changes in the systems or in the user profiles, in order to compare only the direct consequences of the design choices and limit the variables.

With particular reference to a case study in the Municipality of Peristeri, where real involvement of the municipality is possible, this approach aims at searching design tools' sets as variable formal components in the (re)design of urban blocks, to combine the complementary nature of the standard and planned design with the variability required by the inhabitants; the resulting urban design will tend to overcome the current standard distinction between informal settlements and planned developments.

To meet these objectives, a detailed ethnographic analysis of the whole urban context has been developed in order to try to understand and "anticipate" urban dwellers' expectations and use them as a very important input in the redesign process of the buildings. Peristeri is located in the area of Attica, northwest of Athens; as previously indicated, although hit by winds, because of the conformation of the territory and its geographical position, Peristeri is the town with the highest summer temperatures.

The workers' houses' urban compound (Fig. 8.1) is characterized by the presence of massive volumes with different building types and building geometry, as outlined in Fig. 8.2.

In particular, the Peristeri urban compound consists of 12 buildings:

1. Three "towers" (T11);
2. Three north-south oriented double blocks (A7);
3. Four east-west oriented blocks (T7);
4. Two north-south oriented blocks (B6).

All these buildings are residential, with the sole exception of the B6 building type, where small businesses shops, offices, and retail establishments on the ground floor level on the south side are located.

The urban compound extension is 37,820 m^2, of which 25,713 (68 %) is occupied by building construction and impermeable surfaces (parking areas and

Fig. 8.1 Peristeri urban compound selected as reference case study

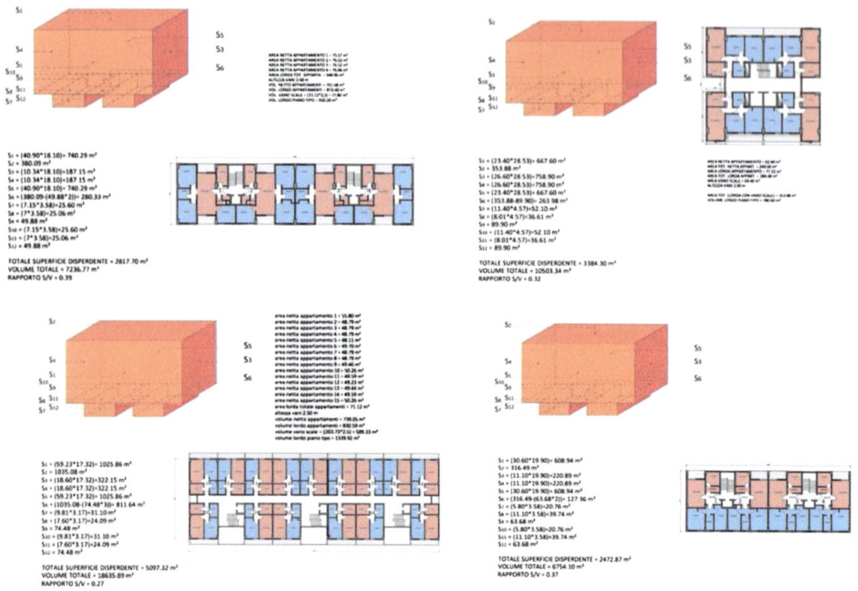

Fig. 8.2 Main building type's geometry in Peristeri urban compound and various surface/volume ratios

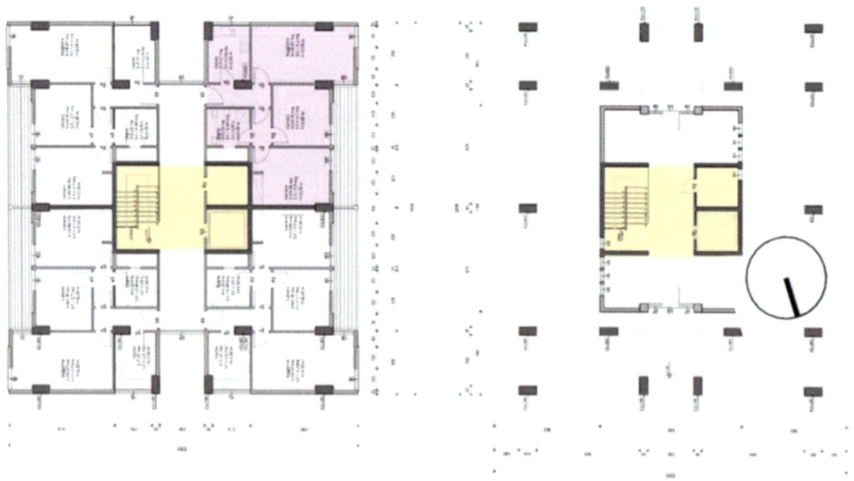

Fig. 8.3 Typical floor (*left*) and ground floor (*right*) of the tower building block

streets); the remaining (32 %) is a green open area. The built area represents 29 % (7504 m^2) of the total impermeable surfaces. In Fig. 8.3, the tower blocks' floors are depicted.

Real façade reconstruction of the various building types have been developed (Figs. 8.4, 8.5, 8.6, 8.7, 8.8, 8.9, 8.10, 8.11, 8.12, 8.13, 8.14, 8.15, 8.16, 8.17, 8.18, 8.19, 8.20, 8.21, 8.22, 8.23, 8.24, 8.25, 8.26, 8.27, 8.28, 8.29, 8.30, 8.31, 8.32, 8.33, 8.34 and 8.35), in order to understand all current modifications/appropriations of the building spaces; this analysis has been developed as a first step in the ethnographic analysis in the whole urban context, to predict urban dwellers' expectations and use them as very important input in the redesign process of the buildings.

In order to support the development of EE solutions in socio-oriented RUEs in the use of appropriate social engagement methods and user-driven design technologies, it is necessary to involve users, build competences, and create/implement a real demand for EE solutions. The social engagement methods include: (1) a functional analysis of socio-technical systems; (2) human and organizational factors' best practices; and (3) user requirements' elicitation and ethnographic research. The use of ethnographic research and the real representation of the buildings through photos is a key passage to inform the technical aspects of design with proper consideration of the end-users' needs and expectations.

The ethnographic observation of end-users' appropriations that are visible on the buildings' envelopes can be summarized in the following four types of modifications introduced by urban dwellers in the specific case study of the Peristeri urban compound:

1. Shading devices' additions;
2. HWAC and heat pump components;

Fig. 8.4 East façades (real photographic reconstruction on the *left*) of the tower building block in the Peristeri urban compound

Fig. 8.5 South and north façades (real photographic reconstruction on the *left*) of the tower building block in the Peristeri urban compound

3. Windows' modifications in size;
4. Space appropriation of balconies and loggias as internal volumes.

The presence of modification is higher in the southern, western and eastern façades of the buildings.

These modifications will be used to drive the redesign process related to the energy retrofitting of the buildings, tackling them by means of an interdisciplinary, socio-technical and physical-engineering approach.

8.3.1 Energy Performance Evaluation in the Buildings as Built

The structural system of the buildings is very simple, made up by a regular grid of beams and pillars, with presumably prefabricated slabs as horizontal elements. The main thermal parameters are provided for the horizontal ground, intermediate, and top flat floors as well as for the external building envelope have been considered the same as in the previous case study. Simulations have been performed by using DesignBuilder, a fully equipped interface using an EnergyPlus platform to detect energy, carbon, lighting, and comfort performance of buildings. EnergyPlus is the U.S. Department of Energy's third-generation dynamic building energy

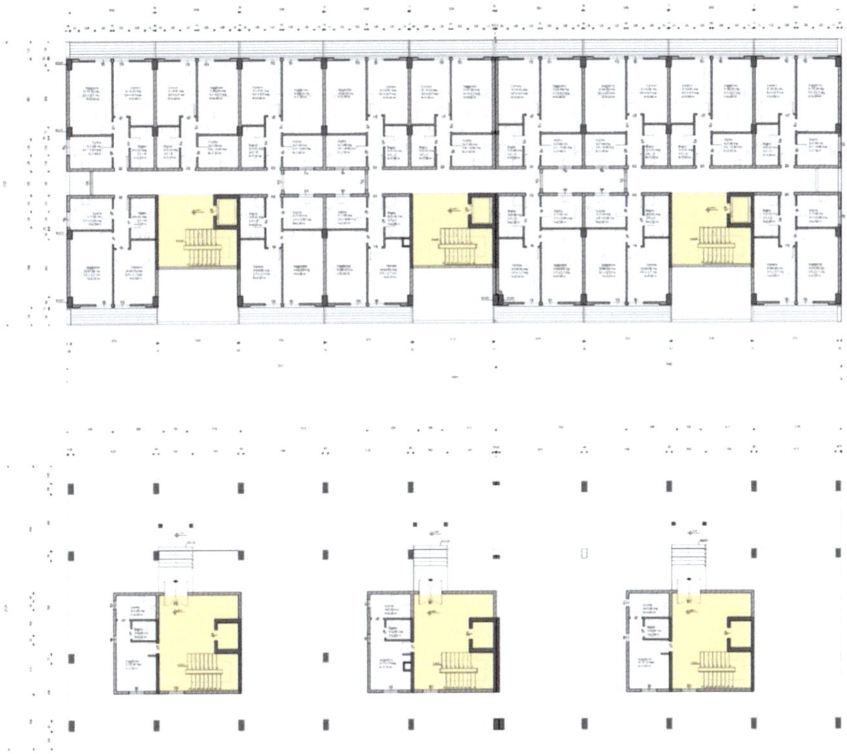

Fig. 8.6 Type floor (*up*) and ground floor (*down*) of one of the three double building blocks

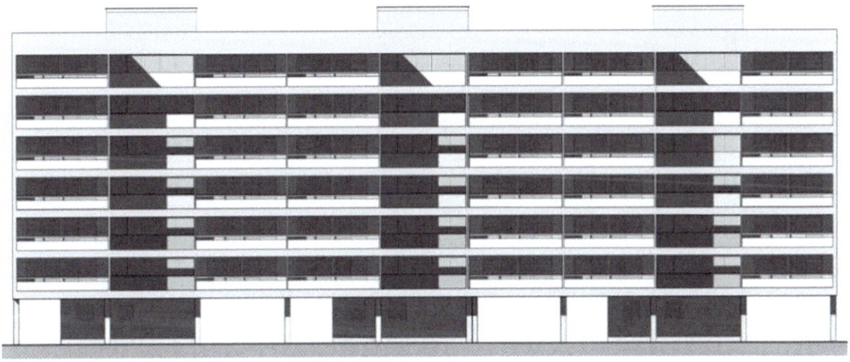

Fig. 8.7 East façade of one of the three double building blocks in the Peristeri urban compound

simulation engine for modelling building, heating, cooling, lighting, ventilation, and other energy flows. The thermal analysis of the as-it-is condition has been performed for each building type in the urban context of Peristeri.

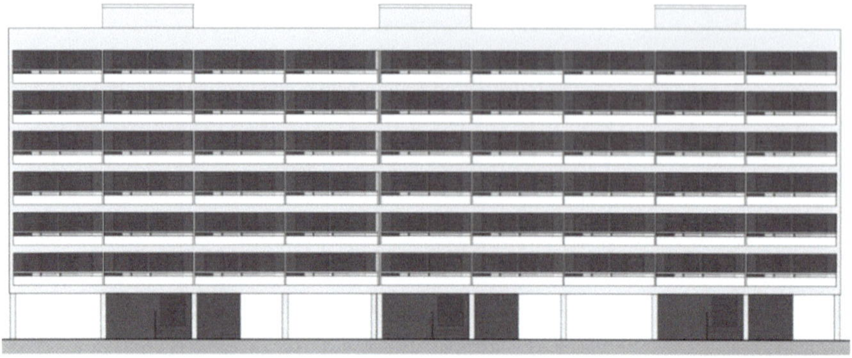

Fig. 8.8 West façade of one of the three double building blocks in the Peristeri urban compound

Fig. 8.9 West façade (real photographic reconstruction) of one of the three double building blocks in the Peristeri urban compound

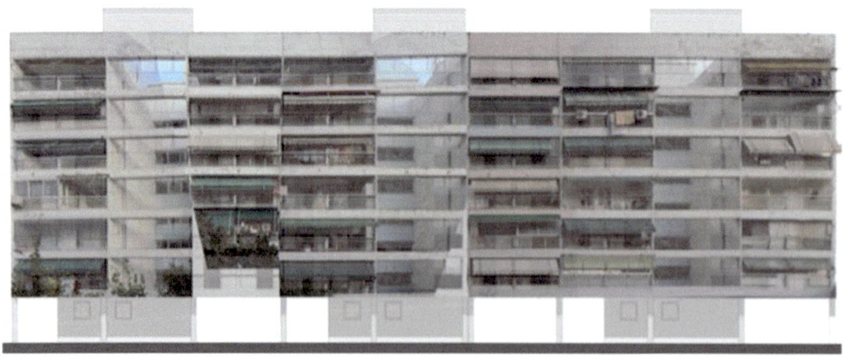

Fig. 8.10 East façade (real photographic reconstruction) of one of the three double building blocks in the Peristeri urban compound

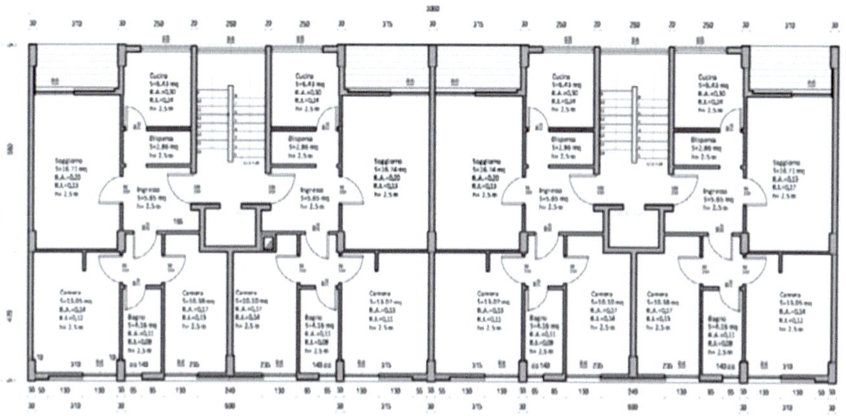

Fig. 8.11 Type floor of one of the four building blocks

Fig. 8.12 East façade of one of the three double building blocks in the Peristeri urban compound

Fig. 8.13 East façade (real photographic reconstruction) of one of the three double building blocks in the Peristeri urban compound

Fig. 8.14 Apartment as thermal zones selected for energy simulation in the single building block in Peristeri

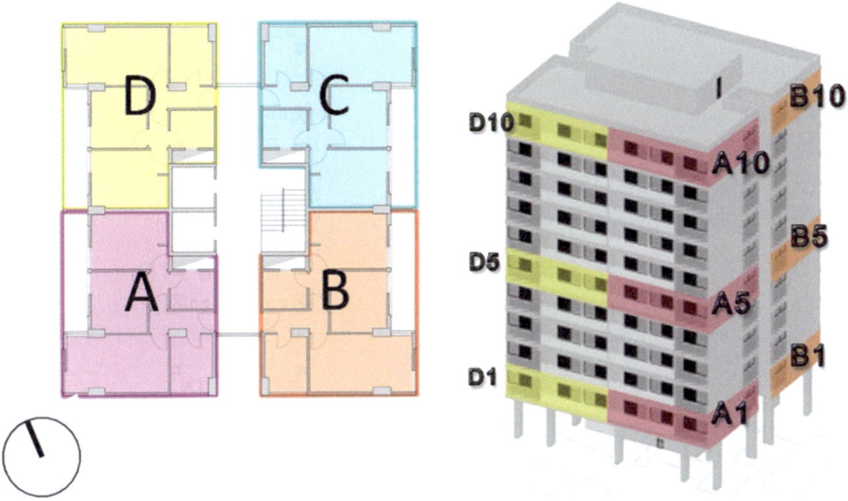

Fig. 8.15 Apartment as thermal zones selected for energy simulation in the tower building in Peristeri

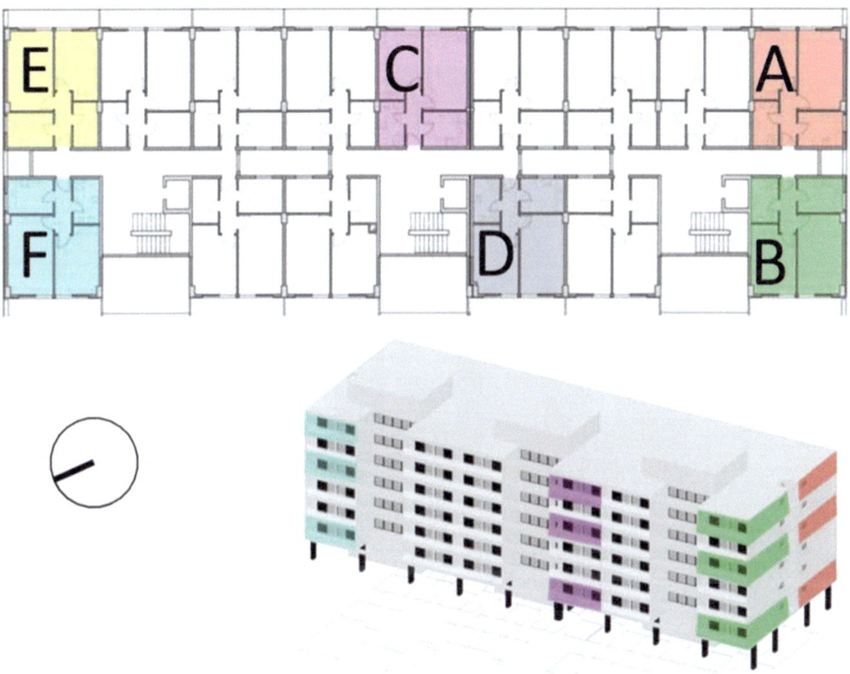

Fig. 8.16 Apartment as thermal zones selected for energy simulation in the double block building in Peristeri

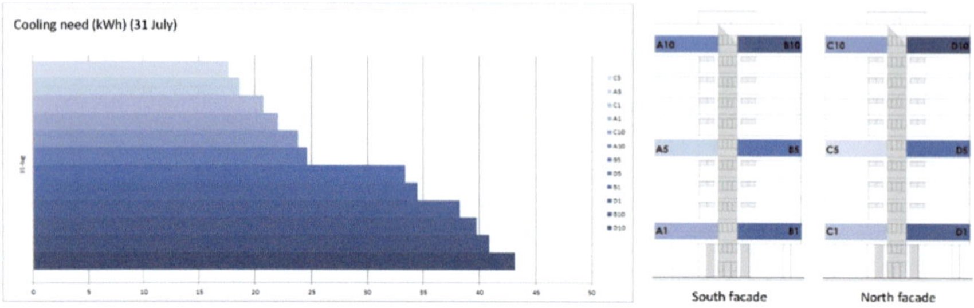

Fig. 8.17 These graphs highlight the different energy demands of the reference units in the summer as a function of both the position of the unit within the building block and the solar orientation

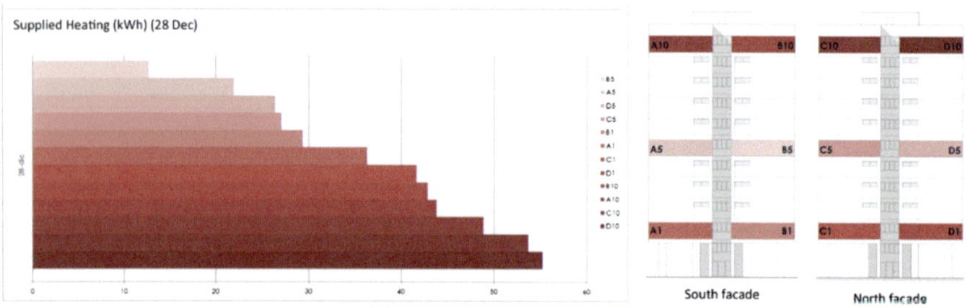

Fig. 8.18 These graphs highlight the different energy demands in the winter of the reference units, as a function of both the position of the unit within the building block and the solar orientation

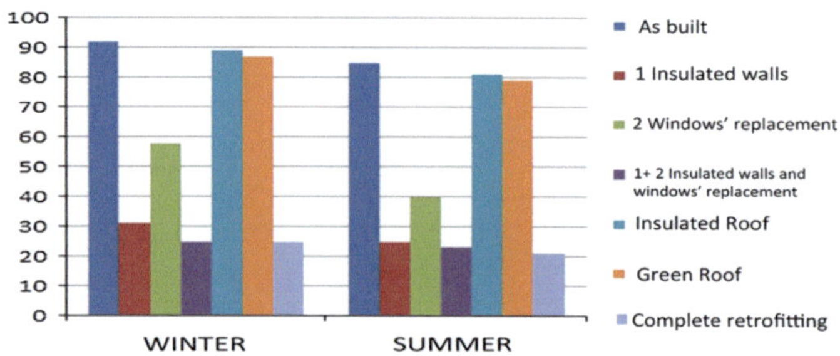

Fig. 8.19 Possible energy retrofitting options in the single block building and corresponding savings in terms of energy performance indexes (kWh/m^2 * y)

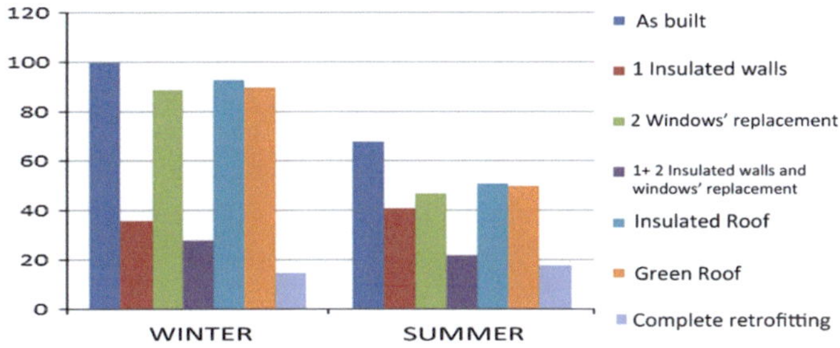

Fig. 8.20 Possible energy retrofitting options in the tower building and corresponding savings in terms of energy performance indexes (kWh/m^2 * y)

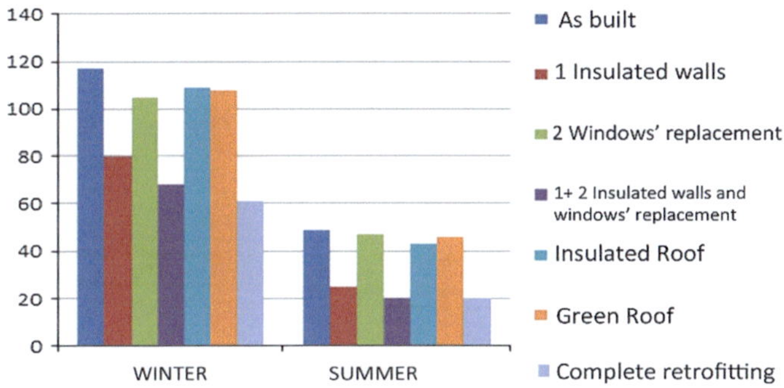

Fig. 8.21 Possible energy retrofitting options in the double block building and corresponding savings in terms of energy performance indexes (kWh/m^2 * y)

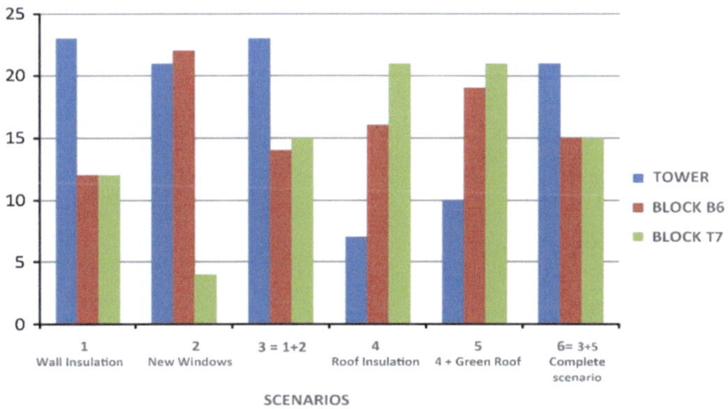

Fig. 8.22 Payback times of the investments range widely: a variation between 23 and 7 years can be seen as a function of the various hypothesized scenarios in the different building types (B6 = double building block; T6 = single building block)

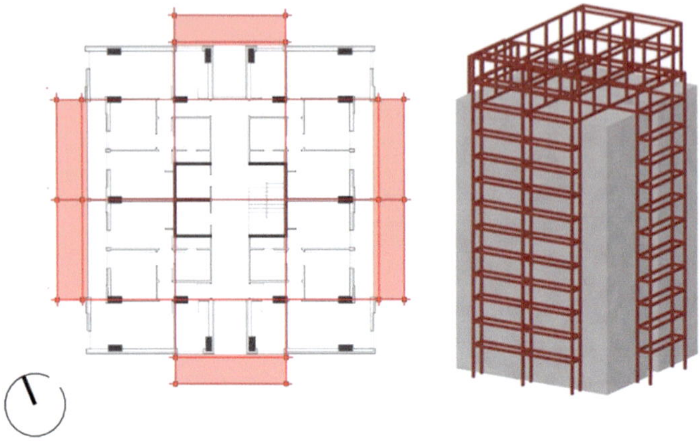

Fig. 8.23 Structure frame in the Peristeri case study for the tower typology. Technical study of the structural feasibility to combine additions on the tops of the buildings (and thus new residential units) with façade additions

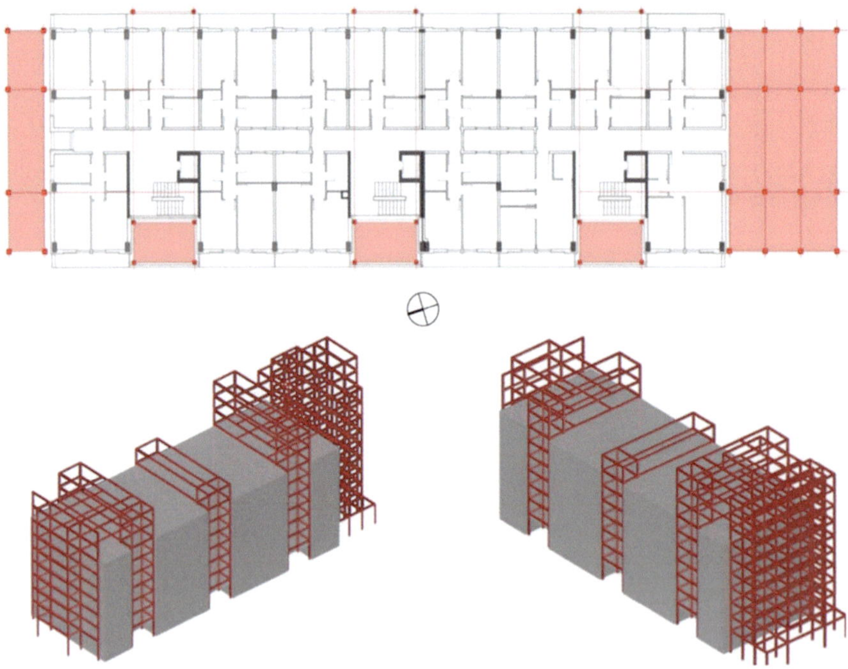

Fig. 8.24 Structure frame for the B6 block type. Technical study of the structural feasibility to combine additions on the *top* and *sides* of the buildings (and thus new residential units) with façade addictions

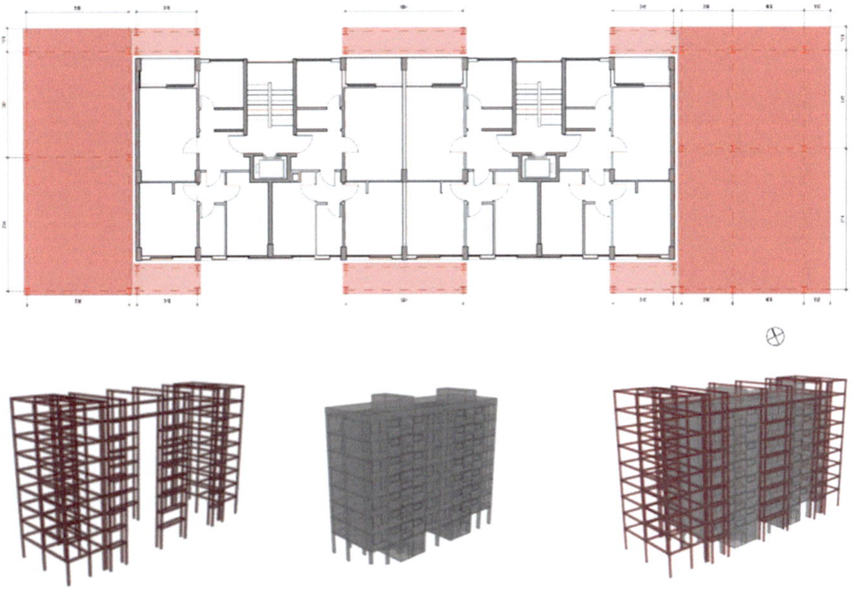

Fig. 8.25 Structure frame for the T7 block type. Technical study of the structural feasibility of combining additions on the tops and sides of the buildings (and thus new residential units) with façade additions

Fig. 8.26 New façade solutions for the existing residential units

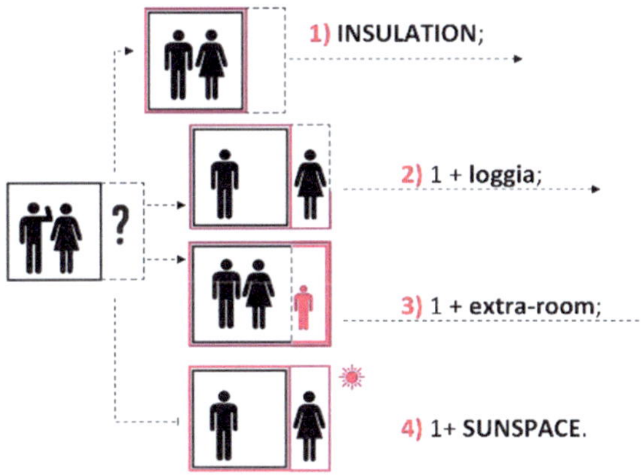

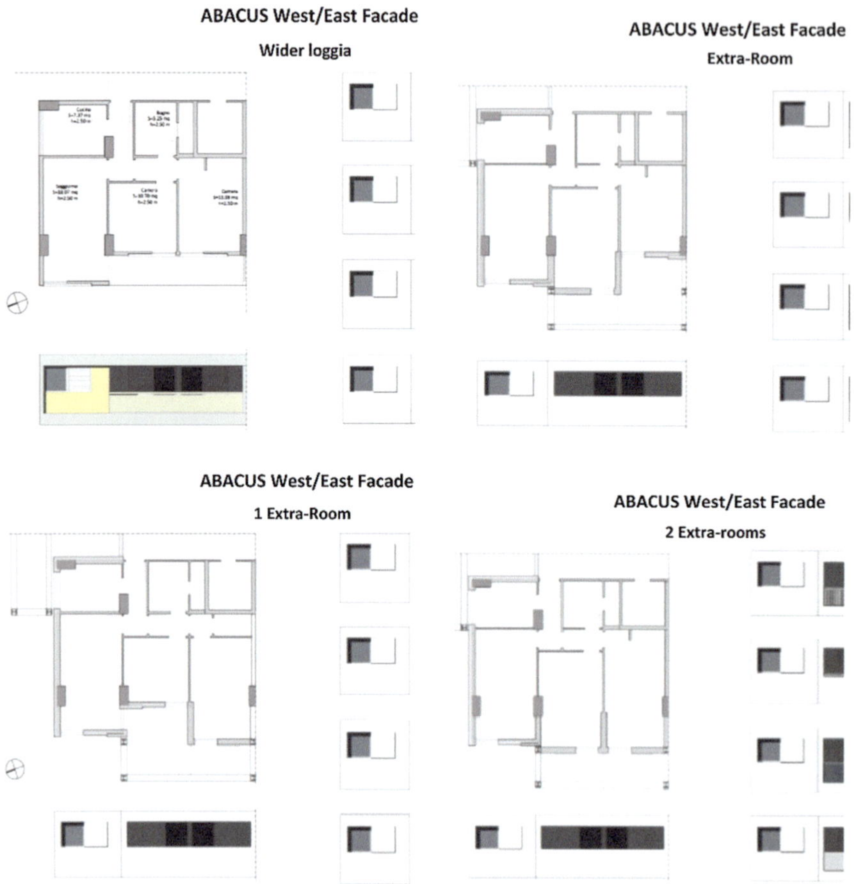

Fig. 8.27 Abacus of modular options as a function of possible users' expectations and requirements in the tower building type (west and east orientation)

In this case study, the following research tasks have been designed and developed:

- Evaluation of the overall energy performance and sensitivity analysis of its variability as a function of the different orientations of the buildings as well as of its different apartments; simulations have been performed for each representative unit type.
- Results' simulation and comparison in terms of comfort, energy contributions and overall heat balance, in order to: (1) identify the main critical problems affecting the building types; (2) classify them according to the their effects on the overall energy balance of the building; and (3) propose targeted solutions.
- Design of building refurbishment interventions and options in the selected reference buildings.

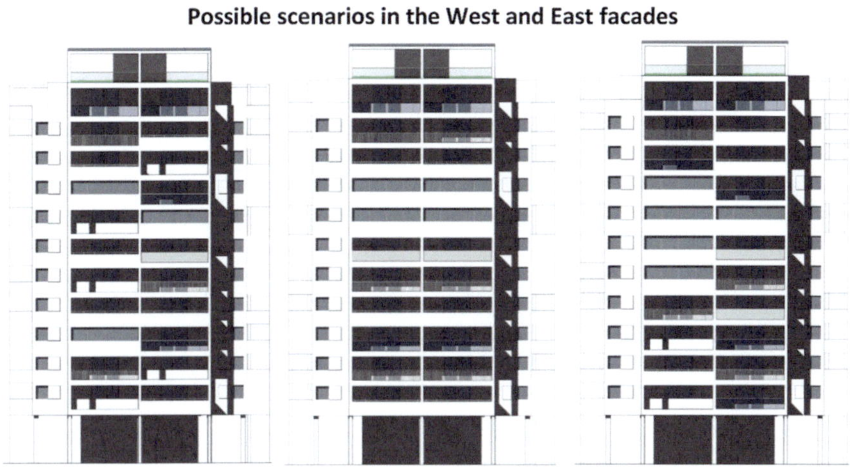

Fig. 8.28 Combination of additions on the *top* (new residential units) and on the elevation of the buildings with new façade solutions selected from the abacus for the north and west façades in the tower typology

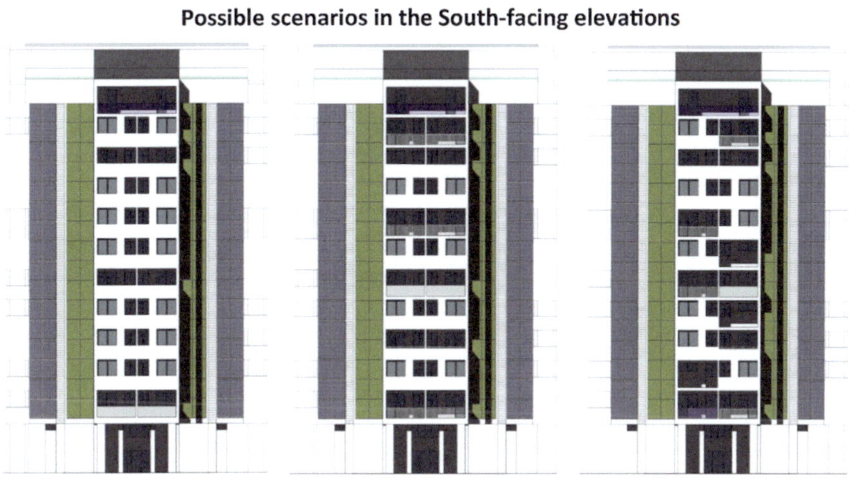

Fig. 8.29 Combination of additions on the *top* (new residential units) and on the elevation of the buildings with new façade solutions selected from the abacus for the south-facing façades in the tower typology

- Run of the simulation of the effects that the alternative design scenarios would produce on the overall energy balance of the building, so as to understand the effectiveness of the various retrofitting options in terms of energy response.
- Final evaluation of the results in terms of cost-effectiveness and quality.

Fig. 8.30 New residential units on the south side of block building B6 combined with add-ons and new façade solutions in the existing residential units

Fig. 8.31 New residential units both on the south and north sides of block building B6 combined with add-ons and new façade solutions in the existing residential units

Each building type has been divided into different representative thermal zones corresponding to the real apartments of each building block. The selection of the representative units (apartments) has been made according to both the position of the unit within the building block and the solar orientation, as shown in Figs. 8.14, 8.15, and 8.16.

Energy simulations have been performed for each building type, both in the cold winter and hot summer seasons. The artificial results for the single building block (type t7) are shown in Table 8.1.

The artificial results for the tower building are shown in Table 8.2.

The artificial results for the double building block are reported in Table 8.3.

Fig. 8.32 South-oriented façades of the Block T7: new residential units combined with add-ons and new façade solutions in the existing residential units

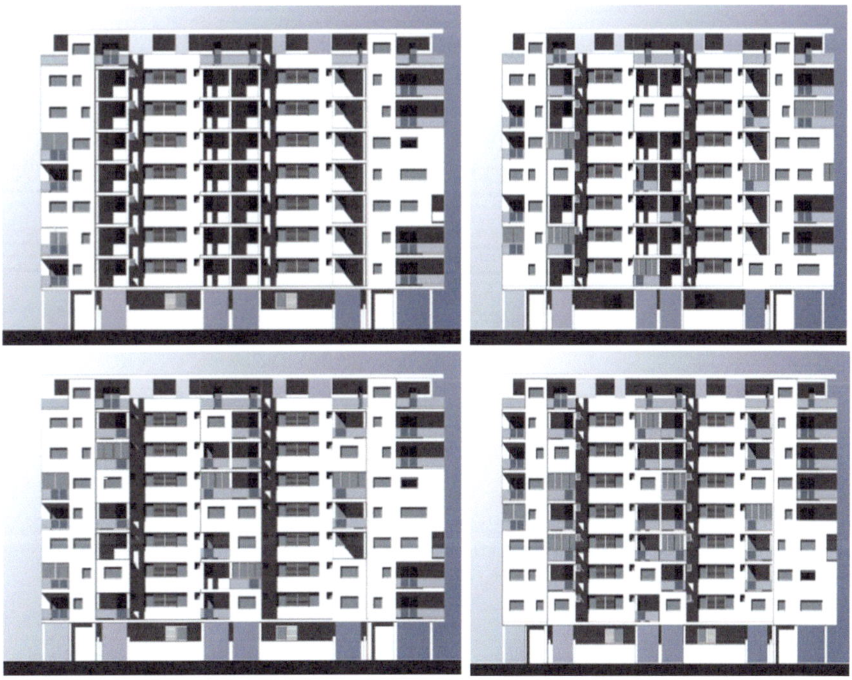

Fig. 8.33 North façade of Block T7: new residential units combined with add-ons and new façade solutions in the existing residential units

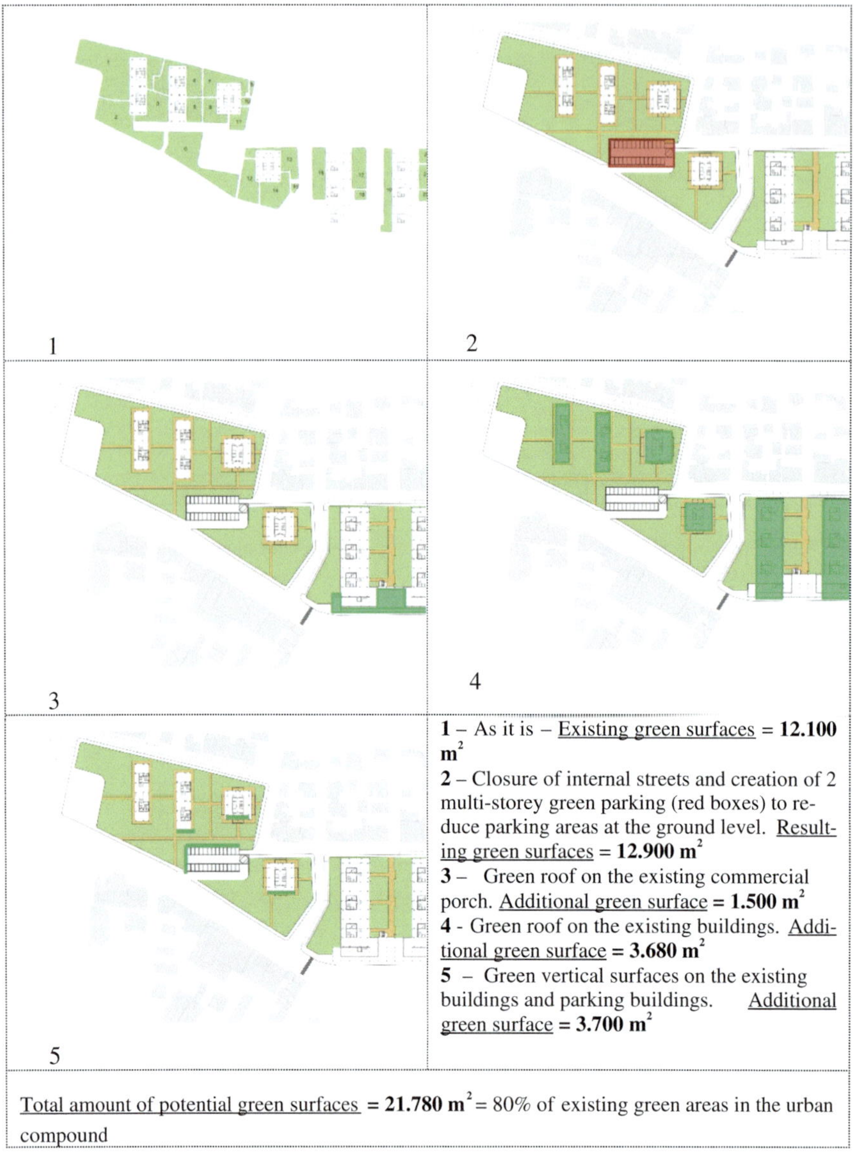

Fig. 8.34 Five steps have been hypothesized for the alternative scenario in the outdoor spaces of the urban compound in Peristeri. The results show how the potential increase of green surfaces in the area spreads to the 80 % of the existing permeable areas

To highlight the various energy demands of the single reference units as a function of both the position of the unit within the building block and the solar orientation, the graphs, as shown in Fig. 8.17, have been drawn. As expected, the most

Fig. 8.35 Final scenario in the urban compound in Peristeri

energy-consuming units (both in summer and winter) are the apartments located in the upper building's floor, just underneath the flat roof of the building. As also expected, the less consuming units (both in summer and winter) are the apartments located in the building's middle floors. A substantial difference in the energy demand may also be observed by comparing the residential units in a same floor location, but with different solar orientations.

As shown on the right side of Figs. 8.17 and 8.18, south-facing apartments are less energy-demanding than the corresponding north-facing ones, both in summer and winter.

Table 8.1 This table quantitatively evaluates the energy performance indexes of the building's residential units as a function of different locations within the building block (see Fig. 8.14)

Apartment/thermal zone	Epi (kWh/m^2 year)	Epe (kWh/m^2 year)
A1	113	89
A4	99	96
A7	167	106
B1	59	84
B4	71	89
B7	145	102
C1	84	95
C4	81	92
C7	170	102

Table 8.2 This table shows the energy performance indexes building residential units as a function of different locations within the building block (see Fig. 8.15)

Apartment/thermal zone	Epi (kWh/m^2 year)	Epe (kWh/ m^2 year)
A1	89	64
A5	75	54
A10	101	72
B1	100	67
B5	94	61
B10	113	73
C1	89	49
C5	75	43
C10	101	57
D1	99	52
D5	87	45
D10	107	62

Table 8.3 This table shows the energy performance indexes in the building's residential units as a function of different locations within the building block (see Fig. 8.16)

Apartment/thermal zone	Epi (kWh/m^2 year)	Epe (kWh/ m^2 year)
A1	111	36
A4	89	42
A6	124	45
B1	115	40
B4	94	47
B6	129	49
C1	103	30
C4	82	34
C6	116	39
D1	107	33
D4	94	37
D6	129	43
E1	112	31
E4	90	31
E6	126	37
F1	107	29
F4	90	33
F6	125	38

8.3.2 *Energy Retrofitting Scenarios of Existing Buildings in the Peristeri Urban Compound*

To evaluate the technical and cost-effective feasibility of the possible energy retrofitting options in this case study, a series of transformations—gradually increasing, from the sectorial to the far-reaching—have been considered on the different buildings' types. These transformations produce the following possible scenarios: Scenario1—wall insulation (walls and lower floor on "pilotis"); Scenario 2—window replacement; Scenario 3—1 + 2 combined; Scenario 4—roof insulation; Scenario 5—roof insulation and green roof; Scenario 6—3 + 5 combined complete scenario. Results of the simulations performed are synthetized in the graphs shown in Figs. 8.19, 8.20, and 8.21.

8.3.3 *Cost-Benefit Analysis*

To identify the economic feasibility of the energy retrofit scenarios, a cost-benefit analysis was conducted by means of a market survey, determining, for each particular design option, the evaluation in terms of energy performance improvement and the related cost estimates. Thus, a cost-effective analysis has been developed for each building type and for each scenario. Simple payback time of the investments has been calculated, and the artificial results of these calculations appear in Fig. 8.22.

Generally, the tower building type presents the higher payback times in all the retrofitting options with the sole exception of the intervention of the roof (Scenarios 4 and 5—roof insulation and green roof), given the limited extension of the roof surface in this building type. Evaluations on cost-benefit assessments of these interventions generally show excessive payback times (up to 10–15 years without incentives). Furthermore, these costs do not yet include the installation of an RES plant and the plant interventions that are in any case necessary to set to zero the energy balance of the selected buildings.

Thus, the energy retrofitting in the existing building of the Peristeri urban compound is proved to be a technically feasible goal, but it is necessary to face the problem of very high costs due to the need to operate both active and passive interventions; amortized over a relatively long time, they are generally comparable with the life of the systems of energy production from RES.

8.3.4 First Conclusions on the Peristeri Urban Compound and Further Design Scenarios

A sample of about 12 different building blocks in the Peristeri urban compound in Athens have been analyzed with different levels of simulations quantifying all the encountered variants in buildings and apartment types, in order to identify the most appropriate retrofitting actions to achieve energy consumption's reduction up to nZEBs in existing building blocks; the detailed diagnosis of the individual residential units have shown that it is possible to reach an average Energy Performance (EP) down to about 20 $kWh/m^2 * y$, for winter and summer respectively, by insulating opaque surfaces –roofs and walls—and replacing existing windows (a complete retrofitting scenario).

Passive retrofitting interventions and the production of energy from renewable sources are necessary to achieve nZEBs in the retrofitting of highly energy-consuming building stock; this is exceptionally challenging, considering the specific context of the local urban compound and the general situation of global economic crisis that the Mediterranean areas and societies are currently experiencing.

Nonetheless, the effective reduction of energy consumption towards zero energy buildings and districts is and remains an unavoidable objective, especially in the Mediterranean urban areas, where the dramatic combination of HI phenomena, fuel poverty, and global overheating severely threatens both humans and the environment.

The reflection on the estimated costs of proposed energy retrofitting options clearly underlines, once again, the need for additional strategies and measures to be developed at the social, public, legislative and market levels, to counterbalance the large payback times of energy retrofitting measures aimed at achieving nZEBs in existing urban buildings of the Mediterranean areas. In particular, it is necessary to activate new forms of incentives in order to attract potential investors as well as dynamically involve the social participation to motivate urban dwellers in the energy retrofitting of their urban environments.

According to these preliminary results and to the general goal of comparing various urban contexts' analysis and design scenarios, the study has envisioned the possibility of a higher transformation of the existing buildings. Therefore, higher transformation scenarios have been conceived and developed for the energy retrofit towards zero energy in the urban compound of Peristeri, as will be shown in sections.

8.3.5 Energy and Cost Benefits of Volumetric Addition in Energy Retrofitting Actions

The necessary integration between the need for retrofitting and the opportunity to assume a wider rehabilitation process scenario have found further pilot experiments in the energy retrofit design of the public residential district in the Peristeri urban compound; within this urban context, various reference buildings have been used to test and compare four possible energy retrofit options according to a socio-oriented approach.

The different structures that have been studied in this case study show a very high degree of possibilities and combinations: the design solutions have not been restrained to a simple prefabricated structure; indeed, the need to develop modular and variable solutions has lead to a wide set of possible urban configurations. The improvement in terms of energy performance and structural safety of the intervention has been investigated and have proved the high potential of expandable architecture also in a cost-effective analysis.

The research goal of the case studies is to study the possible evolution of the buildings in these specific post-war districts, by envisioning and simulating how inhabitants could take over the architectural role in the future and possibly transform their own houses. In this context, ethnography is not considered to be a mere techniques' toolbox directly borrowed from social sciences; on the contrary, design rhetoric refers to the ability and the purpose of design thinking to engage and foster transformations.

In Peristeri, there is a high percentage of elderly people and a continuously increasing number of young couple becoming part of the community. The traditional approach for a multicultural society based on an unsubstantiated and narrow perspective would be ineffective in these urban contexts; therefore, inhabitants should be given structural and technological solutions on which they may base their own personal changes and design processes. By re-elaborating on the real photographic environments, the appropriations took place and the users' modification in the buildings' façades of the urban compound, where a large abacus of possibilities has been studied, starting from the organization of the ground floor plan in different building types. Therefore, the expendable spaces are thought of as extensions of the existing internal spaces: whether occupants desire a bigger balcony rather than a loggia or an extension of the existing room, this system will allow people the possibility of choosing the use of their indoor and outdoor spaces.

A variety of façade solutions are proposed: the "abacus" of the possible technological and formal solutions is the main instrument that should be used to give the inhabitants the possibility to express their needs within a common order, respecting the guidelines of the project. Since the "letters" of this alphabet have been studied as a modular system, they can be joined together in multiple combinations, creating the "discourse" of the urban dwellers.

Forecasting the possibility and studying the pattern as a modular system and keeping in mind the development of various combinations, it is possible to let everyone do what he or she wants with their homes while still leaving open the possibilities for self expression, without clashing with construction codes and regulations, according to a metabolic approach. Concepts such as durability, adaptability, and energy efficiency have been increasingly taken into account in the fields of building and urban research. Especially when considering the high standards and requirements for the retrofitting actions, one of the main goals consists of the development of long-lasting building products that can adapt to various needs and environmental conditions through their entire life cycle and that could guarantee a high level of a building envelope's energy performance.

Our suburbs, and the current condition of recent buildings, provide clear evidence that the building techniques and market have not been able to take into account the further development and changes both in the environment and in the building itself. The approach suggested here proposes a reverse methodology: defining the main design principles and parameters according to their variability, it is possible to develop a modular system that incorporates multiple design solutions where the single components that can be interwoven, changed, or transformed still guarantee the coherence of the urban configuration in its evolving process. Thus, the potential add-ons will be considered to be a result of combined ethnographic and end-user needs' involvement aimed at delivering forms of customized and variable components of self-expression in operations of energy retrofitting of existing building stock, with respect to urban dwellers' and users' expectations.

These additions have been grouped in an abacus of possibilities that will become one of the main design tools for planners and professionals involved in the process. The abacus may represent the communication instrument between technicians and inhabitants enriching the possibilities of synergies between the technical and sociological strategies proposed. To achieve this, it is important to consider participatory and socio-integrated policies already during the abacus's definition phase. Up to six modifications of building envelope for each span have been considered in the reference case studies.

The abacus, whose technical and structural feasibility relies on the hypothesis of an external supporting frame (Figs. 8.23, 8.24 and 8.25), have been tailored and customized as a function of different constructive elements and architectural types (Figs. 8.26 and 8.27); the volumetric additions and options (including the zero-option "one = no intervention") have been further combined with various options of materials, according to the construction and types of the different reference buildings.

The hypothesis of buildings' transformation has been based on the assumption of creating both additional spaces for users and inhabitants in the current residential units, and new additional units for potential investors. Regarding the existing residential units, the hypothetical scenarios are summarized in Fig. 8.26.

Similar abacuses have been hypothesized, assessed, and verified for all the façades in all the different building types of the urban compound in Peristeri. The abacus can be used as leverage to incentivize inhabitants in their decision to renovate their apartments. It also represents a possible user-driven design scheme to let inhabitants adopt tailored solutions according to their needs and requirements.

The overall strategy may represent a different design approach. The result is far to be a common project—it represents to a greater degree the projection of a large set of possible projects, at the same time all different but similar, and therefore closer to human uses, habits and kind. The result is not the design of a model, rather the design of a process, as shown in the following different options, as envisaged for the different building types in the urban compound (Figs. 8.28 and 8.29).

Furthermore, the possibility of combining add-ons on the tops and sides of the buildings (and thus new residential units) with façade solutions have been explored (Figs. 8.30, 8.31, 8.32, 8.33, 8.34 and 8.35) for the block building types

The abacus tool could even be conceived as a "real-time, interactive tool" to help inhabitants and owners in the selection process of the required solutions, giving life to creative and variable, yet controlled, self-expressed possibilities within the retrofitting process of urban buildings. The same strategy has been finally applied for the case study of Block T7, and some of the possible configurations are shown in Figs. 8.32 and 8.33.

8.3.6 Low Versus High Transformation Retrofitting Options Towards Near Zero Energy in Existing Buildings

Further economic analysis has been developed to test the feasibility of the previously envisaged scenarios. Results reported in Tables 8.4, 8.5, and 8.6 show that in each building type, the option of incremental units and add-ons represents an effective strategy for decreasing the long payback time of the energy retrofitting.

As shown in Tables 8.4, 8.5 and 8.6, the potential gains related to the construction of new residential units greatly decrease the payback time of all the initial costs of investment, including the RES (Row C in the tables) required to set the energy balance of the buildings to zero. Furthermore, if we consider the hypothetic investment of developers in add-ons, the gains obtained by sales of the new flats would be close to counterbalancing both the standard energy retrofit and the cost of RES (PV and solar panels) to set the energy demand of the whole building to zero.

Further studies are under way to understand whether it is possible to achieve complete compensation between investments and gains in the initial state. The studies have developed, performed and evaluated various possible solutions for achieving a sustainable and spatially localized distribution of energy via technologies to be diffusely integrated in the urban buildings.

Thus, we have determined the energy-saving potential of real urban environments in a representative case study by designing:

1. Alternative urban scenarios by means of technological building components/materials and volumes aimed at reducing the buildings' heating and cooling loads; and
2. Alternative urban scenarios integrating renewable energy sources in the urban building envelope.

Nonetheless, to properly address the objective of low-carbon, zero energy in existing urban environments, we need to consider the buildings and the related

Table 8.4 Cost-benefit assessment of "deep retrofitting," including the creation of new units in the tower building type T11

T11 (40 units)			
Retrofitting options	Costs/gains related to the option	Saving	Payback time
	€	€/year	year
A—Complete retrofitting	−371.575	18.120	20.5
B—Metal structural frame	−173.910		
C—Solar and PV plants	−196.000	16.500	11.9
D—Construction of new residential units	−264.000		
Total costs/investments	−1.005.485		
G—Gain from new units	600.000		
Total	−405.485	34.620	11.7

Table 8.5 Cost-benefit assessment of "deep retrofitting," including the creation of new units in the double block building type B6

B6 (90 units)			
Retrofitting options	Costs/gains related to the option	Saving	Payback time
	€	€/year	year
A—Complete retrofitting	−558.559	39.015	14.3
B—Metal structural frame	−323.897		
C—Solar and PV plants	−440.000	30.800	14.3
D—Construction of new residential units	−1.331.000		
Total costs/investments	−2.653.456		
G—Gain from new units	2.420.000		
Total	−233.456	69.815	3.3

Table 8.6 Cost-benefit assessment of "deep retrofitting," including the creation of new units in the block building type T7

T7 (28 units)			
Retrofitting options	Costs/gains related to the option	Saving	Payback time
	€	€/year	year
A—Complete retrofitting	−280.717	18.704	15.0
B—Metal structural frame	−174.319		
C—Solar and PV plants	−236.800	30.800	7.7
D—Construction of new residential units	−1.582.000		
Total costs/investments	−2.273.836		
G—Gain from new units	1.924.000		
Total	−349.836	49.504	7.1

open areas; this is especially true in the Mediterranean warm regions of the EU, where the snowballing effects of all the buildings and open areas is very important, especially in the summer. Furthermore, a more holistic vision of the "urban energy demand" considering outdoor spaces and buildings as a whole, can better drive urban planning decisions and bottom-up, grass-roots initiatives towards the adoption of alternative residential configurations/redevelopment of existing urban areas, thus motivating main stakeholders such as public bodies, policy-makers, urban dwellers, and administrative bodies in the adoption of participative processes or green behavior.

This is particularly challenging in urban contexts of "modernism," such as in the district of Peristeri, where most of the blocks are currently characterized by the presence of pillars and a narrow, dark, basement level. As Gehl points out, most social/residential areas were built according to principles and ideologies that gave low priority to outdoor activities, to the connection between distance, intensity, or to the closeness in various contact settings; this has an interesting parallel in decoding and experiencing cities and city spaces (Gehl 2010). Cars are everywhere and block the possible fruition of the street/garden/park level from the pedestrian perspective. A proper renewal action has to be developed to guarantee a higher degree of freedom for the inhabitants that will be able to choose the most fitting solution according to their needs. This approach should also be applied to the ground floor level, incorporating various levels of privacy of the collective areas, thus starting from a direct and active participation of the inhabitants.

A new spatial organization for the outdoor spaces in the Peristeri urban compound has been outlined so far. The master plan feasibility tries to increase the green and permeable surfaces as much as possible. The entry is reorganized to let people enter the area from the renewed retail and commercial areas, into "semi-private" green courtyards, to give a warming and welcoming feeling that would enrich the quality of the communal spaces. Thus, a five-step progressive design scenario has been developed for the transformation of the existing urban area of Peristeri, keeping in mind both the need for a more inclusive environment for urban dwellers and the cooling potential of increased green and permeable surfaces (Fig. 8.34(1,2,3,4,5)).

8.4 Conclusions

A significant real urban environment (RUE) has been selected for investigation in the Athens Metropolitan area. In particular, the energy analysis on a large set of different building types in the large urban complex of the Workers' Houses in Peristeri, Western Athens, has demonstrated the energy-saving potential of various "standard" retrofitting operations (insulation and coatings of opaque elements and envelope surfaces, window replacement, green surfaces). To identify the economical feasibility of the standard energy retrofit, a cost-benefit analysis was conducted by means of a market survey, determining, for each particular design option, the evaluation in terms of energy performance improvement and the related cost estimations.

Simulations showed that it is technically possible to reach an average Energy Performance (EP) of around 35–60 to kWh/m^2 * y, by the insulation of opaque surfaces and the replacement of existing windows. However, the cost-benefit assessments of these interventions have always shown excessive payback times (up to 10–15 years without incentives). Moreover, these operations do not fully compensate the various energy performances of the residential units within the building; this can produce additional barriers to the adoption of monitoring systems to control energy waste.

Thus, the research design simulation has considered the costs of other energy retrofitting operations envisaging the building "densification" by the use of new residential units, new building additions such as "sunspaces," buffer zones and extra rooms (add-ons) in the existing units, and the integration with RES. The various design assumptions and the consequent variation in energy performance indices (the building's primary energy demand for heating and producing hot water) have been compared.

These more radical transformation hypotheses drastically reduce the energy performance indexes of the buildings, up to the target of a "passive house," which can easily achieve the target of nZEB by a reduced use of RES. Furthermore, the different energy performance indexes (kWh/m^2 * y) of the "deep renovation" options have been compared to the relative options' costs. As expected, the higher the building's transformation, the higher the costs. Thus, a feasibility study of these volumetric additions have been developed, concentrating them in the most dispersive parts of the building (roofs and blind façades). Results have shown that in case of a hypothetical investment by developers or the same inhabitants in add-ons, the gains obtained by sales of the new flats would be close to counterbalancing both the standard energy retrofit and the cost of RES (PV and solar panels) to set to zero the energy demand of the whole building. Further studies are under way to understand if it is possible to achieve complete compensation between investments and gains at the initial state.

It can be concluded that according to the hypothesis of a higher transformation of selected building blocks, the volumetric additions may be conceived as powerful, energy-generating and insulating buffer zones able to set the energy demand of the original building to zero. Furthermore, this effective strategy may help balance the different EPs of the units within the building blocks. On the side of economical feasibility, the incremental transformation may produce an interesting opportunity to counterbalance the high investment costs of energy transformation.

The strategy of add-ons has been tested for several scenarios in different building blocks, and cost-benefit analyses' results, as well as architectural enhancements and technical feasibility, are very encouraging. Thus, in the search for additional forms of compensation and incentives for nZEB in existing buildings, volumetric addition, densification, and/or "infill" may represent crucial tools to enhance the technical and economic feasibility of energy retrofitting operations. These interventions' costs are not lower than the standard solutions, but their payback (considering roof extensions and construction of new units), together with the direct benefit of potential investors which gain space and money by possible add-ons, have a very positive effect on the technical and economic feasibility of buildings retrofit towards nZEBs.

Three other advantages—not always quantifiable and not related to technical-economic estimations—confirm the strategic role of the compensation of the volumes' additions. First of all, there is the possibility of using the adds-on to optimize the thermal operating conditions in summer conditions (due to the reduction of incident solar radiation, for vertical air flow, etc.); furthermore, these new structures can also accommodate or act as support for new/renewed ducts for plants and lighting, etc. Secondly, there is the possibility of using the support structures of the additions' volumetric also for the purpose of improving the conditions of the static and mechanical performance of existing structures. These "structural invariants," in fact, if properly sized and affixed to the frames of existing structures (often made up by reinforced concrete), provide an additional element of construction quality and durability. Last but not least, the use of these invariants generates a higher degree of adaptability and variability that facilitates the transition from standardized solutions towards custom-made and tailored designs responses.

This highly flexible and adaptive character of the add-ons gives an extra chance to the nZEB process: various possibilities of add-ons can be studied to encounter the multiple needs and respond to the heterogeneous demand of users/tenants/dwellers. The connecting theme of the proposed research path is that the crises of energy supply and global warming need to be tackled with interdisciplinary, socio-technical and engineering approaches. In particular, with respect to the previous studies already developed on the topic, the design study and the already performed technical-economical evaluation demonstrate that energy efficiency in residential urban complexes can be an extraordinary opportunity to restore environmental, social, and urban quality. The techno-economical feasibility assessment, the proper identification of the types of intervention and their combination in possible scenarios, must be investigated and estimated on a case-by-case basis, with an effective and interdisciplinary design approach integrating a whole system of the socio-technical aspects into the feasibility study of economical and architectural issues.

As already stated, environmental sustainability and energy efficiency in urban settings are more than just technical problems. In this context, a renewed role of architects and planners is what is needed for a real shift in the building practice. In fact, instead of the architectural production based on authorship, architects should consider the users' perspective and their needs as self-organized processes of negotiation. A renewed role is what is needed for a real shift in the building practice. "The architect will have to act not as a 'prima donna,', but as a human being willing to donate his/her energy and his/her ideas to society."

- The difference is in the "how": while in modernity he is in the center as a form-giver, now he/she becomes the interpreter of a process.
- Instead of trying to structure the informal through the architectural production based on authorship, architects should consider both the technical requirements

and the users' perspectives in a participative process of negotiation. Even though the approach is based on specific case studies, the underlying principles will be applicable to other districts and other cities as well.

- This kind of approach can help engage a real shift from current practice towards a social sustainable process where inhabitants and designers work together to find effective and real solutions to social and technical questions. The urban and technological strategies presented here suggest a multifaceted approach that could stimulate the process of energy renewal according to a socio-oriented use of architectural tools in urban environments.

Acknowledgments This paper is part of the project "URBAN RECREATION: Energy efficient retrofit for carbon zero and socio-oriented urban environments, project n. 326060 FP7-PEOPLE-2012-IEF," funded by the European Commission within the framework of the funding scheme of Marie Curie Actions—Intra-European Fellowships (IEF); fellow: Annarita Ferrante. The author also acknowledges the masters students in engineering and architecture in the Department of Architecture of Bologna Lorenza Cavasino, Rachele Iannone e Nicoletta Salvi, for the drawings, tables and data provided in this paper, and the colleague Luca Boiardi for his support in the design builder simulations. The author also acknowledges the Municipality of Peristeri and, in particular, Dimitris Lagaris and Nikos Venetas, for their invaluable help in the production of data, and Prof. Mat Santamouris and the researchers of the Group Building Environmental Research UOA, National University of Athens, Physics Department, for their support.

References

Akbari H, Menon S, Rosenfeld A (2009) Global cooling: Increasing world-wide urban albedos to offset CO_2. Clim Change 94:275–286. doi:10.1007/s10584-008r-r9515-9

Bitan A (1992) The high climatic quality of the city of future. Atmos Environ 26B(1992):313–329

Brown HS, Vergragt PJ (2008) Bounded socio-technical experiments as agents of systemic change: the case of a zero-energy residential building. Technol Forecast Soc Chang 75:107–130

Bulkeley H (2010) Cities and the governing of climate change. Annu Rev Environ Resour 229–253

Bürer MJ, Wüstenhagen R (2009) Which renewable energy policy is a venture capitalist's best friend? Empirical evidence from a survey of international cleantech investors. Energy Policy 37:4997–5006

Buttstädt M, Sachsen T, Ketzler G, Merbitz H, Schneider C (2010) Urban temperature distribution and detection of influencing factors in urban structure. ISUF; International Seminar on Urban Form, Hamburg

Cecodhas, POWER HOUSE, Nearly ZERO Energy CHALLENGE (2013) Progress Report March 2013

Copenhagen V (2009) City of Copenhagen, Copenhagen Climate Plan, The Technical and Environmental Administration City Hall, 1599

Doulos L, Santamouris M, Livada I (2001). Passive cooling of outdoor urban spaces. The role of materials. Sol Energy 77:231–249

EU report, Cities of tomorrow: challenges, visions, ways forward (2011). European Commission—Directorate General for Regional Policy, Luxembourg: Publications Office of the European Union, 2011, 112 pp. ISBN: 978-92-79-21307-6

EU (Commission of) Communities (2009) Design as a driver of user-centred innovation. Brussels, 7.4.2009, SEC(2009)501
Energy Cities (2012) Background paper related to Energy Cities' position paper on the European Commission's legislative proposals for the EU Cohesion Policy 2014–2020
EU Report (2010) World and European sustainable cities. Insights from EU research, Directorate-General for Research. Directorate L, Science, economy and society, Unit L.2. Research in the Economic, Social sciences and Humanities. European Commission, Bureau SDME 07/34, B-1049 Brussels, 2010. Socio-economic Sciences and Humanities; EUR 24353 EN
EU Parliament (2009) Report on the proposal for a Directive of the European Parliament and of the Council on the Energy Performance of Buildings (recast) (COM(2008)0780-C6-0413/2008-2008/0223(COD)), 2009
EU Parliament, Directive 2010/31/EU of the EU P of 19 May 2010 on the energy performance of buildings
EU Communication (2011a) Energy Efficiency Plan 2011, Communication from the Commission to the European Parliament, the Council, the EU Economic and Social Committee and the Committee of the Regions. Brussels, 8.3.2011 COM(2011) 109 final
EU Communication (2011b) A resource-efficient Europe—Flagship initiative under the Europe 2020 Strategy. Communication from the Commission to the EU Parliament, the Council, the EU Economic and Social Committee and the Committee of the Regions. Brussels, 26.1.2011 COM(2011) 21
EU Report, Housing statistics, 2010
Energy 2020. A strategy for competitive, sustainable and secure energy. Communication from the Commission to the European Parliament, the Council, the EU Economic and Social Committee and the Committee of the Regions. Brussels, 10.11.2010 COM(2010) 639 final
EU Communication (2009) Investing in the Development of Low Carbon Technologies (SET-Plan), Communication from the Commission to the European Parliament, the Council, the European Economic and Social Committee and the Committee of the Regions, Brussels, 7.10.2009, COM(2009) 519 final
Ferrante A, Semprini G (2011) Building energy retrofitting in urban areas. Procedia Eng 21:968–975
Ferrante A (2014) Energy retrofit to nearly zero and socio-oriented urban environments in the Mediterranean climate. Sustainable Cities and Society, Available online 12 Feb 2014
Ferrante A (2012) Zero- and low-energy housing for the Mediterranean climate. Adv Build Energy Res 6(No. 1):81–118
Ferrante A, Cascella MT (2011) Zero energy balance and zero on-site CO_2 emission housing development in the Mediterranean climate. Energy Build 43(2011):2002–2010
Ferrante A, Mihalakakou G (2001) The influence of water, green and selected passive techniques on the rehabilitation of historical industrial buildings in urban areas. J Sol Energy 70(3):245–253
Ferrante A, Mihalakakou G, Santamouris M (1999) Natural devices in the urban spaces to improve indoor air climate and air quality of existing buildings. In: International conference of indoor Air 99, Edinburgh, Scotland, 8–13 Aug 1999
Ferrante A, Santamouris M, Koronaki I, Mihalakakou G, Papanikolau N (1998) The design parameters' contribution to improve urban microclimate: an extensive analysis within the frame of POLIS Research Project in Athens, The joint meeting of the 2nd Eu conference on energy performance and indoor climate in buildings and 3rd international conference IAQ, ventilation and energy conservation. Lyon, 19–21 Nov 1998
Ferrante A (2014b) Energy retrofit to nearly zero and socio-oriented urban environments. Sustain Cities Soc Sustain Cities Soc 13:237–253
Ferrante A, Boiardi L, Fotopoulou A (2014) On the viability of nearly zero energy buildings in the Mediterranean urban contexts. Adv Build Energ Res. doi:10.1080/17512549.2014.9410 08

Gehl J (2010) Cities for people. Island Press, 15 Aug 2010
Giannopoulou K, Livada I, Santamouris M, Saliari M, Assimakopoulos M, Caouris YG (2011) On the characteristics of the summer urban heat island in Athens, Greece. Sustain Cities Soc 1(2011):16–28
Grammenos Fanis (2011) European urbanism: lessons from a city without suburbs. http://www.planetizen.com/node/48065. Accessed on April 2014
Guy S (2006) Designing urban knowledge: competing perspectives on energy and buildings. Environ Plann C: Gov Policy 24:645–659
Heiskanen E, Johnson M, Robinson S, Vadovics E, Saastamoinen M (2010) Low-carbon communities as a context for individual behavioural change. Energy Policy 38:7586–7595
Hernandez P, Kenny P (2010) From net energy to zero energy buildings: Defining life cycle zero energy buildings (LC-ZEB). Energy Build 42(2010):815–821
Hassid S, Santamouris M, Papanikolaou M, Linardi A, Klitsikas N, Georgakis C et al (2000) The effect of the Athens heat island on air conditioning load. Energy Build 32:131–141
JWG (2013) Towards assisting EU Member States on developing long term strategies for mobilising investment in building energy renovation (per EU Energy Efficiency Directive Article 4), Composite Document of the Joint Working Group of CA EED, CA EPBD and CA RES, November 2013. http://www.ca-eed.eu/reports/art-4-guidance-document/eed-article-4-assistance-document
Karlessi T, Santamouris M, Apostolakis K, Synnefa A, Livada I (2009a) Development and testing of thermo chromic coatings for buildings and urban structures. Sol Energy 83(4):538–551
Karlessi T, Santamouris M, Apostolakis K, Synefa A, Livada I (2009b) Development and testing of thermochromic coatings for buildings and urban structures. Sol Energy 83(2009):538–551
Leontidou L (1990) Spontaneous urban development: in search of a theory for the Mediterranean city. In: The Mediterranean city in transition: social change and urban development. Cambridge University Press, Cambridge, pp. 7–35
Livada I, Santamouris M, Niachou K, Papanikolaou N, Mihalakakou G (2002) Determination of places in the great Athens area where the heat island effect is observed. Theoret Appl Climatol 71:219–230
Livada I, Santamouris M, Assimakopoulos MN (2007) On the variability of summer air temperature during the last 28 years in Athens. J Geophys Res 112:D12103
Leschke J (2013) Labour market impacts of the global economic crisis and policy responses in Europe, Janine and Andrew Watt, European Trade Union Institute. http://www.socialwatch.eu/wcm/documents/labour_market_impacts_and_policy_responses.pdf. Accessed in August 2013
Marsh G (2002) Zero Energy Buildings, Key Role for RE at UK Housing Development, May–June 2002, Refocus, pp 58–61
Masini A, Menichetti E (2010) The impact of behavioural factors in the renewable energy investment decision making process: Conceptual framework and empirical findings. Energy Policy 1–10
Mihalakakou P, Santamouris M, Papanikolaou N, Cartalis C, Tsangrassoulis A (2004) Simulation of the urban heat island phenomenon in Mediterranean climates. J Pure Appl Geophys 161:429–451
Meyer A. et al., A Copenhagen Climate Treaty, Version 1.0, 2010 Karlsruhe
Mulugetta Y, Jackson T, van der Horst D (2010) Carbon reduction at community scale. Energy Policy 38:7541–7545
Niachou K, Papakonstantinou K, Santamouris M, Tsangrassoulis A, Mihalakakou G (2001) Analysis of the green roof thermal properties and investigation of its energy performance. Energy Build 33:719–729
Papanikolaou, Livada, Santamouris, Niachou K (2008) The influence of wind speed on heat island phenomenon in Athens, Greece. Int J Vent 6(4):337–348
Peattie K (2010) Green consumption: behaviour and norms. Annu Rev Environ Resour 35(2010):195–228

Ricciardelli F, Polimeno S (2006) Some characteristics of the wind flow in the lower Urban Boundary Layer. J Wind Eng Ind Aerodyn 94:815–832
Synnefa A, Santamouris M, Apostolakis K (2007) On the development, optical properties and thermal performance of cool coloured coatings for the urban environment. Sol Energy 81:488–497
Santamouris M (ed) (2001a) Energy and climate in the urban built environment. James and James Science Publishers, London
Santamouris M, Mihalakakou G, Papanikolaou N, Assimakopoulos DN (1999a) A neural network approach for modelling the heat island phenomenon in urban areas during the summer period. Geophys Res Lett 26(3):337–340
Santamouris M, Papanikolau N, Koronakis I, Livada I, Asimakopoulos D (1999b) Thermal and air flow characteristics in a deep pedestrian canyon under hot weather conditions. J. Atmospheric Environment 33:4503–4521
Santamouris M, Papanikolaou N, Livada I, Koronakis I, Georgakis C, Argiriou A et al (2001) On the impact of urban climate to the energy consumption of buildings. Sol Energy 70(3):201–216
Santamouris M, Sfakianaki A, Pavlou K (2010) On the efficiency of night ventilation techniques applied to residential buildings. Energy Build 42:1309–1313
Sfakianaki A, Pagalou E, Pavlou K, Santamouris M, Assimakopoulos MN (2009) Theoretical and experimental analysis of the thermal behaviour of a green roof system installed in two residential buildings in Athens, Greece. Int J Energy Res 33(12):1059–1069
Stathopoulou M, Synnefa A, Cartalis C, Santamouris M, Karlessi T, Akbari H (2009) A surface heat island study of Athens using high-resolution satellite imagery and measurements of the optical and thermal properties of commonly used building and paving materials. Int J Sustain Energ 28(1):59–76
Santamouris M (2001b) Energy and climate in the urban built environment. James & James Science Publishers Ltd, London
Santamouris M, Kapsis K, Korres D, Livada I, Pavlou C, Assimakopoulos MN (2007) On the relation between the energy and social characteristics of the residential sector. J. Energy Build 39(2007):893–905
Webler T, Tuler SP (2010) Getting the engineering right is not always enough: researching the human dimensions of the new energy technologies. Energy Policy 38:2690–2691
Yamashita S (1996) Detailed structure of heat island phenomena from moving observations from electric tram-cars in Metropolitan Tokyo. Atmos Environ 33(3):429–435

Part II
The Built Environment

Chapter 9
Households: Trends and Perspectives

Antonio Serra

Abstract The household final energy consumption in the EU-28 was 26.2 % of the total (with Serbia achieving almost 37 %) in 2012. According to the International Energy Agency's (IEA) statistics for energy balance for 2004–2005, the residential sector accounted for 27 % of the final energy use throughout the world. Starting with these average data, a more detailed study is necessary to show the dissimilarities of performance and energy consumption in Mediterranean households, considering concepts such as energy poverty, housing deprivation, and the rate of overcrowding. The analysis is completed by taking into account a wider range of environmental indicators related to the household sector, such as CO_2 emissions and concentration, etc. In addition, the energy efficiency requirements in building codes and policies to achieve net zero energy buildings (nZEBs)—is a technical concept accepted worldwide, put the focus on new building technologies and on the refurbishment of existing buildings. Upgrading the energy performance of homes offers immediate benefits and helps reduce energy poverty. It also expands the potential for providing additional homes in existing communities while saving energy, land, and materials. These topics will be analyzed according to the temperate climatic conditions of the Mediterranean area.

9.1 Introduction

Household energy consumption (also called "residential sector energy consumption") covers all the energy consumed in households for space heating, water heating, lighting, cooking, and electricity for appliances, including large appliances and small plug loads. Energy consumption trends in the residential sector and the

A. Serra (✉)
Department of Thermal Engineering, University of the Basque Country,
Paseo Rafael Moreno "Pitxitxi" 3, 48013 Bilbao, Spain
e-mail: antonio.serra@ehu.eus

S.-N. Boemi et al. (eds.), *Energy Performance of Buildings*,
DOI 10.1007/978-3-319-20831-2_9

different end uses are driven by a wide range of factors, including overall energy efficiency improvements, changes in population, energy mix, urbanization rates, number of occupied dwellings, inhabitants per household, dwelling size, dwelling type, building characteristics, and age profile, income level and growth, consumer preferences and behavior, energy availability, climatic conditions, and appliances and equipment penetration rate and standards (OECD/IEA 2014). Because of this, energy efficiency policies in the household sector, among others, should address social inequalities, such as energy poverty, housing deprivation, and the over-crowding rate for certain sectors of the population.

This section analyzes the latest available data on energy consumption in the residential sector for some Mediterranean countries. It has to be taken into consideration that national data are not always available, and that sometimes data are not completely reliable or accurate. In addition, social differences between countries, sharpened by the financial crisis, do not permit a simple comparison and interpretation of housing consumption data. Moreover, even in the so-called Mediterranean countries, climate conditions may vary greatly in the same country, i.e., the case of the south and the north of Italy or France. However, the study and analysis of data is important to build a "state-of-the-art" knowledge of the household consumption of Mediterranean countries, looking for common patterns and common frames to achieve more realistic and socially acceptable measures and policies for energy saving and efficiency in households.

9.2 Analysis of Data in the Crisis Period

9.2.1 Household Energy Consumption

The countries analyzed in this section are Greece, Spain, France, Croatia, Italy, Cyprus, Malta, Portugal, Slovenia, Montenegro, Serbia, and Turkey. The period of analysis covers 10 years, from 2003 until 2012 (in many cases the last available data).

Final energy consumption is the total energy consumed by end-users, such as households, industry, and agriculture. It is the energy that reaches the final consumer's door and excludes that used by the energy sector itself. The trend of final energy consumption of a country depends on many factors, such as economic development, population growth, efficiency of its use, etc.

Almost all the countries show a decrease in the final energy consumption coinciding with the financial crisis (Table 9.1). Energy consumption grew between 2003 and 2007–2008, in many cases with 2007 and 2008 a turning point in this tendency (Fig. 9.1). Considering the difference between 2003 and 2012 (see Table 9.2), the countries with the sharpest decrease were Greece (−17.16 %) and Portugal (−13.16 %), while countries with the greatest increase were Turkey (+42.35) and Malta (+16.32 %).

Table 9.1 Final energy consumption in Mediterranean countries

Geo/Year	2003	2004	2005	2006	2007	2008	2009	2010	2011	2012
Greece	20.67	20.45	20.95	21.55	22.06	21.37	20.53	19.00	18.87	17.12
Spain	90.43	94.70	97.76	95.47	98.12	94.63	87.76	89.08	86.67	83.12
France	161.6	163.3	162.8	161.4	158.2	160.0	153.1	158.4	147.2	150.7
Croatia	5.98	6.18	6.34	6.46	6.48	6.60	6.35	6.34	6.19	5.90
Italy	131.0	132.7	134.5	132.6	129.5	127.9	120.9	124.7	122.0	119.0
Cyprus	1.81	1.82	1.82	1.85	1.91	1.96	1.92	1.91	1.91	1.75
Malta	0.39	0.44	0.39	0.38	0.38	0.49	0.44	0.45	0.46	0.46
Portugal	18.59	18.93	19.00	18.78	18.90	18.39	18.18	18.10	17.29	16.15
Slovenia	4.71	4.81	4.89	4.94	4.88	5.24	4.74	4.92	4.96	4.85
Montenegro	n.a.	n.a.	0.81	0.86	0.91	0.92	0.78	0.76	0.72	0.71
Serbia	9.14	10.33	9.57	9.70	10.18	9.47	8.48	9.01	9.26	8.48
Turkey	59.12	61.11	63.44	69.06	73.28	72.15	69.69	73.84	78.49	84.16
EU-28	1173.6	1186.5	1188.5	1190.0	1170.3	1174.5	1107.8	1159.8	1107.9	1104.4

Period 2003–2012. *Unit* Million tons of oil equivalent (TOE). *Data Eurostat* [nrg_100a]

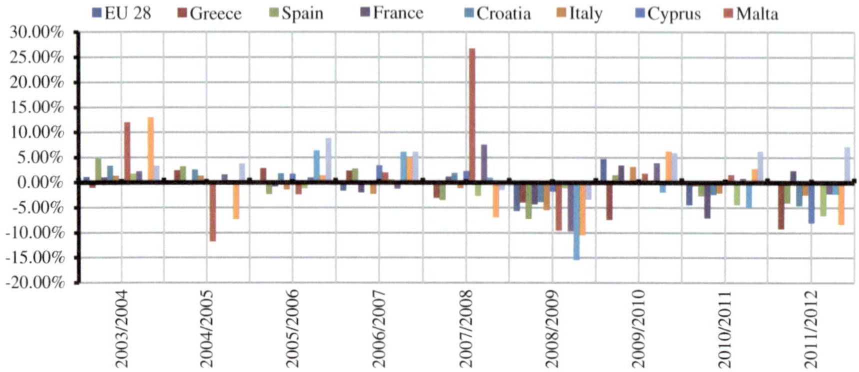

Fig. 9.1 Trends in final energy consumption. Data expressed as percentage change compared to same period in previous year. *Source* Elaboration of data Eurostat

Table 9.2 Difference of final energy consumption in Mediterranean countries expressed as percentage change compared to same period in other years: 2003–2007, 2007–2012, and 2003–2012

Geo/Year	2003–2007 (%)	2007–2012 (%)	2003–2012 (%)
EU-28	−0.28	−5.63	−5.90
Greece	6.69	−22.35	−17.16
Spain	8.50	−15.29	−8.08
France	−2.13	−4.71	−6.74
Croatia	8.37	−8.88	−1.25
Italy	1.14	−8.11	−9.16
Cyprus	5.95	−8.44	−2.99
Malta	−1.72	18.35	16.32
Portugal	1.67	−14.58	−13.16
Slovenia	3.61	−0.56	3.03
Montenegro[a]	12.88	−22.37	−12.37
Serbia	11.41	−16.69	−7.18
Turkey	23.95	14.85	42.35

Source Elaboration of data Eurostat
[a]Due to lack of data, the reference year for Montenegro is 2005 instead of 2003

Greece, Spain, Croatia, Cyprus, Portugal, Slovenia and Montenegro show a positive trend, considering the difference between 2003 and 2007, but a great decrease in final energy consumption between 2007 and 2012. Italy and France show a negative trend (decrease) in both periods. Greece, Spain, Croatia, Italia, Cyprus, Portugal, and Montenegro reached the minimum of total energy consumption in the 2012, France in 2011, Serbia in 2009, and Malta in 2006.

Trends in residential final energy consumption (Table 9.3) are quite different from the data of final energy consumption. Spain and Turkey reached their

Table 9.3 Final energy consumption in the residential sector, 2003–2012

Geo/Year	2003	2004	2005	2006	2007	2008	2009	2010	2011	2012
Greece	5.51	5.42	5.51	5.51	5.39	5.23	4.83	4.61	5.47	5.04
Spain	13.89	14.67	15.13	15.57	15.62	15.49	15,92	16,91	15,62	15,49
France	42.54	44.18	43.30	42.84	39.55	42.94	42.58	43.77	38.24	42.07
Croatia	1.87	1.88	1.92	1.85	1.72	1.78	1.80	1.89	1.85	1.80
Italy	29.33	30.45	31.31	29.45	27.24	27.32	28.81	31.66	31.32	31.33
Cyprus	0.20	0.19	0.31	0.32	0.33	0.33	0.35	0.33	0.34	0.34
Malta	0.08	0.08	0.07	0.08	0.81	0.08	0.06	0.06	0.07	0.07
Portugal	3.11	3.21	3.22	3.21	3.22	3.12	3.20	2.97	2.78	2.71
Slovenia	1.25	1.24	1.18	1.15	1.04	1.11	1.21	1.23	1.21	1.18
Montenegro	0.0	0.0	0.29	0.30	0.31	0.32	0.33	0.33	0.26	0.28
Serbia	3.25	3.12	3.11	2.83	3.24	2.91	3.05	3.09	3.16	3.13
Turkey	17.50	18.07	19.30	19.89	20.72	22.60	21.40	22.45	23.53	20.90
EU-28	303.75	305.32	304.52	300.92	279.18	291.34	288.88	311.05	277.59	289.15

Unit Million tons of oil equivalent (TOE). *Data Eurostat* [nrg_100a]

Table 9.4 Trends of residential energy consumption in Mediterranean countries expressed as percentage change compared to same period in other years: 2003–2007, 2007–2012, and 2003–2012

Geo/Year	2003–2007 (%)	2007–2012 (%)	2003–2012 (%)
EU-28	−8.09	3.57	−4.81
Greece	−2.09	−6.53	−8.48
Spain	12.42	−0.80	11.52
France	−7.04	6.37	−1.11
Croatia	−8.02	4.61	−3.78
Italy	−7.13	15.01	6.81
Cyprus	63.32	2.28	67.04
Malta	−8.56	−4.56	−12.73
Portugal	3.54	−15.88	−12.90
Slovenia	−16.33	13.26	−5.24
Montenegro[a]	4.31	−10.15	−6.28
Serbia	−0.14	−3.47	−3.60
Turkey	18.44	0.85	19.44

Difference of final energy consumption in residential sector. *Source* Elaboration of data Eurostat
[a]Due to lack of data, the reference year for Montenegro is 2005 instead of 2003

minimum residential consumption in 2003, Cyprus in 2004, Serbia in 2006, Croatia, Slovenia, and Italy in 2007, Greece and Malta in 2010, France and Montenegro in 2011, and Portugal in 2012.

Malta, Serbia, and Slovenia reached their maximum final energy consumption in 2003, France in 2004, Greece and Croatia in 2005, Portugal in 2007, Spain, Montenegro and Italy in 2010, and Turkey in 2011. Considering the difference between 2003 and 2012 (Table 9.4), Spain (+11.52 %), Italy (6.81 %), Cyprus (67.04 %), and Turkey (19.44 %) are the only countries where the energy consumption grew; in all the other countries there was a decrease, in accordance with the tendency of Euro 28 countries. Considering the difference between 2007 and 2012, Portugal (−15.88 %), Montenegro (−10.15 %), and Greece (−6.53 %) are the countries with the greatest decrease, while Italy (15.01 %), Slovenia (13.26 %), and France (6.37 %) are the countries with the greatest increases in energy consumption.

9.2.2 Population Change

To better understand the evolution of the residential final energy consumption of a country (at least, in a short period of time such as the one considered here), it is necessary to evaluate the demography changes (i.e., total population living in the country). Considering the difference between 2003 and 2012, the populations increased in all the countries with the exception of Croatia (−0.68 %) and Serbia (−3.66 %).

Spain (11.93 %) and Cyprus (20.78 %) are the countries with the greatest population increases (Tables 9.5 and 9.6). Greece shows a negative trend in the last three years (2010–2012), Croatia in the last five years, and Portugal in the last two. Serbia lost population during the entire period studied.

9.2.3 Building Stock

An increase in population should produce a greater necessity for housing and, consequently, more energy consumption in that sector. For this reason it is necessary to analyze data on the building stock (Table 9.7) and on the floor area of dwellings. Data are available for Greece, Spain, Croatia, Cyprus, Malta, Portugal, and Slovenia.

The trend is an increase in the stock of dwellings until 2012 (last available data). Analyzing the difference between 2003 and 2012 (Table 9.8), following the tendency in the energy consumption, Spain, Cyprus, Croatia, and Malta show the largest increase in the stock of dwellings. From 2007 until 2012, Italy (2.39 %) and Greece (2.95 %) registered a lower rate of dwelling construction. In the three periods considered (2003–2007/2007–2012/2003–2012), there is an increase of the stock in all the countries.

Among the very important data to consider in this analysis is the number of dwellings permanently occupied[1] (see Table 9.9).

It is interesting to evaluate the percentage of dwellings permanently occupied compared to the total (see Table 9.10; data for France not available). In Slovenia, it is over 90 %, in Italy, Cyprus and Croatia between 80 and 90 %, and less than 70 % in Greece, Spain and Portugal.

The number of occupied dwellings increased between 2003 and 2012 (Table 9.11) in all countries, and the trend is quite similar to the total dwelling stock data. Exceptions are Italy, where the number of dwellings permanently occupied grew more (in percentages) than the total stock of dwellings (11.58–5.57 %), with Cyprus (33.69–40.98 %) and Malta (24.63–41.14 %), where the percentage of occupied dwellings became less than the total dwelling stock.

In conclusion, in Greece, Malta, Portugal, and Slovenia, the increase in population and in the dwelling stock is associated with a remarkable decrease in the energy consumption in the residential sector. In Spain there was a notable proliferation of dwellings and increase of household energy consumption, but the growth of the population was not so large.

In Italy, the three indicators follow a comparable trend, while in Cyprus, the population and the total dwelling increase do not explain the percentage variation of energy consumption in dwellings (+67 % between 2003 and 2012). In Croatia, the population decreased and the number of dwellings increased, but the final energy consumption of dwellings decreased. In Table 9.12, the trend of the dwelling stock expressed by the total floor area in square meters can be seen.

[1]The number of permanently occupied dwellings excludes vacant dwellings and summer houses.

Table 9.5 Population change in Mediterranean countries

Geo/Year	2003	2004	2005	2006	2007	2008	2009	2010	2011	2012
Greece	10.9	11.0	11.0	11.1	11.1	11.1	11.1	11.1	11.1	11.1
Spain	41.8	42.5	43.2	44.0	44.7	45.6	46.2	46.4	46.6	46.8
France[a]	60.1	60.5	60.9	61.3	61.7	62.1	62.4	62.7	63.0	63.3
Croatia	4.30	4.30	4.31	4.31	4.31	4.31	4.30	4.30	4.28	4.27
Italy	57.1	57.4	57.8	58.0	58.2	58.6	59.0	59.1	59.3	59.3
Cyprus	0.71	0.72	0.73	0.74	0.75	0.77	0.79	0.81	0.83	0.86
Malta	0.39	0.39	0.40	0.40	0.40	0.40	0.41	0.41	0.41	0.41
Portugal	10.4	10.4	10.4	10.5	10.5	10.5	10.5	10.5	10.5	10.5
Slovenia	1.99	1.99	1.99	2.00	2.01	2.01	2.03	2.04	2.05	2.05
Montenegro	0.61	0.62	0.62	0.62	0.62	0.62	0.63	0.61	0.61	0.62
Serbia	7.49	7.47	7.45	7.42	7.39	7.36	7.33	7.30	7.25	7.21
Turkey	69.7	70.6	71.6	72.5	69.6	70.5	71.5	72.5	73.7	74.7
EU-28	490.81	492.70	494.77	496.63	498.40	500.41	502.18	503.37	504.96	506.09

Unit Millions. *Data Eurostat* [demo_gind]
[a]metropolitan

Table 9.6 Population trends in Mediterranean countries expressed as a percentage change compared to the same period in other years: 2003–2007, 2007–2012, and 2003–2012

Geo/Year	2003–2007 (%)	2007–2012 (%)	2003–2012 (%)
EU-28	1.55	1.54	3.11
Greece	1.32	−0.19	1.13
Spain	7.07	4.54	11.93
France	2.88	2.58	5.53
Croatia	0.19	−0.87	−0.68
Italy	1.91	2.01	3.96
Cyprus	6.19	13.73	20.78
Malta	2.09	2.94	5.10
Portugal	0.84	0.09	0.94
Slovenia	0.77	2.24	3.03
Montenegro	0.31	−0.59	−0.28
Serbia	−1.25	−2.45	−3.66
Turkey	−0.12	7.22	7.10

Population difference. *Source* Elaboration of data Eurostat

Table 9.7 Stock of dwellings

Geo/Year	2003	2004	2005	2006	2007	2008	2009	2010	2011	2012
Greece	5.99	6.11	6.30	6.43	6.53	6.61	6.67	6.72	6.75	6.72
Spain	21.9	22.5	23.1	23.7	24.4	25.0	25.4	25.7	25.9	26.0
Croatia	1.64	1.64	1.65	1.67	1.69	1.71	1.84	1.86	1.87	1.89
Italy	27.6	27.8	28.0	28.2	28.4	28.6	28.8	28.9	29.0	29.1
Cyprus	0.30	0.31	0.32	0.34	0.35	0.37	0.39	0.40	0.42	0.43
Malta	0.15	0.15	0.15	0.16	0.16	0.16	0.17	0.17	0.22	n.a.
Portugal	5.39	5.46	5.46	5.51	5.60	5.66	5.72	5.75	5.88	5.91
Slovenia	0.79	0.79	0.80	0.81	0.82	0.83	0.83	0.84	0.84	0.85

Data expressed in millions. *Source* Odyssee

Table 9.8 Trends of dwelling stock in Mediterranean countries expressed as percentage change compared to same period in other years: 2003–2007, 2007–2012, and 2003–2012

Geo/Year	2003–2007 (%)	2007–2012 (%)	2003–2012 (%)
Greece	9.02	2.95	12.24
Spain	11.19	6.63	18.57
Croatia	3.17	11.74	15.29
Italy	3.10	2.39	5.57
Cyprus	17.38	20.11	40.98
Malta[a]	3.47	36.41	41.14
Portugal	3.78	5.50	9.49
Slovenia	3.68	4.08	7.91

Dwelling stock difference. *Source* Elaboration of data Odyssee
[a]Reference year for Malta: 2011

Table 9.9 Stock of dwellings permanently occupied

Geo/Year	2003	2004	2005	2006	2007	2008	2009	2010	2011	2012
Greece	4.07	4.15	4.28	4.37	4.44	4.49	4.53	4.57	4.59	4.57
Spain	14.49	14.90	15.32	16.09	16.60	17.00	17.36	17.62	17.81	18.05
France	25.19	25.52	25.84	26.20	26.48	26.74	26.98	27.24	27.51	27.78
Croatia	1.42	1.42	1.42	1.44	1.47	1.48	1.59	1.61	1.62	1.64
Italy	22.87	23.31	23.60	23.90	24.28	24.64	24.90	25.17	25.40	25.87
Cyprus	0.23	0.24	0.25	0.26	0.28	0.29	0.315	0.33	0.344	0.35
Malta	0.12	0.12	0.12	0.13	0.13	0.14	0.13	0.13	0.15	0.16
Portugal	3.73	3.84	3.97	4.03	3.78	3.79	3.93	3.93	3.96	3.98
Slovenia	0.71	0.72	0.72	0.73	0.74	0.75	0.76	0.76	0.77	0.77

Data expressed in millions. *Source* Odyssee

Table 9.10 Amount of dwellings permanently occupied compared to the total

Geo/Year	2003	2004	2005	2006	2007	2008	2009	2010	2011	2012
Greece	68.0	68.0	68.0	68.0	68.0	68.0	68.0	68.0	68.0	68.0
Spain	65.9	66.1	66.2	67.6	67.9	67.8	68.1	68.4	68.7	69.3
Croatia	86.5	86.5	86.5	86.5	86.5	86.5	86.5	86.5	86.5	86.5
Italy	82.7	83.7	84.1	84.5	85.2	85.8	86.2	86.8	87.3	88.6
Cyprus	76.6	77.0	77.6	78.6	79.1	79.7	80.5	81.2	81.7	82.0
Malta	79.3	79.6	80.0	82.4	83.5	84.5	79.7	75.2	68.2	n.a
Portugal	69.2	70.4	72.7	73.0	67.6	66.9	68.8	68.3	67.3	67.5
Slovenia	90.1	90.2	90.3	90.4	90.5	90.6	90.6	90.7	90.7	90.7

Data expressed as percentage. *Source* Odyssee

Table 9.11 Trends of occupied dwellings by country expressed as percentage change compared to same period in other years: 2003–2007, 2007–2012, and 2003–2012

Geo/Year	2003–2007 (%)	2007–2012 (%)	2003–2012 (%)
Greece	8.28	2.86	10.90
Spain	12.69	8.05	19.72
France	4.89	4.69	9.34
Croatia	3.08	10.51	13.26
Italy	5.79	6.15	11.58
Cyprus	17.49	19.63	33.69
Malta	8.21	17.90	24.63
Portugal	1.33	5.03	6.29
Slovenia	3.91	4.22	7.96

Dwelling permanently occupied difference. *Source* Odyssee

Table 9.12 Total floor area of the dwelling stock

Geo/Year	2003	2004	2005	2006	2007	2008	2009	2010	2011	2012
Greece	0.50	0.51	0.53	0.54	0.55	0.56	0.56	0.57	0.57	0.57
Spain	1.93	1.99	2.04	2.10	2.16	2.22	2.26	2.29	2.31	2.32
Croatia	0.12	0.12	0.12	0.12	0.12	0.13	0.14	0.14	0.14	0.14
Italy	2.64	2.66	2.67	2.69	2.71	2.72	2.73	2.74	2.75	2.76
Cyprus	0.04	0.04	0.04	0.04	0.05	0.05	0.05	0.05	0.60	0.62
Malta	0.01	0.01	0.01	0.01	0.01	0.01	0.01	0.01	0.02	n.a.
Portugal	0.48	0.49	0.49	0.50	0.52	0.53	0.53	0.61	0.64	0.64
Slovenia	0.06	0.06	0.06	0.06	0.06	0.06	0.06	0.06	0.06	0.06

Data expressed in thousand millions. *Source* Elaboration of data Odyssee

Table 9.13 Trends of total floor area of dwellings in Mediterranean countries expressed as percentage change compared to same period in other years: 2003–2012, 2007–2012, and 2003–2007

Geo/Year	2003–2007 (%)	2007–2012 (%)	2003–2012 (%)
Greece	9.02	2.95	12.24
Spain	11.88	7.49	20.26
Croatia	5.34	12.31	18.30
Italy	2.42	1.87	4.33
Cyprus	20.01	22.97	47.57
Malta*	3.47	36.41	41.14
Portugal	8.58	23.60	34.20
Slovenia	6.84	6.45	13.74

Reference year for Malta: 2011. Dwelling floor area difference. *Source* Elaboration of data Odyssee

In Croatia, Cyprus, Malta, and Portugal (Table 9.13), the greatest increases are concentrated in the period between 2007 and 2012, while in Greece, Spain, and Italy, the biggest difference was between 2003 and 2007. Considering the difference between 2003 and 2012, Italy is the only country with an increase of less than 10 %.

9.2.4 Greenhouse Gas Emissions

Until the end of 2006, construction output in Europe had increased rather steadily, but with the economic and financial crisis, the output began to decline quite dramatically. Since spring 2008, the level of total construction in the EU-28 has been in constant decline. The most recent level of this index (May 2014) for the EU-28, is more than 25 % lower than the pre-crisis highest.[2] In Portugal, the variation in

[2]Construction production (volume) index overview. Main statistical findings. European Commission. Eurostat.

the total floor area of dwellings is mainly due to the variation of the average floor area that changed from 89 m^2 in 2003 to 109.09 in 2012. In countries like Greece, Italy, Spain, France, and Malta, the change was almost nil.

A useful indicator for better understanding the building trends in the housing sector is the greenhouse gas emission associated with the sector. Emissions of carbon dioxide (CO_2), nitrous oxide (N_2O), and methane (CH_4), resulting from the activities of various industries and households in the EU-27 in 2011, stood at 4.66 billion tons of carbon dioxide (CO_2) equivalents. Table 9.14 shows the trend of greenhouse gas emissions (tons of CO_2 equivalents expressed as the sum of carbon dioxide, carbon dioxide from biomass used as a fuel, methane nitrous oxide, hydrofluorocarbons, perfluorocarbons and sulphur hexafluoride) due to total activities by households (heating/cooling activities, transport activities, other activities).[3]

Considering the period from 2003 until 2012 (Table 9.15), the emissions in Italy and Portugal decreased, respectively, by 11 and 8 %, while in Turkey they increased more than 100 %. Between 2008 and 2012, Croatia, Cyprus, Malta, Slovenia, and Turkey increased their emissions, while in Greece there was a decrease of almost 24 %, with 13.56 % in Spain, and 7.63 % in Italy.

Data on heating/cooling activities by households (no data were available for Portugal) follows the same tendencies of the total emissions. The exceptions are Italy, by an increase in greenhouse gas emissions in the two periods under analysis, and Croatia, with greenhouse gas emissions for heating and cooling 10 times higher (data expressed in percentages) than the total emissions of the sector.

Table 9.16 shows the amount of greenhouse gas emissions for heating and cooling compared to the total emissions of the sector. The volume of data does not permit defining a common range of variations. France (52.52 %), Italy (53.30 %), and Slovenia (55.28 %) are the countries where the contribution of heating and cooling to greenhouse gas emission was highest, while in Cyprus (19.29 %) and Spain (27.49 %) it was the lowest.

9.2.5 Discussion of Data

As a short summary of all the previous analyses, data show that in most of the countries, the total final energy consumption decreased, while the population increased. Moreover, the total stock and the floor area of households also increased, while the residential final energy consumption decreased, mainly due to the financial crisis and a general condition of poverty. Nevertheless, to complete the analysis, it is necessary to evaluate the energy savings achieved by energy efficiency progress in the household sector.

---

[3]Data for Italy, Portugal and Turkey are complete for the whole period of analysis 2002–2012, while for the other countries are only available for the period 2008–2012.

Table 9.14 Greenhouse gas emission due to total activities by households (heating/cooling activities, transport activities, other activities)

Geo/Year	2002	2003	2004	2005	2006	2007	2008	2009	2010	2011	2012
EU-28	n.a.	n.a	n.a	n.a	n.a	n.a	950.1	929.1	961.3	870.6	870.6
Greece	n.a	n.a	n.a	n.a	n.a	n.a	22.5	22.4	21.1	20.1	17.1
Spain	n.a	n.a	n.a	n.a	n.a	n.a	74.2	72.4	73.3	67.0	64.2
France	n.a	n.a	n.a	n.a	n.a	n.a	165.9	165.7	172.3	154.3	161.6
Croatia	n.a	n.a	n.a	n.a	n.a	n.a	6.88	7.01	7.12	7.17	7.03
Italy	128.4	133.3	135.8	136.5	130.7	126.6	128.4	128.2	129.9	123.8	118.6
Cyprus	n.a	n.a	n.a	n.a	n.a	n.a	1.96	2.03	2.05	2.11	2.15
Malta	n.a	n.a	n.a	n.a	n.a	n.a	0.33	0.34	0.35	0.35	0.34
Portugal	18.9	18.8	18.6	18.2	18.1	17.6	17.5	17.5	18.2	17.5	17.3
Slovenia	n.a	n.a	n.a	n.a	n.a	n.a	5.48	5.71	5.76	5.82	5.79
Turkey	43.8	47.4	51.6	54.2	56.6	62.7	76.0	81.3	78.2	80.8	99.0

Data expressed in million tons of CO_2 equivalents. *Source* Enerdata [env_ac_ainah_r2]

Table 9.15 Trends of greenhouse gas emissions due to all activities by households and to heating and cooling activities expressed as percentage change compared to same period in other years: 2008–2012, 2002–2012

Geo/Year	2008–2012 (%)	2002–2012 (%)	2008–2012 (%)	2002–2012 (%)
EU-28	−8.36		−8.50	
Greece	−23.97		−25.15	
Spain	−13.56		−14.06	
France	−2.60		−3.13	
Croatia	2.05		11.86	
Italy	−7.62	−11.01	2.78	2.07
Cyprus	9.38		4.68	
Malta	2.90		19.65	
Portugal	−1.01	−7.99		
Slovenia	5.72		5.22	
Turkey	30.17	108.70	26.75	136.80

Difference of greenhouse gas emissions in residential sector. *Source* Enerdata

To do this, it can be useful to analyze the main results of the project ODYSSEE-MURE.[4] In accordance with this project, the variation of the households' energy consumption is influenced by:

- Climatic difference between years t and t0 ("climatic effect")
- Change in number of occupied dwellings ("more dwellings")
- "More appliances per dwelling" (electrical appliances, central heating)
- Change in floor area of dwelling for space heating ("larger homes")
- Energy savings, as measured from ODEX[5]
- Other effects (mainly change in heating behaviors).

In the period studied (2003–2012), the energy savings in many countries of Europe were estimated; among them were Spain, France, Italy, Portugal, and Slovenia (Table 9.17). The energy savings achieved in Slovenia were more than three times (3.47) the decrease of total household energy consumption[6] in that

[4]The ODYSSEE-MURE project is co-ordinated by ADEME with the technical support of Enerdata, Fraunhofer, ISIS and ECN. It is supported by the Intelligent Energy Europe program and is part of the activity of the EnR Club.

[5]ODEX is the index used in the ODYSSEE-MURE project to measure the energy efficiency progress by main sector (industry, transport, households) and for the whole economy (all final consumers). For each sector, the index is calculated as a weighted average of sub-sectoral indices of energy efficiency progress; sub-sectors being industrial or service sector branches or end-uses for households or transport modes. For more information check the document "Definition of ODEX indicators in ODYSSEE data base. March 2010" http://www.indicators.odyssee-mure.eu/energy-efficiency-database.html.

[6]It has to be noted that in the project ODYSSEE-MURE the definition of final consumption is the same as the "final consumption for energy uses" from EUROSTAT . It differs however from IEA who includes the non-energy uses in the final consumption. Non-energy uses are excluded in ODYSSEE, as their utilization is not related to energy efficiency considerations, but rather to materials management.

Table 9.16 Amount of greenhouse gas emission for heating and cooling compared to the total emissions of the sector, expressed as percentage change compared to same period of the previous year

Geo/Year	2002	2003	2004	2005	2006	2007	2008	2009	2010	2011	2012
EU-28	n.a.	n.a.	n.a	n.a	n.a	n.a	46.8	46.2	48.6	44.8	46.8
Greece	n.a.	n.a.	n.a.	n.a.	n.a.	n.a.	36.2	31.0	32.9	34.3	35.6
Spain	n.a.	n.a.	n.a.	n.a.	n.a.	n.a.	27.6	27.8	29.3	26.8	27.4
France	n.a.	n.a.	n.a.	n.a.	n.a.	n.a.	52.8	52.7	54.2	49.7	52.5
Croatia	n.a.	n.a.	n.a.	n.a.	n.a.	n.a.	33.5	34.1	35.9	37.0	36.8
Italy	44.5	46.4	47.4	48.9	47.5	45.6	47.9	49.5	51.3	50.0	53.3
Cyprus	n.a.	n.a.	n.a.	n.a.	n.a.	n.a.	20.1	21.3	18.0	19.9	19.2
Malta	n.a.	n.a.	n.a.	n.a.	n.a.	n.a.	8.63	9.21	8.05	9.69	10.0
Portugal	n.a.	n.a.	n.a.	n.a.	n.a.	n.a.	n.a.	n.a.	n.a.	n.a.	n.a.
Slovenia	n.a.	n.a.	n.a.	n.a.	n.a.	n.a.	55.5	56.3	55.9	55.8	55.2
Turkey	39.7	40.8	41.7	42.7	42.0	40.7	47.5	49.2	48.5	48.3	46.3

Unit Percentage. *Source* Enerdata

Table 9.17 Decomposition of factors influencing household energy consumption in the period 2003–2012

Households—(2003–2012) ktoe/country	Slovenia	France	Portugal	Spain	Italy
Variation consumption	−0.085294	−2994.4	−411.43	1600.803	2116.866
Climate	−0.068144	−243.839	52.331	403.137	−1338.358
More dwellings	0.099636	4324.27	206.079	3433.532	3480.32
More appliances per dwelling	0.015357	−98.219	n.a.	n.a.	1001.263
Larger homes	0.074428	889.47	0	194.747	−328.632
Energy savings	−0.295907	−8742.146	−713.731	−206.114	−1822.322
Other	0.089335	876.065	43.893	−2224.5	1124.598

Data expressed in ktoe. *Source* ODYSEE_MURE project

period, i.e., 2.91 times in France and 1.73 times in Portugal. However, in Spain and Italy, the total household energy consumption increased—7.76 times the contribution of energy savings in Spain, and 1.16 in Italy. In other words, it seems that in Slovenia, France, and Portugal the energy savings were the major responsibility of the variations in the households' energy consumption.

The contribution to the variation in changes in heating behaviors (the row named "other" in the table) constitutes an important phenomenon, mainly associated with the so-called "energy poverty," that will be analyzed in the next section.

9.3 Housing and Living Quality

Access to good quality and affordable accommodations is both a fundamental need and a right, but meeting this need is still a significant challenge in a number of European Union countries. Poor housing conditions point to the risk of poverty and social exclusion.

9.3.1 Overcrowding Rate

One of the key dimensions in assessing the quality of housing conditions is the availability of sufficient space in the dwelling. The overcrowding rate[7] is the proportion of people living in an overcrowded dwelling, as defined by the number of

[7]The overcrowding rate is defined as the percentage of the population living in an overcrowded household. A person is considered as living in an overcrowded household if the household does not have at its disposal a minimum number of rooms equal to: one room for the household; one room per couple in the household; one room for each single person aged 18 or more; one room per pair of single people of the same gender between 12 and 17 years of age; one room for each single person between 12 and 17 years of age and not included in the previous category; one room per pair of children under 12 years of age. Definition: Glossary Eurostat. European Comission.

rooms available to the household and the household's size, as well as its members' ages and their family situation (Table 9.18).

In 2012 the average value of the overcrowding rate in Europe was 17 %. Croatia, Greece, and Italy with 44.4, 27.3, and 26.2 % respectively, registered the highest overcrowding rates that year, while Malta, Cyprus, and Spain with 4, 2.8, and 5.5 %, registered the lowest. Serbia, in 2013, had a rate of 54.3 %. The largest decrease between 2005 and 2012 in the share of the population living in overcrowded dwellings was reported by Slovenia, decreasing by 26.4 % points.

Within the population at risk of poverty[8] (Table 9.19), the overcrowding rate in the EU-28 was 29.3 % in 2012, 12.2 % points above the entire population rate; in Greece it was 12.9 points above the rate, 15.2 % in France, and 12.6 % in Italy. This difference increased in 2013 for Greece (14.7 %) and Italy (14.4 %), and decreased in France (14.3 %). Spain moved from 6.5 to 5.8 %, Croatia from 4.2 to 4 %, Cyprus from 4.7 to 2.5 %, Portugal from 7.1 to 2.3 %, and Slovenia from 10.5 to 9.6 %. In 2013, the overcrowding rate for the population at risk of poverty in Serbia was 62.3 %.

9.3.2 Severe Housing Deprivation Rate

In order to complete the framework of the quality of housing, it is useful to consider the severe housing deprivation rate (Table 9.20). This indicator is defined as the percentage of persons living in a dwelling that is considered to be overcrowded, while having some other aspects of housing deprivation—such as the lack of a bath or a toilet, a leaking roof, or a dwelling considered to be too dark.[9]

Across the EU-28 as a whole, 5.1 % of the population suffered from severe housing deprivation in 2012. That same year, Croatia (9.6 %), Italy (8.4 %), Slovenia (8.1 %), and Greece (7 %) registered the highest rates, while Malta (1 %), Cyprus (1.2 %), and Spain (1.3 %) registered the lowest rates. Slovenia moved from 12.3 % (more than one out of 10 people) in 2005 to 6.5 % in 2013. In the same period, Portugal decreased from 7.7 to 5.6 %. In 2013 the Serbian population suffering from severe housing deprivation was 16.4 %.

Within the population at risk of poverty (Table 9.21) across the EU-28 as a whole, 12.6 % of the population suffered from severe housing deprivation in 2012. That same year, Slovenia, registered the highest rate (16.9 %, with a clear decrease in the trend of the whole period of analysis), while Italy registered 16.7 % in 2012 and 17.3 % in 2013 (increasing since 2010), Croatia 13.7 % and Greece 13.2 %.

The lowest rate in 2012 was registered in Malta (1.6 %). In Serbia, the data of 2013 show that almost 3 out of 10 people at risk of poverty suffered from severe housing deprivation.

[8] People living in households where equivalized disposable income per person was below 60 % of the national median.

[9] European Commission. Eurostat. Statistics Explained Glossary.

Table 9.18 Overcrowding rate, expressed as percentage of total population, calculated considering age, gender, and poverty status

Geo/Year	2003	2004	2005	2006	2007	2008	2009	2010	2011	2012	2013
EU-28	n.a.	n.a.	n.a.	n.a.	n.a.	n.a.	n.a.	17.9	17.3	17.0	17.4
Greece	31.6	29.2	29.2	29.3	29.2	26.7	25.0	25.5	25.9	26.5	27.3
Spain	n.a.	13.6	8.4	6.5	5.8	5.6	5.2	5.0	6.6	5.6	5.2
France	n.a.	10.3	9.4	8.1	10.1	9.7	9.6	9.2	8.0	8.1	7.6
Croatia	n.a.	n.a.	n.a.	n.a.	n.a.	n.a.	n.a.	43.7	44.6	44.4	42.8
Italy	n.a.	25.7	24.2	24.3	24.4	24.2	23.3	23.9	25.0	26.2	27.3
Cyprus	n.a.	n.a.	2.2	1.9	1.6	3.3	2.6	3.5	2.9	2.8	2.4
Malta	n.a.	n.a.	3.8	3.5	4.2	3.9	3.8	4.0	4.4	4.0	3.6
Portugal	n.a.	15.3	16.5	15.8	16.1	15.7	14.1	14.6	11.0	10.1	11.4
Slovenia	n.a.	n.a.	42.0	40.3	39.9	39.5	38.0	34.9	17.1	16.6	15.6
Serbia	n.a.	n.a.	n.a.	n.a.	n.a.	n.a.	n.a.	n.a.	n.a.	n.a.	54.3

Source Eurostat [ilc_lvho05a]

Table 9.19 Overcrowding rate, expressed as percentage of total population, calculated considering all age, gender, and the population at risk of poverty

Geo/Year	2003	2004	2005	2006	2007	2008	2009	2010	2011	2012	2013
EU-28	n.a.	n.a.	n.a.	n.a.	n.a.	n.a.	n.a.	29.8	29.3	29.3	30.4
Greece	38.4	30.1	34.1	37.3	37.1	35.2	32.3	34.7	35.8	39.4	42.0
Spain	n.a.	16.1	13.2	11.7	11.5	10.8	10.3	9.5	10.8	12.1	11.0
France	n.a.	25.0	21.4	18.5	25.7	26.4	27.2	25.6	23.5	23.3	21.9
Croatia	n.a.	n.a.	n.a.	n.a.	n.a.	n.a.	n.a.	46.4	47.3	48.6	46.8
Italy	n.a.	37.9	35.8	36.4	37.9	35.3	35.9	36.2	38.8	38.8	41.7
Cyprus	n.a.	n.a.	6.0	5.0	4.8	8.5	5.2	7.5	7.8	7.5	4.9
Malta	n.a.	n.a.	4.7	3.4	5.4	4.2	5.1	6.6	9.2	6.6	6.3
Portugal	n.a.	21.8	24.7	22.2	21.1	25.1	23.4	22.3	20.6	17.2	19.7
Slovenia	n.a.	n.a.	50.7	47.1	48.9	47.2	44.4	46.3	26.4	27.1	25.2
Serbia	n.a.	n.a.	n.a.	n.a.	n.a.	n.a.	n.a.	n.a.	n.a.	n.a.	62.3

Source Eurostat—SILC [ilc_lvho05a]

Table 9.20 Severe housing deprivation rate by age, sex and poverty status expressed as percentage of total population

Geo/Year	2003	2004	2005	2006	2007	2008	2009	2010	2011	2012	2013
EU-28	n.a.	n.a.	n.a.	n.a.	n.a.	n.a.	n.a.	5.8	5.5	5.1	5.2
Greece	9.6	8.2	9.1	9.1	8.5	8.1	7.6	7.6	7.2	7.0	7.0
Spain	n.a.	4.8	2.9	2.6	2.6	1.6	1.8	1.8	2.1	1.3	1.8
France	n.a.	3.4	3.0	2.7	3.3	3.4	3.0	3.0	2.5	2.6	2.2
Croatia	n.a.	n.a.	n.a.	n.a.	n.a.	n.a.	n.a.	12.3	9.9	9.6	9.0
Italy	n.a.	9.0	8.0	7.7	7.2	7.4	7.3	6.7	8.8	8.4	8.9
Cyprus	n.a.	n.a.	1.6	1.3	0.8	1.6	1.1	1.6	1.6	1.2	1.4
Malta	n.a.	n.a.	1.0	0.8	0.9	1.0	1.3	1.4	1.6	1.0	1.1
Portugal	n.a.	7.1	7.7	7.5	7.6	6.9	4.7	5.6	4.0	4.3	5.6
Slovenia	n.a.	n.a.	12.3	13.1	12.3	16.6	17.5	15.4	8.7	8.1	6.5
Serbia	n.a.	n.a.	n.a.	n.a.	n.a.	n.a.	n.a.	n.a.	n.a.	n.a.	16.4

Source Eurostat—SILC [ilc_mdho06a]

Table 9.21 Severe housing deprivation rate by age and gender, considering population at risk of poverty

Geo/Year	2003	2004	2005	2006	2007	2008	2009	2010	2011	2012	2013
EU-28	n.a.	n.a.	n.a.	n.a.	n.a.	n.a.	n.a.	13.7	13.1	12.6	12.9
Greece	15.6	11.3	14.7	14.7	15.0	13.6	13.0	14.0	14.2	13.2	11.8
Spain	n.a.	6.8	6.4	5.5	6.1	2.9	4.6	3.8	4.8	3.6	4.4
France	n.a.	9.6	8.2	6.9	8.7	10.0	10.4	9.8	8.8	8.7	8.4
Croatia	n.a.	n.a.	n.a.	n.a.	n.a.	n.a.	n.a.	19.5	15.4	13.7	14.9
Italy	n.a.	18.2	14.9	15.3	14.1	13.9	14.9	14.4	15.9	16.7	17.3
Cyprus	n.a.	n.a.	4.2	3.7	2.2	2.7	0.8	4.4	5.7	3.6	3.2
Malta	n.a.	n.a.	1.5	0.5	2.3	1.6	3.2	1.7	3.9	1.6	1.9
Portugal	n.a.	13.0	13.8	11.8	13.2	12.4	10.6	10.6	8.2	7.8	11.4
Slovenia	n.a.	n.a.	21.1	19.2	22.6	26.8	24.6	25.4	16.2	16.9	14.1
Serbia	n.a.	n.a.	n.a.	n.a.	n.a.	n.a.	n.a.	n.a.	n.a.	n.a.	27.3

Source Eurostat—SILC [ilc_mdho06a]

9.3.3 Housing Cost Overburden Rate

The housing cost overburden rate (Table 9.22) is the percentage of the population living in households where the total housing costs ("net" of housing allowances) represent more than 40 % of disposable income ("net" of housing allowances).[10]

In 2012 there were significant differences among the member states. Malta (2.6 %) and Cyprus (3.3 %), as well as France and Slovenia (both 5.2 %), had a relatively small proportion of their populations living in households where housing costs exceeded 40 % of their disposable income, while around one-third of the population of Greece (33.1 %) spent more than 40 % of its equivalized disposable income on housing.

In Italy, the housing cost overburden rate was 6.8 % of the population in 2012 and 8.4 % in 2013. In Slovenia, it was 5.2; percentages were registered in Serbia, in 2013, with 28 % of the population living in households where housing costs exceeded 40 % of their disposable income.

9.4 Energy Poverty

Among energy-related problems across Europe, energy poverty is a major problem, as between 50 and 125 million people are unable to afford proper indoor thermal comfort. Energy poverty (or "fuel poverty" as it is also known), occurs when a household is unable to afford the most basic levels of energy for adequate heating, cooking, lighting, and use of appliances in the home due to a combination of factors related to low incomes, energy-inefficient housing, and high energy prices. Fuel poverty is primarily a determinant of three household factors: income, energy prices, and the energy needs of the household (influenced by the energy efficiency status of the property).

The United Kingdom is the only country to have come up with an official definition: "a household is in a situation of fuel poverty when it has to spend more than 10 % of its income on all domestic fuel use, including appliances, to heat the home to a level sufficient for health and comfort." (EPEE 2009)

It can be expressed by the equation:

$$\text{Fuel poverty ratio} = \frac{\text{Energy costs(space heating} + \text{DHW} + \text{electricity)}}{\text{Income}}$$

Fuel poverty is associated with adverse impacts on physical health, mental well-being and quality of life (Liddell 2010), making it a prominent social issue throughout the European Union (Walker 2014). The Buildings Performance Institute Europe (BPIE 2014) measures fuel poverty by the inability of people to keep their home *adequately warm*, to pay their *utility bills*, and to live in a *dwelling without defects* (leaks, damp walls, etc.).

[10]European Commission. Eurostat. Statistics Explained Glossary.

Table 9.22 Housing costs overburden rate by age, gender, and poverty status

Geo/Year	2004	2005	2006	2007	2008	2009	2010	2011	2012	2013
EU-28	24.7	22.7	24.6	15.8	22.2	21.8	18.1	24.2	33.1	36.9
Greece	4.9	5.3	7.4	8.3	10.1	12.8	13.2	13.8	14.3	10.3
Spain	5.4	5.3	6.0	5.7	4.2	4.0	5.1	5.2	5.2	5.0
France	n.a.	n.a.	n.a.	n.a.	n.a.	n.a.	14.1	8.0	6.8	8.4
Croatia	12.8	12.7	12.3	7.7	8.1	7.5	7.5	8.4	7.9	8.7
Italy	n.a.	6.6	3.0	1.7	1.8	2.4	3.1	3.1	3.3	3.3
Cyprus	n.a.	1.7	1.8	2.5	3.3	2.8	3.7	3.0	2.6	2.6
Malta	4.7	4.3	4.5	7.4	7.6	6.1	4.2	7.2	8.3	8.3
Portugal	n.a.	4.7	3.0	5.0	4.4	3.9	4.3	4.7	5.2	6.0
Slovenia	n.a.	n.a.	n.a.	n.a.	n.a.	n.a.	n.a.	n.a.	n.a.	28.0
Serbia	24.7	22.7	24.6	15.8	22.2	21.8	18.1	24.2	33.1	36.9

Source Eurostat—SILC [ilc_lvho07a]

9.4.1 Inability to Keep Homes Adequately Warm

Table 9.23 shows the inability to keep homes adequately warm,[11] expressed as a percentage of the total population, while Table 9.24 shows the inability to keep homes adequately warm as expressed as a percentage of the total population only for people who are below 60 % of the median equivalized income. In 2013 in Greece, 29.5 % of the entire population was not able to heat their homes adequately. In 2004 the data showed 16.8 %, with an increase of more than 75 %. In Malta (23.4 % in 2013) there was an increase of 85.7 % since 2005.

In 2013 the highest percentages were registered in Cyprus (30.5 %) and Portugal (27.9 %), where in both countries the values decreased since 2007. In Italy, the 19.1% of people were not able to warm their homes adequately, 8 % in Spain (both countries had 10.1 % in 2006), 6.8 % in France (increasing in the last years), and 4.9 % in Slovenia. Serbia registered 18.3 % in the only available data for 2013.

Considering people below 60 % of median equivalized income, in 2013 there are four countries where over 40% of people were not able to warm their homes adequately: Cyprus (51 %), Greece (48.4 %), Portugal (44.6 %), and Italy (40.1 %). Among them, Portugal is the only country with a decreasing rate in the last years. The average rate in Europe was 24.1 %. Slovenia reached the minimum value among all the countries with 13.1 %.

9.4.2 People Living in Dwellings with Poor Conditions

Table 9.25 shows the share of the total population living in a dwelling with a leaking roof, damp walls, floors or foundations, or rotting window frames or floors, expressed as a percentage of the total population, while in Table 9.26 we see the same data referring to people at risk of poverty (below 60 % of median equivalized income).

In 2013, the poorest data were registered in Portugal (31.9 %), Cyprus (31.1 %), Slovenia (27 %), and Italy (23.1 %); Greece, Malta, Croatia, and France were below the EU-28 average (15.7 %). Spain reached 16.7 % and Serbia 21.6 % (the only available data). Restricting the analysis to people with low income, there were five countries with a percentage higher than 30: Portugal (40.1 %), Slovenia (39.6 %), Cyprus (35.1 %), Serbia (33.1 %), and Italy (30.5 %). Malta (11.5 %) and Greece (20.8 %) registered the lowest values.

[11]This indicator is about affordability (ability to pay) to keep the home adequately warm, and refers to the perception of the participants (and not to an objective criterion of thermal comfort).

Table 9.23 Inability to keep home adequately warm expressed as percentage of total population

Geo/Year	2003	2004	2005	2006	2007	2008	2009	2010	2011	2012	2013
EU-28	n.a.	n.a.	n.a.	n.a.	n.a.	n.a.	n.a.	9.5	9.8	10.8	10.8
Greece	17.4	16.8	15.7	12.0	13.8	15.4	15.7	15.4	18.6	26.1	29.5
Spain	n.a.	9.5	9.4	10.1	8.0	6.0	7.2	7.5	6.5	9.1	8.0
France	n.a.	5.9	5.3	5.9	4.6	5.3	5.5	5.7	6.0	6.0	6.8
Croatia	n.a.	n.a.	n.a.	n.a.	n.a.	n.a.	n.a.	8.3	9.8	10.2	9.9
Italy	n.a.	10.7	10.6	10.1	10.4	11.3	10.6	11.2	18.0	21.2	19.1
Cyprus	n.a.	n.a.	33.7	33.8	34.6	29.2	21.7	27.3	26.6	30.7	30.5
Malta	n.a.	n.a.	12.6	10.8	10.2	8.8	11.1	14.3	17.6	22.1	23.4
Portugal	n.a.	36.3	40.0	39.9	41.9	34.9	28.5	30.1	26.8	27.0	27.9
Slovenia	n.a.	n.a.	2.6	3.0	4.2	5.6	4.6	4.7	5.4	6.1	4.9
Serbia	n.a.	n.a.	n.a.	n.a.	n.a.	n.a.	n.a.	n.a.	n.a.	n.a.	18.3

Source Eurostat—SILC [ilc_mdes01]

Table 9.24 Inability to keep home adequately warm expressed as percentage of total population for people below 60 % of median equivalized income

Geo/Year	2003	2004	2005	2006	2007	2008	2009	2010	2011	2012	2013
EU-28	n.a.	n.a.	n.a.	n.a.	n.a.	n.a.	n.a.	21.0	21.7	24.3	24.1
Greece	37.4	32.0	30.3	24.7	29.4	29.9	36.8	38.4	38.8	47.6	48.4
Spain	n.a.	17.1	17.1	21.4	17.2	12.4	15.4	15.9	12.3	18.2	15.6
France	n.a.	16.5	12.9	15.7	12.8	11.5	15.0	15.3	16.9	15.2	17.7
Croatia	n.a.	n.a.	n.a.	n.a.	n.a.	n.a.	n.a.	18.9	22.5	23.9	24.0
Italy	n.a.	24.6	25.6	23.4	23.9	25.8	25.7	27.0	35.2	44.1	40.1
Cyprus	n.a.	n.a.	51.3	56.3	63.0	48.1	37.8	40.1	46.3	50.6	51.0
Malta	n.a.	n.a.	17.2	16.8	16.5	13.9	17.5	25.1	28.1	32.1	34.9
Portugal	n.a.	56.9	62.1	60.6	64.9	56.0	44.3	49.7	44.8	43.1	44.6
Slovenia	n.a.	n.a.	6.6	8.7	11.4	14.3	11.5	13.1	12.4	17.3	13.1
Serbia	n.a.	n.a.	n.a.	n.a.	n.a.	n.a.	n.a.	n.a.	n.a.	n.a.	30.0

Source Eurostat—SILC [ilc_mdes01]

Table 9.25 Share of total population living in a dwelling with a leaking roof, damp walls, floors or foundation, or rotten window frames or floor

Geo/Year	2003	2004	2005	2006	2007	2008	2009	2010	2011	2012	2013
EU-28	n.a.	n.a.	n.a.	n.a.	n.a.	n.a.	n.a.	16.1	15.5	15.1	15.7
Greece	21.0	19.7	20.9	20.4	19.4	18.6	17.6	17.1	15.3	14.7	14.0
Spain	n.a.	20.6	17.9	18.5	19.0	16.8	18.3	21.8	16.1	12.0	16.7
France	n.a.	15.0	12.3	12.0	13.7	12.8	12.6	12.5	10.9	12.8	13.2
Croatia	n.a.	n.a.	n.a.	n.a.	n.a.	n.a.	n.a.	19.8	15.2	13.3	13.1
Italy	n.a.	23.1	22.6	21.9	21.1	20.4	20.5	20.0	23.2	21.4	23.1
Cyprus	n.a.	n.a.	36.3	32.9	30.1	26.5	29.6	30.0	29.8	30.0	31.1
Malta	n.a.	n.a.	7.9	7.4	5.4	6.9	9.8	12.1	10.2	10.8	11.8
Portugal	n.a.	20.3	19.4	18.7	19.5	18.9	19.7	21.9	21.3	22.0	31.9
Slovenia	n.a.	n.a.	19.0	21.6	17.5	30.2	30.6	32.4	34.7	31.5	27.0
Serbia	n.a.	n.a.	n.a.	n.a.	n.a.	n.a.	n.a.	n.a.	n.a.	n.a.	21.6

Source Eurostat—SILC [ilc_mdho01]

Table 9.26 Share of population below 60 % of median equivalized income living in a dwelling with a leaking roof, damp walls, floors or foundations, or rotting window frames or floors

Geo/Year	2003	2004	2005	2006	2007	2008	2009	2010	2011	2012	2013
EU-28	n.a.	n.a.	n.a.	n.a.	n.a.	n.a.	n.a.	25.7	24.3	23.5	23.5
Greece	30.1	30.0	29.1	28.6	26.5	26.9	27.4	26.1	21.2	21.0	20.8
Spain	n.a.	26.8	23.6	25.8	28.2	23.4	27.3	29.3	23.3	17.9	21.7
France	n.a.	22.3	20.6	19.6	22.3	22.6	22.8	20.9	19.6	22.1	23.1
Croatia	n.a.	n.a.	n.a.	n.a.	n.a.	n.a.	n.a.	30.2	24.8	19.2	21.5
Italy	n.a.	32.7	30.7	30.3	28.1	26.7	27.9	27.9	29.2	30.1	30.5
Cyprus	n.a.	n.a	44.0	41.5	39.7	33.2	34.2	36.1	40.4	34.6	35.1
Malta	n.a.	n.a	11.9	8.4	7.3	7.9	12.8	15.6	11.5	12.4	11.5
Portugal	n.a.	30.4	29.7	28.6	31.5	26.1	28.6	31.1	28.4	28.5	40.1
Slovenia	n.a.	n.a	30.4	32.7	30.4	44.5	40.9	47.0	47.2	46.1	39.6
Serbia	n.a.	n.a	n.a	n.a	n.a	n.a	n.a	n.a	n.a	n.a	33.1

Source Eurostat—SILC [ilc_mdho01]

9.4.3 Difficulties Paying the Bills

Table 9.27 shows the percentage of the population in arrears on utility bills, Table 9.28 shows the same data restricted to populations below 60 % of the median equivalized income.

In 2013, 35.2 % of the total Greek population had arrears on utility bills, with 30.4 % in Croatia, and 21.9 % in Cyprus. In the same period, Spain (8.3 %), France (6.2 %), and Portugal (8.2 %) were below the average of EU-28 (10.1 %), although Italy (12 %), Malta (11.4 %), and Slovenia (19.7 %) were above. The only data for Serbia show 36.7 %.

Analyzing the trend for 2005 to 2013, Cyprus is the country with the highest increase (125.8 %), followed by Spain (84.4 %), Portugal (64 %), Slovenia (56.3 %), and Malta (44.3 %). Italy increased by 14.3 %, while France was the only country that decreased (−13.9 %).

Considering only people below 60 % of the median equivalized income, in 2013 61.4 % of the Greek population experienced arrears on utility bills, 53.6 % in Serbia, 47.2 % in Croatia, 39.9 % in Cyprus, and 35.7 % in Slovenia. Spain (18.4 %), Malta (19.3 %), Portugal (19.4 %) and France (19.8 %) were below the average of EU-28 (22.9 %); Italy achieved 26.3 %. Between 2005 and 2013, Spain (135.9 %), Cyprus (175.2 %), and Portugal (100 %) registered the highest increase in arrears on utility bills, while France (8.2 %) and Italy (5.6 %) had the lowest.

9.4.4 Population Living in Uncomfortable Dwellings

In order to complete the analysis, it is useful to consider the percentage of the population living in a dwelling not comfortably cool during the summer or warm during the winter, by material deprivation status (Table 9.29).[12]

This indicator is different than the one already discussed about the inability to keep homes adequately warm, because it turns attention to another concept:

[12]The definition of material deprivation is based on the inability to afford a selection of items that are considered to be necessary or desirable, namely: having arrears on mortgage or rent payments, utility bills, hire purchase instalments or other loan payments; not being able to afford one week's annual holiday away from home; not being able to afford a meal with meat, chicken, fish (or vegetarian equivalent) every second day; not being able to face unexpected financial expenses; not being able to buy a telephone (including mobile phone); not being able to buy a color television; not being able to buy a washing machine; not being able to buy a car; or not being able to afford heating to keep the house warm. The material deprivation rate is defined as the proportion of persons who cannot afford to pay for at least three out of the nine items specified above, while those who are unable to afford four or more items are considered to be severely materially deprived. Source: Eurostat. Social inclusion statistics.

Table 9.27 Arrears on utility bills expressed as percentage of total population

Geo/Year	2003	2004	2005	2006	2007	2008	2009	2010	2011	2012	2013
EU-28	n.a.	n.a.	n.a.	n.a.	n.a.	n.a.	n.a.	9.1	9.1	10.0	10.1
Greece	30.5	24.7	26.5	25.0	15.7	15.9	18.9	18.8	23.3	31.8	35.2
Spain	n.a.	5.0	4.5	4.3	4.6	4.6	6.3	7.5	5.7	7.5	8.3
France	n.a.	9.0	7.2	6.6	6.4	6.1	7.5	7.1	7.1	6.7	6.2
Croatia	n.a.	n.a.	n.a.	n.a.	n.a.	n.a.	n.a.	28.0	27.5	28.9	30.4
Italy	n.a.	10.4	10.5	10.9	10.4	13.6	11.0	10.5	12.1	11.7	12.0
Cyprus	n.a.	n.a.	9.7	9.9	10.0	7.5	13.3	16.3	16.9	18.4	21.9
Malta	n.a.	n.a.	7.9	6.5	7.2	7.3	7.9	6.8	8.6	10.1	11.4
Portugal	n.a.	5.3	5.0	4.8	5.2	3.7	6.1	6.4	6.7	6.3	8.2
Slovenia	n.a.	n.a.	12.6	11.3	11.4	14.2	16.9	18.0	17.3	19.3	19.7
Serbia	n.a.	n.a.	n.a.	n.a.	n.a.	n.a.	n.a.	n.a.	n.a.	n.a.	36.7

Source Eurostat—SILC [ilc_mdes07]

Table 9.28 Arrears on utility bills expressed as percentage of population below 60 % of median equivalized income

Geo/Year	2003	2004	2005	2006	2007	2008	2009	2010	2011	2012	2013
EU-28	n.a.	n.a.	n.a.	n.a.	n.a.	n.a.	n.a.	20.1	20.5	22.1	22.9
Greece	48.5	40.8	46.3	49.5	47.3	29.5	37.1	38.0	41.3	54.4	61.4
Spain	n.a.	10.0	7.8	8.2	9.2	8.4	13.2	16.3	12.4	17.9	18.4
France	n.a.	22.5	18.3	17.5	16.6	18.7	20.6	23.3	22.2	17.8	19.8
Croatia	n.a.	n.a.	n.a.	n.a.	n.a.	n.a.	n.a.	32.3	38.0	43.7	47.2
Italy	n.a.	23.8	24.9	24.2	24.3	26.5	23.1	23.5	24.8	24.5	26.3
Cyprus	n.a.	n.a.	14.5	16.7	13.8	10.6	21.6	23.5	21.1	25.9	39.9
Malta	n.a.	n.a.	12.2	13.1	13.4	12.0	16.2	14.4	16.2	19.4	19.3
Portugal	n.a.	9.7	9.7	7.5	10.4	7.5	12.8	13.0	14.6	14.6	19.4
Slovenia	n.a.	n.a.	23.0	21.1	23.3	24.8	28.9	33.6	32.5	37.5	35.7
Serbia	n.a.	n.a.	n.a.	n.a.	n.a.	n.a.	n.a.	n.a.	n.a.	n.a.	53.6

Source Eurostat—SILC [ilc_mdes07]

Table 9.29 Share of population living in a dwelling not comfortably cool during summer and not comfortably warm during winter time by material deprivation status, 2012

	Dwelling not comfortably cool during summer		Dwelling not comfortably warm during winter	
	Total population	Severely materially deprived population	Total population	Severely materially deprived population
EU-28	19.2	38.4	12.9	41.6
Greece	34	67.5	26.2	54.4
Spain	25.6	49.4	17.7	56.4
France	18.9	33.7	17.7	52.5
Croatia	24.2	42.7	7.8	22.7
Italy	26.3	44.9	16	48.9
Cyprus	29.6	57.2	21.9	51.4
Malta	35.4	71.3	28.8	76.7
Portugal	35.7	47.7	46.6	70
Slovenia	17.3	33.6	5.1	27.6

Source Eurostat 2012 ad hoc module "Housing Conditions" (HC070)

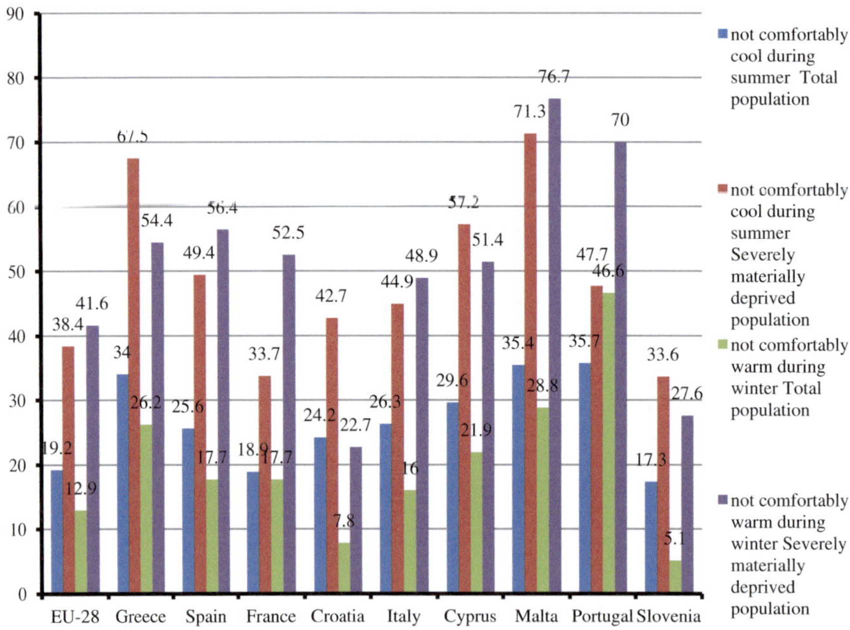

Fig. 9.2 Development of household situation by country. Analysis of comfort conditions

there are more people—at least in the Mediterranean countries considered in this study—who feel discomfort due to overheating problems in summer than underheating problems in winter (with the exception of Portugal). Considering the severely materially deprived population, the same happens in Greece, Croatia, Cyprus, and Slovenia. Of course, it should be necessary to consider the same indicator for the summer (as the indicator of inability to keep the homes adequately warm) in order to build better understanding of the phenomena; nevertheless, it makes sense to affirm that energy poverty in Mediterranean countries has an important component due to overheating problems in dwellings.

Finally, during the last decade there has been a general, severe worsening in the quality of life and housing. Some positive data have come from the share of the total population living in dwellings with a leaking roofs or damp walls, floors or foundations, or rotten window frames or floors, as some countries are registering a general reduction over the last years. As was thought, data for people at risk of poverty are even worse, with the effects of financial crisis striking the weakest social stratum (Fig. 9.2).

9.5 Conclusions

The Energy Performance of Buildings Directive (EPBD, 2002/91/EC) was implemeted in 2002 and it is the main legislative instrument in force at EU level covering the energy efficiency of buildings. The Directive has been recast in 2010 (EPBD recast, 2010/31/EU) with more ambitious provisions. Through the EPBD introduction, requirements for certification, inspections, training or renovation are now imposed in Member States (BPIE 2011). Moreover, the new directive requires all new buildings to be nearly zero-energy by the end of 2020. All new public buildings must be nearly zero-energy by 2018.

The household sector of the Mediterranean countries is weak and needs to be updated to achieve minimum goals requested by the European directives. With the exception of Cyprus and Spain, with, respectively, 36.4 and 20 % of their buildings having been built after 2000, the buildings are too old to achieve minimum comfort and energy consumption standards. Moreover, even in the case of buildings built after 2000, in many cases the national standard did not match the minimum level of performance requested by European directives[13] (Fig. 9.3). These data show the necessity to update national building codes and energy politics, focusing them not only on the performance of new constructed buildings, but mainly on the need to improve existing dwellings' fitness.

[13]For more information about the progresses of national building codes to achieve nZEb, check the pages http://ec.europa.eu/energy/en/topics/energy-efficiency/buildings/nearly-zero-energy-buildings and the document "Implementing the Energy Performance of building Directivee—Featuring Country Reports 2012. Concerted Action Energy performance of Builidngs". Web page: http://www.epbd-ca.org/Medias/Pdf/CA3-BOOK-2012-ebook-201310.pdf.

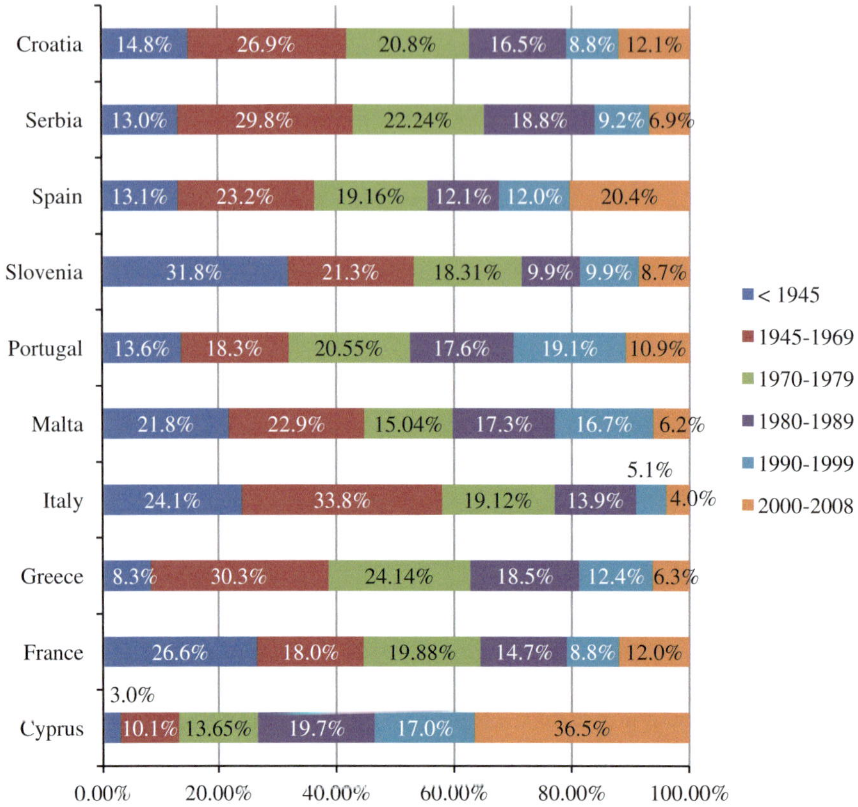

Fig. 9.3 Dwellings according to construction date. *Source* ENTRANZE

The improvement has to work on different concepts: the use of the dwelling, energy performance, environmental protection, use of renewable energies and physical accessibility, and integrating social, energy, and climate strategies. Technical solutions must be adapted to the potential of the temperate climate of Mediterranean countries and its restrictions, must be cost-effective and mainly based on synergies between passive strategies, renewable energy sources and active systems, combining measures to achieve energy performance and indoor environment quality (Gaitani 2014) by a holistic approach. In addition, concepts such as life cycle analysis and embodied energy must be known by the building professionals. Moreover, it is necessary to think of a building as part of a more complex system, its environment and morphology, where each part of that system has a function and mutual influence (Santamouris 2001).

Building retrofitting is strictly related to urban renovation, a complex task that involves many scales (building, neighborhood, city, territory) and many stakeholders (political actors, developers, urban planners, architects, communities, etc.)

(Sauer 2013). To achieve these goals, energy efficiency strategies for the residential sector must be multidimensional, fully synergized with housing policy, and incorporating the principles of equity, access, and balanced geographical development (Golubchikov and Deda 2012). Under these conditions, energy efficiency improvement and energy savings can have several benefits (Ryan 2012):

(a) Impact on health and well-being
(b) Poverty alleviation: energy affordability and access
(c) Increased disposable income.

The effects of the financial crisis, such as energy poverty, housing deprivation, etc., have a greater impact on low-income housing in the Mediterranean area. The decreasing household energy consumption in this case cannot be seen as a positive effect as long as it is linked to a general condition of poverty and marginalization. Because of it, new policies and financial mechanisms (on the European, national, and regional levels) should be developed to counteract these effects and permit wider access to available and affordable financing, improving the environmental quality in low-income households, and as a measure to retrofit the existing stock towards net zero energy buildings (CECODHAS 2011).

References

BPIE (2011) Europe's buildings under the microscope. A country-by-country review of the energy performance of buildings. Study prepared by the Buildings Performance Institute Europe. ISBN: 9789491143014, http://www.bpie.eu/uploads/lib/document/attachment/20/HR_EU_B_under_microscope_study.pdf

BPIE (2014) Alleviating fuel poverty in the EU. Investing in home renovation, a sustainable and inclusive solution. Study prepared by the Buildings Performance Institute Europe. http://bpie.eu/uploads/lib/document/attachment/60/BPIE_Fuel_Poverty_May2014.pdf

CECODHAS (2011) Housing Europe review 2012. The nuts and bolts of European social housing systems. Published by CECODHAS Housing Europe's Observatory, Brussels (Belgium). http://www.housingeurope.eu/file/38/download

Gaitani N (2014) Nearly zero energy buildings (nZEB) status report in Mediterranean countries. ZEMedS Project, Executive summary. http://www.zemeds.eu/sites/default/files/nZEB%20Status%20Report%20in%20MED%20Countries%20ExS_EN_0.pdf

Golubchikov O, Deda P (2012) An integrated policy framework for energy efficient housing. Energy Policy 41:733–741

Liddell C, Morris C (2010) Fuel poverty and human health. Energy Policy 38:2897–2997

OECD/IEA (2014) Energy efficiency indicators: essentials for policy making. OECD/IEA, International Energy Agency. http://www.iea.org/publications/freepublications/publication/IEA_EnergyEfficiencyIndicators_EssentialsforPolicyMaking.pdf

Ryan L, Campbell N (2012) Spreading the net: the multiple benefits of energy efficiency improvements. OECD/IEA, International Energy Agency. http://www.iea.org/publications/insights/ee_improvements.pdf

Santamouris M (ed) (2001) Energy and climate in the urban built environment. James and James Science Publishers, London

Sauer B, Arraiz-Garcia M, Pedrola-Vidal B (2013) Europe: an opportunity to regenerate cities in a holistic manner. Exploring synergies of integrated approaches in urban design. In: Proceedings of the conference "High Performance Buildings—Design and Evaluation Methodologies", 24–26 June 2013, Brussels, EU Sustainable Energy Week. http://iet.jrc.ec.europa.eu/energyefficiency/sites/energyefficiency/files/final_handouts_printer.pdf
Walker R, McKenzie P et al (2014) Estimating fuel poverty at household level: an integrated approach. Energy Build 80:469–479

Chapter 10
Office Buildings/Commercial Buildings: Trends and Perspectives

Dionysia Denia Kolokotsa

Abstract Zero and Positive Energy Buildings are buildings that use their own locally-installed (renewable) energy-generating sources to produce, over a certain period of time (e.g., on an annual basis), more power than they consume. The aim of this chapter is to present the trends and perspectives for office buildings to become zero energy in the coming years. Innovative technologies and energy management practices are presented. The role of case studies and shining examples for the Mediterranean region are revealed.

10.1 Introduction

In recent years, concerns about the shortage of traditional energy reserves, the rising demand for energy (fuelled partly by the development of new economies), and obvious concerns regarding the effect of irrational energy use and human activity on the environment, have made the topic of energy savings and efficiency almost universal.

Buildings today account for 40 % of the world's primary energy use and 24 % of the greenhouse gas emissions (IEA 2011). It is estimated that 26 % of the total final energy consumption in Europe was consumed in residential buildings, and 13 % in non-residential buildings (Berggren et al. 2012). The tertiary sector (non-residential building and agriculture) is one of the fastest-growing energy demand sectors, and it is expected to be 26 % higher in 2030 than it was in 2005, compared to only 12 % higher for residential buildings (Boyano et al. 2013). This estimation makes it necessary to investigate the use of energy in the non-residential buildings sector, but the difficulty is in finding the way to achieve smart consumption strategies without causing negative consequences of people's standard of living and productivity.

D.D. Kolokotsa (✉)
Technical University of Crete, Technical University Campus,
Office:K2.114, Kounoupidiana, GR 73100 Chania, Greece
e-mail: dkolokotsa@enveng.tuc.gr

S.-N. Boemi et al. (eds.), *Energy Performance of Buildings*,
DOI 10.1007/978-3-319-20831-2_10

On the other hand, climate change is imminent in many parts of Europe—in particular, southern Europe is suffering from heat and drought extremes that have a negative impact on energy efficiency. Energy consumption is steadily increasing in various countries of southern European, partly due to the increase of urbanization and use of air conditioning in cities. Various energy effiicient technologies are mature enough and can be used for decrease of energy use and impovement of indoor comfort conditions. These technologies can be put into the following basic categories:

- Measures for the improvement of the building's envelope (addition or improvement of insulation, change of color, placement of heat-insulating door and window frames, increase of thermal mass, building shaping, superinsulated building envelopes, etc.).
- Incorporation of high-efficiency heating and cooling equipment, e.g., air conditioning equipment with higher EER, high-efficiency condensing boilers, etc.
- Use of renewables (solar thermal systems, buildings' integrated photovoltaics (PV), hybrid systems, etc.).
- Use of "intelligent" energy management, i.e., advanced sensors, energy control (zone heating and cooling), and monitoring systems.
- Measures for the improvement of indoor comfort conditions in parallel with minimization of the energy requirements (increase in ventilation rate, use of mechanical ventilation with heat recovery, improvement of boilers and air conditioning efficiency, use of multi-functional equipment, i.e., integrated water heating with space cooling, etc).
- Use of energy-efficient appliances, compact fluorescent lighting, LED, etc.

Furthermore, zero energy buildings (ZEB) have great potential in helping to alleviate the problems related to the deterioration of the environment and the depletion of energy sources.

The aim of this chapter is to analyze the ZEB' perspective in the recent highly demanding environment for the construction and retrofitting sectors. The challenges for ZEB in southern Europe are discussed and revealed. Two offices (ZEB case studies) are analyzed, showing the future research prospects.

10.2 The Zero Energy Buildings' Perspectives in the Mediterranean Region

In the building sector, net energy is considered to be the balance between the energy consumption in a building and the energy produced by its renewable energy systems. In this respect, ZEB refers to buildings that are connected to the energy grids, and the ZEB is more general and may include autonomous buildings (Li et al. 2013). Several countries have proposed future building energy targets to establish ZEBs, such as the Building Technology Program of the US Department of Energy, and the EU Directive on Energy Performance of Buildings (Sartori et al. 2012). Achieving a Net ZEB involves two main strategies: first, by minimizing the required energy through energy-efficient measures, and then by meeting the minimal energy needs by adopting renewable energy.

Following the research activities concerning ZEB, it is well understood that even if the most effective energy-efficient measures are applied, energy will still be required in order to power the daily running of a building. The main difference between energy-efficient buildings and ZEBs is that in ZEBs, the minimized energy is covered by the use of renewable energy or low-carbon energy production technologies. The most widely used energy technologies that are adopted for on-site applications are (Li et al. 2013):

- PV and building-integrated photovoltaic (BIPV)
- Wind turbines integrated in buildings or sites
- Solar thermal (solar water heaters, concentrated solar power)
- Heat pumps

From studying the literature, the following shortcomings have been discovered:

- Large discrepancies in targets and fulfillments among the various countries worldwide. For example in Europe, while the Energy Performance of Buildings Directive sets the framework definition of net zero energy building (nZEBs), member states are responsible for reporting on the detailed application in practice of that definition (i.e., reflecting their national, regional, or local conditions). For residential buildings, the maximam primary energy consumption for ZEB ranges from 33 kWh/m^2/year in Croatia to 95 kWh/m^2/year in Latvia, with most of the countries (Belgium, France) aiming at 45 or 50 kWh/m^2/year.
- Lack of clear definition, since depending on the project's objectives and goals, the proposed definitions have different characteristics. However, through research, exchange of ideas, and discussions during the past few years, a common view has been emerging that a widely accepted definition of nZEB should be a definition framework that contains various elements, such as boundaries, metrics, criteria, etc. (see Fig. 10.1). In order to make a clear balance calculation for the net zero goal, a boundary needs to be clarified for the building system with on-site renewable. Inside this boundary, the building system consumes delivered energy, such as electricity and natural gas, from on-site renewable and energy grids, and output energy back to the grid when the renewable energy power (REP) system generates excess electricity.
- Significant cost burden: Several nZEB buildings and settlements are currently in operation or in development stages. (See the list in http://www.iea-shc.org/data/sites/1/publications/T40A52-DC-TR1-30-Net-ZEBs.pdf, where energy-efficient technologies, combined with solar, wind, and other sources of energy are used to attain "near zero-energy" behavior (Comodi et al. 2014; Groezinger et al. 2014; Noguchi et al. 2008; Pyloudi et al. 2014)). Based on international estimations, the additional initial investment costs for residential buildings range from €200–700/m^2, depending on the passive technologies used and the active systems' sizing and design (Provata et al. 2015).

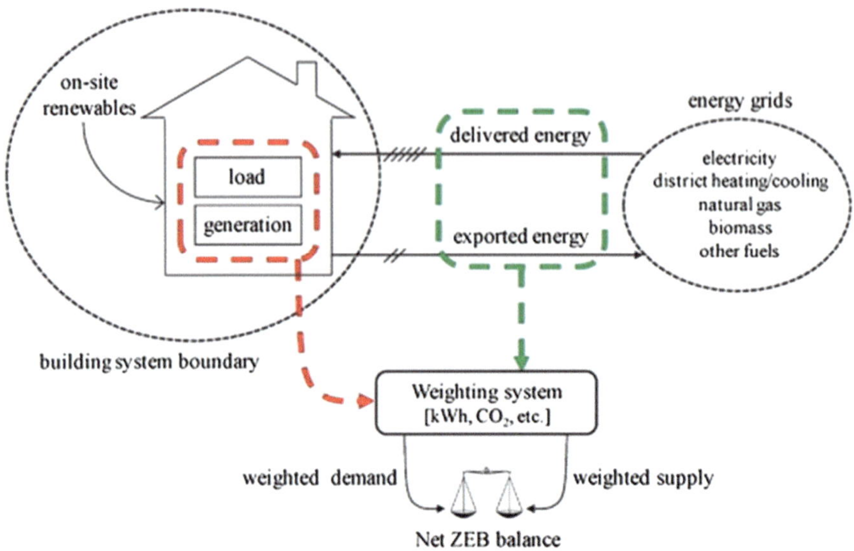

Fig. 10.1 System structure and basic elements of nZEB (Deng et al. 2014)

- Delays in smart/micro grid integration on the neighborhood and district levels. Smart/micro grids can create a revolution in the building sector. The accumulated experience of the last decades has shown that the hierarchical, centrally controlled grid of the twentieth century is ill-suited to the needs of the twenty-first century. The smart micro grid can be considered to be a modern electric power grid infrastructure for enhanced efficiency and reliability, thus ensuring effective optimization of resources.

In addition, Mediterranean countries are facing a number of challenges related to climate change and weather variations. The temperature increase has a serious impact on the energy consumption of the urban buildings, by increasing the energy and the electric power necessary for cooling needs (Kapsomenakis et al. 2013). Forecasts of the future ambient temperatures in the specific geographic area, as well as projections of the expected energy consumption in the building sector, show an important temperature increase followed by a serious increase in energy consumption (Asimakopoulos et al. 2012).

10.3 Office Buildings as ZEB in the Mediterranean Region

In this section, two selected ZEB office buildings in the Mediterranean region are analyzed and discussed.

10.3.1 Office Building in Crete, Greece

The office building of Environmental Engineers at the Technical University of Crete is situated on the university's campus in the suburb of Chania, Greece, and its coordinates are 24°A (longitude) and 35°B (latitude). The building was constructed in 1997 and is a two-story building with a total height of 11 m and total length of 48 m. The ground floor (605 m^2) and the first floor (610 m^2) consist of offices, laboratories, stairways, elevators, toilets, hallways, and PC labs; on the second floor (421 m^2) is the mechanical equipment, as no employees work there. The electricity demands of the building are met by the grid (Pyloudi et al. 2014). The building characteristics are depicted in Fig. 10.2.

The building is occupied by 56 persons performing seated, light work, such as typing. Its operating schedule is from 8:00 a.m. until 6:00 p.m. every weekday. The ventilation rate depends on the size of each zone, the number of persons, and the use of each zone. The infiltration rate is considered a constant value of 0.5 air changes per hour (ach). The energy consumption is approximately 37,600 kWh (29 kWh/m^2) for cooling and 10,400 kWh (8 kWh/m^2) for heating. The electricity consumption for lighting and computers is 23,300 kWh (18 kWh/m^2) and 26,800 kWh (20.7 kWh/m^2), respectively.

In order to reach the ZEB objective, the following energy efficiency measures are proposed:

- Increasing the insulation on the walls and ceilings, replacing windows, and applying cool paint to the external walls (Scenario 1).
- Replacing lights and power supply of personal computers with more efficient ones (Scenario 2).

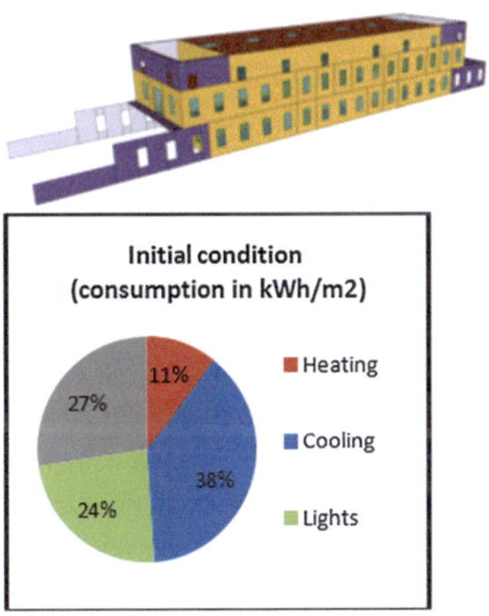

Characteristics of the building	
General information	
Location	Chania. Crete. Greece
Building type	Office building
Floors	3 (The ground and the first floor are in operation and the second floor is used for the installation of HVAC systems
Surface area	1296 m2 (including the surface area of ground and first floor)
Operation hours	08:00 - 18:00 Monday - Friday
Orientation	SE
Building envelope	
Walls	Three-layer (plasterboard, polysterine, concreteboard) outer walls of total U-value = 0.915W/m2K
Roof	Two-layer (concrete iron, polysterene) ceiling of total U-value = 1.449W/m2K
Windows	Double glazing windows with U-value = 2,83 W/m2K)
Floor	Two-layer (polysterene and marble) with U-value = 0,945 W/m2K)
Building services	
Cooling system	VRV system (6.44 kW) with separate mechanical ventilation support
Heating system	Same with cooling system

Fig. 10.2 The details of the building

- Adjusting the ventilation rates in offices based on the technical guidelines (Scenario 3).
- Combining all scenarios (Scenario 4).

The envelope improvements are tabulated in Table 10.1, and the energy efficiency achieved is illustrated in Fig. 10.3.

In the initial condition, (see 2nd collumn of the energy consumption table in Fig. 10.2) the greater proportion (38 %) of the energy is consumed for cooling needs—which is logical, considering that the building is situated in Chania, Greece, which is in the "A category" (TEE 2010). The smaller proportion (11 %) is consumed for heating needs, with approximately the same proportion of the energy used for lighting and computers (24 and 27 %, respectively). The changes in the building envelope (Scenario 1) cause an important reduction in the consumption for cooling, which reaches 69 %.

The combination of all the previous scenarios (presented in Scenario 4) shows that the major reduction can be detected in cooling needs, in which there is a reduction of almost 83 %. The reduction of lighting, heating and computer needs follow, with 47, 38 and 30 % rates, respectively. In general, the total energy consumption is reduced from 76 to 34 kWh/m^2, and the energy saving reaches 55 %.

The minimized energy is covered by the installation of PVs and/or roof-mounted wind turbines. The energy production of the renewables is depicted in Fig. 10.4. The minimized energy, which needs to be covered by renewable energy sources, is 44,000 kWh/year. The PV panels installed on the roof of thc building produce approximately 24,000 kWh/year, and the wind turbines Roof-mounted produce 20,000 kWh annually. Since no energy storage is anticipated, it is necessary to have 23,000 kWh/year grid purchases.

Thus, the energy production that comes from renewable energy sources has the same quantity as the energy consumption of the building. Subsequently, applying the "best-case scenario" (energy-efficient measures in combination with renewable energy sources), the retrofitting of the present building in order to become a nZEB is achieved.

10.3.2 Laboratory Building in Cyprus

The building belongs to the Cyprus Institute and hosts the laboratory and administration spaces of the Energy, Environment, and Water Research Center. It is located in Aglantzia, Nicosia (the capital of Cyprus). It has a total space of 2130 m^2 and includes a mixture of laboratory spaces, offices, conference/computer rooms, and auxiliary common areas. It consists of the basement, ground floor, 1st floor and 2nd floor. A view of the building is in Fig. 10.5. Almost 1450 m^2 are

Table 10.1 Measures of scenario 1 (building envelopes)

		Walls	Roofs	Windows			Walls	Roofs	Windows
Reference building	Polystyrene insulation (cm)	2	1	–	Envelope improvements	Polystyrene insulation (cm)	5	3	–
	Total U-value (W/m^2)	0.915	1.449	2.83		Total U-value (W/m^2)	0.436	0.671	1.04
	g-value	–	–	0.755		g-value	–	–	0.227
	Solar reflectance (%)	10	–	–		Solar reflectance with cool paint (%)	89	–	–

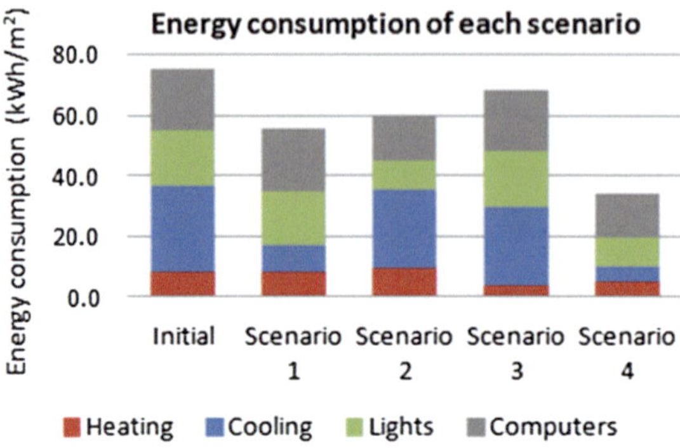

	Energy consumption (kWh/m²)				
	Initial	Scen. 1	Scen. 2	Scen. 3	Scen. 4
Heating	8.0	8.0	9.5	4.0	5.0
Cooling	29.0	9.0	26.0	26.0	5.0
Lights	17.8	17.8	9.4	17.8	9.4
Computers	20.7	20.7	14.5	20.7	14.5
Total	75.5	55.5	59.4	68.5	33.9
Energy Saving		26%	21%	9%	55%

Fig. 10.3 The reduction of energy consumption applying different scenarios

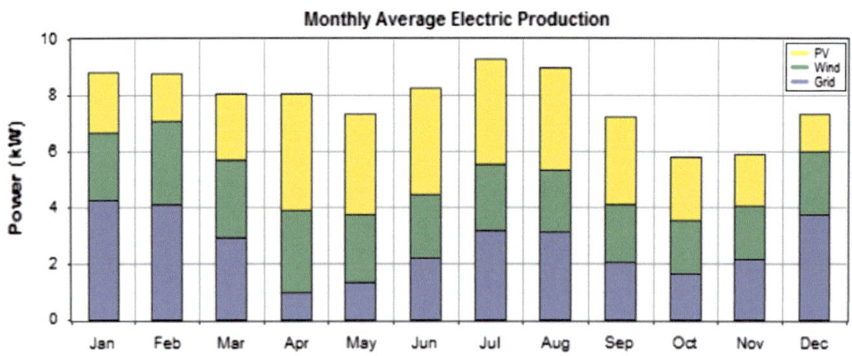

Fig. 10.4 Monthly average electricity production

Fig. 10.5 The EEWRC laboratory building

used as laboratory spaces, while 124 m^2 are offices and seminar rooms. In order to achieve the zero energy goal for the specific building, the following energy efficiency scenarios are followed:

- Various types of low emissivity glazing with and without argon filling (Scenarios 2, 3, 4, and 5) to reduce solar gains and increase the thermal resistance of the building skin.
- Various levels of wall insulation (Scenario 6 with *U* value equal to 0.61 W/m^2 K, Scenario 7 with *U* value equal to 0.43 W/m^2 K, and Scenario 8 with *U* value equal to 0.32 W/m^2 K).
- Increased insulation in horizontal elements (Scenario 9 with *U* value equal to 0.47 W/m^2 K, Scenario 10 with *U* value equal to 0.32 W/m^2 K and Scenario 11 with *U* value equal to 0.23 W/m^2 K).
- Intensive shading to reduce the solar gains (Scenario 12: In the original design, it was done with the provision of external vertical blinds. It simulated the energy performance of the building, assuming that the blinds are placed two-thirds of the distance they had been in the original architectural design. Scenario 13: The placing of the external vertical blinds remains the same as in the original architectural design, but the width has been doubled. Scenario 14: External horizontal blinds to the south façade. Scenario 15: Shading to all east, south and west openings).
- Use of ceiling fans (Scenario 16).

- Waste heat recovery (Scenario 17). In order to minimize ventilation losses, the use of waste heat recovery ventilators with a nominal efficiency of 60 % is studied.
- Free cooling (Scenario 18): In order to minimize the required cooling load, the use of the free cooling technique with a ventilation rate of up to five ach is studied.
- Night ventilation: With 5, 10, and 15 ach corresponding to scenarios 19, 20, and 21, respectively.
- Combined Scenario (Scenario 22): The combined scenario involves low emissivity glazing filled with Ar (Scenario 5), building elements with sufficient thermal insulation (Scenarios 7 and 10), external shading of all south-, east- and west-facing openings (Scenario 15), ceiling fans (Scenario 16), waste heat recovery ventilators as in Scenario 17, and night ventilation with up to five ach (Scenario 19).
- Combined Scenario (Scenario 23): This combined scenario involves low emissivity glazing filled with Ar (Scenario 5), building elements with sufficient thermal insulation (Scenarios 7 and 10), external shading of all south-, east- and west-facing openings (Scenario 15), with shading coefficient 85 %, ceiling fans (Scenario 16), waste heat recovery ventilators as in Scenario 17, and night ventilation of up to five ach (Scenario 19).
- Combined Scenario (Scenario 24): This combined scenario involves low emissivity glazing filled with Ar (Scenario 5), building elements with sufficient thermal insulation (Scenarios 7 and 10), external shading of all south-, east- and west-facing openings (Scenario 15) with shading coefficient 85 %, ceiling fans (Scenario 16), waste heat recovery ventilators as in Scenario 17, night ventilation of up to five ach (Scenario 19), and demand control ventilation.

The entire energy system is monitored through a "Building Energy Management" that controls the heating and cooling devices as well as the ventilation and lighting in the building. Additionally, smart sensors are installed in various parts of the building to monitor the energy performance and indoor comfort conditions. Detailed thermal simulations using the TRNSYS software were carried out to calculate the performance of the previously defined energy conservation scenarios.

The heating load of the base case scenario (conventional building), is calculated close to 57 kWh/m^2/year, and the corresponding cooling load around 87 kWh/m^2/year. The final energy conservation of the building for the combined scenarios ranges from 67 to 77 % for the heating performance and from 73 to 77 % for the cooling. The combination proposed in Combined Scenario 22, without the use of the gas-filled windows, is the one followed and installed in the building (Figs. 10.6 and 10.7).

The renewable energy used to cover the minimized energy demands consist of:

- Monocrystalline PVs placed on the roof of the building. The total installed capacity ranges from 13 to 15 kW.
- A solar cooling system using Fresnel collectors with thermal capacity of the solar concentrating collectors close to 100 kWth.

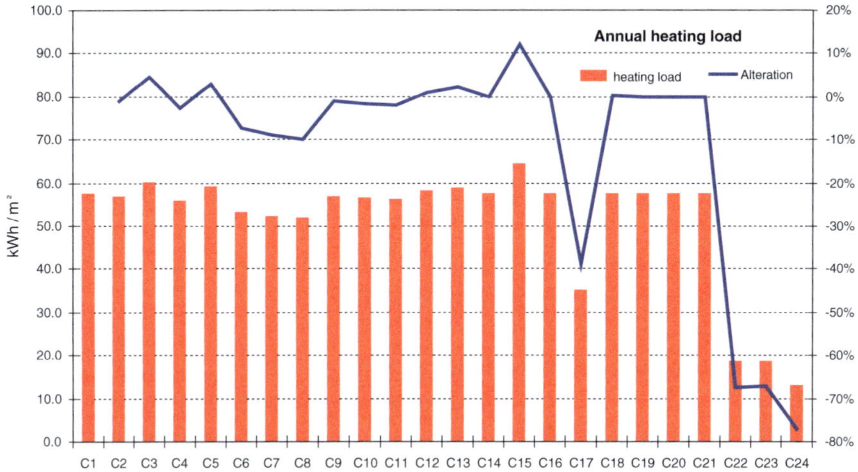

Fig. 10.6 Predicted annual heating load and energy conservation for the scenarios

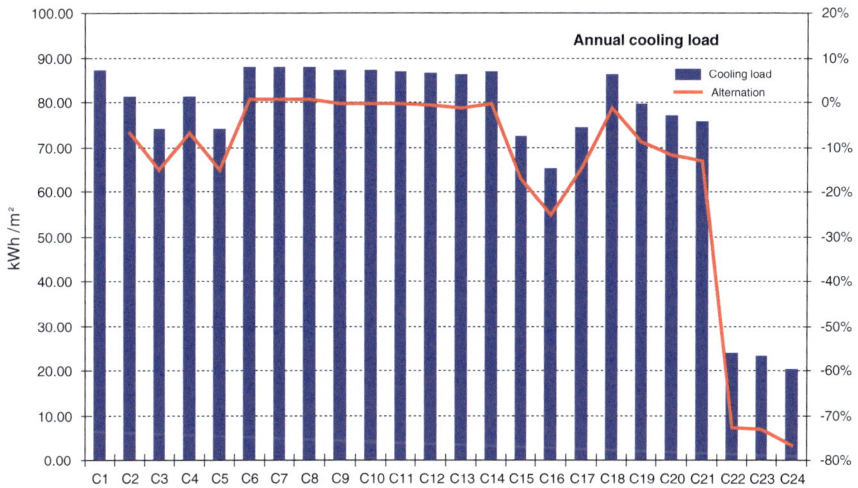

Fig. 10.7 Predicted annual cooling load and energy conservation for all the scenarios

Concerning the Fresnel concentrating collectors, the modeling of the system is done using TRNSYS software. At first, the Fresnel collectors heat up the fluid concentrating solar radiation, then the heating production is stored in storage tanks. Moreover, a heat exchanger transfers heat from one fluid to a second one, which is stored in a secondary tank before being transferred to the absorption chiller as the heating medium. Chilled water leaves the absorption chiller and is stored in a storage tank and used for covering the cooling loads of the building. Finally, a heat dissipation system is used for cooling the water leaving the absorption chiller.

Concerning the operation of the absorption chiller, it is calculated that during the summer, it performs better than in the winter, when solar radiation does not allow the system to start working, since the heat source does not reach the required temperature to start operations (175 °C). As can be seen, the cooling power when the absorption chiller is operating, is at a minimum of 25 kW and can reach up to 41 kW.

10.4 Conclusions and Future Prospects

The energy consumption of the buildings is quite high and may increase considerably in the future because of the improving standards of life and increasing use of air conditioning. For the Mediterranean region, the temperature increase leads to an excessive use of air conditioning, thus increasing peak electricity demands and making the zero energy goal much more difficult to be reached.

The main priority in reaching the ZEB objective in the Mediterranean region is the minimization of the energy demand for cooling as well as the peak load shaving and shedding during the summer. Towards this priority there are numerous energy efficiency technologies that support passive cooling, ranging from superinsulated envelopes with increased thermal mass to solar control and heat dissipation techniques such as ventilative cooling, evaporative cooling and earth cooling (Santamouris and Kolokotsa 2013). Advanced control strategies and predictive optimization techniques can contribute to the load shaving and shedding, especially during heat waves where the possibilities of blackouts are increased (Kolokotsa et al. 2011).

One significant ingredient for ZEB is the proper sizing of renewable energy sources to produce the energy needed by the buildings. It has been observed in design studies that design teams were overly optimistic, as they overestimated the energy produced through RES and they underestimated the operational energy requirements (Crawley et al. 2009). This discrepancy is due, in part, to the inaccuracy of the weather data that are obtained from meteorological stations, usually near airports, and sufficiently far from the building location. Significant efforts are being made for the improvement of the accuracy for the short-term prediction of meteorological conditions, such as temperature (Papantoniou et al. 2014), or solar radiation (Provata et al. 2015).

Moreover, the urban microclimate (e.g., the presence of heat islands) can lead to significant discrepancies in the local weather data, and this can have a significant effect on the actual energy requirements (Santamouris 2007). It is therefore important that there be accurate weather measurements or modeling to yield pragmatic weather data at the site location so they can be used to properly estimate building loads and energy availability through renewable energy. Furthermore, combined integrated solutions that on the one hand reduce the energy and environmental burden of the buildings and on the other contribute to climate change adaptation and mitigation, should be disclosed.

Moreover, significant efforts have been put into quantifying and bridging the gap (energy, financial, and environmental) between the cost of optimal combination of energy technologies' level with the nZEB level (Groezinger et al. 2014; Nolte 2013; Hamdy et al. 2013; Kurnitski 2011). The aforementioned efforts show that although innovative passive and active smart and even low-cost technologies are available, their successful and optimal combination is still missing. In addition, as shown by recent research results, the effective of smart balancing and optimization of ZEB and micro grid management resources can lead to an almost 6 % reduction of operational costs with zero extra investment cost (Provata et al. 2015). Therefore, the focus should be on the successful balancing of ZEB systems and costs, while significantly accelerating their market uptake.

Finally, to realistically attain the ZEB on a larger scale (for buildings and districts), optimal pilot test cases should become the leading examples.

References

Asimakopoulos DA, Santamouris M, Farrou I, Laskari M, Saliari M, Zanis G, Giannakopoulos C (2012) Modelling the energy demand projection of the building sector in Greece in the 21st century. Energy Build 49. doi:10.1016/j.enbuild.2012.02.043

Berggren B, Wall M, Flodberg K, Sandberg E (2012) Net ZEB office in Sweden—a case study, testing the Swedish Net ZEB definition. Int J Sustain Built Environ 1(2):217–226. doi:10.1016/j.ijsbe.2013.05.002

Boyano A, Hernandez P, Wolf O (2013) Energy demands and potential savings in European office buildings: case studies based on energyPlus simulations. Energy Build 65:19–28. doi:10.1016/j.enbuild.2013.05.039

Comodi G, Giantomassi A, Severini M, Squartini S, Ferracuti F, Fonti A, Polonara F (2014) Multi-apartment residential microgrid with electrical and thermal storage devices: experimental analysis and simulation of energy management strategies. Appl Energy. doi:10.1016/j.apenergy.2014.07.068

Crawley D, Pless S, Torcellini P (2009) Getting to net zero. ASHRAE J 51(9):18–25

Deng S, Wang RZ, Dai YJ (2014) How to evaluate performance of net zero energy building—a literature research. Energy 71:1–16. doi:10.1016/j.energy.2014.05.007

Groezinger J, Boermans T, John A, Seehusen J, Wehringer F, Scherberich M (2014) Overview of member states information on NZEBs working version of the progress report—final report. ECOFYS

Hamdy M, Hasan A, Siren K (2013) A multi-stage optimization method for cost-optimal and nearly-zero-energy building solutions in line with the EPBD-recast 2010. Energy Build 56:189–203. doi:10.1016/j.enbuild.2012.08.023

IEA (2011) Towards net zero energy solar buildings—fact sheet

Kapsomenakis J, Kolokotsa D, Nikolaou T, Santamouris M, Zerefos SC (2013) Forty years increase of the air ambient temperature in Greece: the impact on buildings. Energy Convers Manag 74:353–365

Kolokotsa D, Rovas D, Kosmatopoulos E, Kalaitzakis K (2011) A roadmap towards intelligent net zero-and positive-energy buildings. Sol Energy 85(12):3067–3084

Kurnitski J (2011) How to calculate cost optimal nZEB energy performance?. The REHVA European HVAC Journal 48(5):36–41

Li DHW, Yang L, Lam JC (2013) Zero energy buildings and sustainable development implications—a review. Energy 54:1–10. doi:10.1016/j.energy.2013.01.070

Noguchi M, Athienitis A, Delisle V, Ayoub J, Berneche B (2008) Net zero energy homes of the future: a case study of the ÉcoTerra house in Canada. Renewable Energy Congress

Nolte I (2013) Implementing the cost optimal methodology in EU Countries. BPIE

Papantoniou S, Kolokotsa D, Kalaitzakis K (2014) Building optimization and control algorithms implemented in existing BEMS using a web based energy management and control system. Energy Build. doi:10.1016/j.enbuild.2014.10.083

Provata E, Kolokotsa D, Papantoniou S, Pietrini M, Giovannelli A, Romiti G (2015) Development of optimization algorithms for the leaf community microgrid. Renewable Energy 74:782–795. doi:10.1016/j.renene.2014.08.080

Pyloudi E, Papantoniou S, Kolokotsa D (2014) Retrofitting an office building towards a net zero energy building. Adv Build Energy Res, 1–14. doi:10.1080/17512549.2014.917985

Santamouris M (2007) Heat Island research in Europe—the state of the art. Adv Build Energy Res 1(1):123–150

Santamouris M, Kolokotsa D (2013) Passive cooling dissipation techniques for buildings and other structures: the state of the art. Energy Build 57(0):74–94. doi:http://dx.doi.org/10.1016/j.enbuild.2012.11.002

Sartori I, Napolitano A, Voss K (2012) Net zero energy buildings: a consistent definition framework. Energy Build 48:220–232. doi:10.1016/j.enbuild.2012.01.032

TEE (2010) Analytic national standards of parameters for the calculation of energy performance of buildings and issuing of energy performance certificate. In Technical Committee of Greece (ed) TOTEE 20701-1

Chapter 11
Energy Efficiency in Hospitals: Historical Development, Trends and Perspectives

Agis M. Papadopoulos

Abstract Hospitals are interesting buildings: they are highly complex, as they comprise a wide range of services and functional units. The prevailing indoor conditions should ensure thermal comfort, air quality, and visual comfort, in order to have a healing effect on patients. At the same time, hygienic regulations are getting tighter and are of obvious importance. Meeting those requirements leads, not unreasonably, to the fact that the energy demand of hospitals is among the highest of non-residential buildings. This was accepted for many years as an unpleasant but inevitable "side effect," but increased energy costs, cuts in public health budgets, the competition in the private health sector, and growing sustainability concerns have changed this approach. It is therefore interesting to discuss the way in which this complex problem has been addressed over the last few decades and, even more so, the way it is expected to be solved in the coming decades.

11.1 Introduction: On the Evolution of Hospital Buildings

Hospitals as buildings have been part of the European urban landscape for quite a few centuries. Over time, they have developed from rather primitive nursing centers into highly complex organizations, providing health care on all levels—from acute to chronic—and from general to highly specialized treatment. Even small hospitals, with fewer than 100 beds, can host centers for such treatment, requiring sophisticated infrastructures. Furthermore, hospitals operate as incubators for research and development in medicine and pharmaceutical sciences, and as educational and training centers. Hospital buildings therefore have to be functional and supportive of the various functions taking place. It is unnecessary to mention that because hospitals are public buildings *par excellence*, they are also

A.M. Papadopoulos (✉)
School of Engineering, Aristotle University Thessaloniki, 54124 Thessaloniki, Greece
e-mail: agis@eng.auth.gr

S.-N. Boemi et al. (eds.), *Energy Performance of Buildings*,
DOI 10.1007/978-3-319-20831-2_11

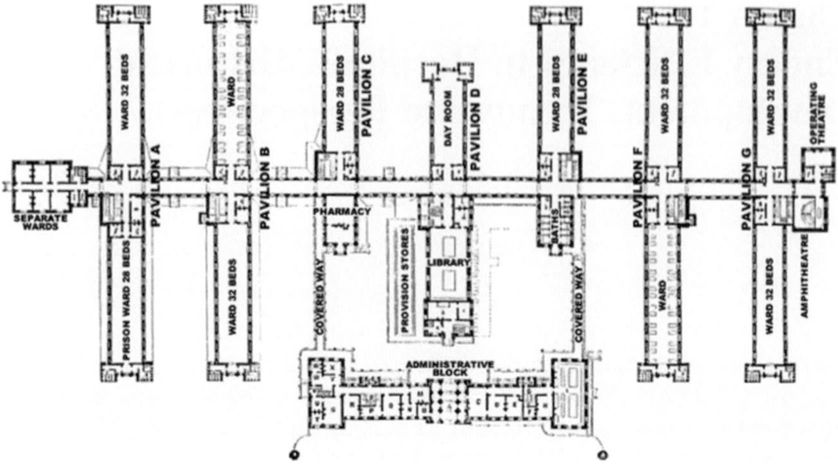

Fig. 11.1 Plan of the Royal Herbert Hospital in Woolwich (London) (http://www.royalherbert.co.uk/history.php)

representative ones and demonstrate architectural trends, social attitudes, and political approaches.

The necessity for, and the benefits of, high indoor environmental quality was perceived earlier than one would think. Responding to a mandate from King Louis XV of France at the end of the eighteenth century, a committee detailed two main types of hospital buildings—the radial and the pavilion. In particular, the latter enabled a significant improvement in ventilation and natural lighting, which became very popular.

Florence Nightingale encapsulated this approach in her "Notes on Hospitals" paper in 1863, underlining the necessity to provide patients with natural light, air, and landscape. Those principles were to influence the design of hospitals until World War II. A typical implementation of the pavilion style is the St. Thomas Hospital of London, completed in 1865, with six pavilions arranged side by side along the riverfront. The "comb" or "rack" design, like the Royal Herbert Military Hospital in Woolwich depicted in Fig. 11.1, solved problems of communication for the medical staff by connecting the pavilions by means of a gallery.

Two significant changes took place in the interwar period that affected hospital buildings in a most dramatic way: The evolution and technological developments in medicine led to a much more "technological" and "industrialized" approach to health care, leading to the provision of integrated services, with patients being inducted to primary care, i.e., the starting point for entry to the health system, and proceeding to secondary care (i.e., referral services that are intermediate in intensity), and on to tertiary, where patients are referred for very intensive subspecialty care, all in the same building. Such an approach made the idea of a single-block, deep plan arrangement building very appealing, as this type of building enables a very efficient, "all-in-one" process, as a rule by means of vertical transportation of the patient.

This type of building can obviously provide only very limited access to natural ventilation and light. The evolution of this kind of building's typology would not have been possible had it not coincided with the introduction and rapid development of air conditioning technology, which provided the answer to establishing and maintaining high standards of indoor air quality and of thermal comfort. One should not forget to mention the development of deep span bearing load frames, which enabled the design and construction of those large plan-view areas buildings, leading to what became globally known as the typical deep plan, block hospitals of the post-World War II era.

Since the 1990s, questions have been raised on the wisdom, but also on the feasibility, of mega-block hospitals, considering the estrangement of patients and staff from the environment, but also the long-term costs of those buildings' operation, which may outweigh their lower capital investment costs. The deep plan, block-type hospitals require lower capital investment, but their life cycle cost is eventually higher, due to higher operational expenses compared to narrower types of buildings (Glanville and Nedin 2009). A more layered typology appears as a reasonable alternative, leading to the so-called "podium on a platform" type of hospitals. This development is shown in Fig. 10.2.

The evolution in the typology of hospitals was also necessitated by the development in the provision for and the perception of health care after World War II: the establishment of the National Health Service system in the U.K., which acted as a model for most European countries and ensured access to health and medical care for entire populations, leading to an exponential increase of required services. Combined with the rapid advance of the services provided, this has led to the contemporary, highly sophisticated *modus operandi* of hospitals, a consequence of the latter being the quest for very specific quality features, focusing on the integrated design of the buildings, the reliability of their services, the flexibility in order to adapt to medical and technical changes and, last but not least, the economics. Still, in order to ensure those qualities, contemporary health care buildings have a very high energy and environmental impact, therefore making the call for sustainability imperative (Fig. 11.2).

Eventually it is also a matter of energy and economics: there are some 15,000 hospitals in the European Union. There are no pan-European data, but most member-state hospital buildings account for between 4 and 6 % of the building sector's energy consumption, although they represent less than 0.5 % of the built area. The fact that those buildings' operational costs account for 25 to 60 % of overall health expenditures, depending on the country, should not be left unobserved (HOPE 2007; BPIE 2011).

11.2 On the Use of Energy in Hospitals

The use of energy, and in that sense energy efficiency, is the major component of sustainability, waste management probably being the second most important one. Practically every study carried out over the last 30 years seems to confirm

Fig. 11.2 Evolution of health care, from "hope" to "podium on platform" typebuildings (Burpee 2009)

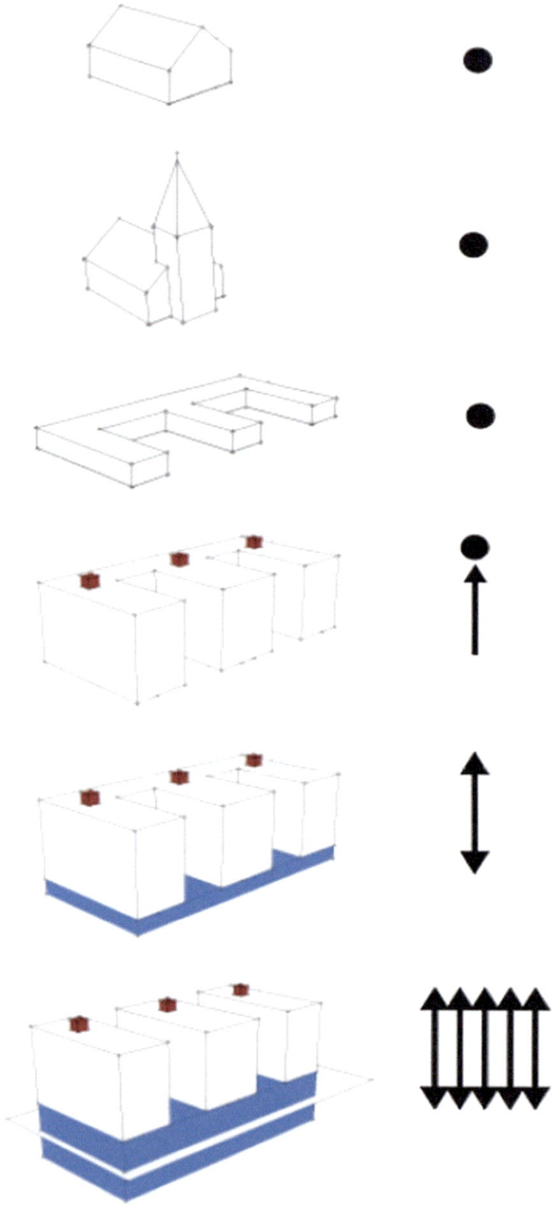

that from Slovenia to the U.K., and from Bulgaria to Finland, hospitals are among the biggest energy consumers of all non-residential building types, together with hotels and restaurants. Specific energy consumption values vary between 250 and 600 kWh/m^2, depending on the hospital type, its size, its functions, its location and, of course, the state of its building envelope and HVAC systems (Santamouris

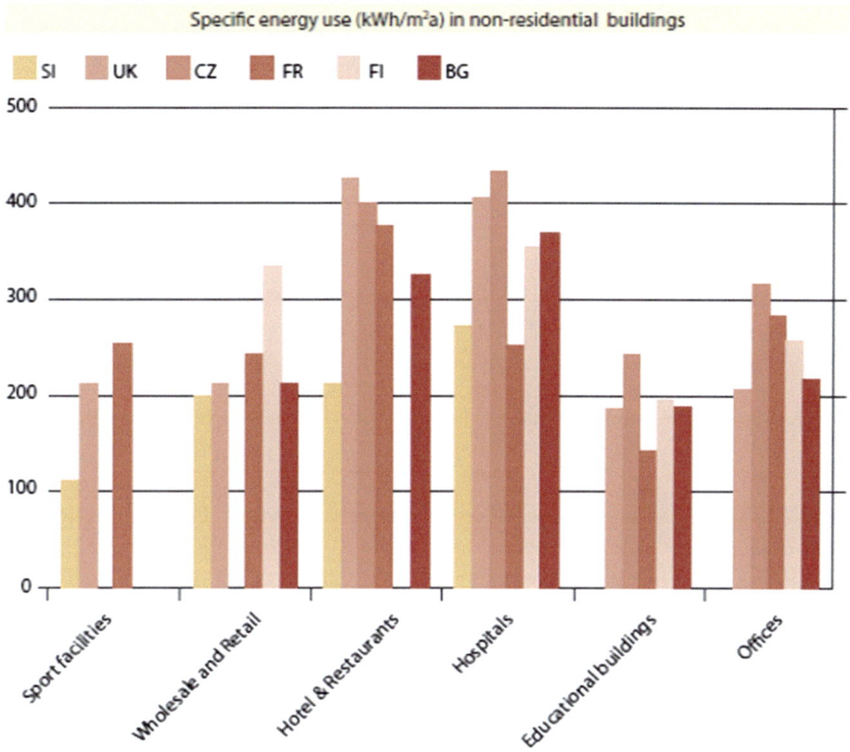

Fig. 11.3 Specific energy use in hotels, as compared to other non-residential buildings, for specific European countries (BPIE 2011)

and Asimakopoulos 1996; BPIE 2011). Figure 11.3 depicts some interesting data on the specific use of energy in hotels, as compared to other non-residential buildings, for some European countries.

A series of studies has shown that there is a correlation between the size of the hospital and its functional costs; unfortunately, this cannot be limited to energy costs only, as it also includes operation and maintenance. Still, based on the analysis of most of the studies published over the last 20 years, Posnett concludes that there is a range of between 300 and 600 beds, where functional costs are minimized (Posnett 2002). The assumption that this trend also applies to energy consumption and cost is reasonable, as it is in line with the results from energy audits carried out in hospitals that have been published (Tzikopoulos et al. 2005; Garcia-Sanz-Calcedo et al. 2014) and with the statistical data published by EIA for the USA (EIA 2007).

The explanation for this seemingly paradoxical result becomes quite clear, considering the breakdown of the energy consumption of two comparable-in-size, large university hospitals, in Thessaloniki, Greece, and in Oslo, Norway, which have a comparable final energy consumption: 430 kWh/m^2 for the Oslo hospital,

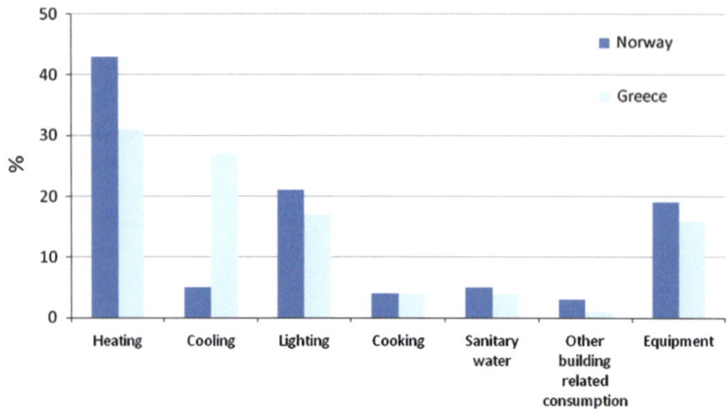

Fig. 11.4 Energy consumption breakdown for comparable hospitals in Norway and Greece

390 kWh/m^2 for the one in Thessaloniki. The breakdown of the consumption values is depicted in Fig. 11.4 (Shchuchenko et al. 2013; Christodoulidis 2012).

Oslo has 4714 heating degree days (HDD), and 9 cooling degree days (CDD), while Thessaloniki has 2012 HDD and 710 CDD, respectively (EURIMA 2011). It is therefore reasonable to expect that the heating and cooling loads are correlated to the climatic conditions. The hospital in Thessaloniki has higher energy consumption for heating than one would expect, which is due to the building's age: it was constructed in the mid-1950s, so a thorough renovation is quite necessary.

Still, one cannot fail to notice that in both hospitals, some 20 % of the energy is used by the equipment and another 15–20 % by the lighting. This observation is of particular interest, since one of the major points to be addressed in the energy performance of hospitals is their rather modest improvement over the last decades.

Energy consumption in hospitals has not been reduced significantly over the last 30 years, at least not proportionally to the significant progress made in the design and construction of the buildings' envelopes and the HVAC systems, and to the implementation of the Energy Performance of Buildings Directive after 2003. If one compares the findings of field studies from the 1980s (Adderley et al. 1988, 1989; Santamouris et al. 1994) to the most recent ones (Garcia-Sanz-Calcedo et al. 2014; Rohde et al. 2014), they seem to confirm this impression, which has also been verified by the findings of a series of EU-funded projects presented in the last few years, including among other RES-Hospitals, LCB Healthcare, EMAS and Information Technology in Hospitals, Green@hospitals, Hospilot, etc.

There are a few reasons for this development: (1) there has been a very significant increase in the medical equipment used, considering both the plethora of apparatuses and their power requirements; (2) there has been a trend towards single-patient rooms, as part of the policy to improve health care services; and (3) requirements for indoor air quality, and for visual and thermal comfort as well, have become stricter.

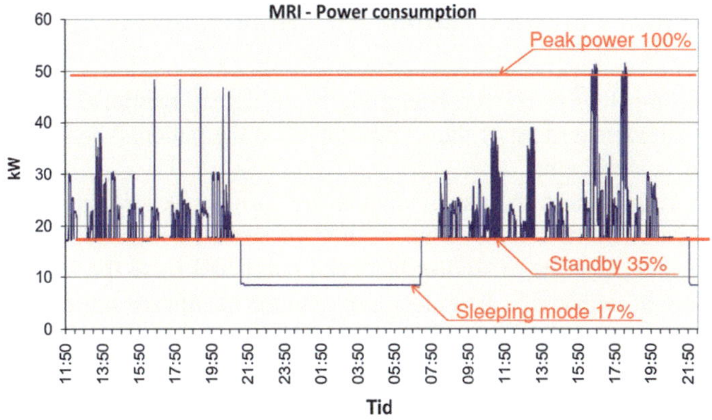

Fig. 11.5 Typical daily profile of an MRI's power consumption (Harsem 2011)

To name only a few examples of the impact of equipment, computed tomography imaging (CT) units have a plug load of 30–60 kW, and an angiography calls for loads of 15–30 kW, while the introduction of magnetic resonance imaging (MRI) in the 1990s led to a load of some 30–50 kW per MRI unit. Such a unit is in a service and diagnostic mode for 2 h daily on average, and for 10 h in a standby/ sleep mode for 10 h (COCIR 2011); however, even in its sleep mode, it uses a not insignificant amount of energy, as can be seen in Fig. 11.5.

Those large medical imaging pieces of equipment may account for up to 50 % of a hospital's energy utilization for healthcare requirements, leaving all the other small equipment responsible for the remaining 50 %.

The impact of providing more single patient rooms cannot be discounted: According to international design practice, average values of between 130 and 170 m^2 per bed can be considered reasonable for medium to large teaching hospitals. In the 1960s and 1970s, single- and double-bed rooms accounted for 50 % of the total capacity, the other 50 % being 4- to 6-bed rooms. As there is a shift towards single- and double-bed rooms, this will inevitably lead to larger hospitals and therefore to higher energy consumption, even though the specific consumption will not be affected (Bobrow and Thomas 2000; Healy and McKee 2002; Maben et al. 2012).

A final, but very important point that has to be made has to do with the age of hospital buildings: only 25 % of the health care facilities in the European Union were constructed after 1980, with figures varying from 22 % in Germany to 24 % in Hungary, 45 % in Sweden, and 70 % in Spain (which has the newest building stock) (ECOFYS 2011). The age of the health care facilities underlines the necessity to emphasize the renovation and refurbishment of the existing building stock, so as to improve the energy performance of the buildings on the one hand, and to comply with the new increased requirements, especially considering indoor environmental issues, on the other.

11.3 Thermal Comfort, Indoor Air Quality, and Hygiene

Indoor environmental issues are becoming more tightly regulated since there has been a better understanding of their importance on the patients' healing and on the medical staff's well-being.

Thermal comfort and indoor air quality are, of course, strongly linked due to the role of ventilation, yet the same amount of attention was not always paid to both parameters. A series of studies has been published since the late 1980s, considering thermal comfort in hospitals and smaller healthcare buildings. The biggest part of those studies, especially in the 1990s, dealt with the environmental parameters—namely, indoor air temperature, relative humidity, and air velocity, as those are the physical parameters determining thermal comfort, based on the PMV method. However, in the last two decades, research has also focused on the thermal comfort of patients and hospital staff, especially with the adaptation of the latest ASHRAE 55 and the EN 7730 standards (ASHRAE 55-2010; ISO-EN 7730 2009) that adopt the adaptive thermal comfort approach.

The ASHRAE 55 adaptive comfort standard applies only to naturally cooled buildings and therefore cannot be used in hospitals that feature central HVAC systems. Furthermore, there are many hospitals, especially older ones in the Mediterranean area, with a combination of natural ventilation, by means of operable windows, and local refrigeration, by means of room air conditioners. In this case, the new EN 15251 standard can be used, although it is also not without difficulties (EN 15251 2012). As Nicol and Wilson pointed out, the approach taken in EN 15251 foresees a prediction of the PMV index within very tight limits with respect to the six input parameters (air temperature, radiant temperature, relative humidity, air velocity, clothing insulation, and metabolic rate), where each parameter must be measured with great accuracy. In the case of the "personal" variables—clothing insulation and metabolic rate—this is not practically possible (Nicol and Wilson 2011).

Furthermore, it has to be emphasized that all those standards apply for healthy people. As a rule, sick patients require higher air temperatures, as they have lower metabolic rates, due to both medications and the lack of physical activity. Using the ISO/TS 14415:2005 standard, which focuses on the ergonomics of the thermal environment for people with special requirements, may be helpful, but considering that there are also healthy people in the same rooms, the problem remains complicated. Medical staff, especially in surgery and examination rooms, needs lower air temperatures, due to the intensity of their activities. At the same time, the higher air temperature and relative humidity values are a potential problem, as they favor infections, bacterial growth, and transfer of airborne bacteria (Khodakarami and Nasrollahi 2012).

There are therefore contradictions that have to be solved, and this can be done mainly by compromising with respect to the air temperature and by direct control of the fresh air requirements that ensures a minimization of hygienic problems. One of the very characteristic features of hospital buildings is that they include a wide range of services and functional units. There are the medical care areas, for the diagnostic and treatment units, including clinical laboratories, imaging, emergency rooms, and operating rooms. The inpatient care or bed-related functions probably account for the biggest surface percentage of the building and, as will be discussed, also for energy consumption.

Furthermore, there are the hospitality functions, including the food services and housekeeping units, and, finally, there is the administrative part of the buildings, with offices, archives, and supply stores. This broad range of functions is reflected in the regulations, standards, guidelines and codes that apply to hospital construction and operations, which foresee different requirements for the various uses. Table 11.1 presents a synopsis of the major parameters that determine both indoor air quality and thermal comfort, based on the recommendations by ASHRAE (Khodakarami and Nasrollahi 2012).

Operating rooms are a very particular and demanding type of rooms, with important requirements regarding fresh air and recirculated air requirements, both in quantitative terms, i.e., how many air changes are need, and in qualitative terms, i.e., the air flow type (laminar or turbulent) and its distribution in the operating room. In the literature, one can find recommendations for between 10 and 150 air changes per hour, while laminar, downward air flows are considered to be the most effective solution to maximize the exhaust of contaminated air without having dispersion in the room (Balaras et al. 2007). Table 11.2 shows a synopsis of the main parameters, as recommended by ASHRAE, DIN, and the American Institute of Architects.

Guaranteeing that those levels of thermal comfort and indoor air quality will be established, and at the same time doing it with the least possible energy consumption, is probably the key issue of any energy design and energy conservation exercise in the field. Furthermore, considering the number and ages of hospitals in Europe, which was discussed previously, it is clear that the priorities lie in the renovation and refurbishment of the existing building stock rather than in the design and construction of new hospitals. This trend is enhanced by the growing pressure to reduce costs in the health care sector, as part of the austerity policy applied in most European countries in the 2010s. It is therefore imperative to focus on strategies, tools, and means to improve the energy performance of the existing hospitals and reduce their operational costs while simultaneously ensuring a high level of indoor environmental quality and, not forgetting the prevailing trends, improving their overall index of sustainability—and all this on tightening budgets. A challenging task indeed, presenting analogies to the squaring of the circle.

11.4 Improving Energy Efficiency and Reducing Energy Costs: Energy Optimization

The methodological approach to improving the energy efficiency of hospitals is, in principle, not different from that adopted for any other type of building. It is based on an action plan that calls for a monitoring, analyzing, and optimizing sequence of actions, as depicted in Fig. 11.6.

Table 11.1 Indoor air quality and thermal comfort parameters and their values as recommended by ASHRAE

Function	Pressure relationship to adjacent area	Minimum air changes of fresh air per hour	Minimum total air changes per hour	All air exhausted directly to outside	Air recirculated within room units	Relative humidity (%)	Design temperature (°C)
Patient room	[a]	2	6[b]	–	–	30 (W), 50 (S)	24 ± 1
Newborn nursery suite	[a]	2	6	–	No	30–60	22–26
Labor/delivery/recovery	[a]	2	6[b]	–	–	30 (W), 50 (S)	24 ± 1
Patient corridor	[a]	2	4	–	–	–	–
General inpatient area not covered by ASHRAE	[a]	2	4	–	–	30–60	≤24

[a]Continuous directional control required
[b]Total air changes per room may be reduced to 4 when using supplemental heating and/or cooling systems, like radiant heating and cooling, baseboard heating, etc.

Table 11.2 Air temperature, relative humidity and air-change requirements for operating rooms

Temperature[a] (°C)	Relative humidity (%)	Ventilation	Source/References
20–23	30–60	Positive pressurization. Minimum 15 ACH, at least 3 ACH outdoor air. Filter all recirculated and fresh air through minimum 90 % efficient filters	American Institute of Architects
		In rooms not engineered for horizontal laminar airflow, introduce air at the ceiling and exhaust air near the floor	
17–27	45–55	Positive pressurization. Minimum 25 ACH, at least 5 ACH outdoor air	ASHRAE
20–24	30–60	Positive pressurization (at least 2.5 Pa). Primary supply diffusers, non-aspirating. Minimum 20 ACH, minimum 4 ACH outdoor air	ASHRAE
22–26[b]	50–60	Positive pressurization. The air changes should be 60 $m^3/m^2/h$, if the room height is 3 m, or else 20 ACH	German Institute for Standardization

[a]Continuous directional control required
[b]Total air changes per room may be reduced to 4 when using supplemental heating and/or cooling systems, like radiant heating and cooling, baseboard heating, etc.

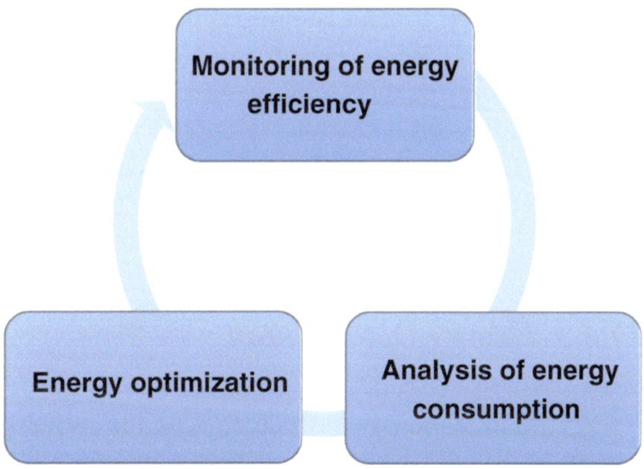

Fig. 11.6 Methodological approach for the improvement of energy efficiency

One point that has to be made in advance is that given the high energy consumption and the tightening of hospital budgets, emphasis has to be put on the reduction of energy costs, which is not always achieved only by improving energy efficiency; it is also a matter of optimizing the use of energy and taking into consideration energy policies and tariffs as well as the overall economic appraisal of energy investments, including their initial financing options. The three steps of the methodological approach can be briefly described as follows:

11.4.1 Monitoring of Energy Efficiency

In order to be able to reduce energy consumption in a hospital, this consumption first has to be measured. This may sound self-evident, but in many older buildings, where building management systems are not available, it may prove a daunting and tedious task. The measured data subsequently has to be validated and consolidated, so that it can provide useful consumption values. This must be done, considering the weather conditions for the time period of the measurements, the operational patterns and conditions of the hospital, as well as the indoor environmental quality conditions that have prevailed for the same time period. In addition, data are needed for the cost of the various energy forms, since they are needed for making the long-term optimal decisions.

11.4.2 Analysis of Energy Consumption

This is the classical exercise of a detailed, in-depth energy audit, which is required to analyze a hospital's energy results in a meaningful manner so as to be able to compare values to generate benchmarks and use those values as the foundation for optimization. It is paramount to being able to determine the energy breakdown with respect to its use, especially considering electrical loads, as they are more difficult to monitor. It must be noticed that the energy audit and certification process foreseen by the Energy Performance of Buildings Directive is, in most cases, an inadequate tool for this purpose, as it serves to benchmark and compare the reference building on a calculative approach (Panayiotou et al. 2010; Theodoridou et al. 2011). What is needed are analytical energy audits, according to the very recent EN 16247-2 standard, which do provide the necessary detailed data.

11.4.3 Energy Optimization

Energy optimization is the outcome of the process, based on the results of the energy audit and subsequent analysis consisting of proposed actions on three different levels: (1) small-scale improvements as part of the operation and maintenance procedure, expected to reduce energy consumption by perhaps up to 5 % and reduce energy costs due to smarter energy source and tariff choices; (2) standard energy conservation and renewable energy sources utilization measures, expected to achieve energy savings ranging between 20 and 30 %; and (3) energy refurbishment measures as part of an extensive renovation of the hospital, expected to reduce the building's consumption drastically—and to yield an almost new building. It must be noted that there is no uniform definition for "extensive renovation": in Europe this implies a minimum of 75 % energy savings, whereas in the U.S. it means of improving the energy efficienct by 30–50 % (GBPN 2013).

It becomes quite clear that the key to the successful implementation of a strategy to improve the hospital's energy efficiency lies in the implementation of an integrated strategy that will consider technical and non-technical parameters, in order to meet the overall optimization goals. One possible form of this strategy, albeit surely not the only one, is shown in Fig. 11.7.

There is an abundance of literature that has been published on the technical aspects of the strategy. It is beyond the scope of this chapter to go into this discussion, and it would also be beyond the space allocated. What is important, however, is to focus on a few basic points that make hospitals different from other non-residential buildings, when it comes to applying energy conservation and renewable energy utilization measures.

In most European countries, the development of national regulations on thermal insulation since the 1970s has quite drastically reduced heat losses to the buildings' envelopes (Papadopoulos 2007). This trend has been intensified by the implementation of the Energy Performance of Buildings Directive, since 2003 (Papadopoulos et al. 2008; Böttcher 2012). At the same time, internal loads have been steadily increasing, as discussed in the previous chapters, but also extensively in literature (Congradac et al. 2012). Furthermore, those loads are quite evenly distributed throughout the day and the year. It is therefore not uncommon in practice to need heat rejection rather than heat addition over large periods of the year for most European climatic conditions. It is also interesting to note that this trend is also being monitored in the U.S., where the thermal protection of the buildings' envelope is as a rule not on the European level (Burpee et al. 2009).

This problem is exacerbated in southern Europe by solar loads and high ambient temperatures in summer. The use of sun-protection and cool materials provides some relief for the former (Synnefa et al. 2012; Khodakarami et al. 2009). *The latter is a more difficult problem, which has to be tackled by a combination of high efficiency air conditioning, and passive, low-energy cooling techniques* (Avgelis and Papadopoulos 2009; Santamouris and Kolokotsa 2013).

A comment must be made about the conservation potential that can be exploited by means of state-of-the-art building automation and control (BAC) systems. Energy-optimized operations require the comprehensive application of occupancy sensor control for lighting and HVAC systems in order to control and regulate those systems. Modern sensors, such as gas mixture sensors for room air quality and actuating elements to compensate natural and artificial lighting, by dimming the latter instead of switching on and off, provide solutions that can be integrated in the building management system. The EN 15232 standard provides the framework for the consideration of those systems' contribution within the assessment of the buildings' energy performance.

A comment also has to be made regarding the generation part of the energy balance. The use of renewable energy sources must definitely be considered for new

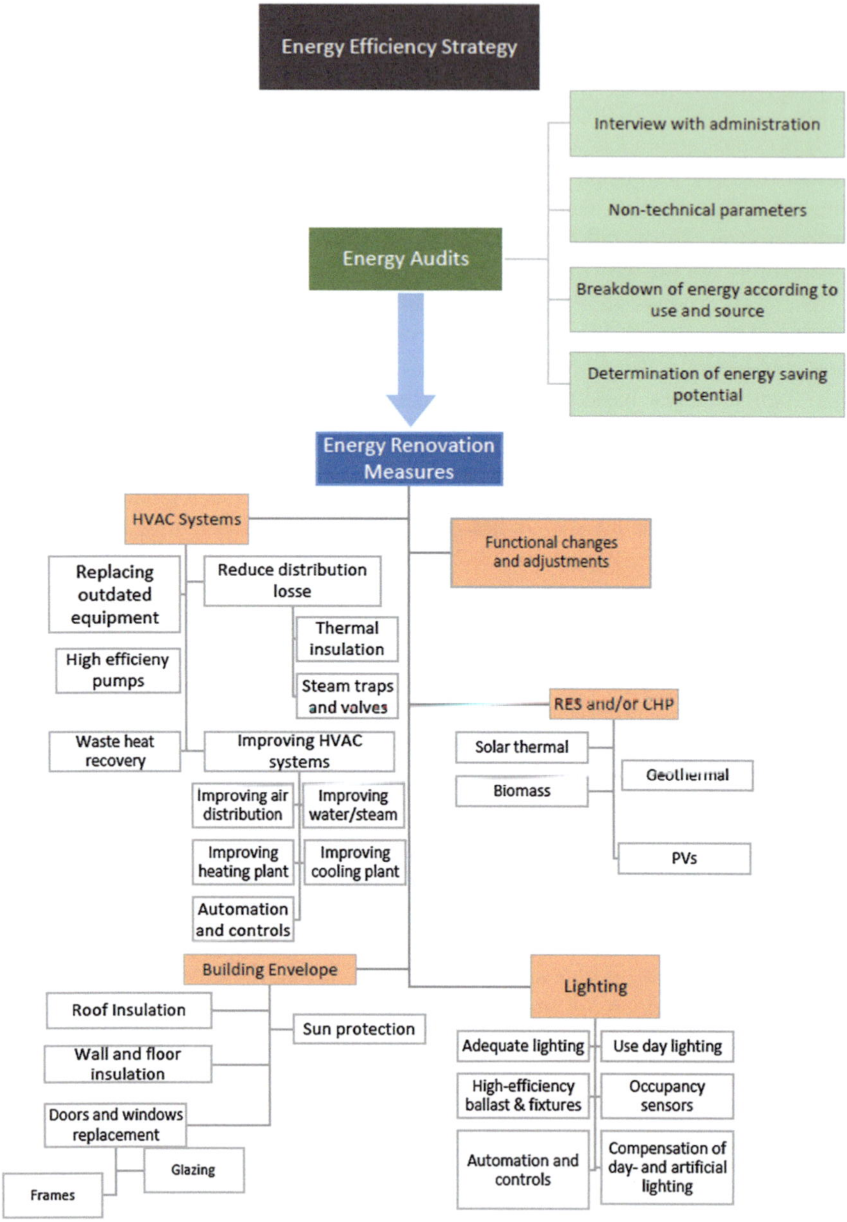

Fig. 11.7 A strategy for promoting energy efficiency in hospitals

hospital buildings: geothermal energy can be utilized to cover part of the heating and cooling requirements; solar thermal systems can be used for sanitary water, but also for solar air conditioning by means of absorption cooling; photovoltaics can be integrated in the buildings' façades; and the use of biomass is, at least in some regions of Europe, an interesting option. In existing buildings, however, the options are more limited, especially for hospital buildings located in densely built urban environments. Still, there are options, mainly solar ones, to cover at least a fraction of the requirements (Tsoutsos et al. 2010; Karteris et al. 2013). Perhaps the most attractive option, however, is the use of "combined heat and power" (CHP) systems: As heat, whether for heating, cooling or sanitary water, is frequently needed at the same time as electricity, it leads to a simultaneity between the electric and thermal loads. Furthermore, the heat requirements are evenly spread over the year, and natural gas provides an excellent primary source for power generation. These are ideal conditions for the combined generation of heat and power, which has become quite popular over the last 15–20 years (Kayo et al. 2014).

11.5 Conclusions

Hospitals are particularly interesting buildings, considering their historical evolution, their architectural features, their HVAC installations, and, of course, their energy behavior. Furthermore, given their importance for public health and well-being, it is only reasonable to apply state-of-the-art technological solutions so as to ensure that the buildings will perform in the expected way—namely, to serve as true health care facilities—and not just as public, non-residential buildings. The introduction of the Energy Performance of Buildings Directive provided, also for hospitals, for the first time a commonly accepted methodological approach to evaluate their efficiency, allowing us to produce comparable indices.

At the same time, significant changes have begun to affect the public sector, both public and private, with a common denominator being the pressure to cut operational expenses—in our case, energy costs—much more intensively than has been the case in the 40 years since the first oil crisis. Solutions therefore have to be elaborated, which will ensure the indoor environmental quality needed while reducing energy costs.

Technological developments in the fields of building materials, of HVAC systems, and, of course, of building design have been rapid, providing tools and means to the designer and the builder. Research activity has been intense and fertile, with many studies and publications available, enabling the dissemination of precious expertise. Still, there are challenges that have to be met in the near future. Some of them are of a purely technical nature; others have more to do with the approach we adopt when trying to solve the technical aspects of the problem.

It is true that hospitals have indoor environmental quality requirements. However, this does not necessarily imply that all the areas in a hospital have to be "clean rooms." It is also true that hospitals operate on a 24/7 basis, but this does not mean that all the areas in a hospital operate in this mode; as a matter of fact, only a small part of a hospital does so—and this is the one with the more conventional requirements.

It is true as well that hospitals are big energy consumers—a fact that is often cited by governmental and other authorities when it comes to budget discussions. However, even the most energy-intensive hospital consumes less than a shopping mall. This is not an excuse to stop the efforts to improve a hospital's energy performance, but it had to be said in order to set the record straight.

Finally, one should also notice that it is not only a matter of technology or of "hardware," but also of the way in which the hospital, a highly complex and sophisticated organism, is managed on a day-to-day basis. The implementation of energy management systems should not be considered a luxury.

References

Adderley AE, O'Callaghan PW, Probert SD (1988) Optimising the choice of energy-thrift measures for hospitals. Appl Energy 30(2):153–160

Adderley AE, O'Callaghan PW, Probert SD (1989) Energy-signature characteristic of a hospital. Appl Energy 34:125–153

ASHRAE Standard 55R (2010) Thermal environmental conditions for human occupancy. American Society of Heating, Refrigerating and Air-Conditioning Engineers, Inc, Atlanta, GA

Avgelis A, Papadopoulos AM (2009) Application of multicriteria analysis in designing HVAC systems. Energy Build 41:774–780

Balaras C, Daskalaki E, Gaglia A (2007) HVAC and indoor thermal conditions in hospital operating rooms. Energy Build 39:454–470

Bobrow M, Thomas J (2000) Inpatient care facilities. In: Kobus R et al (eds) Building type basics for healthcare facilities. Wiley, New York, pp 131–192

Böttcher O (2012) Energy efficient and sustainable—federal buildings in Germany. REHVA J 49(3):41–45

BPIE report (2011) Europe's buildings under the microscope. Buildings Performance Institute Europe, pp 58–52

Burpee H, Loveland J, Hatten M, Price S (2009) High performance hospital partnerships: reaching the 2030 challenge and improving the health and healing environment. White Paper for the American Society of Healthcare Engineering

Christodoulidis C (2012) Feasibility study for a CHP plant for the AHEPA hospital in Thessaloniki. Dissertation at the Department of Mechanical Engineering, Aristotle University Thessaloniki (in Greek)

COCIR report (2011) MRI—Measurement of energy consumption. Draft report for discussion with the Ecodesign Consultation Forum

Congradac V, Prebiracevic BJorgovanovic N, Stanisic D (2012) Assessing the energy consumption for heating and cooling in hospitals. Energy Build 48:146–154

ECOFYS (2011) Panorama of the European non-residential construction sector, Schimschar S et al (2011)

EIA report (2007) Commercial buildings energy consumption survey: large hospital tables, Energy Information Administration, 2012

EN 15251 (2012) Indoor environmental input parameters for design and assessment of energy performance of buildings addressing indoor air quality, thermal environment, lighting and acoustics, CEN 2012

EURIMA (2011) U-values for better energy performance of buildings, Annex 1, ECOFYS, 2011

Garcia-Sanz-Calcedo J, Lopez-Rodriguez F, Cuadros F (2014) Quantitative analysis on energy efficiency of health centers according to their size. Energy Build 73:7–12

GBPN report (2013) What is a deep renovation, Technical report, Green Buildings Performance Network

Glanville R, Nedin P (2009) Sustainable design for health. In McKee M, Healy SJ (eds) Investing in hospitals of the future, European Observatory on Health Systems and Policies, pp 229–246
Harsem TT (2011) Reduction of total energy consumption in hospitals. In 4th European conference on healthcare engineering, 1 June 2011
Healy J, McKee M (2002) Improving performance within the hospital. In Rechel B, Wright S, Edwards N, Dowdeswell B, McKee M (eds) Hospitals in a changing Europe, European Observatory on Health Systems and Policies, pp 205–225
HOPE report (2007) The Crisis, Hospitals and Healthcare, European Hospital and Healthcare Federation, Brussels
ISO EN 7730 (2009) Moderate thermal environments: determination of PMV and PPD indices and specification of the conditions for thermal comfort
Karteris M, Slini T, Papadopoulos AM (2013) Urban solar energy potential in Greece: a statistical calculation model of suitable built roof areas for photovoltaics. Energy Build 62:459–468
Kayo G, Hasan A, Siren K (2014) Energy sharing and matching in different combinations of buildings, CHP capacities and operation strategy. Energy Build 82:685–695
Khodakarami J, Nasrollahi N (2012) Thermal comfort in hospitals—a literature review. Renew Sustain Energy Rev 16:4071–4077
Khodakarami J, Knight I, Nasrollahi N (2009) Reducing the demands of heating and cooling in Iranian hospitals. Renew Energy 34(4):1162–1168
Maben J, Penfold C, Robert G, Griffiths P (2012) Evaluating a major innovation in hospital design: workforce implications and impact on patient and staff experiences of all single room hospital accommodation. Report for HaCIRIC, National Nursing Research Unit, Department of Health Policy and Management, King's College London, pp 116–119
Nicol F, Wilson M (2011) A critique of European standard EN 15251: strengths, weaknesses and lessons for future standards. Build Res Inf 39(2):183–193
Panayiotou GP, Kalogirou SA, Florides GA, Maxoulis CN, Papadopoulos AM, Neophytou M, Fokaides P, Georgiou G, Symeou A, Georgakis G (2010) The characteristics and the energy behaviour of the residential buildings stock of Cyprus in view of Directive 2002/91/EC. Energy Build 42(11):2083–2089
Papadopoulos AM (2007) Energy cost and its impact on regulating the buildings' energy behaviour. Adv Build Energy Res 1:105–121
Papadopoulos AM, Oxizidis S, Papathanasiou L (2008) Developing a new library of materials and structural elements for the simulative evaluation of buildings' energy performance. Build Environ 43(5):710–719
Posnett J (2002) Are bigger hospitals better? In Rechel B, Wright S, Edwards N, Dowdeswell B, McKee M (eds) Hospitals in a changing Europe, European Observatory on Health Systems and Policies, pp 100–118
Rohde T, Martinez R, Mysen M (2014) Activity modeling for energy-efficient design of new hospitals. Int J Facility Manage 5(1)
Santamouris M, Asimakopoulos D (eds) (1996) Passive cooling of buildings, James and James Publication, pp 12–18
Santamouris M, Kolokotsa D (2013) Passive cooling dissipation techniques for buildings and other structures: the state of the art. Energy Build 57:74–94
Santamouris M, Dascalaki E, Balaras C, Argiriou A, Gaglia A (1994) Energy performance and energy conservation in health care buildings in Hellas. Energy Convers Manage 35(4):293–305
Shchuchenko V, Lie B, Harsem T (2013) Influence of building form of hospital on its energy performance. Eur Sci J 3:286–295
Synnefa A, Saliari M, Santamouris M (2012) Experimental and numerical assessment of the impact of increased roof reflectance on a school building in Athens. Energy Build 55:7–15
Theodoridou I, Papadopoulos AM, Hegger M (2011) A typological classification of the Greek building stock. Energy Build 43(10):2779–2787
Tsoutsos T, Aloumpi A, Gkouskos Z, Karagiorgas M (2010) Design of a solar absorption cooling system in a Greek hospital. Energy Build 42:265–272
Tzikopoulos AF, Karatza MC, Paravantis JA (2005) Modeling energy efficiency of bioclimatic buildings. Energy Build 37:529–544

Chapter 12
The Hotel Industry: Current Situation and Its Steps Beyond Sustainability

Sofia-Natalia Boemi and Olatz Irulegi

Abstract The Mediterranean area attracts almost 20 % of the world's tourism (over 150,000,000 tourists each year) due to its mild and pleasant climate and famous cultural heritage. Tourism-related energy studies generally focus on estimating total energy use and efficiency and comparing the energy consumption regarding various facets of tourism-related activities, mostly hotels. This chapter presents a review of EU projects and papers. There is no doubt that research on tourism and climate change has developed substantially. Recent publications have begun focusing on sustainability—specifically, how climate change will impact tourism and how destinations can be adapted. Considerable attention is also paid to tourism's role as a contributor to greenhouse gas emissions and how these can be mitigated, due to the fact that the tourism sector has been transformed into an increasing environmentally conscious marketplace where consumers have realized the impact of the behaviors that they have paid for, which are strongly associated with environmental problems. This is the main achievement of recent years, and it certainly represents a qualitative difference in the way that the activity is considered.

12.1 Introduction

The tourism industry, and particularly the hotel sector, is becoming increasingly competitive and dynamic, motivated by the pressures of globalized supply and demand (Oliveira et al. 2013). In fact, over the past six decades, despite occasional

S.-N. Boemi (✉)
Process Equipment Design Laboratory, Aristotle University of Thessaloniki, Box 487, GR-54124 Thessaloniki, Greece
e-mail: nboemi@gmail.com

O. Irulegi
Architecture Department, University of the Basque Country, Plaza Oñate 2, 20018 San Sebastian, Spain
e-mail: o.irulegi@ehu.eus

S.-N. Boemi et al. (eds.), *Energy Performance of Buildings*,
DOI 10.1007/978-3-319-20831-2_12

shocks, tourism has experienced almost uninterrupted growth and diversification, becoming one of the largest and fastest-growing economic sectors in the world.

This pattern has continued in recent years, despite the global financial and economic crisis, with tourism having the potential to be one of the main engines of recovery in the EU.

According to the United Nations World Tourism Organization, in 2012—for the first time in history—there were more than one billion international tourist arrivals. Europe remained the most frequently visited region in the world, accounting for over half of all international tourist arrivals in 2012. The wealth of European cultures, the variety of its landscapes, and the quality of its tourist infrastructure are likely to be among the reasons that tourists choose to take their holidays in Europe.

But the tourism industry is under pressure to reduce its greenhouse gas emissions (Dalton et al. 2008). Therefore, one of the main trends dominating the sector is the search for a continuous and systematic improvement of processes and resources toward efficiency (Chen 2007; Pestana Barros 2005). Differences in markets, tradable products, quality, facilities, location, and differentiations in prices, among other aspects, can generate the critical factors of success and viability of these enterprises (Farrou et al. 2012). Thus, this is slowly responding to the task with the implementation of energy efficiency initiatives and the adoption of renewable energy supplies (RESs) (Karagiorgas et al. 2004). Specifically, the building sector in Greece accounts for about one-third of carbon dioxide (CO_2) emissions, and for about 36 % of the total energy consumption. CO_2 emissions from the building sector have an annual growth rate of around 4 %, constantly inflating the energy consumption of buildings (Theodoridou et al. 2011). Even so, there is no specific legislative requirement or even a comprehensive energy efficiency strategy at present for the tourism industry to reduce its energy use and to adopt RES.

The only legislative measure that applies for any kind of building is the implementation of the European Buildings' Directive EPBD (Energy Performance of Buildings), which began in 2002 and was completed in the summer of 2010. Along the lines of EPBD, several regulations for the EU countries have been launched; they have set specific limitations and minimum requirements for energy efficiency concerning the design and construction of various building types. However, emphasis is put on improving the energy performance of the residential building stock; therefore, in order to evaluate the existing building stock of hotels and to determine their energy consumption, a thorough investigation has to be carried out (Boemi et al. 2011).

This chapter aims to identify the gaps between energy efficiency and sustainability and to provide an overview of the current situation of the tourism accommodation sector, mostly hotel buildings.

12.2 An Overview of Energy Performance in Hotels

A review of the literature has shown that the performance of hotels has been a topic of long-time interest to academics who adopt a plethora of approaches, such as finance (Phillips and Sipahioglu 2004), economics (Chen and Dimou 2005), international business (Quer et al. 2007) and energy demand in terms of the energy use index or energy intensity (defined as the site energy consumption per unit of gross floor area) (Yu and Lee 2009).

In most cases, a number of international studies on resource consumption in hotels and other accommodation facilities have been performed. Furthermore, some consumption indicators are also included in the environmental reports of hotel companies. However, only a few cases actually investigated the influence of various operational characteristics of energy and water consumption. Most investigations focused on determining the consumption indicators for various regions (Bohdanowicz and Martinac 2007).

In general, the subject of energy consumption in the hotel industry has been developing since the 1990s and has focused on the energy use index (energy intensity), which is defined as the building energy consumption per unit of gross floor area. The first study was reported in 1991 where the average annual energy intensity of hotel buildings in Ottawa (Canada) was reported to be 688.7 kWh/m^2 (Bohdanowicz and Martinac 2007). Since then, the average consumption of hotel buildings has been reported in several countries.

Worldwide, a study of the Australian hotel industry (Becken et al. 2001), analyzed five different accommodation categories (15 hotels, 22 bed-and-breakfast, 20 motels, 35 hostels, and 13 campgrounds) on the west coast of New Zealand, and reported an energy demand of 277.78 kWh/m^2 or 34 m^2/bed (Becken et al. 2001). In Ottawa, researchers investigated energy use in 16 hotels, and the average energy consumption was found to be 688.7 kWh/m^2 (Deng and Burnett 2000). In Singapore, the energy use intensity of 29 hotels was 427.0 kWh/m^2 (Priyadarsini et al. 2009). Deng and Burnett reported an average energy use intensity of 564.0 kWh/m^2 when studying the energy performance of 16 hotels in Hong Kong; in another study in 2003, the same researchers presented an energy use intensity of 542.0 kWh/m^2 in 36 Hong Kong hotels. Önüt and Soner (2006) studied the energy breakdown of 32 five-star hotels in the Antalya region of Turkey and found that the energy use intensity was 400.0 kWh/m^2.

In the EU, Bohdanowicz and Martinac (2007) reported that the average energy use for European hotels in the 1990s was between 239.0 and 300.0 kWh/m^2 year. More analytically, Santamouris et al. (1996) collected energy consumption data from 158 Greek hotels and estimated the energy savings potential of various retrofitting scenarios. The annual average total energy consumption was measured as 273.0 kWh/m^2. This is similar to what was reported by Markis and Paravantis (2007), who found that the hotels' energy use is equal to 297.0 kWh/m^2. In addition, Farrou et al. (2012) wrote that average electricity and thermal energy consumption is calculated to be approximately 290 kWh/m^2 year for hotels that operate all year round, and approximately 200 kWh/m^2 year for hotels with seasonal operation. In the other Mediterranean countries, hotel energy use was reported to be 215.0 kWh/m^2 for Italy, 278.0 kWh/m^2 for Spain, and 420.0 kWh/m^2 for France (Farrou et al. 2012). A more clear view of the energy consumption per country is provided in the next paragraph and in Table 12.2.

The aforementioned research on hotel efficiency gives a clear understanding of the fluctuation of the energy use intensity across major tourism places. But it has to be mentioned that a great fluctuation appears in the value of the annual average consumption not only among various counties, but also among regions of the same country and even for the same country over different periods of time, which is a result of the different climatic conditions and operational characteristics of hotels globally, as was reported by Pieri et al. (2015).

12.2.1 Tourism in Countries with Temperate Climates

Most of the European countries with temperate climates are those that surround the Mediterranean Sea. In recent years, these countries have been top tourist destinations, due to their mild, pleasant climates and their famous cultural heritages, attracting almost 20 % of world tourism, which means more than 150,000,000 tourists each year (SETE 2014). The number of accommodation establishments in Europe is estimated at about 200,000, with almost half of the building stock located in the Mediterranean countries (Hotel Energy Solutions 2011).

Within the next 20 years, the number of tourist arrivals is expected to double, increasing the need for suitable hotel accommodations and facilities. Most of the existing establishments in the Mediterranean basin were built between the 1960s and 1980s; they are buildings of low quality with regard to energy consumption or users' comfort according to today's standards. This explains why most of them need a partial or complete renovation to compete with the market demands in 20 years (on average) after their construction.

However, the tourism sector has been resilient despite the financial crisis of the last few years, since it has had a steady augmentative development with an average growth of 3.2 % from 2005 until 2013 in southern Europe. Only in 2013, 11,000,000 more international arrivals were reported, presenting an increase of 6 % in comparison to 2012 (UNWTO 2014).

As the number of overnight stays reflects both the length of stay and the number of visitors, it is considered to be a key indicator within tourism accommodation statistics. Therefore, the nights spent in tourism accommodations located in countries with temperate climates are presented in Table 12.1. Specifically, there were 2.58 billion nights spent in tourist accommodation establishments—composed of hotels and similar accommodations (NACE Group 55.1), holiday and other short-stay accommodations (NACE Group 55.2), and camping grounds, recreational vehicle parks, and trailer parks (NACE Group 55.3) across the EU-28 in 2012.

12.2.2 Basic Figures for the Greek Sector

Tourism is one of the most dynamic branches within the service sector of the Greek economy (Boemi et al. 2011). It has always been a significant development

Table 12.1 Nights spent at tourist accommodation establishments by NUTS 2 regions [tour_occ_nin2]

Country	2009	2010	2011	2012	2013
Greece	64,292,443	65,059,095	69,138,050	62,887,010	70,089,017
Spain	250,984,815	267,147,471	286,743,027	280,659,549	286,030,160
France	191,741,154	195,906,121	201,970,247	201,900,781	201,402,749
Croatia	18,606,977	19,344,700	20,466,922	21,137,979	21,093,536
Italy	246,618,107	251,098,476	259,910,852	255,610,143	254,759,348
Cyprus	12,807,922	13,599,262	14,087,844	14,546,723	14,022,197
Malta	6,740,299	7,416,254	7,529,165	7,676,434	8,265,375
Portugal	36,457,069	37,391,291	39,440,315	39,681,040	42,507,098
Romania	16,514,083	15,417,560	17,367,337	16,502,576	16,537,621
Slovenia	5,918,377	5,853,329	6,184,598	6,195,576	6,174,738
Montenegro	–	–	2,968,719	2,976,279	–
Serbia	–	–	–	3,945,932	3,985,666
Total EU-28	1,529,561,249	1,571,309,389	1,638,335,337	1,664,371,686	1,701,135,711

Hotels and similar accommodations. *Source* Eurostat (tour_occ_nin2)

tool and the key driver of economic growth, holding a high position in world rankings throughout the years. Based on a recent statistical survey, the Greek hotel industry follows a steady growth with more than 10,000 million euro direct receipts, resulting in the contribution of 16.4 % to the Gross Domestic Product (GDP) (SETE 2014). In 2012, the international tourist arrivals reached over 15,500,000 people, bringing Greece to 17th place in the world rankings, while 2013 was a year of strong growth, with arrivals almost 10 % higher than in the previous peaks of 2007 and 2011.

Specifically, during the summer of 2014, an impressive increase in the arrivals was reported, leading to almost 1,500,000 more tourists in the last eight months compared to 2013. During the following years, despite the worldwide economic crisis, it is estimated that tourism in Greece will reach 21,500,000 million—15.8 % more than in 2014.

But tourism affects other sectors as well. In fact, tourism has been one of the few sectors since 2000 that has been contributing to the creation of employment opportunities. In 2012, about 700,000 people were working in the hospitality industry, representing 20 % of the work force in Greece. Moreover, it appears that there is a widespread underemployment of hotel infrastructure and a hotel overconcentration capacity in only four areas (Crete, North Aegean Islands, Cyclades, and Attica). Specifically, 47 % of the total hotel building stock is situated north of Greece, including all Aegean islands.

An interesting feature of Greek tourism is that the great majority—around 80 %—visit Greece for leisure, and only a small percentage travel for business or other reasons. This easily explains the strong seasonality of arrivals, where 85 % of the total tourist traffic corresponds to the May-October period, and more than half between July and September (SETE 2014).

Greece is on the way to transform itself to an upmarket tourism destination, so the need for an increase in high-quality, resort accommodations becomes apparent. Specifically, hotels represent 0.82 % of the Greek building stock, with more than 9000 hotels, 32,806 buildings, and 773,445 rooms (SETE 2014).

In 1990 the official licenses issued by the National Tourist Organization totalled more than 6400 hotel buildings with 423,660 beds. Thus, it is obvious that a significant increase of more than 50 % in the hotel units has been reported in the last 20 years. This development follows a steady trend and has an average annual growth of 1.9 %. So despite the financial crisis, the hotel capacity of the country is growing, while the average size of hotels in Greece continues to increase.

Most of the hotels are 2-star hotels with a low budget infrastructure. Therefore, a strategic plan for spatial development is needed in order to achieve a balanced tourist growth and high-quality tourism facilities development. Aiming for the improvement of the existing infrastructure, the lengthening of the operational season, and the improvement of alternative forms of tourism, a series of structural, financial, and administrative measures were adopted by the state.

Other data show that most (68.2 %) of the total hotel building stock was constructed after the application of the first Greek regulation for buildings' insulation, which means that most of the hotels are properly insulated (Boemi et al. 2011). When energy demand is analyzed with respect to its time variation, it is obvious that the peak values are recorded in the summer season, due to the use of air conditioning for space cooling, a fact that is enhanced by the seasonal character of Greek tourism. Therefore, the improvement of the existing infrastructure, the lengthening of the operational season, and the improvement of alternative forms of tourism, combined with a series of structural, financial and administrative measures, should be adopted by the State. With respect to sustainability, the measures are based on the EU's encouragement of improving the environmental performance of services and products. Also, in order to improve the energy performance of the tourism sector, one cannot fail to emphasize the role of voluntary environmental quality labelling schemes, such as Eco-label and ISO 14001.

12.3 Features of the Hotel Industry in Countries with Temperate Climates

As defined by the World Trade Organization (WTO 1998): "A hotel is any facility that regularly or occasionally provides overnight accommodation for tourists. Hotels and similar establishments are typified as being arranged in rooms, in number exceeding a specified minimum; as coming under a common management; as providing certain services including room service, daily bed-making and cleaning of sanitary facilities; as grouped in classes and categories according to the facilities and services provided; and as not falling in the category of specialized establishments." In general, hotels are classified, depending on their facilities and services provided, into five star categories: 5*, 4*, 3*, 2*, 1*.

Moreover, hotels are designed to provide high levels of comfort for guests; however, frequent complaints related to uncomfortable thermal environment and inadequate indoor air quality (IAQ) appear. In Portugal, for example, Asadi et al. (2011) used a 4-star hotel building to audit the IAQ. The study reveals four main problems: insufficient ventilation rate; too-high particle concentration in some rooms; contamination with Legionella on the sanitary hot-water circuit; and poor filtration effectiveness in all air-handling units.

Also in Portugal, Gonçalves et al. (2012) studied a hotel building located in Coimbra (Portugal). From an individual analysis, the electrical equipment was found to be the main contributor to the primary energy consumption in the hotel. However, they present the highest exergy efficiency when compared to processes related to space air conditioning. In other studies (Oliveira et al. 2013; Pestana Barros 2005), the efficiency of individual hotels was the issue; specifically, they studied the efficiency of hotels in terms of star rating, location, and training activities. The results pointed out the important role of the operational environment.

A similar study conducted in Spain examined the link between environmental practices and firm business performance in the Spanish hotel industry (Barberán et al. 2013). The results revealed that hotels showing a stronger commitment to environmental practices reached higher performance levels (Molina-Azorín et al. 2009). Using the same approach, Roselló-Batle et al. (2010) identified the processes that have the greatest impact on the life cycle of a tourist building. The results show that the operating phase, which represents between 70 and 80 % of the total energy use, is the one with the greatest impact.

But in order to understand the correlation between the operational phase, energy efficiency, and energy costs on the operational costs in a tourist accommodation building, an overview of the available data on energy use (mostly derived from projects) of existing hotels is needed. Therefore, data from past EU projects have been gathered and presented below. Also, Table 12.2 presents data from all the accumulated literature.

VAROLEN was a Greek project aimed at collecting data from 158 hotels. Most of the hotels (140) were in Athens, and the average annual energy consumption was reported at 273 kWh/m^2 year (Santamouris et al. 1996). Within the EU CHOSE (2001) project, data were collected from 10 hotels (6 on the mainland and 2 on islands) in Greece. Their size ranged from 3500 m^2 to 35,110 m^2, with an annual average energy consumption of 289.9 kWh/m^2 year. The energy consumption in the hotel sector consisted of approximately 48 % for heating and air conditioning, 13 % for domestic hot water supply, 7 % for lighting, 25 % for catering (kitchen facilities), and 7 % for other electrical appliances. The Hotel Energy Solutions (HES 2015) project reported significant variations in energy use in facility types within the hotel sector. However, it concluded that regarding climatic conditions, overall energy use levels can be relatively constant (energy needs for cooling and heating balance out), but with significant differences in the necessary technologies to reduce energy use in different climate zones.

In addition, HES provided average energy use levels according to available certification schemes for the energy performance of hotels (e.g., Accor, Nordic

Table 12.2 Energy use in the hospitality sector from past EU projects and literature

Country	Year	Hotels and restaurants energy index (kWh/ m^2 year)	Sample	Source
France	2008	398	–	nZEB (2014)
France	1975–2005	292–375	–	BPIE (2011)
Croatia	2008	397.8	–	BPIE (2011)
Greece	2013	273	90	Farrou et al. (2014)
Greece	2011	300	92	Boemi (2011)
Greece	2008	418	–	nZEB (2014)
Greece	2002	134.67	1	Xuchao et al. (2010)
Greece	1996	273	158	Santamouris et al. (1996)
Italy	2008	222	–	BPIE (2011)
Italy	2001	249–436	–	CHOSE (2001)
Spain	2008	204	–	BPIE (2011)
Spain	2003	Seasonal hotels: 88.4–122.0 Annual hotels: 99.8–179.6		Roselló-Batle et al. (2010)
Spain, Majorca, Zanzibar	1999	61.4–254.4	30	Gössling (2001)
Turkey, Antalya	2005	194.28–733.77	32	Önüt and Soner (2006)
Portugal	2001	99–444.6	–	CHOSE (2001)

Swan, LowE, WWF/IBLF, Thermie) delivered energy use range 200–400 kWh/(m^2 a), with average energy use 305–330 kWh/(m^2 a).

The HOTRES project was aimed at promoting renewable technologies (solar, solar passive, thermal, solar PV, geothermal and biomass). Therefore, data from 100 hotels (specifically, for eight hotels of three geographical types: mountain, city, and coastal) were derived from fuel and electricity bills. Based on the results of this audit, electricity was reported as the main fuel intake.

Another project focusing not only on reducing the energy consumption of existing buildings of the hospitality sector into nZEB but on promoting the frontrunners was Nearly Zero Energy Hotels (nZEH). The project, with a consortium of seven European countries (Croatia, France, Greece, Italy, Romania, Spain, and Sweden), showed an average primary energy consumption of between 221.5 and 591.8 kWh/m^2 year.

During the last few decades, several tourism projects were implemented; all of them forced tourism and hoteliers to get interested in how they can turn their business "green," how to apply renewable systems, and how energy reduction can also have a serious impact in their operational costs. The most important projects, as reported by the EU (CORDIS 2015), which set the roots for the next project that had the greater impact in the tourism accommodation industry, are presented in Table 12.3.

Table 12.3 The most important hotel projects of the last 15 years (CORDIS 2015)

Project	Year	Participating countries	Objective
REST: Renewable energy and sustainable tourism	2002–2004	Italy, Portugal, France, Spain, German, UK	The aim was to encourage the hotel industry to look at its energy performance and take action (participants: Italy, Portugal, France, Spain, etc.)
HOTRES: Technical support to the tourism industry with renewable energy technologies. Phase 1: hotel sector	2001–2003	France, Portugal, Italy, Spain, Greece	The aim was to devise and implement a strategic methodology for the promotion of RET (Renewable Energy Technologies)
GREEN HOTEL: Integrating self-supply into end use for sustainable tourism—target action C	2003–2012	Portugal, Spain, Italy, Greece, Denmark, Sweden, Switzerland, Belgium	Specific aims: to develop and apply a new concept of sustainable community
Optimized solar system for hotels in Algarve	1987–1989	Portugal, Spain	The aim was to demonstrate the possibility of energy saving within a nationwide group of hotels through centralized management as a function of level of room occupancy, climatological conditions and energy consumption under normal conditions
Feasibility studies of solar heating systems for the production of sanitary hot water (SHW) in the tourist industry	1995–1997	Spain, France	The aim of the project is to provide the ICAEN and the ADEME with the structure necessary to carry out a significant number of feasibility studies for solar heating systems for the production of sanitary hot water in tourist establishments, principally hotels and camping sites
Integrated small-scale solar heating and cooling systems for sustainable air conditioning of buildings	2007–2011	France, Italy, Spain, Germany, Egypt, Austria	The project aims to develop highly integrated solar heating and cooling systems for small- and medium-capacity applications that are easily installed and economically and socially sustainable. The envisioned applications are residential houses, small office buildings and hotels

(continued)

Table 12.3 (continued)

Project	Year	Participating countries	Objective
XENIOS: Development of an audit tool for hotel buildings and the promotion of RUE and RES	2002–2003	Greece, Spain, Italy and France	The project is aimed at the hotel sector. Its main objectives are to provide a methodology for a preliminary audit with the necessary tools for a first assessment of where and how to integrate the most cost-effective, energy-efficient renovation practices, technologies, and systems
Tourism hotel with solar passive architecture	1988–1992	France	The main aim of the project is to attain a high comfort level and save energy in the type of hotels where per definition, the seasonal occupation and energy use are largely correlated
ShMILE: Sustainable hotels in the Mediterranean area	2004–2007	France, Greece, Italy, Austria	The main aim of the project was to develop tools, boosting implementation, training and education actions for EU-wide implementation of EU ecolabel on tourist accommodation service

12.4 Beyond Energy: Hotels and Sustainability

Ever since the oil crisis in 1973, it has been understoodthat a large proportion of energy is consumed in buildings for heating, cooling, and lighting; moreover, it is directly related to the way the buildings are designed (Boemi et al. 2011; Karkanias et al. 2010). At the same time, it has also been realized that the densely built urban environment creates a microclimate of its own, affecting the energy's balance. Therefore, several studies in the past focused on the energy balance of buildings, and especially in the case of hotels (Boemi et al. 2011; Pestana Barros and Dieke 2008; Han et al. 2009; Markis and Paravantis 2007; Santamouris et al. 1996). But recently, there has been a major turnaround, and an increasing number of studies have focused on the hotels' environmental performance and their sustainability.

Since the early 1980s, tourism studies have begun focusing on service and product quality (Sharpley and Forster 2003). The nature of tourism consumption has undergone a significant transformation, from the traditional "3-S" (sun-sea-sand), to an increasingly competitive environment. The hoteliers are striving to move towards a new model based on a competitive advantage based on quality, while finally, greater legislation and regulation imposed on the industry is obliging many organizations to focus on quality (Sharpley and Forster 2003).

This is due to the fact that the tourism sector has been transformed into an increasingly environment-conscious marketplace where consumers have realized the impact of their purchased behaviors, which are strongly associated with environmental problems (Laroche and Parsa 2001). In fact, according to Eurobarometer (2009), almost half of EU citizens said that ecolabelling plays an important role in their purchasing decisions; the percentage believing that this is important ranged from 22 in the Czech Republic to 64 in Greece. With a growing number of customers seeking green destinations, environmentally friendly practices can help not only in characterizing a hotel as "green," but in providing a basis for a better marketing strategy to improve their presence in the competitive arena (Manaktola and Jauhari 2007).

In that marketplace, the influence of environmental management is a competitive advantage that will be analyzed through its impact on costs and differentiation. Environmental management may allow the hotel to reduce its operational costs and its energy consumption, and to re-use materials through recycling (Boemi et al. 2011). In this context, the tourism sector and hoteliers understood that environmental improvement in terms of resource productivity is attractive as a product (Logar 2010).

An environmentally friendly hotel is a lodging property that applies and follows ecologically sound programs and practices, such as water- and energy-saving measures, reduction and management of solid waste, and structural measures in order to promote sustainability. The UNEP and WTO (2002) have defined sustainable tourism as a form of tourism that leads to management of all resources in such a way that economic, social, and aesthetic needs can be fulfilled while maintaining cultural integrity, essential ecological processes, and biological diversity and life support systems. Also, it describes the development of sustainable tourism as a process that meets the needs of the current tourists and host communities while protecting and enhancing the requirements for future generations.

In further research, trying to link energy consumption, climate change, and economic growth in energy, economic studies have shown that international tourism is one of the specific segments or sectors of the economy that deserves further attention (Park and Hong 2013; Lu et al. 2013; Jayanthakumaran et al. 2012). However, it is also likely that tourism development might have indirect effects on climate change through economic growth and energy capacity expansion. For example, an increase in tourism activities implies an increased demand for energy at various functions, such as transportation, catering, accommodations, and the management of tourist attractions (Katircioglu 2014; Becken 2011; Gössling 2002), which is also likely to lead to environmental pollution and degradation. In fact, a study suggests (Tsagarakis et al. 2011) that tourists from countries with higher energy awareness prove to be more willing to choose hotels with energy-saving installations and renewable energy sources. They argue that citizens from those countries (i.e., Canada, Japan, Sweden, and Finland) that successfully adopt energy-saving policies, are more likely to prefer hotels with energy-saving installations in destination countries. In this respect, an investigation of the relationship among international tourism, energy consumption, and climate change will be of interest to both policy-makers and practitioners.

Moreover, the existing literature of energy economics generally focuses on the link among economic growth, energy, and carbon emissions as a proxy of environmental pollution, but results are still inconclusive (Katircioglu 2014). A relatively small sampling of the literature investigates the issue of energy consumption related to the tourism sector, mainly due to its implications for environmental issues, such as its contribution to climate change (Gössling et al. 2002; Becken 2005; Gössling et al. 2002). Only a few studies, on the other hand, focus on the link between tourism and electricity consumption as it was analyzed before. The link among the energy sector, the environment, and international tourism has rarely been considered from different perspectives in the relevant literature.

12.5 Conclusions

Most European countries with temperate climates are located in the Mediterranean area. In the last few years, these countries have been top tourist destinations, due to their mild and pleasant climates and their famous cultural heritage, attracting almost 20 % of world tourism, which comes to more than 150,000,000 tourists annually (SETE 2014). Within the next 20 years in those countries, the number of tourist arrivals is expected to double, increasing the need for suitable hotel accommodations and facilities.

Therefore, tourism-related energy studies will play a leading role in research; these studies generally focus on estimating total energy use and efficiency (Becken 2005) and comparing energy consumption across various parts of tourism-related activities (Nepal 2008). But there has been a lack of validated and verified data on energy behavior and performance of hotels. Thus, more studies, like the one mentioned above, focus on statistical approximation in order to be safe and to present study cases with a small number of hotel buildings that cannot be generalized on a country level.

Yet the hotel sector is of vital importance not only for the economy, but also for its energy performance for the building sector. The best information on the hotel building stock has been culled from several EU projects and papers, which have been described above. In fact, there is no doubt that research on tourism and climate change have developed substantially. As Becken (2013) reported, the number of publications between 1986 and 2005, compared with 2006–2012, increased by 368 %, with 108 publications in 2011 and 2012. But the analysis of energy efficiency in hotels is limited not only by the awareness of the technologies and their benefits in hotels, but due to the main difficulty in reporting the huge number of buildings, the management capacity and commitment, lack of staff resources, and concerns about financing and performance; these limitations were confirmed by several study cases.

However, recent publications focus on sustainability—specifically, how climate change will impact tourism and how destinations can be adapted. Considerable attention is also paid to tourism's role as a contributor to greenhouse gas emissions and how these can be mitigated; this is due to the fact that the tourism sector has been transformed into an increasingly environment-conscious marketplace where consumers have realized the impact of their purchased behaviors, which are strongly associated with environmental problems.

Moreover, it is easier for researchers to export data from environmentally conscious hotels, because they use it as an attraction and way of promoting their facilities. But despite any limitations, it is undeniable that research on tourism is driven towards sustainability, due to the extent that it is now inseparable from the notion of contemporary tourism. Any initiative that aims to influence tourism in the present context must consider sustainability as a factor in achieving the kind of tourism desired. This is the main achievement of recent years, and it certainly represents a qualitative difference in the way that the tourism activity is conducted

References

Asadi E, Costa JJ, Gameiro da Silva M (2011) Indoor air quality audit implementation in a hotel building in Portugal. Build Environ 46:1617–1623

Barberán R, Egea P, Gracia-de-Rentería P, Salvador M (2013) Evaluation of water saving measures in hotels: a Spanish case study. Int J Hosp Manag 34:181–191

Becken S (2005) Harmonising climate change adaptation and mitigation: the case of tourist resorts in Fiji. Glob Environ Change 15:381–393

Becken S (2011) A critical review of tourism and oil. Ann Tourism Res 38:359–379
Becken S (2013) Developing a framework for assessing resilience of tourism sub-systems to climatic factors. Ann Tourism Res 43:506–528
Becken S, Frampton C, Simmons D (2001) Energy consumption patterns in the accommodation sector—the New Zealand case. Ecol Econ 39:371–386
Boemi SN (2011) Energy and environmental management of the greek hotel building stock. Dissertation in University of Ioannina, Greece
Boemi SN, Slini T, Papadopoulos AM, Mihalakakou G (2011) A statistical approach to the prediction of the hotel stock's energy performance. Int J Vent 10:163–172
Bohdanowicz P, Martinac I (2007) Determinants and benchmarking of resource consumption in hotels—case study of Hilton International and Scandic in Europe. Energy Build 39:82–95
BPIE—Buildings Performance Institute Europe (2011) Europe's buildings under the microscope. A country-by-country review of the energy. Available via DIALOG. http://www.europeanclimate.org/documents/LR_%20CbC_study.pdf. Cited 21 Apr 2015
Chen C-F (2007) Applying the stochastic frontier approach to measure hotel managerial efficiency in Taiwan. Tour Manag 28:696–702
Chen JJ, Dimou I (2005) Expansion strategy of international hotel firms. J Bus Res 58:1730–1740
CHOSE, Energy savings by combined heat cooling and power plants (CHCP) (2001) In the hotel sector. In: Report of the commission of the European communities, directorate general for energy, Stockholm
CORDIS–Community Research and Development Service (2015) Available via DIALOG. http://cordis.europa.eu/projects/home_en.html
Dalton GJ, Lockington DA, Baldock TE (2008) A survey of tourist attitudes to renewable energy supply in Australian hotel accommodation. Renew Energy 33:2174–2185
Deng S-M, Burnett J (2000) A study of energy performance of hotel buildings in Hong Kong. Energy Build 31:7–12
Eurobarometer (2009) Survey on the attitudes of Europeans towards tourism. Available via DIALOG. http://ec.europa.eu/public_opinion/flash/fl_258_en.pdf
Farrou I, Kolokotroni M, Santamouris M (2012) A method for energy classification of hotels: a case-study of Greece. Energy Build 55:553–562
Farrou I, Kolokotroni M, Santamouris M (2014) Building envelope design for climate change mitigation: a case study of hotels in Greece. Int J Sustain Energy (in press)
Gonçalves P, Rodrigues Gaspar A, Gameiro da Silva M (2012) Energy and exergy-based indicators for the energy performance assessment of a hotel building. Energy Build 52:181–188
Gössling S (2001) The consequences of tourism for sustainable water use on a tropical island: Zanzibar, Tanzania. J Environ Manag 61:179–191
Gössling S (2002) Global environmental consequences of tourism. Glob Environ Change 12:283–302
Gössling S, Borgström Hansson C, Hörstmeier O, Saggel S (2002) Ecological footprint analysis as a tool to assess tourism sustainability. Ecol Econ 43:199–211
Han H, Hsu L-T, Lee J-S (2009) Empirical investigation of the roles of attitudes toward green behaviors, overall image, gender, and age in hotel customers' eco-friendly decision-making process. Int J Hosp Manag 28:519–528
Hotel Energy Solutions (2011) Publications. Available via DIALOG. http://hotelenergysolutions.net/publications. Cited 10 Jan 2015
Hotel Energy solutions (HES) (2015). EU project, Available via DIALOG. http://www.buildup.eu/links/27664
Jayanthakumaran K, Verma R, Liu Y (2012) CO2 emissions, energy consumption, trade and income: a comparative analysis of China and India. Energy Policy 42:450–460
Karagiorgas M, Mendrinos D, Karytsas C (2004) Solar and geothermal heating and cooling of the European Centre for Public Law building in Greece. Renew Energy 29:461–470

Karkanias C, Boemi SN, Papadopoulos A, Karagiannidis A (2010) Energy efficiency in the Hellenic building sector: an assessment of the restrictions and perspectives of the market. Energy Policy 38:2776–2784
Katircioglu ST (2014) International tourism, energy consumption, and environmental pollution: the case of Turkey. Renew Sustain Energy Rev 36:180–187
Laroche M, Parsa HG (2001) Brand management in hospitality: an empirical test of the Brisoux-Laroche model. J Hosp Tourism Res 24:199–222
Logar I (2010) Sustainable tourism management in Crikvenica, Croatia: an assessment of policy instruments. Tourism Manag 31:125–135
Lu S, Wei S, Ke Zhang, Kong X, Wu W (2013) Investigation and analysis on the energy consumption of starred hotel buildings in Hainan Province, the tropical region of China. Energy Convers Manag 75:570–580
Manaktola K, Jauhari V (2007) Exploring consumer attitude and behaviour towards green practices in the lodging industry in India. Int J Contemp Hosp Manag 19:364–377
Markis T, Paravantis JA (2007) Energy conservation in small enterprises. Energy Build 39:404–415
Molina-Azorín JF, Claver-Cortés E, Pereira-Moliner J, Tarí JJ (2009) Environmental practices and firm performance: an empirical analysis in the Spanish hotel industry. J Clean Prod 17:516–524
Nepal SK (2008) Tourism-induced rural energy consumption in the Annapurna region of Nepal. Tourism Manag 29:89–100
nzEB: Nearly Zero Energy Hotels (2014) REHVA journal. Available via DIALOG. http://www.buildup.eu/news/41058. Cited 15 Jan 2015
Oliveira R, Pedro MI, Cunha Marques R (2013) Efficiency performance of the Algarve hotels using a revenue function. Int J Hosp Manag 35:59–67
Önüt S, Soner S (2006) Energy efficiency assessment for the Antalya Region hotels in Turkey. Energy Build 38:964–971
Park J, Hong T (2013) Analysis of South Korea's economic growth, carbon dioxide emission, and energy consumption using the Markov switching model. Renew Sustain Energy Rev 18:543–551
Pestana Barros C (2005) Measuring efficiency in the hotel sector. Annals of Tourism 32:456–477
Pestana Barros C, Dieke PUC (2008) Technical efficiency of African hotels. Int J Hosp Manag 27:438–447
Phillips PA, Sipahioglu MA (2004) Performance implications of capital structure: evidence from quoted UK organisations with hotel interests. Serv Ind J 24:31–51
Pieri SP, Tzouvadakis I, Santamouris M (2015) Identifying energy consumption patterns in the Attica hotel sector using cluster analysis techniques with the aim of reducing hotels' CO_2 footprint. Energy Build 94:252–262
Priyadarsini R, Xuchao W, Siew Eang L (2009) A study on energy performance of hotel buildings in Singapore. Energy Build 41:1319–1324
Quer D, Claver E, Andreu R (2007) Foreign market entry mode in the hotel industry: the impact of country- and firm-specific factors. Int Bus Rev 16:362–376
Roselló-Batle B, Moià A, Cladera A, Martínez V (2010) Energy use, CO2 emissions and waste throughout the life cycle of a sample of hotels in the Balearic Islands. Energy Build 42:547–558
Santamouris M, Balaras CA, Dascalaki E, Argiriou A, Gaglia A (1996) Energy conservation and retrofitting potential in Hellenic hotels. Energy Build 24:65–75
SETE (2014) The contribution of tourism to the Greek economy for 2014—a snapshot. Available via DIALOG. http://sete.gr/media/2005/simasia_tourismou_sete_intelligence_report.pdf. Cited 20 Mar 2015
Sharpley R, Forster G (2003) The implications of hotel employee attitudes for the development of quality tourism: the case of Cyprus. Tourism Manag 6:687–697

Theodoridou I, Papadopoulos AM, Hegger M (2011) Statistical analysis of the Greek residential building stock. Energy Build 43:2422–2428

Tsagarakis KP, Bounialetou F, Gillas K, Profylienou M, Pollaki A, Zografakis N (2011) Tourists' attitudes for selecting accommodation with investments in renewable energy and energy saving systems. Renew Sustain Energy Rev 15:1335–1342

UNEP—United Nations Environment Programme, WTO—World Tourism Organization (2002) Making tourism more sustainable. A guide for policy makers. Available via DIALOG. http://www.unep.fr/shared/publications/pdf/DTIx0592xPA-TourismPolicyEN.pdf. Cited 10 Jan 2015

UNWTO—World Tourim Organisation (2014) Tourism highlights, 2014 edition. Available via DIALOG. http://mkt.unwto.org/publication/unwto-tourism-highlights-2014-edition. Cited 20 Mar 2015

WTO—World Trade Organisation (1998) Tourism services, background note by the secretariat. Available via DIALOG. https://www.wto.org/english/tratop_e/serv_e/w51.doc. Cited 10 Jan 2015

Xuchao W, Priyadarsini R, Siew Eang L (2010) Benchmarking energy use and greenhouse gas emissions in Singapore's hotel industry. Energy Policy 38:4520–4527

Yu M-M, Lee BCY (2009) Efficiency and effectiveness of service business: evidence from international tourist hotels in Taiwan. Tourism Manag 30:571–580

Chapter 13
Schools: Trends and Perspectives

Martha C. Katafygiotou and Despoina K. Serghides

Abstract This chapter's emphasis is on existing school buildings, with an overview of the European and, more specifically, the Mediterranean region. Following the overview of schools in the Mediterranean region, the study focuses on secondary schools in Cyprus. It identifies the prevailing building practices in school construction with specific reference to the schools in Cyprus. The construction and energy consumption details of the secondary school buildings in Cyprus are also presented. Indoor comfort and energy efficiency are analyzed through questionnaires, surveys, interviews, and simulations on specific pilot school buildings. The field studies are conducted to evaluate the indoor thermal conditions during the students' classes. Further investigation of the energy efficiency of schools is carried out through building simulations. Existing situations, current trends and tendencies of schools provide essential information to facilitate the energy performance assessment of the building stock and to highlight the potential of energy savings and the upgrading of their indoor comfort.

13.1 Introduction

Energy efficiency in buildings is controlled by a number of directives, regulations, and strategies on both international and national levels. In December 2008, the European Commission implemented a strategy for "climate action," according to which, the EU has set an indicative objective to reduce its primary energy consumption by 20 % in relation to the projected 2020 energy consumption (Directive

M.C. Katafygiotou (✉) · D.K. Serghides
Department of Environmental Science and Technology, Cyprus University of Technology,
33 Anexartisias Street, 1st Floor, PC 3036 Limassol, Cyprus
e-mail: martha.katafygiotou@cut.ac.cy

D.K. Serghides
e-mail: despina.serghides@cut.ac.cy

S.-N. Boemi et al. (eds.), *Energy Performance of Buildings*,
DOI 10.1007/978-3-319-20831-2_13

2012/27/EU). The directive on energy performance of buildings (Directive 2010/31/EU) outlines measures that require member states to set minimum requirements and develop methods for determining the energy performance of buildings. Cities play a key role in mitigating climate change, as most of the construction and transportation occurs in urban areas.

The increase of complexity in school buildings to ensure users' comfort and social and functional demands are changing energy consumption patterns. In the Mediterranean context, the rise in energy consumption, and especially in electricity demand, confirms the need to improve energy management in schools in order to reduce their environmental impact (Mumovic et al. 2009; Santamouris et al. 2008). Also, the compliance with legal ventilation rates is a factor potentially increasing pressure on energy demand in schools. This is a current important research topic that often reveals the poor air quality conditions in the classrooms (Pegas et al. 2011).

Energy consumption in schools and day-care centers is usually high, and it has an impact on the communities' energy consumption and hence the energy bills (EU 2008; IPCC 2001; Directive 2002/91/EC; SET-Plan 2007; Itard et al. 2009; EuroACE 2004). Some reports, specifically on the context of the southern Europe and the Mediterranean, focus on the buildings' envelope and systems' features in order to propose adequate strategies for energy efficiency and indoor air quality in schools (Katafygiotou and Serghides 2014; Desideri and Proietti 2002; Becker et al. 2007; Dascalaki and Sermpetzoglou 2011). Therefore, benchmarking standards for the energy performance of schools already exist in many EU countries. Several studies also examined the typical yearly use of heating in school buildings (Santamouris et al. 2007; Jones et at. 2000; Corgnati et al. 2008). An investigation in Italy states that there is a considerable amount of oil consumption that can be significantly reduced if appropriate energy-saving techniques and renewable sources are used (Desideri and Proietti 2002). The school sector consumes high amounts of energy for heating and electricity, and therefore energy-saving measures are imperative. An average of 13 % of the total energy use in USA, 4 % in Spain and 10 % in the UK is consumed by schools (Perez-Lombard et al. 2008).

It must be mentioned that besides building design and systems, users are also a determining factor impacting energy consumption in buildings. However, behavioral issues are still among the areas least covered in the scientific literature. Understanding the interaction of users and buildings is therefore relevant to promoting better design strategies to enhance buildings' sustainability (De Santoli et al. 2014). According to Christina et al. (2014), organizational, social, and behavioral issues are still among the areas least covered. Becker et al. (2007) mentioned also that information on the energy consumption of school buildings is still very limited, especially in the Mediterranean region.

This chapter focuses on the Mediterranean island of Cyprus and it will initially identify the prevailing building practices of schools there. The average energy consumption for heating and electricity is calculated, and the typical secondary school building in Cyprus is described in order to examine the building stock principal techniques. The chapter provides useful information on the energy categorization of schools and the efficient refurbishment scenarios needed in order to decide on the most appropriate strategies and techniques to reduce their energy consumption. One of the greatest challenges of modern architecture is creating a thermally comfortable environment inside buildings, and studying in detail all the factors that contribute to it. The aim in each case is to achieve indoor air quality and thermal comfort at the minimum energy cost possible.

13.2 Methodology

The investigation of the schools' characteristics and their indoor comfort in Cyprus schools in parallel with the comparative study of their energy efficiency is the main theme of this chapter. Energy efficiency in buildings is always inseparable from the main internal thermal comfort. In obtaining certain indoor conditions, the cost in terms of energy must be taken into consideration. Humans, from the beginning of life, tend to create thermally neutral environments, in which they live and feel comfortable. In previous research by the authors, studies of schools were conducted over the last five years and they are presented as a comprehensive chapter today.

As a first step, the structural characteristics of the Mediterranean schools were investigated. A questionnaire campaign was carried out in all the secondary schools in Cyprus in order to define the most common structural practices and the average energy consumption of schools (Katafygioutou and Serghides 2014). The aim of the survey was to define the typical secondary school and present the most commonly applied construction techniques, building services, electricity and oil consumption, etc. Cyprus is a very good example of a Mediterranean country because it consists of three different climatic zones—inland, coastal, and mountainous—and it can effectively describe the Mediterranean trends and tendencies. Therefore, questionnaires were sent to all 156 secondary schools on the island, and a high percentage of them responded. The typical secondary school building was described through this research and will help extract important results regarding the principal practices and techniques that exist in the school buildings of the Mediterranean region.

Following this step, some pilot schools were selected and an investigation was carried out to examine the perception of indoor, environmental, and energy parameters of schools in Cyprus. Part of that study was developed within the framework of the Mediterranean (MED) Program TEENERGY SCHOOLS (high-energy efficiency schools in the Mediterranean area), which dealt with the energy efficiency of Mediterranean school buildings (Katafygiotou and Serghides 2013). For the analysis of the derived results, a statistical tool has been developed that is based on Excel spreadsheets, and remarkable results described the indoor comfort of schools in correlation with their energy consumption.

The users' opinion concerning the indoor, and especially the thermal, comfort of schools is also supported with in situ measurements in a typical school building on the island. During that time, climatic parameters, such as air temperature and relative humidity per minute, are recorded by data loggers inside and outside the building, and the monitoring was done for seven continuous days during each of the seasons. The field measurements were recorded by sensors and specifically by the CENTER 340 Thermo Recorder, Extech RHT10 Humidity and Temperature USB Dataloggers. The sensors' accuracy is ±1 °C for air temperature and ±3.5 % for relative humidity. Indoor environmental conditions are recorded in the most representative spaces of the building. It is worth noting that schools in Cyprus, and in almost all the Mediterranean countries, have a similar time schedule—five days a week from September until the end of June, with two major two-week holiday breaks (at Christmas and Easter). Classes are held from 7:30 until 13:35, and in many cases this changes from country to country and can be up to 14:30.

The school overview is concluded in this chapter with a case study in which energy simulations were carried out in order to define the energy performance of the representative/typical school building that was selected. The building design, and specifically its construction characteristics, were selected through field inspections and examined through parametric simulations using iSBEMcy software, which is the official governmental software for the categorization of energy efficiency in buildings and the calculation of CO_2 emissions according to the European Directive 2002/91/EC (Infotrend 2009). The software calculates the energy consumption of the building and concludes its energy performance certificate, which states the energy categorization of the building. All the European member countries have similar software for examining the energy classification of their buildings. Energy ratings of a building can provide useful information on its energy consumption and it is done through standard measurements under a specific experimental protocol. The energy auditing investigates the aspects that affect energy efficiency and the conservation potential of schools. As the energy consumption of a building is affected from its construction elements, alternative strategies and techniques for energy efficiency must be used to create a comfortable indoor environment that at the same time will achieve energy conservation. Therefore, based on the energy categorization of the building, various retrofitting scenarios are studied in simulations (Katafygiotou and Serghides 2014).

13.3 Schools' Building Stock Data

The Cyprus schools' building stock is described in this section in order to fully understand the school typologies and characteristics. Cyprus, although small in size, consists of three different temperate climatic zones: coastal, inland and mountainous (see Fig. 13.1). The map below was developed by the Meteorological Service of Cyprus and the three climatic zones have the following altitudes: coastal 0–100 m; inland 100–500 m; and mountainous >500 m.

The studied sample of school buildings includes a variety of morphological and architectural characteristics. The construction details provide information about the techniques that are followed in the different chronological periods, since each

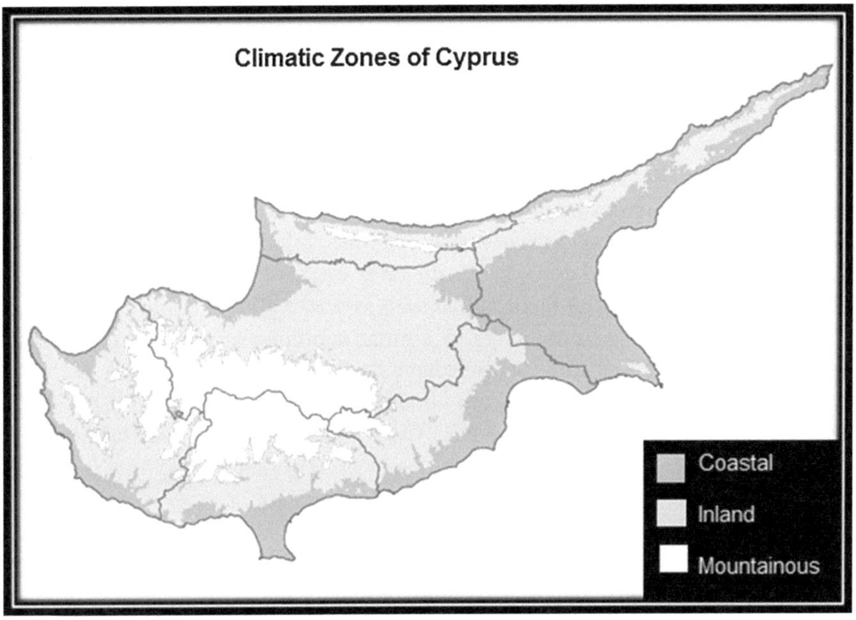

Fig. 13.1 The three climatic zones of Cyprus and the boundaries of each district

period is characterized by different influences, especially in the architectural sector. As was seen in the survey, most schools in Cyprus were built before the establishment of thermal insulation regulations and the same situation exists in all the European and Mediterranean countries (Katafygiotou and Serghides 2014). The height, the shape, and the orientation of the school buildings directly influence their energy behavior. Concerning the height of the buildings; most of the schools consist of two floors with a typical floor height of 3–3.5 m. The density of pupils and the equipment in classrooms also directly affect the internal gains and indoor comfort. The average area of a classroom is 46 m^2 and the density ranges from 2–3 m^2/pupil, which is satisfactory according to ASHRAE standards (Handbook 2001). Specifically, the average area for each pupil in the classroom is 2.03 m^2/pupil while in the entire school building it is 8.56 m^2/pupil; rectangular and Π shapes are the most common in school buildings. North is the prevailing orientation of school buildings, possibly due to the pleasant visual comfort that it gives the classrooms, and south and east orientation followed in preferences. West orientation is not preferred, except in individual cases with northwest or southwest orientation. Furthermore, all the schools in the sample have outside courtyards, since the Mediterranean climate offers the possibility of utilizing it almost all through the year. Usually a large part of the courtyard is surrounded by the school buildings, and therefore atriums are often created. The main corridors of schools are divided into internal and external ones.

There are also some schools that have both types of corridors. Outside corridors enhance natural ventilation and air changes in the classrooms during the summer but have increased heat loss during the cold winter if the building envelope is not insulated. In Cyprus, the outside corridors are often seen in schools, and this is again common in most Mediterranean areas. Concerning roofs, 83 % of school buildings consist of horizontal roofs in schools; pitched roofs with tiles are mainly found in the mountainous areas of the island and are similar to the roofs of northern European schools, which makes snow removal easier during winter. Regarding the thermal shield of the envelope, with only one exception among the public schools that were built after the establishment of thermal regulations, all the other schools have single-glazed openings. A similar condition is also found regarding the envelope's insulation. Most schools have no insulation and only in a few cases are schools partially insulated, and only one school of the whole sample in Cyprus has insulation on both its roof and walls. The walls of the schools are mostly made of brick, and in case they are damaged, 5 cm of extruded polystyrene is usually used. Further analysis of the sample concludes with the average thermal resistance of the building envelopes. The typical construction of walls, roofs, floors, and openings are studied and the usual average thermal conductivity of the building components is calculated and presented in Table 13.1.

Concerning the heating, ventilation, and air conditioning of schools (HVAC), they all have central thermostats that are adjusted daily by the school director or the maintenance employees. The heating systems are very old in most of the schools, and most have split units for cooling—but only in the staff offices and never in the classrooms. Exceptions are the private schools that have installed split units for cooling in most of the school's areas. In most cases, in order to control the cooling system, thermostats are autonomous in each room. From the answers to the questionnaires, it seems that the cooling systems are younger than the heating systems, and most of the split unit systems for cooling have energy classification A or B. Also, occasionally during the winter, split units operate for heating purposes. In almost half of the sample schools, a hot water system is installed, but only a few of them have solar panels for water heating, while most have central heating by oil or electric heaters for hot water use.

Regarding illumination, all schools have artificial lighting with fluorescents in classrooms and offices, and conventional or economic lamps in corridors and restrooms. The power of the lighting usually ranges from 50 to 80 W. Many schools

Table 13.1 Thermal conductivity of the buildings' components

Building envelope	Components structure	Insulated (yes/no)	U-Value (W/m^2 K)	Building envelope
External walls	Brick	No	1.37–1.39	External walls
	Concrete	No	3–3.5	
Ground floor	Reinforced concrete	No	0.6–0.8	Ground floor
Horizontal roof/ exposed floors	Reinforced concrete	No	2–3.5	Horizontal roof/ exposed floors

have already replaced almost all the conventional lamps with economical ones. Furthermore, the new office devices (e.g., computers, copy machines, scanners) in schools have the "Energy Star" mark.

The energy consumption of schools is mainly from electricity and heating oil. The lighting, the cooling, the electrical heating of hot water, and use of office equipment are the main expenses that affect the consumption of electricity. During the winter months, the consumption of heating oil is also an important factor of energy consumption. In the questionnaires, schools are asked to supplement the monthly consumption of the past three years for oil and electricity in order to examine their energy consumption levels. As is derived from the statistical results, the annual average consumption in Cyprus schools is 62.75 kWh/m^2. It is seen that in coastal areas, schools have increased needs for electricity, while the mountainous areas have increased needs for heating.

On the other hand, the most energy-intensive schools are in the mountainous areas, with an increased need for heating. In Table 13.2, the characteristics of the

Table 13.2 Characteristics of typical school buildings in each climatic zone

	Inland	Coastal	Mountainous
Type of school	Public gymnasiums	Public gymnasiums	Public gymnasiums
Number of users	491	516	347
Total area (m^2)	2755	3119	3640
Area of a typical classroom	43	46	55
Density of students in classrooms	1.82	1.99	2.59
Age of schools	21.4	28.5	68.3
Number of floors	2	2	3
Orientation	South	North	Northwest
Shape	Shape Π	Rectangular	Combination of shapes
Main corridor	Exterior	Exterior	Exterior
Insulation	No	No	No
Windows	Single glazed	Single glazed	Single glazed
Heating system	Oil central heating	Oil central heating	Oil central heating
Cooling systems	Split units	Split units	Split units
Hot water system	No system	No system	Through oil central heating
Illumination	Fluoresces	Fluoresces	Fluoresces
Equipment	With "energy star"	With "energy star"	N/A
Electricity consumption (kWh/m^2/year)	22.03	25.56	24.51
Oil consumption (kWh/m^2/year)	33.64	22.88	83.48
Total consumption (kWh/m^2/year)	55.67	48.43	107.99

Table 13.3 Typical Cypriot school building components

Type of school	Public
Climatic zone	Coastal
Area of a typical classroom	46 m^2
Density of students in classrooms	2–3 m^2/pupil
Number of pupils	488
Building period	1974–2000
Number of floors	2
Roof	Horizontal
Orientation	North
Shape	Rectangular or Π-shape
Courtyard	External
Main corridor	External
Insulation	No
Windows	Single glazed
Total energy consumption	62.75 kWh/m^2

typical school building per climatic zone are presented. Furthermore, these results conclude to the average typical school in Cyprus (a combination of the three climatic zones), Table 13.3.

13.4 Pilot Schools' Comfort and Energy Performance Investigation

The next step that this chapter presents is the investigation of the indoor comfort in selected Cyprus' pilot schools in parallel with the comparative study of their energy efficiency. Energy efficiency in buildings is always inseparable from the main internal thermal comfort. To obtain certain indoor conditions, the cost in terms of energy must be taken into consideration. In the survey, nine pilot schools were selected, all of which are located in the three climatic zones of Cyprus (coastal, inland and mountainous). The total of 185 students correspond in the survey. The questionnaires included questions on the general characteristics of the school environment, thermal, visual, and acoustic comfort, safety and security, psycho-physical wellness, and some pupils' suggestions for indoor comfort improvement. They were submitted directly to the pupils during classes and the completion was with the assistance of the teacher in charge of the project at each school. The questions on environment include the perception of thermal climate, ventilation, and various aspects of air quality. The questionnaire campaign was begun within the framework of the European-funded MED-TEENERGY (2011) program, in which eight partners from Italy, Spain, Greece, and Cyprus participated (Katafygiotou and Serghides 2013).

13.4.1 Indoor Environmental Conditions

The first overview of the general indoor environmental conditions is presented in Fig. 13.2. Illumination, indoor temperature, acoustic comfort, building construction condition, quality of rooms and furniture, available space per person, quality of other areas (restrooms, locker rooms, etc.), and level of cleaning and maintenance. Pupils were asked to rank the above-mentioned areas on a scale from "insufficient" to "sufficient" conditions. The overall average of nine schools and the general image of the schools rank at a little higher value than the average. The indoor air temperature, the acoustics, the quality of space and furniture, as well as the cleaning and maintenance, do not seem to satisfy the pupils.

A closer look at thermal comfort, which is the most correlated aspect of the energy efficiency of schools, shows that thermal conditions are unsatisfactory. The energy performance of buildings was directly affected by the thermal conditions and vice versa. Thermal comfort of schools is significant, since it promotes the building energy efficiency in the most effective way.

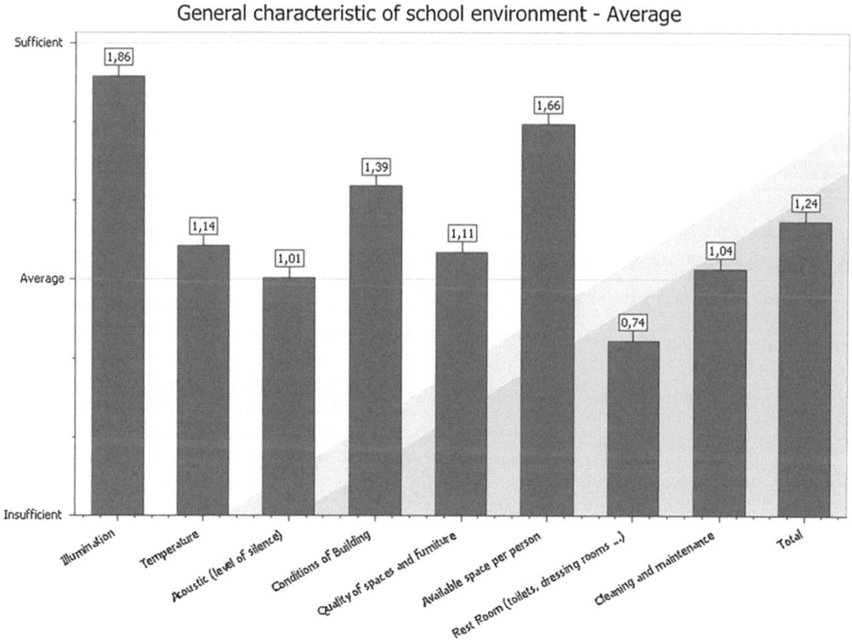

Fig. 13.2 Average profile of nine schools

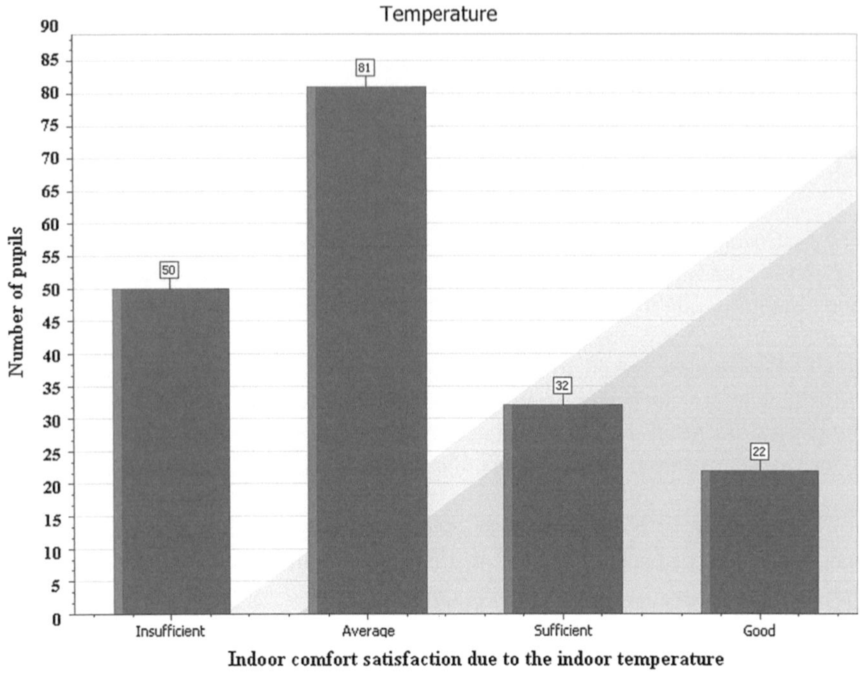

Fig. 13.3 Thermal comfort in Cypriot schools

The results are almost the same in all three climatic zones (Fig. 13.3). The percentage of 70.8 of pupils shows that thermal comfort is at an insufficient or average level. This directly demonstrates the problem of thermal comfort in most of the existing buildings, and particularly in schools. It is important to note that the heating system is the same in all schools in Cyprus and consists of central oil heating. However, the degree of thermal discomfort expressed by the pupils is related to the windows' orientation, combined with any passive heating strategies used and the reduced heat losses.

13.4.1.1 Studies Comparing Thermal Comfort and Energy Efficiency

Continuing the research, a comparative assessment was developed that shows the relation between poor indoor quality conditions and low energy efficiency of buildings. This may occur due to the mismanagement of the HVAC systems in schools and the inefficient insulation of buildings. The relationship between indoor comfort, especially thermal, and energy efficiency is essential because these two issues are inextricably intertwined. The energy efficiency of the pilot school was simulated through iSBEMcy, the official government software, and the energy categorization of the school buildings was derived. The general results from the

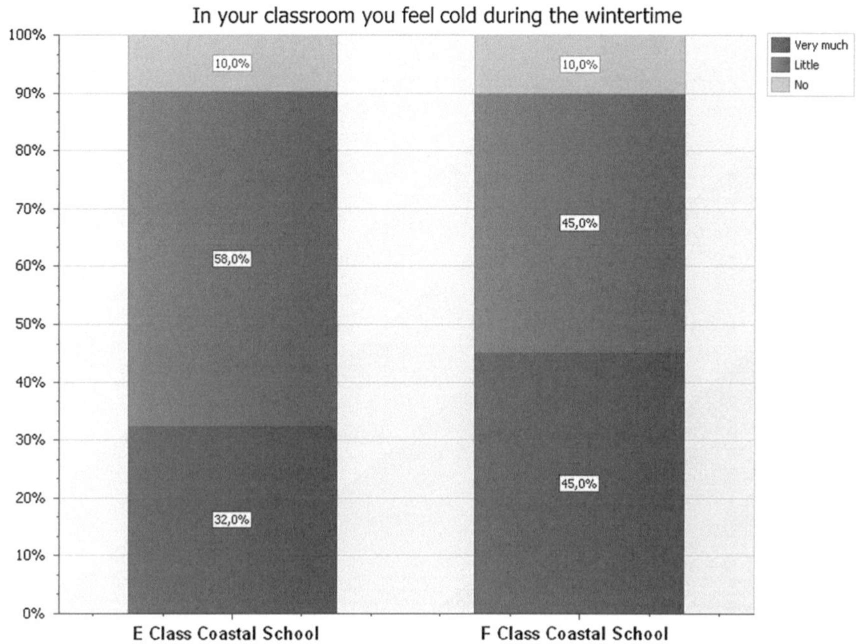

Fig. 13.4 Comparison between "E" and "F" energy class schools in coastal areas during winter

research show that the quality of structures and energy performance of schools are very low—even for schools that were built during the last decade. Briefly, the energy categorization of the pilot schools is "D" classification or lower, and therefore this indicates that the school stock needs urgent upgrading and large amounts of energy conservation (Katafygiotou and Serghides 2013).

As a typical and representative sample of the research, the coastal area is presented. In coastal areas, the energy classification of two schools is inadequate, as they are classified in the "E" and "F" energy classes. The percentage of pupils that do not feel cold during winter in the classroom is the same in both schools and is only 10 % (see Fig. 13.4). On the other hand, the percentage of pupils that feels a little or very cold is higher. The school with the lower energy classification has 13 % more pupils feeling too cold. This shows better indoor thermal comfort in the school with a better energy class.

The thermal comfort of pupils in the same schools' buildings during summer is also presented in Fig. 13.5. In contrast with wintertime, it appears that the "F" energy class school has better internal conditions for summer, and 5 % of the pupils answered that they do not feel hot during the summer in their classrooms.

Nevertheless, the percentage of pupils that feel too hot is more in the "F" class school than in the "E" class school. While it seems that in winter the two schools have differences in thermal comfort, during the summer the thermal comfort

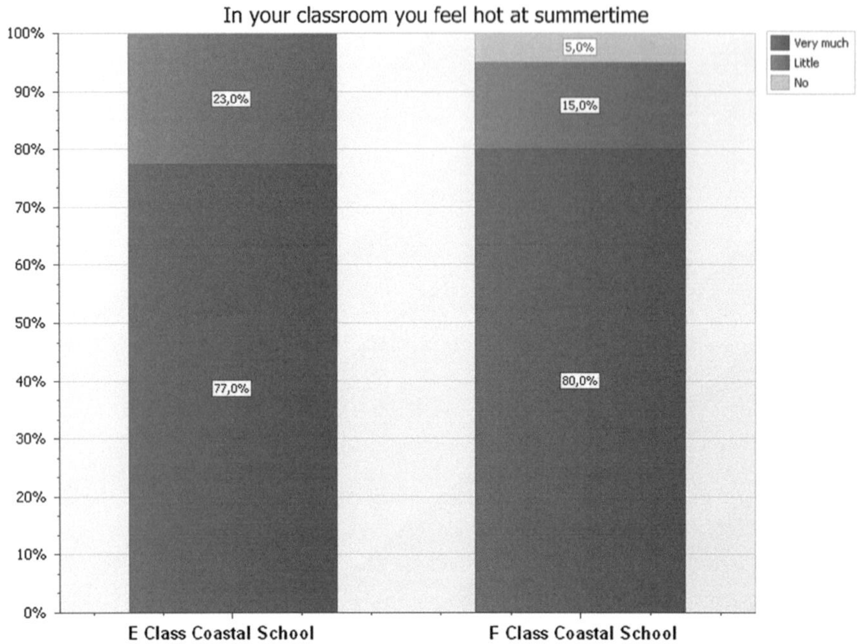

Fig. 13.5 Comparison between "E" and "F" energy class schools in coastal areas during summer

conditions are very similar in both schools. In summer, neither of the schools seems to satisfy the pupils since there is no conventional cooling. The thermal comfort results of the two schools are compatible and this might be connected with their energy classification which is similar.

The results of the comparative studies are significant because most of the pilot schools provide better indoor comfort conditions in buildings with better energy classification. This indicates that users consume less energy if the indoor conditions of a building are better. The results show that the energy efficiency of schools in Cyprus is poor, and therefore the indoor quality conditions are inferior, especially thermal comfort. In a building where bioclimatic design or passive techniques are absent and the construction is poor—without any envelope insulation—there is a lack of thermal comfort. Furthermore, due to the Mediterranean climate of Cyprus, hot months last more than the cold ones and therefore focusing on cooling strategies and techniques is essential.

13.5 Field Measurements of Climatic Parameters in a Typical School Building

The field study was conducted in a selected secondary school building in Cyprus, to assess the indoor thermal conditions during the students' classes. The Neapolis Gymnasium was selected since it is a typical Cypriot school located in the coastal

Fig. 13.6 Neapolis Gymnasium, coastal area of Cyprus

city of Limassol (Fig. 13.6). Briefly, this school was built in 1985 and consists of two floors and a horizontal roof with no insulation; neither is there insulation on the walls, which are made of brick, and the windows are single-glazed. The Neapolis School also has a combination of a rectangular and Π-shape, with north and south orientation. Moreover, the courtyard is outside and the building has mostly exterior corridors, and an old central oil heating system with a central thermostat. It also has split units for cooling in the offices and in the computer labs, and the hot water system operates through the central heating system. The lighting consists of fluorescents and some of the equipment has "Energy Star" marking.

We learned from the study that this "typical" school building does not provide the required thermal comfort conditions for its users. Through detailed field measurements, we confirmed that the climatic conditions and, specifically, the air temperature and relative humidity are in many cases unsatisfactory. Mainly during winter and summer, the majority of users feel discomfort, and this is verified by the measurements from sensors. The temperature differences between outdoor and indoor spaces are small when the air conditioning is not working, and this happens because the building is non-insulated (Serghides and Georgakis 2012).

Through this research, useful conclusions have been reached regarding the thermal comfort of school buildings in Cyprus. An extensive monitoring was take place which lasted a whole academic year. It provides sufficient information about the climatic conditions in school buildings, especially those without insulation. As seen in Fig. 13.7, in the spring the indoor conditions are mostly comfortable, but again the high humidity levels cause discomfort to the students. During the summer, the internal conditions are hot and humid, and in the autumn, the prevailing conditions are mostly comfortable and partially warm and humid. The warmth is especially strong from September until the middle of October and afterwards the temperatures decrease significantly. During the winter, results show a comfortable feeling and although the outdoor conditions are cold inside the school the temperature is higher than the desire one, due of the misuse of the heating system. Personal interviews of students led to the conclusion that pupils who are seated

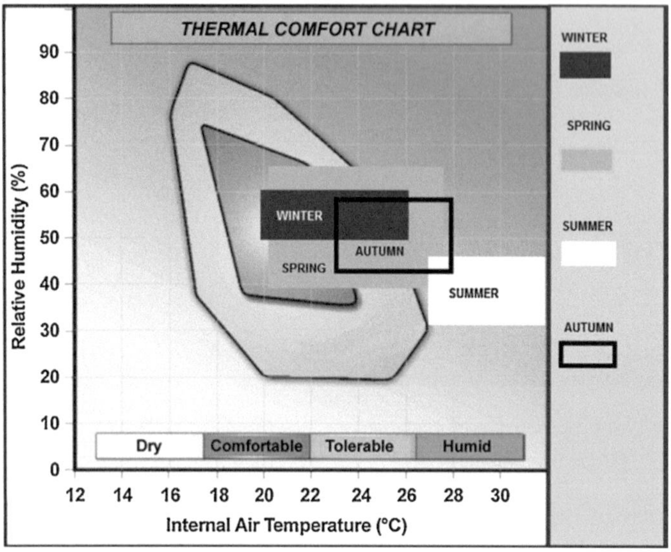

Fig. 13.7 Thermal comfort chart with indoor conditions of the different seasons

near windows and doors feel cold, while those in the middle of the classroom or near the heaters feel hot. The lack of insulation in combination with the misuse of heating system, which is set on higher than the appropriate temperatures, concludes on those results.

13.6 Energy Simulations and Upgrade Scenarios of a Typical School

A case study with simulations for energy performance of the typical school building (Neapolis Gymnasium) has also been developed (Katafygiotou and Serghides 2014). Variables of the building elements are examined through parametric simulations using the iSBEMcy software. Based on the energy categorization of the building, five retrofitting scenarios were studied via simulations. The software calculates the energy consumption of the building and concludes its energy performance with a certificate that states the energy categorization of the building. The energy auditing of the building under study investigates the aspects that affect its energy efficiency and its conservation. Initially, the simulations conclude with the primary energy consumption of the particular building (Fig. 13.8).

The school building was categorized in energy class "D" with an annual primary energy consumption of 442 kWh/m^2/year. The consumption that is derived from the iSBEMcy simulations is more than the actual consumption as calculated via the annual bills, since the software concludes to the primary energy consumption. Primary energy is the energy form found in nature with the

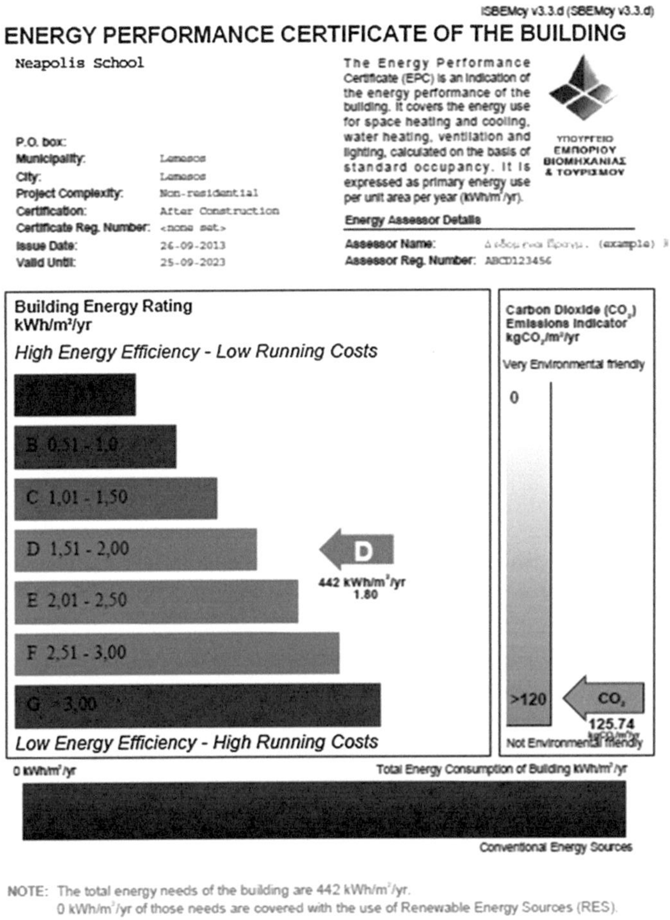

Fig. 13.8 Energy categorization of the existing school building

conversion or transformation process to electricity or oil for heating; this is the reason that it is so high.

Based on the energy categorization of the building, various retrofitting scenarios are studied in simulations. The scenarios concern building insulation, window replacement, and other scenarios for the operation of heating and cooling systems. Following these, and through comparative studies, the most energy-efficient techniques are presented (Table 13.4).

From the results of the simulations, it was deduced that the most efficient retrofitting scenario with a 47.5 % savings is the removal of all the split air-conditioning units except the one in the lobby and two more in the director's and administrative offices. All the other cooling systems, either in the assistant director's offices, in the music room, or in computer labs, can be replaced with fans that consume considerably less energy. The cost for this scenario would maybe be the

Table 13.4 Energy savings from various retrofitting scenarios

Retrofitting scenarios	Energy conservation (%)	Energy classification
Scenario A: Addition of 5 cm extruded polystyrene on roof	16.5	D
Scenario B: Addition of 5 cm extruded polystyrene on walls (external site) and roof	31.9	C
Scenario C: Replacement of single-glazed with double ones	3.6	D
Scenario D: Remove all the split air-conditioning units (except those in management, administrative office and lobby)	47.5	B
Scenario E: Increase of the efficiency of the heating and cooling system	23.5	C

lowest of all the other retrofitting scenarios, and it improves the energy class of the building from D to B.

The second most efficient scenario, with a 31.9 % savings, is the addition of 5 cm of extruded polystyrene on walls and on the horizontal roof. Although this technique (wall insulation) is fundamental for preserving heat loss during the winter, it is also very expensive, difficult to be installed in existing buildings and usually needs a long payback period. With this scenario, the energy class of the building would be upgraded from D to C. Therefore, the first scenario (with the addition of 5 cm of extruded polystyrene only to the roof of the school) seems more realistic and applicable.

The replacement of all the single-glazed windows with double ones is not a very efficient strategy since the energy savings is only 3.6 %, and the energy class of the building remains the same. Nevertheless, in cases where the frames and windows will remain single-glazed, they should be airtight in order to prevent heat loss during the winter.

Finally, upgrading the heating and cooling systems provides energy savings of 23.5 % and an energy class upgrade from D to C. This can be achieved either by maintenance of the heating system or replacement of the old cooling split units with new A energy class units.

13.7 Conclusions

From this chapter it is clearly seen that the Mediterranean school building stock needs a substantial energy upgrade. The construction characteristics of schools directly affect their energy performance. It is concluded that the most appropriate design element for an energy-efficient school building is the insulation of the envelope and the proper use of HVAC systems especially for cooling. The shape, in combination with the orientation of the buildings, also affects indoor thermal comfort. The right building orientation may serve very well and it seems to consume less total energy. Moreover, since the schools are closed during the summer, the cooling system must be limited to office spaces, and the energy classification of the split units must be in the A or B class.

The results also show that pupils are not satisfied with the indoor comfort conditions of their school buildings. In particular, the analysis of the thermal comfort depicts inadequate conditions in all three climatic zones. Through the detailed field measurements, it was also confirmed and validated that the indoor climatic conditions, and specifically air temperature and relative humidity, are in many cases unsatisfactory for the users. Mainly during the winter and summer, the majority of users feel discomfort, and in most cases this is confirmed by the measurements from sensors. The studies show that the problems arise from the old heating systems and the non-insulated building envelopes of schools.

Significant also were the results of the comparative study between the indoor comfort conditions and the energy efficiency of the buildings. Most of the pilot schools reveal better indoor comfort conditions in buildings with better energy classifications; this indicates that users consume less energy if the indoor conditions of a building are better. The outcomes show that the energy efficiency of schools in Cyprus is bad, and therefore the indoor quality conditions are poor—especially the thermal comfort. In a building where bioclimatic design or passive techniques are absent and the construction is poor without any envelope insulation, there is a lack of thermal comfort. Furthermore, due to the Mediterranean climate of Cyprus, there is more hot weather than cold, so focusing on cooling strategies and techniques is essential.

Energy awareness among pupils and teachers is very important in order to upgrade the energy performance of schools. Public buildings, and especially schools, should be on top of the list for improving energy efficiency. In this way they will act as exemplars, and they will raise energy and environmental awareness in schoolchildren. The next goal will be the "nearly zero energy" school building. This will be a challenge in European Union as well as in the Mediterranean region since the summertime needs demand an alternative perspective of design for hot climates.

References

A European Strategic Energy Technology Plan (SET-PLAN); COM (2007) 723 Final. Commission of the European Communities, Brussels

Becker R, Goldberger I, Paciuk M (2007) Improving energy performance of school buildings while ensuring indoor air quality ventilation. Build Environ 42:3261–3276

Christina S, Dainty A, Daniels K, Waterson P (2014) How organisational behaviour and attitudes can impact building energy use in the UK retail environment: a theoretical framework. Archit Eng Des Manag 10:164–179

Corgnati SP, Corrado V, Filippi M (2008) A method for heating consumption assessment in existing buildings: a field survey concerning 120 Italian schools. Energy Build 40(5):801–809

Dascalaki EG, Sermpetzoglou VG (2011) Energy performance and indoor environmental quality in Hellenic schools. Energy Build 43:718–727

De Santoli L, Fraticelli F, Fornari F, Calice C (2014) Energy performance assessment and a retrofit strategies in public school buildings in Rome. Energy Build 68:196–202
Desideri U, Proietti S (2002) Analysis of energy consumption in the high schools of a province in central Italy. Energy Build 34:1003–1016
EU Energy and Transport in Figures, Statistical Pocket Book 2007/2008 (2008) European Communities, Brussels, Belgium
Handbook A (2001) Fundamentals, American society of heating, refrigerating and air conditioning engineers. Atlanta
Infotrend Innovations/BRE for the Ministry of Commerce, Industry and Tourism (2009) Methodology for assessing the energy performance of buildings, 1sted. Cyprus Energy Service, Nicosia, Cyprus
Intergovernmental Panel on Climate Change (IPCC) (2001) Climate change 2001: Mitigation; third assessment report, working group III. IPCC, New York
Itard L, Meijer F (2009) Towards a sustainable Northern European housing stock: figures, facts and future. IOS Press, Amsterdam
Jones P, Turner R, Browne D, Illingworth P (2000) Energy benchmarks for public sec-tor buildings in northern Ireland. In: Proceedings of CIBSE national conference, Dublin
Katafygiotou MC, Serghides DK (2013) Indoor comfort and energy performance of buildings in relation to occupants' satisfaction: investigation in secondary schools of Cyprus. Adv Build Energy Res 1–25
Katafygiotou MC, Serghides DK (2014) Analysis of structural elements and energy consumption of school building stock in Cyprus: energy simulations and upgrade scenarios of a typical school. Energy Build 72:8–16
Katafygiotou MC, Serghides DK (2014) Thermal comfort of a typical secondary school building in cyprus. Sustain Cities Soc 13:303–312
Mumovic D, Palmer J, Davies M, Orme M, Ridley I, Oreszczyn T et al (2009) Winter indoor air quality, thermal comfort and acoustic performance of newly built secondary schools in England. Build Environ 44:1466–1477
Official Journal of the European Union (2010) Directive 2010/31/EU of the European Parliament and of the council
Official Journal of the European Union (2012) Directive 2012/27/EU of the European Parliament and of the council
Pegas PN, Alves CA, Evtyugina MG, Nunes T, Cerqueira M, Franchi M et al (2011) Indoor air quality in elementary schools of Lisbon in spring. Environ Geochem Health 33:455–468
Perez-Lombard L, Ortiz J, Pout C (2008) A review on buildings energy consumption information. Energy Build 40(3):394–398
Recast of the Energy Performance of Buildings Directive (2002/91/EC); COM (2010) 755/ SEC(2010) 2821. Commission of the European Communities, Brussels
Santamouris M, Mihalakakou G, Patargias P, Gaitani N, Sfakianaki K, Papaglastra M, Geros V (2007) Using intelligent clustering techniques to classify the energy performance of school buildings. Energy Build 39(1):45–51
Santamouris M, Synnefa A, Asssimakopoulos M, Livada I, Pavlou K, Papaglastra M et al (2008) Experimental investigation of the air flow and indoor carbon dioxide concentration in classrooms with intermittent natural ventilation. Energy Build 40:1833–1843
Serghides DK, Georgakis CG (2012) The building envelope of Mediterranean houses: optimization of mass and insulation. J Build Phys 36(1):83–98
Towards Energy Efficient Buildings in Europe (2004) Final report. EuroACE, London

Part III
Building's Design and Systems

Chapter 14
New Challenges in Covering Buildings' Thermal Load

Simeon Oxizidis

Abstract This chapter identifies the major features of the emerging heat-power nexus in the built environment and suggests the specific energy technologies that will most likely successfully address the new challenges in covering buildings' thermal loads. These challenges are related in the supply side with the anticipated smart decarbonized grid, and in the demand side with the strict building performance regulations that require Zero Energy Buildings (ZEBs). Heat pumps, solar thermal collectors, photovoltaics and combined heat and power and thermal energy storage (TES) display the energy performance characteristics that can allow buildings to fulfill the ZEB agenda and integrate seamlessly with the smart grid. In the temperate climates of the Mediterranean region, those particular building energy technologies, either for new buildings or retrofits, can increase the flexibility provided by buildings and offer demand response services. For this demand response potential to be significant in high-performance buildings, it needs to be augmented by energy storage technologies. However, to realize the holistic concept of the demand-responsive/smart-grid-ready building, advanced sensoring, controlling, and connectivity infrastructure is required.

14.1 Introduction

The electrification of the modern world and the availability of energy resources during the last century laid the groundwork for the technological development of building energy systems to meet the thermal and visual comfort requirements of building occupants. Even the particular definitions of the concepts of comfort are in one sense derivatives of the features and performance characteristics of those

---

S. Oxizidis (✉)
International Energy Research Center, Tyndall National Institute,
Lee Maltings, Dyke Parade, Cork, Ireland
e-mail: simeon.oxizidis@ierc.ie

S.-N. Boemi et al. (eds.), *Energy Performance of Buildings*,
DOI 10.1007/978-3-319-20831-2_14

building energy systems when covering building thermal loads. Their availability and technical capabilities have in turn shaped the way buildings are conceptualized, designed, and eventually constructed.

Nowadays, buildings are the major end-users of any energy network, and it is anticipated that this century will see another paradigm shift in buildings' energy supply and the relationship between the electric grid and the buildings. The decarbonization of the former by great penetration of intermittent renewable energy sources (RES) and the deployment of interconnectivity, instrumentation, and intelligence features in its structure, will most definitely change the way we perceive, design, and advance both building energy supply systems and the buildings themselves as energy entities. However, another imperative, with respect to buildings and their energy systems, can also be identified in the regulatory framework that sets very strict limits for the energy consumption of buildings.

Thus, on the supply side, the emerging smart decarbonized grid, and on the demand side, the stringent building performance regulations that require zero energy buildings (ZEBs) are formulating a very challenging environment for addressing building thermal loads. The increased participation of RES in the electricity generation fuel mix deprive the supply side of its traditional flexibility, which can only be counterbalanced by increased flexibility on the demand side (i.e., buildings).

Those constraints send signals that the era of prescribing building energy systems to exclusively and timely address a heating or cooling load is no longer the sole aim of heating, ventilation, and air conditioning (HVAC) design. The evolution of the building as a dynamic energy system is more true today than ever. From now on, the focus is not just on addressing thermal loads but designing systems that comply with the ZEB regulations when providing thermal comfort, while at the same time are a best fit with the dynamics of their wider heat-power nexus. The twofold role of building energy systems anticipates the close interdependency of electrical and thermal energy in the built environment, and already suggests specific technologies that can serve that purpose.

RES are in the epicenter of the heat-power nexus; ZEBs need to rely on them and need to adjust to their intermittent nature, both on-site and on the grid level. The lack of flexibility on the supply side (of either distributed or central renewable generation) should be balanced by increased flexibility on the demand (buildings) side. The only way for buildings to achieve that flexibility is by energy storage. And since the major energy uses in buildings are heat-related, it is more than necessary to deploy some means of thermal energy storage (TES)—which, as energy storage goes, is probably the most effective technology available—either active (e.g., hot water tanks), or passive (e.g., high thermal mass building fabrics).

In the following, the building energy supply systems are defined, along with their relationship to climate and their importance in building retrofitting. In Sect. 14.2, the specific imperatives of the contemporary and near future regulatory and technological environment are explored, and the link between building energy systems and thermal comfort is described. Section 14.3 introduces the heat-power nexus in the built environment, and identifies the particular energy technologies that are best suited to address all the challenges that are emerging in this nexus. The chapter concludes with a vision of the future buildings and provides the prescription for the specifications of those buildings in order to be both smart and zero-energy.

14.1.1 Defining Building Energy Supply Technologies

Building energy systems can be described as engineered systems that deliver thermal energy in order to control and maintain the conditions of the indoor built environment. Most often, they are used to convert some other form (primary or renewable or other) of energy into thermal energy delivered in a building. We usually refer to them as HVAC systems, a term that includes both their supply and demand sides.

Figure 14.1 provides a rough sketch of the HVAC main components and modes of operation. Assuming a heating mode, at the top left corner are the energy resources (fossil fuels, electricity, renewable energy, or district heat). To the lower and right are the thermal energy production systems (boilers, solar thermal collectors, etc.), followed by the distribution system (pipes or ducts) that deliver the heating energy to the emission systems (radiators, fan coils, VAV boxes, etc.) in the lower right-hand corner. We can describe any heating system using this taxonomy, where by converting an energy source to heat in the plant (or primary) equipment we can move the produced thermal energy through the distribution system to the inside of the building where it is transmitted to the different building zones by our emission systems to maintain comfort. The reverse case occurs in the cooling mode where we remove heat from the buildings' zones and transfer it to our plant equipment (chiller) to adjust it to the particular conditions suitable for disposal in our heat environmental sink (e.g., ambient air).

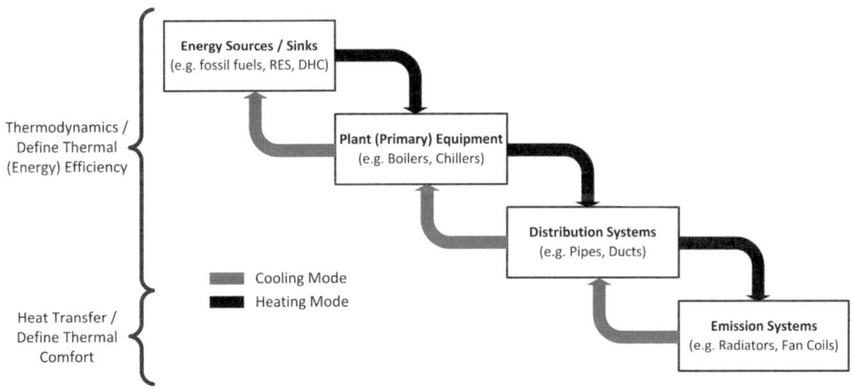

Fig. 14.1 Components of building energy systems

However, all the components shown in Fig. 14.1 may not be physically present in a real system where some of them can be omitted or combined in a single physical entity (packaged systems). A typical example is an electrical infrared heater that does not require a heat distribution system and combines the plant equipment and the emitter in a single piece of equipment. Another point worth mentioning—that is omitted in the graph—is the probable presence of a TES system between the plant equipment and the distribution system to provide a buffer or longer term storage of heat.

Regarding the performance properties of the components in Fig. 14.1, emission systems are usually characterized by their heat transfer features (radiant, convective, or hybrid systems), while plant equipment is characterized by its thermodynamic features, most importantly efficiency. The scope of this chapter includes only the plant equipment that refers to building energy supply technologies. A categorization of the most commonly used plant equipment can be found in Fig. 14.2, where we can distinguish between heating and cooling plant equipment and heat pumps that can reverse their mode so they can perform both functions.

Certain building systems like roof—mounted PVs lie outside the above definition of plant equipment, since usually in grid-connected buildings they just transmit all their produced energy to the grid and do not directly address the thermal or electrical loads of the buildings. However, they can be considered building energy supply technologies and as such they will be treated in this chapter.

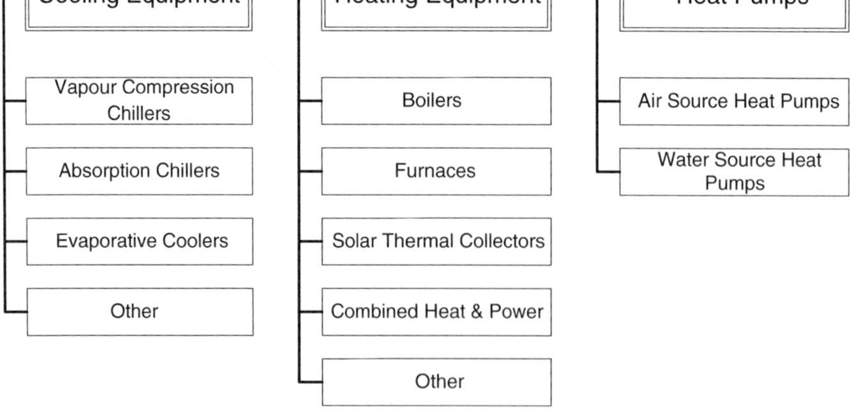

Fig. 14.2 Plant equipment categorization

14.1.2 Building Energy Supply Technologies in Temperate Climates

The climate responsive design of buildings has been exclusively linked to the architectural aspects of buildings. Terms such as "bioclimatic architecture," or "passive" or "solar buildings" suggest that exact notion. However, building energy supply systems themselves are affected by the climate, and their performance is strongly dependent upon specific climatic variables.

Ambient air temperature strongly affects the performance of several energy systems—most importantly of air source heat pumps, whose performance depends on the heat source (ambient air) temperature during heating operation, or the heat sink (ambient air) during the cooling operation. The thermal behavior of several other systems (solar thermal collectors, photovoltaic panels) is determined by their heat losses to the ambient air and are subsequently affected by the air temperature.

Solar radiation magnitude and intensity plays a vital role in the performance of several active solar systems (solar thermal collectors, photovoltaic panels, etc.). Other climatic parameters have a secondary effect on the performance of specific systems. Ambient humidity influences the operational features of cooling towers or evaporative cooling systems, since it limits the potential of cooling by evaporating water, while wind affects the heat losses of energy systems exposed to it (e.g., solar collectors).

It is thus valid to claim that specific building energy supply systems are climate-responsive technologies, which is why climate should be an important factor when selecting and designing building energy supply systems. In that sense, the temperate climate of the countries in the Mediterranean belt already suggests specific building energy supply technologies that are most appropriate for the buildings of those areas. The presence of both a heating and a cooling period, combined with rather mild, long winters, promotes the use of reversible air source heat pumps to cover both loads. Good insolation, on the other hand, makes the installation of either solar thermal collectors of photovoltaics panels or both on the Mediterranean buildings ideal.

14.1.3 Building Energy Systems in Retrofitting

Since the typical lifetime of any building is considered to be between 50 and 100 years, during that time it is envisaged that few changes will occur to its envelope—which will usually be highly intrusive and disruptive, and very expensive. Buildings' energy systems, on the other hand, have a far shorter lifetime (usually no more than 20–25 years), and during a building's lifetime will be replaced or upgraded several times. Thus, modernizing those systems offers great opportunities not only to achieve energy efficiency but also to adjust to all the developments and advancements that are taking place with respect to building technology and regulations.

Nowadays, such developments refer mainly to the emerging "smart grid." Upgrading building energy systems in this environment should make provisions for their interconnection with the anticipated smart grid in order to be ready to exploit the possible capabilities that will arise from the deployment of smart meters. Such energy systems are the ones that can offer demand response (DR), potential, and mainly refer to technologies that can address building thermal loads and are either electrically driven (e.g., heat pumps) or generate electricity (e.g., microCHP or PVs), and, most commonly, utilize some type of energy storage in addition (e.g., hot or cold water tanks).

The same notion regarding upgrading and retrofitting is also valid for the control and automation infrastructure of building energy systems, which usually have an even shorter life span, especially as digital technologies advance at a much faster pace. For buildings to become interconnected and integrated to the smart grid, the upgrading of their ICT (information and communication technology) infrastructure is crucial, but in parallel does not require either huge capital investments or highly disruptive interventions.

It is very important to emphasize that upgrading building energy supply technologies and their ICT infrastructure for providing energy services to the grid (demand response) opens a new market for energy services companies (ESCOs) and can potentially introduce new financing schemes for retrofitting buildings with shorter payback periods. Such a possibility can increase, in the near future, the market for building retrofitting, which is still stale and not in line with the ambitious goals set by the EU.

14.2 Shifting the Paradigm

In order to achieve the EU-set goal of "near zero energy buildings," (nZEB) technological innovations and regulatory measures are required on many levels. The EU has already set up the legislative framework that can facilitate the nZEB agenda; that agenda is presented schematically in Fig. 14.3.

In the next paragraphs the main EU regulations (directives) that are affecting the energy performance of buildings and their systems are being discussed, followed by the advancements that are shaping the technological background of the future buildings with respect to their energy features.

14.2.1 The Zero Energy Building Agenda and the Regulatory Environment

In the EU there is a bouquet of directives that directly or indirectly affect the energy performance of buildings and subsequently promote the nZEB agenda. The most important are:

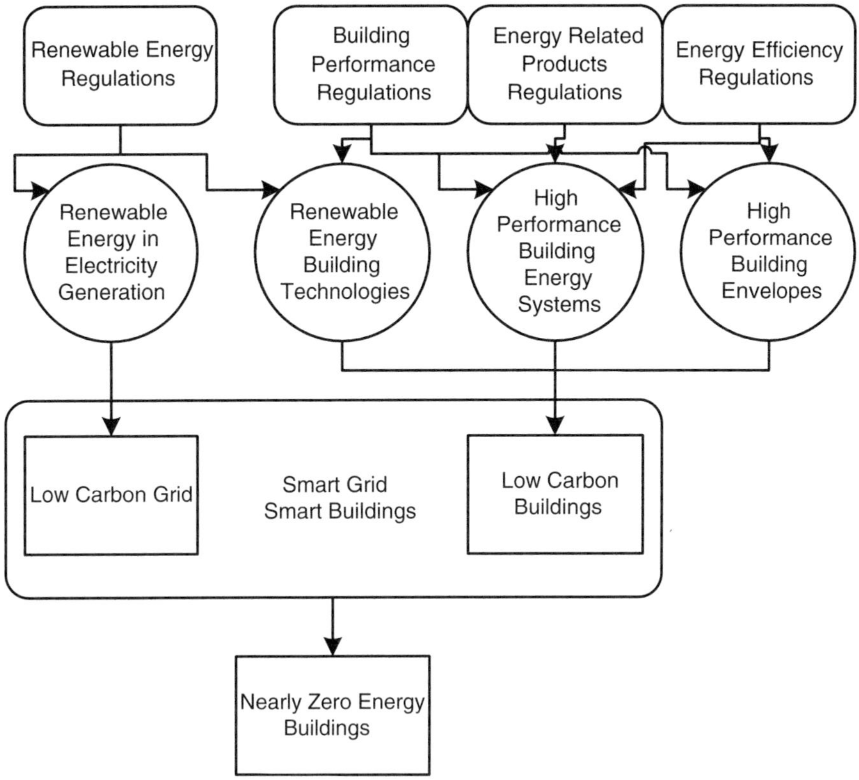

Fig. 14.3 Realization of the nZEB agenda

- EU Directive on the energy performance of buildings (EPBD)—Recast (EC 2010b)
- EU Directive on the promotion of the use of energy from renewable sources (EC 2009a)
- EU Directive on energy efficiency (EC 2012)
- EU Directive establishing a framework for the setting of ecodesign requirements for energy-related products—Recast (EC 2009b), and
- EU Directive on the indication by labelling and standard product information of the consumption of energy and other resources by energy-related products (EC 2010a).

The EPBD, since its first version (EC 2002), was a powerful tool for promoting both policy and technology in the EU with respect to energy-efficient buildings. The recast (EC 2010b) of this directive imposes even stricter building energy standards, which can only be implemented by the use of specific technologies such as heat pumps, solar thermal collectors, and PVs. Probably the single most important requirement is that all new buildings, as of 1.1.2021, are to be built as nZEBs. That can only be achieved with a significant share of renewable energy.

The directive that specifically targets renewables (EC 2009a) states that 20 % of the EU's final energy demand in 2020 should be provided by RES. This directive is important in the built environment not only from the supply (grid) point of view, but also from the demand (buildings) point of view, since it explicitly mentions building energy supply technologies for heating and cooling from RES—namely, solar thermal collectors and heat pumps, and, to a lesser extent, biomass. However, both on-site and off-site RES are strongly related to the buildings' energy features, since the penetration of RES in the grid above a specific threshold can only be achieved by appropriate demand management from the buildings' side.

The energy efficiency directive (EC 2012) itself specifically mentions cogeneration and district heating and cooling as a means for promoting efficiency in the heating and cooling of buildings. However, another critical and very interesting component of the directive is the provision of measures for opening the energy services market; thus, the directive aims to give access and space to aggregators to actively participate and operate in the European market in order to help guarantee grid security and ensure operational stability (SEDC 2014).

Energy services from buildings are also included in the most recent environmental ratings schemes, which are another type of regulatory provisions that are still not mandatory in Europe but are becoming more and more popular. The most commonly used environmental rating scheme (the US-originated "LEED") in its latest version (LEED v4.) includes a provision for DR, so if a building is wired to provide DR, it can claim additional credits. The overall goal and intent of that provision is to make energy generation and distribution systems more efficient and reduce greenhouse gas emissions, but it is very interesting that a green building initiative such as LEED moves towards an integrated approach where a building can be considered green when it uses energy in accordance with the performance features of the grid (USGBC 2015).

The Energy Labelling (EC 2010a) and Ecodesign (EC 2009b) directives refer to product groups that use, generate, transfer, or measure energy. Both establish policy frameworks that are to be implemented by subsequent measures and actions. The Ecodesign directive sets minimum requirements for products that must be fulfilled to minimize their environmental impact. This is of special interest to vendors of energy supply systems, since it includes heaters, boilers, and air conditioning units. The Energy Labelling directive sets the labelling requirements according to the provisions of that of the Ecodesign.

14.2.2 The Smart Decarbonized Grid Landscape and the Connected Building

Modern buildings depend on unconstrained and reliable supply of energy as it is facilitated from the instantaneous functionality provided by today's electricity grid. However, if electricity generation is about to be proliferated by RES contribution, a big challenge emerges. RES are intermittent energy sources, and when used for electricity generation cannot provide the same level of service as the fully dispatchable generation. Thus, the flexibility lost when RESs increase their share in the electricity generation fuel mix should be replaced elsewhere in the supply/demand loop. That non-generation-based flexibility should be accommodated by the buildings themselves since they are the main end-users.

Beyond the vast deployment of RES generation, in order for the smart grid concept to be truly realized, the integration of sensing and monitoring equipment, along with control, automation, and communications infrastructure is required. When all the pieces of the puzzle are in place, a transformational change—mainly to the distribution grid—will be possible. The traditional notion of a one- directional distribution grid where electricity is passively transferred to the buildings will become obsolete, and the flow of electricity will become two-directional, allowing further deployment of building-integrated distributed generation (micro-CHP, PVs).

However, a smart decarbonized grid does not encompass the same features and operational characteristics globally. For each region is being conceptualized locally, and to a great extent follows the climatic (solar and wind resources availability) and geographical (hydro and marine energy resources availability), and even geopolitical (international connections with other electricity and natural gas networks) particularities of the supply side along with the cultural and climate-dependent features of the demand side (buildings loads). Especially with regard to the countries of the Mediterranean belt, most of them deal with a bouquet of RES generation, inclusive of all three main types (wind, solar and hydro).

EEGI (2013) has recognized that one of the most important barriers to be overcome for the realization of the smart grid is the development of all possible sources of flexibility on the electricity networks, which may be provided by generation (large and small), by demand response, and/or by storage resources. Small (distributed) generation, DR and TES are all resources directly linked to buildings' energy systems.

14.2.3 Thermal Comfort and Energy Supply

Thermal comfort—as defined today—is not a catholic, global variable deterministically defined by human physiology and heat transfer physics. The engineering view of thermal comfort regards it as the mental state of a person during which he or she expresses satisfaction with his or her thermal environment and does not want any change to occur in the thermal conditions (ASHRAE 2013). But even that definition implicitly presupposes the ability of occupants to change their thermal environment, thus having at their disposal building energy supply technologies to deliver or remove heat accordingly. CIBSE (2006) takes that notion a step even further, by requiring that the indoor environment be designed and controlled so that occupants' comfort and health are assured. Subsequently, it is not a paradox to describe comfort as the endpoint of a technological request driven by advances in engineering (Roberts 1997). That historical perspective of thermal comfort more aptly describes its dynamic features, connects comfort with social development, and views it not only as a scientific theory but as a cultural phenomenon as well.

Following that line of thought, the expectations of people regarding thermal comfort change continually, succeeding the evolutions in architecture, building physics, and the technological advancements of energy systems and their controls that determine the thermal characteristics of the indoor environment and, consequently, the way in which heat is transferred to and from people. In this context, the presence of a system in a space, delivering or abstracting heat, and the heat transfer mode and features utilized to achieve it, should be an issue clarified and properly quantified when evaluating thermal comfort perception (Oxizidis and Papadopoulos 2013).

Nowadays there are two distinct ways to quantify comfort in a space: (1) The deterministic route of Fanger's PMV-PPD model (Fanger 1970) and all its successors, which was derived from laboratory- based research, and (2) the stochastic approach described by the adaptive comfort models that arose from field-based research (de Dear and Brager 1998). The latter is more suited to evaluating thermal comfort conditions in free-running, naturally ventilated buildings, especially for the cooling period where no mechanical cooling is used. Thermal comfort zones for both methods are depicted schematically for cooling conditions in the charts of Fig. 14.4. The indoor climatic parameter used to determine comfort zones in both charts is the operative temperature. It is obvious that the adoptive model considers a far broader range of conditions as comfortable, which can be attributed to shifting expectations and preferences as a result of occupants having a greater degree of personal control over their thermal environment (e.g., opening

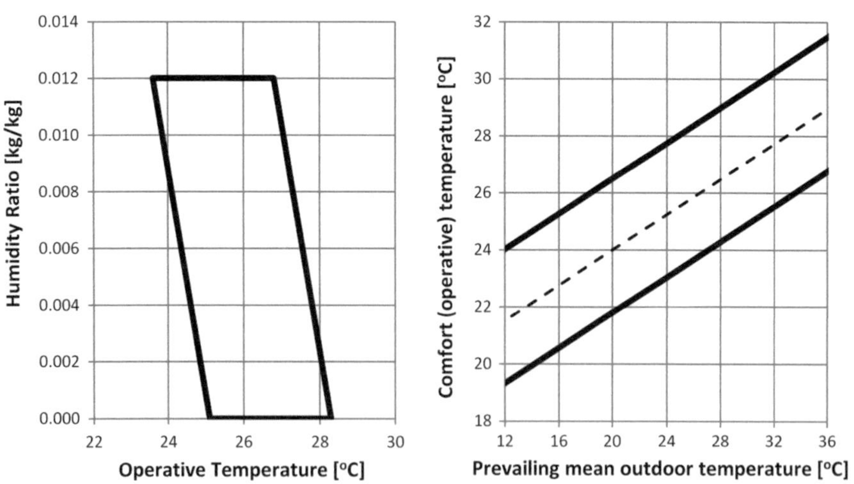

Fig. 14.4 Thermal comfort zones in cooling conditions for deterministic (*left*) and stochastic (*right*) models (90 % acceptance) (ASHRAE 2013)

windows) and becoming more accustomed to variable conditions that closely reflect the natural rhythms of outdoor climate patterns (Brager et al. 2015).

Even the deterministic method prescribes a bandwidth of conditions that can be described as comfortable; thus, it provides some latitude for DR potential if conditions require it. In addition, comfort standards (ASHRAE 2013; ISO EN 2007) allow for small durations of time where indoor conditions can be found outside of the comfort zones (either as a total number of hours, such as 100–150, or as weighted hours, where the level of deviation is also taken into account). All those features do increase the DR potential but are doing it by considering tight comfort constraints based on steady state heat balances and imposed by the availability and operation of HVAC systems.

However, if the future buildings are going to be adaptable and responsive to the fluctuations of energy supply, then their occupants will not have unrestricted, continuous, direct control over the HVAC output and their indoor spaces could not be characterized as fully steady state environments, it is possible that the thermal comfort conceptualization will move towards an approach similar to the adoptive model provisions. In that sense, a new thermal comfort narrative could emerge that fits more closely to older, historical notions of comfort, as was the case in less-engineered (with respect to heating and cooling systems) buildings before the proliferation of HVAC systems.

14.3 Energy Technologies for Building Supply

To increase the energy efficiency of buildings and minimize the environmental impact derived by the operation of energy systems covering heating and cooling loads, it is important to use RES in addressing those loads. Solar thermal collectors or heat pumps driven by RES-generated electricity (locally, on-site, or centrally) offer that possibility. Indeed, nowadays the commercialization of plant equipment, such as solar thermal collectors and air-to-water heat pumps, is very high.

Especially for air-to-water heat pumps, all major manufacturers have introduced to the market very efficient systems, anticipating that electrification of thermal loads—such as space heating and domestic hot water production—will most certainly get a huge boost when electricity will be generated by a high share of RES and will be transmitted and distributed through smart grids. Though, as in the case of solar thermal collectors, due to the intermittent nature of those resources TES is a most important parameter and a key component for the achievement of the maximum energy saving as well as good economic performance.

Additionally, although solar thermal collectors and air-to-water heat pumps require less energy than conventional heating systems, their operational characteristics, and consequently their energy saving potential, are greatly determined by the particular terminal heating and cooling units used to deliver the produced thermal—heating and cooling—energy inside the conditioned space. Those types of plant equipment—in order to maximize their efficiency—require the utilization of low-temperature heating and high-temperature cooling (low exergy) terminal equipment (emitters).

In the next paragraphs the interdependency of electrical and thermal loads in the built environment is explored and the particular technologies that rise up to address that challenge are being named and shortly described.

14.3.1 The Heat-Power Nexus (Interdependency of Electrical and Thermal Energy in the Built Environment)

The heat-power nexus in the built environment can be simplistically portrayed as the electrification of the thermal loads and the storing of electricity as thermal energy. The electrification of buildings' thermal loads in final energy demand is already a dominant trend. The need to increase regulated electrical loads in buildings to better absorb RES-generated electricity and the market proliferation of very efficient heat pumps are the main drivers leading that trend.

Figure 14.5 provides a schematic of the heat-power nexus in the built environment. The dotted line delineates the boundaries of the buildings, while outside that frame are the various energy networks, most notably the electric grid that is the one that most buildings are connected to. At the left side of the schematic are the commonly available primary energy resources for buildings, while at the right side are the final end uses (i.e., the demand side) of energy in buildings.

Using building supply technologies, those primary resources can be converted either to heat or power; this heat or power can be directly used to address the thermal or electrical loads of buildings. However, with particular building energy systems, e.g., electrically driven or thermally driven heat pumps, we can further relocate the heat and power to heating or cooling thermal energy ready either to be stored or used instantly to address the heating and cooling loads. In all cases, the buildings are interconnected to energy networks (electrical grid, district heating and cooling) with which they can exchange energy.

Future buildings should utilize zero carbon energy resources, either locally available (on the buildings sites) or centrally available (from the grid) to achieve ZEB status. One such possible—among several others—path of virtue is highlighted in the schematic of Fig. 14.5 by shaded boxes and dotted lines. A particular dimension that is only implicitly present in this schematic is time. The availability of RES does not always coincide with the buildings' loads; that mismatch can be mitigated by the use of thermal—either heating or cooling—energy storage technologies.

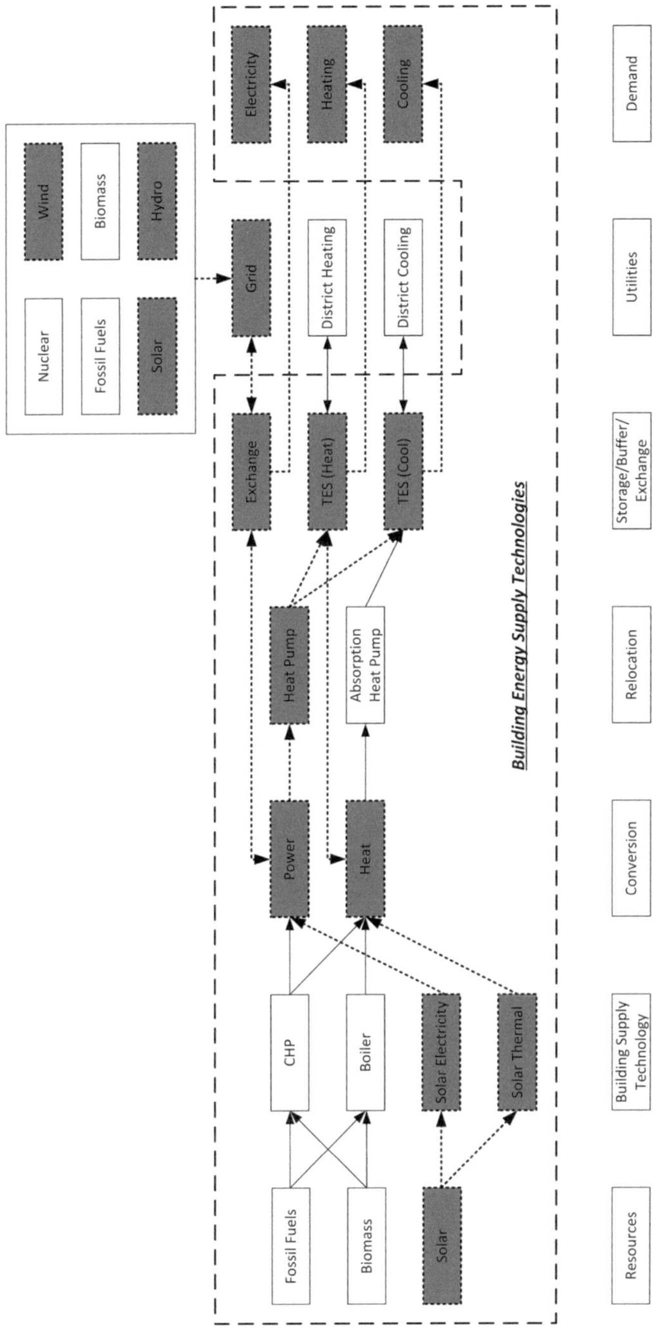

Fig. 14.5 The heat-power nexus (the shaded boxes and dotted arrows depict one possible path of virtue for ZEBs) in the built environment

14.3.2 Emerging Building Energy Systems

IEA (2011) identified the key technology options for heating and cooling in buildings that have the greatest long-term potential for reducing CO_2 emissions. They are:

- Active solar thermal
- Combined heat and power (CHP)
- Heat pumps for space heating and cooling, and hot water, and
- Thermal storage

It is not a coincidence that those particular technologies have been highlighted throughout this chapter as the appropriate ones for dealing with the various challenges emerging in the built environment. They are the technologies that can play the facilitator role of the heat-power nexus by introducing flexibility to the smart grid while exploiting and responding to the particular features of temperate climates.

CER (2011) has recognized heat pumps, microgeneration, and TES as the particular building energy technologies with merit to the DR. It is envisaged that the electrification of heat (and transport) by utilizing such technologies would play a significant role in the decarbonization of the entire energy system, thereby allowing the facilitation of high levels of electricity generation from RES. That will be possible by balancing the output of the variable sources of generation with demand flexibility provided by buildings.

In temperate climates, the potential of those technologies to reduce CO_2 emissions from buildings is even higher. Reversible heat pumps can perform with greater efficiency under mild conditions and can provide both space heating and cooling as required in the buildings of the Mediterranean belt. Microgeneration provided by PVs can contribute to a significant fraction of the buildings' overall energy consumption.

In the following paragraphs the main features of those systems are presented.

14.3.2.1 Microgeneration (or the Distributed Generation Narrative)

Microgeneration refers to on-building site generation of electricity, usually facilitated by CHP technologies, hence with heat production as well. However, microgeneration or distributed generation (because the generation plant is connected to a distribution network rather than the transmission network) on the buildings' side can be accomplished by PVs as well. Especially in the case of PVTs (Hybrid photovoltaic/thermal solar systems)combined heat and power generation driven by solar energy is possible.

CHP units are generally more efficient, from a systems perspective, than the separate, centralized generation and distribution of electricity and local production of heat. Beyond their energy performance advantages, CHP units also reduce transmission and distribution losses and they improve energy security and the reliability of energy supplies, particularly when combined with TES. When CHP units are connected to thermally driven chillers to provide cooling as well, the term "tri-generation" is used.

14.3.2.2 Heat Pumps

Heat pumps can address all the common thermal loads (space heating, cooling, and DHW) in buildings, and they are one of the most effective ways to cover those loads since they utilize renewable energy (i.e., aerothermal, hydrothermal, and geothermal, according to the terminology used in EU directives to promote the use of renewable energy). However, to capitalize on a comparatively low CO_2 intensity, they need to be driven by electricity that is generated by a fuel mix that is not heavily reliant on fossil fuels—especially coal.

14.3.2.3 Solar Thermal Collectors

Solar thermal collectors can address most of the thermal loads met in the built environment. They are extremely well suited to cover DHW and space heating loads (especially when feeding hydronic low temperature heat emitters like floor heating systems), whereas if they are connected to thermally activated chillers (mainly absorption), they can provide cooling as well.

The particular technologies of solar thermal collectors that are appropriate to the built environment are flat plate, evacuated (or vacuum) tubes, and air collectors. However, the latter cannot have a significant share of the market in Mediterranean buildings since they require both opaque (vs glazed) south façades, and mechanical ventilation systems—which is a very rare combination. In general, flat plate collectors are more suited for building applications in the Mediterranean countries since the required temperature of the heating loads are not high enough to justify the significantly higher cost of evacuated tube collectors.

All the above types of active solar thermal collectors are mature products from a technological and commercial perspective. However, commercially are popular for covering mainly DHW loads and less so for space heating. The vast potential of solar thermal collectors in space heating and cooling applications, especially in temperate climates, has not yet been realized (combisystems provide both space heating and DHW, but they only have a significant market penetration in central European countries). This has to do mainly with the cost, installation and operation complexity that have been identified as priorities to overcome in the near future (RHC-ETP 2014). In parallel, developments in materials research (super-insulation materials and advanced glazings) and new ICT technologies seem to be very promising for increasing performance and reliability.

14.3.2.4 Energy Storage (Thermal)

For centuries, fossil fuels have addressed—successfully and in a timely way—thermal loads in buildings by being not just forms of energy able to be converted in plant equipment to heating or cooling thermal energy, but also by being stored energy ready to be used when a need arose.

Thus, replacing fossil fuels in both the built environment and electricity generation fuel mix with RES, energy storage technologies are of crucial importance. RES are just forms of energy and they do not provide any storage capacity, while electricity grids have essentially no storage as well.

Since the major energy uses in buildings are heat-related, it is essential to deploy some means of TES—which as energy storage goes is probably the best technology available—either active (e.g., hot water tanks) or passive (e.g., high thermal mass building fabric). And since all primary energy sources can be used to produce heat, and the latter is used in almost every energy conversion process, TES is probably the way forward—especially for distributed storage in buildings.

A schematic depicting the operation principle of a TES system is presented in Fig. 14.6.

The main advantages of TES deployment in the built environment are:

- Increased RES utilization both on distributed and centralized generation. The energy provided by RES is usually intermittent and often not synchronous with the corresponding demand. TES directly increases the utilization of heat-generating RES technologies like solar thermal collectors, and indirectly of the electricity generation technologies such as PVs and wind turbines when electrically driven technologies—e.g., heat pumps—address the buildings' thermal loads.

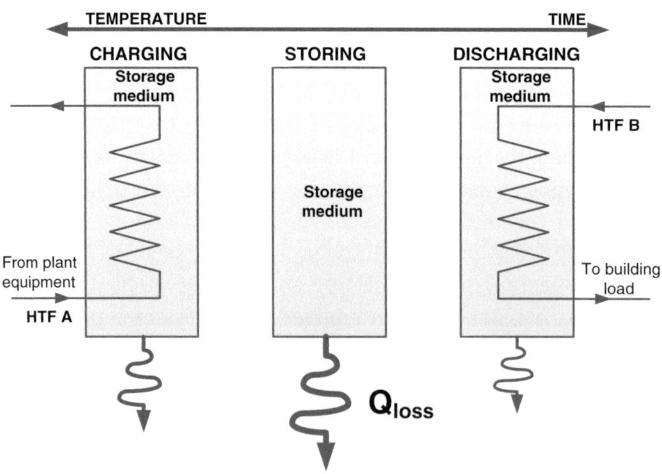

Fig. 14.6 Operational principle of a TES system

- Reduced maintenance costs. TES deployment results in reduced wearing down of equipment due to rapid cycling, which is very important for increasing the lifetime of both heat pumps and CHP units.
- Increased equipment efficiency. TES allows the operation of plant equipment during favorable environmental conditions to increase efficiency (e.g., chillers operating at night, when ambient temperatures are lower and more susceptible to heat disposal). In the same manner, TES can be used to avoid low-efficiency partial load operation, which is especially true for heat pumps.
- Decreased operation costs. TES can be used to avoid the operation of plant equipment during peak tariffs or during DR events, thus minimizing utility bills.
- Smaller size equipment—lower capital costs. TES can be sized to complement (by shaving peak loads) conventional plant equipment, thus reducing its overall sizing and cost.

The main types of TES technologies are sensible heat storage (e.g., water tanks or masonry walls for increased thermal mass in buildings); latent heat storage (PCM tanks or ice storage or even microencapsulated PCMs in wall boards for increased building thermal mass); and thermochemical heat storage. Currently, the most commonly used TES technology is sensible heat storage, either by water tanks or building fabric thermal mass. PCM products are commercially available, but for the time being they have only captured a niche of the market. There are several other criteria for categorizing TES, such as temperature (high temperature, low to medium temperature, cool storage), duration (seasonal, daily, buffer), size (centralized or distributed), and functionality (active or passive).

Passive heat storage is a crucial fixture of the thermal mass of any building and it is expected to become even more important in the future. CER (2011) has identified it as a technology feature of the highest value for the implementation of DR. The report went even further, suggesting a change in building regulations to ensure that new housing stock is equipped with heat storage. It is especially crucial to point out that thermal mass, combined with thermally resistant building elements, broadens the duration and capacity of the DR potential since it significantly attenuates and delays indoor air temperature fluctuations, thereby allowing increased periods of occupants' thermal tolerance to the non-availability of heating and cooling systems. Thermal mass heat storage can additionally provide DR, not only from the plant equipment—as with active TES—but from the circulation pump as well (if hydronic systems are being used).

Increased thermal mass in buildings in temperate climates has been a must-have feature of vernacular architecture to mitigate the high solar radiation and high daytime temperatures (by exploiting the variations in daytime and nocturnal temperatures and in environmental thermal radiation between sunny skies and starry night skies) in these climates during summer conditions. It is very interesting that the notion of high thermal mass in the era of smart grid ready buildings resurfaces to address supply load variations.

14.3.3 Building Energy Management Systems

If TES makes the deployment of DR schemes in buildings feasible, then advancements in ICT make them possible. The latter, in combination with building energy management systems (BEMS), can provide the automated communication and control of HVAC systems to offer services to the grid.

The traditional role of BEMS has been to optimize levels of service by HVAC equipment and assure comfort provision in an energy-efficient manner. But in the near future, it is expected that BAS will not only provide building occupants with expected comfort services, but will also make optimal decisions about storing, exporting, or importing energy resources depending on expected weather and occupancy patterns, and in response to signals from the grid (Salom et al. 2014).

So as buildings are beginning to be equipped with smart technologies, their interaction with utilities will intensify, allowing the deployment of numerous applications for innovation that will offer new opportunities to improve energy performance and control the energy demand/supply balance. These developments will make the building the cornerstone of a future energy system that integrates RES technologies, leading to a more efficient use of available energy resources (JRC 2014). This will take place as heating and cooling systems acquire intelligence features thereby increasing the flexibility available to the energy system. Subsequently, the increased flexibility can enable renewable based heating and cooling systems to satisfy a greater share of the energy demand (RHC-ETP 2013).

However, for that vision to materialize, a new narrative is required with regard to BEMS. It is not just the control or communication functions that need upgrading, but also advanced sensoring and submetering is essential to exploit all the flexibility of the HVAC systems.

Van de Bree et al. (2014) identified and summarized the following most relevant building automation functions for nZEBs:

- Central, concerted control of all energy-related components to tap all internal potentials and to ensure that the whole system can work as efficiently as possible.
- Monitoring and providing feedback to ensure that the demanded and calculated (low) energy demand of nearly zero energy is met and to encourage the users to save energy.
- Load shifting and storage management to increase coverage rates of onsite renewable energy, to increase free cooling potential in central and southern European regions, and to increase grid stability.
- Ensuring thermal comfort.

14.4 Envisioning the Building of the Future (Is the All-Electric Building the Future?)

It is clear that high-energy performance buildings will play an important role in the future energy system in balancing the demand and supply of energy. For the vision to be realized, of buildings not only being receivers but also providers of energy services, there are a series of conditions that need to be met.

The first of these conditions is an increase of the buildings' electrical loads capacity to allow higher penetration—much above the current thresholds—of RES in the electricity generation fuel mix. Since a high performance building is by definition very energy-efficient, the electrification of all its regulated thermal loads is a prerequisite for displaying any DR potential. Electrifying thermal loads allows the deployment of TES systems on buildings either inherent in their construction, like thermal mass, or as additional HVAC equipment (water storage tanks) that will significantly augment the DR potential of electrically driven heating and cooling systems. The last condition concerns the ability of buildings to gauge their thermal status, and receive and exchange information with the grid and control their HVAC and local RES systems in an optimal way that will require an order of magnitude upgrade of the current ICT infrastructure used in buildings.

Such a building will most probably be an all-electric one. It will use electrically driven systems (heat pumps) to produce thermal energy for its needs or systems that generate electricity in parallel (CHP) when addressing thermal loads, or it will just generate electricity locally (PVs)—or a combination of the two. Solar thermal technologies and techniques, whether passive or active, will enhance the thermal performance of the buildings towards the realization of the ZEB agenda. In that heat-power nexus, the future connected building will emerge as an indispensable component of the decarbonized energy system.

References

ASHRAE (2013) Standard 55-2004: thermal environment conditions for human occupancy. American Society of Heating Refrigeration and Air conditioning Engineers, Atlanta

Brager G, Zhang H, Arens E (2015) Evolving opportunities for providing thermal comfort. Build Res Inf 43(3):274–287

CER (2011) Demand side vision for 2020—decision paper. Commission for Energy Regulation, Dublin

CIBSE (2006) Guide a: environmental design. Chartered Institute of Building Services Engineers, London

de Dear RJ, Brager G (1998) Developing an adaptive model of thermal comfort and preference. ASHRAE Trans 104:145–167

EC (2002) Directive 2002/91/EC on the energy performance of buildings. European Commission, Official Journal of the European Union L 001

EC (2009a) Directive 2009/28/EC on the promotion of the use of energy from renewable sources. European Commission, Official Journal of the European Union L 140/16

EC (2009b) Directive 2009/125/EC establishing a framework for the setting of ecodesign requirements for energy-related products (recast). European Commission, Official Journal of the European Union L 285/10
EC (2010a) Directive 2010/30/EU on the indication by labelling and standard product information of the consumption of energy and other resources by energy-related products. European Commission, Official Journal of the European Union L 153/1
EC (2010b) Directive 2010/31/EU on the energy performance of buildings. European Commission, Official Journal of the European Union L 153/13
EC (2012) Directive 2012/27/EU on energy efficiency. European Commission, Official Journal of the European Union L 315/1
EEGI (2013) European electricity grid initiative: research and innovation roadmap 2013–2022, Jan
Fanger PO (1970) Thermal comfort—analysis and applications in environmental engineering. Danish Technical Press, Copenhagen
IEA (2011) Technology roadmap: energy-efficient buildings: heating and cooling equipment. International Energy Agency, France
ISO EN (2007) CEN Standard EN15251. Indoor environmental input parameters for design and assessment of energy performance of buildings—addressing indoor air quality, thermal environment, lighting and acoustics. Comite Europeen de Normalisation, Brussels
JRC (2014) 2013 Technology map of the European strategic energy technology plan—report EUR 26345 EN, technology descriptions. Joint Research Centre, Luxembourg, Publications Office of the European Union
Oxizidis S, Papadopoulos AM (2013) Performance of radiant cooling surfaces with respect to energy consumption and thermal comfort. Energy Build 57:199–209
RHC-ETP (2013) Strategic research and innovation agenda for renewable heating and cooling. European Technology Platform on Renewable Heating and Cooling, Brussels, Belgium
RHC-ETP (2014) Solar heating and cooling technology roadmap. European Technology Platform on Renewable Heating and Cooling, Brussels, Belgium
Roberts B (1997) The quest for comfort. Chartered Institute of Building Services Engineers, London
Salom J, Marszal AJ, Candanedo JA, Widén J, Lindberg KB, Sartori I (2014) Analysis of load match and grid interaction indicators in Net ZEB with high resolution data. IEA SHC Task 40 and ECB Annex 52, Subtask A Report
SEDC (2014) Mapping demand response in Europe today. Smart Energy Demand Coalition, Apr
USGBC (2015) LEED O + M: existing buildings v4—LEED v4: demand response. http://www.usgbc.org/node/2613007?view=guide. cited at 16 Feb 2015
Van de Bree A, Von Manteuffel B, Ramaekers L, Offermann M (2014) Role of building automation related to renewable energy in nZEB's. ECOFYS, Utrecht

Chapter 15
Energy Technologies for Building Supply Systems: MCHP

Sergio Sibilio and Antonio Rosato

Abstract Micro-cogeneration is an emerging technology with the potential to—if designed and operated correctly—reduce both the primary energy consumption and the associated greenhouse gas emissions, when compared to traditional energy supply systems. The distributed nature of this generation of technology has the additional advantages of (1) reducing electrical transmission and distribution losses; (2) alleviating the peak demands on the central power plants; and (3) diversifying the electrical energy production, thus improving the security of energy supply. The micro-cogeneration devices are used to meeting the electrical and heating demands of buildings for space heating/hot water production, as well as potentially (mainly for temperate and hot climates) absorption/adsorption cooling systems. Currently, the use of commercial micro-cogeneration units in applications such as hospitals, leisure facilities, hotels, or institutional buildings is well established. The residential cogeneration industry is in a rapid state of development and flux, and the market remains undeveloped, but interest in the technologies by manufacturers, energy utilities, and government agencies remains strong.

15.1 Introduction

The electricity demand of buildings is usually satisfied by large central power plants using combustion-based energy conversion; one waste by-product of this conventional electricity power generation is heat. "Cogeneration" (CHP = combined heat and power) is a proven technology (more than 100 years old) that is

S. Sibilio (✉) · A. Rosato
Department of Architecture and Industrial Design "Luigi Vanvitelli",
Second University of Naples, via San Lorenzo, 81031 Aversa, CE, Italy
e-mail: sergio.sibilio@unina2.it

A. Rosato
e-mail: antonio.rosato@unina2.it

S.-N. Boemi et al. (eds.), *Energy Performance of Buildings*,
DOI 10.1007/978-3-319-20831-2_15

able to recover and use this otherwise wasted heat. It is usually defined as the simultaneous generation in one process of thermal energy and electrical and/or mechanical energy from a single stream such as oil, coal, natural or liquefied gas, biomass, solar, etc. (ASHRAE Handbook 2000). It allows developing the power and heat transfer using an integrated system that achieves a larger overall utilization efficiency if compared to the separate energy production, thanks to the recovery and use of waste heat in addition to electricity.

In principle, the concept of cogeneration can be applied to power plants of various sizes, ranging from small-scale for residential buildings to large-scale cogeneration systems for industrial purposes, to fully grid-connected utility generating stations. In the past, mostly because of economy-of-scale reasons, cogeneration was limited to large-sized power plants operating in central locations. These stations are usually characterized by significant electric losses due to the transmission and distribution to the final user through high voltage lines and the transformers; in addition, large cogeneration systems generally utilize the waste heat by piping hot water into the buildings of the surrounding community, a process that involves significant heat loss due to transportation of hot water over long distances, along with the relevant investment costs for the pipes. For these reasons, in recent years a great deal of attention has been focused on the transition from centralized to decentralized systems with an increasing diffusion of the so-called "micro-cogeneration" (MCHP, micro combined heat and power) aiming to produce electricity and heat near where both energy flows can be used. According to Directive 2004/8/EC, this process is usually defined as the local combined production of electrical and thermal energy from a single fuel source with an electric output lower than 50 kW_{el}. Some authors use the term "small-scale cogeneration" for the combined heat and power generation systems with electrical power less than 100 kW_{el} (Angrisani et al. 2012; Maghanki et al. 2013), while the term "micro-cogeneration" is sometimes used to denote cogeneration units with an electric capacity smaller than 15 kW_{el} (Angrisani et al. 2012; Maghanki et al. 2013).

Micro-cogeneration may potentially change the traditional roles attributed to the private consumer. Typically, households purchase and consume electricity from the grid and produce heat with a heating unit owned by them. With a micro-cogeneration unit installed in their houses, they become electricity producers and may sell electricity to the grid. Figure 15.1 (Evangelisti et al. 2015) shows the concept of MCHP applied to a home: waste heat is used for space heating and domestic hot water production, while electricity is used within the building (for lighting, consumer electronics, or any other electrical needs the house may have) or exported to the grid.

Micro-cogeneration applications have to satisfy either the electrical and thermal demands, or satisfy the thermal demand and part of the electrical demand, or satisfy the electrical demand and part of the thermal demand. If the electric power export to the central electric grid is allowed, residential cogeneration units can operate in response to variations in heat demand. It is called "heat demand following operation;" in this case, the MCHP device is sized to meet the heating needs, while electricity is either used internally or exported to the grid. However, in a

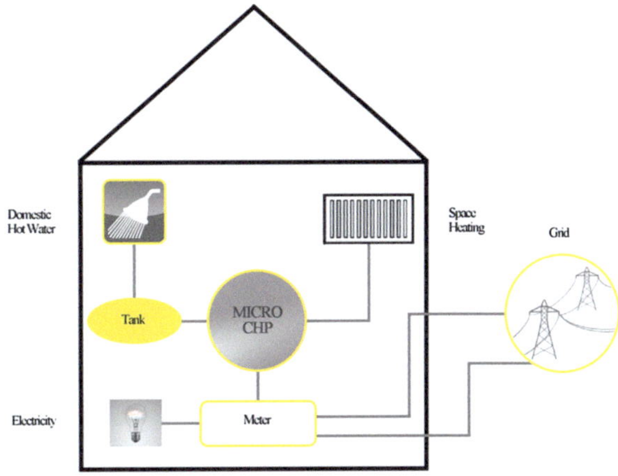

Fig. 15.1 Schematic representation of an MCHP unit for residential use (Evangelisti et al. 2015)

residential cogeneration unit without electric power export, the unit is designed to satisfy the electricity demand of the customer, and heat is used to contribute to water and space heating; its heat output varies in response to its electric power output that follows the electric power demand. In this case ("electricity demand following operation"), a supplementary peak boiler may be required to meet the total heat demand. Depending on the magnitude of the electrical and thermal loads, whether they match or not (as well as the operating strategy), the MCHP system may have to be run at partial load conditions, the surplus energy (electricity or heat) may have to be stored or sold, and deficiencies may have to be made up by purchasing electricity or heat from other sources, such as the central electric grid or a boiler plant; the surplus heat produced can be stored in a thermal storage device such as a water tank, or in phase change materials, while surplus electricity can be stored in electrical storage devices, such as batteries or capacitors (Onovwiona and Ugursal 2006).

Micro-cogeneration is emerging as a fast-growing technique to reduce the primary energy consumption in small- or medium-scale applications (Angrisani et al. 2012; Maghanki et al. 2013; Onovwiona and Ugursal 2006; Wu and Wang 2006; Chicco and Mancarella 2009), thanks to the fact that, if designed and managed properly, the efficiency of energy conversion in cogeneration systems increases to over 80 % when compared to an average of 30–40 % in conventional fossil fuel-fired electricity generation systems. Figure 15.2 (Onovwiona and Ugursal 2006) illustrates how the chemical energy from the fuel is converted into useful thermal energy and electrical energy for a conventional fossil fuel-fired electricity generation and a micro-cogeneration system.

In this figure, α_E is the electric efficiency (ratio between electric output and fuel power input) of the micro-cogeneration unit, α_Q is the thermal efficiency (ratio between thermal output and fuel power input) of the MCHP device, η_E is the electric efficiency of the electrical power plant (production of electricity only), η_Q is the thermal efficiency of the boiler (production of heat only), E is the electricity

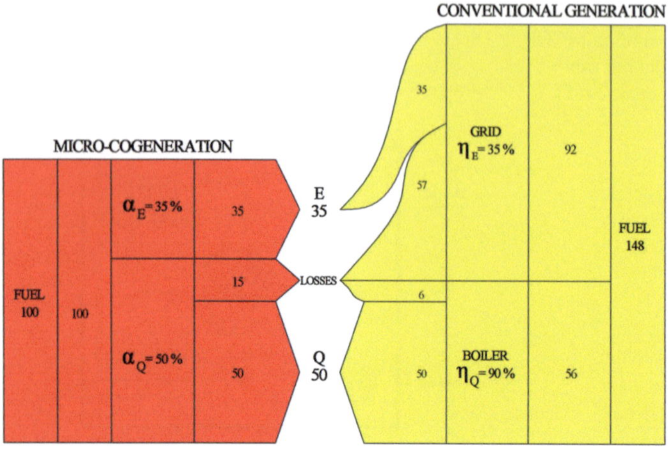

Fig. 15.2 Micro-cogeneration versus conventional separate generation (Onovwiona and Ugursal 2006)

demand, and Q is the heat demand (for space heating and domestic hot water production). This figure highlights how micro-cogeneration technology is potentially able to satisfy the electric and thermal needs of end-users with a lower primary energy consumption when compared to the conventional methods of generating heat (in boilers) and electricity separately (in centralized power plants), as recognized also by the European Community (Directive 2004/8/EC, 2004); in addition, it should be stressed that the increase in energy efficiency with micro-cogeneration can result in lower operating costs and reduced greenhouse gas emissions [assisting in meeting Kyoto targets (United Nations, Framework Convention on Climate Change, 2005)]. The distributed generation nature of this technology also has the potential to reduce losses due to electrical transmission and distribution inefficiencies as well as alleviate peak utility demand problems; these systems are even more attractive for remote communities where a lack of central generation stations and costly connection to the grid is neither an affordable nor a preferable option. MCHP systems have the additional advantage of diversifying electrical energy production (they can use alternate fuels), thus potentially improving the security of the energy supply in the event of problems occurring with the main electricity grid.

MCHP systems could also provide new commercial opportunities for manufacturers as well as for partnerships between manufacturers, energy suppliers, financiers, and others. In addition, a widespread exploitation of micro-cogeneration units could defer huge investments in new, large generation plants, substations, or infrastructures. Finally, it could develop competitive electricity markets and increase the customer participation in the market, providing solutions to exploit the price elasticity to the electrical demand.

In order to obtain that include energy savings, reduction of pollutant emission, and short payback-periods, residential cogeneration units must be appropriately designed and managed in response to variations in residential energy demands. First of all, the effective utilization of the cogeneration device's thermal output for space heating as well as for domestic hot water production is crucial for obtaining high levels of overall energy efficiency, along with the associated environmental benefits (IEA/ECBCS Annex 42 2007). When designing an MCHP system, it should be also taken into account that the efficiency of a cogeneration unit is strongly related to the type of the prime mover, its size, and the temperature at which the recovered heat can be utilized. In addition, taking into account that the economic feasibility of a micro-cogeneration scheme is strongly related to the operating time (Angrisani et al. 2012), the utilization level of the unit typically has to be more than 4000 running hours per year (Onovwiona and Ugursal 2006).

Together with a lot of potential benefits, some drawbacks still limit the large diffusion of micro-cogeneration systems (Angrisani et al. 2012; Sonar et al. 2014; DECENT project 2002; Mancarella and Chicco 2009; Angrisani et al. 2014a):

- The design of systems poses a significant technical challenge due to the non-coincidence of thermal and electrical loads, necessitating the need for electrical/thermal storage or connection in parallel to the electrical grid
- High first-capital costs
- Lack of financial supporting actions to ensure a suitable payback period (possibility to obtain funds as well as to sell the electric surplus to the grid at a good price)
- Insufficient political support mechanisms and administrative hurdles, such as electric network grid connection
- Components that are in the R&D phase or in a pre-selling phase
- Lack of trial data to fill the gap of optimization between the systems and the user load profile as well as to define several operating strategies and their impact on optimum system performances
- The diffusion of distributed cogeneration within urban areas, where air quality standards are quite stringent, brings about environmental concerns regarding local emissions (such as NO_x, CO, SO_x, particulate matter, unburned hydrocarbons, etc.).

Today, leading governments from the countries of Europe, Japan, and the United States have taken roles in promoting and advancing this technology (IEA/ECBCS Annex 42 2007; IEA/ECBCS Annex 54 2013) and micro-cogeneration has a significant market potential. MCHP devices are especially interesting for small and medium family houses, buildings and enterprises, hotels, hospitals, university campuses, etc. The Micro-Map Project (MICRO-MAP 2002) reported that in Europe between 5 and 12.5 million homes could have MCHP systems installed by 2020, which would result in a CO_2 emissions savings of between 3.3 and 7.8 million tons per year.

With respect to the potential utilization of micro-cogeneration technology, it is very important to emphasize that in many countries in the last few years, there has been an increasing demand for cooling energy during the warm season, generally satisfied by electrically-driven units, with this trend having contributed to electrical load-peaking and subsequent network congestion and failure events in different power systems worldwide. This has strengthened the awareness of governments, manufacturers, and communities about energy and environmental issues, pushing forward the search for efficiently combining micro-cogeneration units with various technologies currently available for local cooling generation. A combination of MCHP systems with various thermally fed or electrically driven cooling systems allows setting up a "micro-trigeneration system" (MCCHP = micro combined cooling heat and power) (Angrisani et al. 2012; Maghanki et al. 2013; Onovwiona and Ugursal 2006; Wu and Wang 2006; Chicco and Mancarella 2009), that represents the production in situ of a threefold energy vector requested by the user from a unique source of fuel with significant potential benefits from energy and environmental points of view. MCCHP is an upgrade of the micro-cogeneration unit where thermal or electric energy is further utilized to provide space or process cooling capacity; in this way, the overall energy efficiency increases and the economic payback decreases, due to the larger amount of operating hours per annum (Angrisani et al. 2012).

In order to advance the design, operation, and analysis of micro-cogeneration and micro-trigeneration systems, two consecutive research projects have been sponsored and completed by the International Energy Agency (IEA) (IEA/ECBCS Annex 42 2007; IEA/ECBCS Annex 54 2013).

15.2 Prime Mover Technologies and Market Survey

Today, there are several technologies that are capable of providing cogeneration devices. The conversion process can be based on combustion and subsequent conversion of heat into mechanical energy, which then drives a generator to produce electricity. Alternatively, it can be based on direct electrochemical conversion from chemical energy to electrical energy. Other processes include the photovoltaic conversion of radiation.

There are five main technologies being developed for micro-cogeneration units including:

1. Reciprocating internal combustion engines (ICE)-based MCHP systems
2. Reciprocating external combustion Stirling engines (SE)-based MCHP systems
3. Fuel cells (FC)-based MCHP systems
4. Gas and steam micro-turbines (MT)-based MCHP systems
5. Photovoltaic thermal (PVT) MCHP systems

Table 15.1 shows the most important scientific papers focused on the above-listed micro-cogeneration technologies. The references reported in this table highlight and describe many significant examples referred on them.

At the moment, micro-cogenerators based on reciprocating internal combustion and Stirling engines are already available on the market for the single- and multi-family residential building market; small-scale commercial applications and

Table 15.1 Manufacturer's data on main ICE-based micro-cogenerators

	Proposed literature
Reciprocating internal combustion engines (ICE)-based MCHP systems	Onovwiona and Ugursal (2006), Wu and Wang (2006), Angrisani et al. (2012), Maghanki et al. (2013), Sonar et al. (2014).
Reciprocating external combustion Stirling engines (SE)-based MCHP systems	Onovwiona and Ugursal (2006), Wu and Wang (2006), Angrisani et al. (2012), Maghanki et al., (2013), Sonar et al. (2014)
Fuel Cells (FC)-based MCHP systems	Onovwiona and Ugursal (2006), Wu and Wang (2006), San Martin et al. (2010),Angrisani et al. (2012),Maghanki et al. (2013), Sonar et al. (2014).
Gas and steam micro-turbines (MT)-based MCHP systems	Onovwiona and Ugursal (2006), Wu and Wang (2006), Maghanki et al. (2013), Sonar et al. (2014).
Photovoltaic thermal (PVT) MCHP systems	Chow (2010), Bianchi et al. (2012), Maghanki et al. (2013), Ferrari et al. (2014), Kumar et al. 2015

a large R&D operation that aims at producing, in the medium and long run, small, commercially available units based on fuel cells, gas, and steam micro-turbines, are already in progress (Angrisani et al. 2012). When selecting micro-cogeneration systems, one should consider some important technical parameters that assist in defining the type and operating scheme of the various alternative cogeneration systems to be selected:

- Electric and thermal load patterns of the end-users
- Heat-to-power ratio of the end-users
- Fuels available
- System reliability
- Permission of power export to the central electric grid
- Temperature levels
- Local environmental regulations

In the following sections, the various technologies suitable for micro-cogeneration are described and compared in terms of capacity range, fuel, electricity efficiency, thermal efficiency, overall efficiency (sum of electric and thermal efficiencies), noise level, dimensions, weight, life service, pollutant emissions, capital, and maintenance costs.

15.2.1 Reciprocating Internal Combustion Engines (ICE)

The reciprocating internal combustion engines are coupled with an electricity generator and heat exchangers to recover the heat of the exhaust gases, coolants and oil. Internal combustion engines are the most well-established technology for small and medium MCHP applications. They are a robust and proven technology with long-life service [up to 80,000 h (Angrisani et al. 2012)], a capital cost of between €2000 and 6000/kW$_{el}$ depending on the size (Angrisani et al. 2012), and a typical maintenance cost from 0.010 to 0.015 €/kWh (Onovwiona and Ugursal 2006). ICE-based MCHP systems occupy small installation spaces and also have satisfactory electric (20–34 %) and thermal (50–65 %) efficiencies. In addition to fast start-up capability and good operating reliability, they are characterized by high efficiency at partial load operation and can be fired on a broad variety of fuels with excellent availability, allowing for a range of various energy applications, especially emergency or standby power supplies.

Although they are a mature technology, reciprocating internal combustion engines have obvious drawbacks: (1) relatively high vibrations that require shock absorption and shielding measures to reduce acoustic noise [<60 dB(A) at a 1 m distance (Angrisani et al. 2012)]; (2) a large number of moving parts and frequent maintenance intervals that increase maintenance costs; and (3) high emissions, particularly nitrogen oxides [NO_x emissions are less than 100 ppm with a stable shaft power output in an engine speed range between 1200–3000 rpm (Angrisani et al. 2012)], which are the underlying aspects of this technology and need to be improved. Major manufacturers around the world continuously develop new engines with lower emissions; at the same time, emissions that control options, such as selective catalytic reduction, have been utilized.

Presently, a number of internal combustion engine-based micro-cogeneration units are commercially available. Table 15.2 describes the main ICE-based MCHP systems on the market (with electric output lower than 15 kW$_{el}$) in terms of input power, electric output, thermal output, electric efficiency (ratio of electric power to fuel input power), thermal efficiency (ratio of thermal power to fuel input power),

Table 15.2 Manufacturers' data on main ICE-based micro-cogenerators

	SENERTEC Dachs G 5.5	YANMAR CP5WN-SN	YANMAR CP10WN-SN
Input power (kW)	20.5	17.8	31.5
Electric power (kW)	5.5	5.0	9.9
Thermal power (kW)	12.5	10.0	16.8
Electric efficiency (%)	26.8	28.1	31.4
Thermal efficiency (%)	61.0	56.2	53.3
Overall efficiency (%)	87.8	84.3	84.8
Fuel	Natural gas	Natural gas	Natural gas
Weight (kg)	530	400	756
L (mm)	720	1100	1470
H (mm)	1000	1500	1790
D (mm)	1060	500	800
No. of cylinders	1	3	3
Displacement (cm^3)	579	699	1642
Noise (dB(A))	56	53	56

(continued)

Table 15.2 (continued)

	TEDOM micro T7 AP	COGENGREEN ecoGEN-12AG	EC POWER XRGI 9
Input power (kW)	25.9	43.0	31.0
Electric power (kW)	7.0	12.0	9.0
Thermal power (kW)	17.2	28.0	20.0
Electric efficiency (%)	27.0	27.9	29.0
Thermal efficiency (%)	66.4	65.1	64.5
Overall efficiency (%)	93.4	93.0	93.5
Fuel	Natural gas	Natural gas, LPG	Natural gas, propane, butane
Weight (kg)	645	700	440
L (mm)	1315	1340	920
H (mm)	1480	1218	960
D (mm)	700	780	640
No. of cylinders	3	4	4
Displacement (cm^3)	962	1600	2237
Noise (dB(A))	58	55	49

	HONDA Ecowill	AISIN SEIKI GECC 46 A2	AISIN SEIKI GECC 60 A2
Input power (kW)	3.8	18.0	20.8
Electric power (kW)	1.0	4.6	6.0
Thermal power (kW)	2.5	11.7	11.7
Electric efficiency (%)	26.3	25.6	28.8
Thermal efficiency (%)	65.8	65.0	56.3
Overall efficiency (%)	92.1	90.6	85.1
Fuel	Natural gas, LPG	Natural gas, LPG	Natural gas, LPG
Weight (kg)	71	465	465
L (mm)	580	1100	1100
H (mm)	750	1500	1500
D (mm)	298	660	660
No. of cylinders	1	3	3
Displacement (cm^3)	163	952	952
Noise (dB(A))	43	54	54

	VAILLANT Ecopower e4.7	SENERTEC Dachs G 5.0
Input power (kW)	18.9	19.6
Electric power (kW)	4.7	5.0
Thermal power (kW)	12.5	12.3
Electric efficiency (%)	24.9	25.5
Thermal efficiency (%)	66.1	62.8
Overall efficiency (%)	91.0	88.3

(continued)

Table 15.2 (continued)

	VAILLANT Ecopower e4.7	SENERTEC Dachs G 5.0
Fuel	Natural gas, propane	Natural gas
Weight (kg)	390	530
L (mm)	760	720
H (mm)	1080	1000
D (mm)	1370	1070
No. of cylinders	1	1
Displacement (cm^3)	270	579
Noise (dB(A))	56	56

overall efficiency (sum of electric and thermal efficiencies), fuel, weight, dimensions (L = Length, H = Height, D = Depth), number of cylinders, displacement, and noise (at a 1 m distance), based on the manufacturers' nominal data.

15.2.2 Reciprocating External Combustion Stirling Engines (SE)

Unlike reciprocating internal combustion engines, a Stirling engine is an external combustion device. This means that the cycle medium, usually helium or hydrogen (but also oxygen, nitrogen, and carbon dioxide), is not exchanged during each cycle (but within the device), while the energy driving the cycle is applied externally. Stirling engines can operate on almost any fuel (gasoline, alcohol, natural gas or butane) and renewable energy sources (solar or biomass), with external combustion that facilitates the control of the combustion process and results in low emissions (emissions from current Stirling burners can be 10 times lower than those emitted from reciprocating internal combustion engines with a catalytic converter, making the emissions generated from Stirling engines comparable to those from modern gas burner technology (Onovwiona and Ugursal 2006)). Compared to the ICE-based systems, Stirling engines have fewer moving parts and lower vibrations, longer lives service, lower noise levels, and longer maintenance-free operating periods. The global efficiency is higher than 80 %, and it may even go beyond 95 %, with a good performance at partial load. The capital costs depend on the size, ranging from €2,700 to 5500/kW_{el} (Angrisani et al. 2012); an estimated maintenance cost for the unit is around €0.013/kWh (Onovwiona and Ugursal 2006). Despite many advantages, the Stirling engine has not found the expected applications due to low electric efficiency (ranging from 12 to 28 %), difficult power control system because of the presence of various heat exchangers (heater, cooler, regenerator, and auxiliary heat exchangers), high pressure level of working gas, low durability of parts, and long start-up time.

Various small-scale Stirling-based cogenerators are commercially available or under development. Table 15.3 describes the main SE-based MCHP systems available on the market (with electric output lower than 15 kW_{el}) in terms of input power, electric output, thermal output, electric efficiency (ratio of electric power to

Table 15.3 Manufacturers' data on main SE-based micro-cogenerators

	WHISPERGEN	BAXI Ecogen	Qnergy QCHP7500
Input power (kW)	8.3	7.4	38.0
Electric power (kW)	1.0	1.0	7.5
Thermal power (kW)	7.0	6.0	30.0
Electric efficiency (%)	12.0	13.5	19.7
Thermal efficiency (%)	84.3	81.1	78.9
Overall efficiency (%)	96.4	94.6	98.7
Fuel	Natural gas	Natural gas, biogas	Wood pellets, biomass, liquid fuels, natural gas, propane
Weight (kg)	137	110	200
L (mm)	480	450	630
H (mm)	840	950	770
D (mm)	560	426	1380
No. of cylinders	4	–	–
Displacement (cm3)	–	–	–
Working gas	Nitrogen	–	–
Maximum pressure (bar)	–	–	–
Noise (dB(A))	–	45	65

	SUNMACHINE	SOLO 161
Input power (kW)	12.0	38.8
Electric power (kW)	3.0	9.5
Thermal power (kW)	7.8	26.0
Electric efficiency (%)	25.0	24.5
Thermal efficiency (%)	65.0	67.0
Overall efficiency (%)	90.0	91.5
Fuel	Wood pellets	Natural gas, LPG, biogas, biomass
Weight (kg)	410	460
L (mm)	1160	1280
H (mm)	1590	980
D (mm)	760	700
No. of cylinders	1	2
Displacement (cm3)	520	160
Working gas	Nitrogen	Helium, hydrogen
Maximum pressure (bar)	36	150
Noise (dB(A))	–	–

fuel input power), thermal efficiency (ratio of thermal power to fuel input power), overall efficiency (sum of electric and thermal efficiencies), fuel, weight, dimensions (L = Length, H = Height, D = Depth), number of cylinders, displacement, working gas, maximum working pressure, and noise (at a 1 m distance) based on manufacturers' nominal data.

15.2.3 *Fuel Cells (FC)*

Fuel cell cogeneration-based systems have, perhaps, the greatest potential in micro-cogeneration applications, thanks to their ability to produce electricity at a relatively high efficiency with a significant reduction of greenhouse gas emissions.

In a fuel cell, the chemical energy within the fuel is converted directly to electricity (with by-products of heat and water) without any mechanical drive or generator. Currently, most of the fuel cells are either based on the low temperature (80 °C) proton exchange membrane fuel cell (PEMFC) technology, or on the high temperature (800–1000°C) solid oxide fuel cell (SOFC) technology. They normally run on hydrogen, but can also run on natural gas, methanol, or other fuels, by external or internal reforming. SOFC performs better than PEMFC technology, but start-up and cooling phases take longer, which immediately affects the time and costs required for installation, maintenance, repair, and durability of fuel cells (Angrisani et al. 2012). Additional types of fuel cells are available: alkaline fuel cells (AFC), phosphoric acid fuel cells (PAFC), molten carbonate fuel cells (MCFC), and, as of late, direct methanol fuel cells (DMFC).

Fuel cells have several benefits, such as high electric efficiency (30–60 %) and overall efficiency (80–90 %), near zero emissions [due to their lack of a combustion process, FCs have extremely low emissions of NO_x and CO; their CO_2 emissions are also generally lower than other technologies due to their higher efficiency (Onovwiona and Ugursal 2006)], a good match with the residential thermal to power ratio, reliability, quiet operation, potential for low maintenance, and excellent partial load management.

Nevertheless, the high costs [varying from €6700/kW_{el} for PEMFC to €60,000/kW_{el} for SOFC (Angrisani et al. 2012)] and relatively short lifetime of fuel cell systems are their main limitations. Typically, the total cost is represented by the stack subsystem (25–40 %), the fuel processor (25–30 %), the electronics (10–20 %), the thermal management subsystem (10–20 %), and an ancillary (5–15 %) (Angrisani et al. 2012). Fuel cell maintenance costs vary with the type of fuel cell, size, and maturity of the equipment. A major overhaul of fuel cell systems involves shift catalyzer replacement, reformer catalyzer replacement, and stack replacement; for example, the cost of replacing the stack of a 10 kW PEMFC is estimated to be €0.0188/kWh (Onovwiona and Ugursal 2006).

Ongoing research to solve technological problems and develop less expensive materials and mass production processes are expected to result in advances in technology that will reduce the cost of fuel cells. At this moment there are few fuel

Table 15.4 Manufacturers' data on main FC-based micro-cogenerators

	KYOCERA (SOFC)	PANASONIC ENE FARM (PEMFC)	HEXIS Galileo 1000 N (SOFC)
Electric power (kW)	0.70	0.75	1.00
Thermal power (kW)	0.65	1.08	1.80
Electric efficiency (%)	42.0	35.2	30.0
Thermal efficiency (%)	39.2	50.6	62.0
Overall efficiency (%)	81.2	85.8	92.0
Fuel	Natural gas	Natural gas	Natural gas, biogas
Weight (kg)	94	95	170
L (mm)	600	400	620
H (mm)	935	1850	1640
D (mm)	335	400	580
	BlueGEN (SOFC)	VAILLANT FCU 4600 (PEMFC)	VIESSMAN Vitovalor 300-P (PEMFC)
Electric power (kW)	1.50	4.60	0.75
Thermal power (kW)	0.54	7.00	1.00
Electric efficiency (%)	60.0	35.0	37.0
Thermal efficiency (%)	25.0	50.0	53.0
Overall efficiency (%)	85.0	85.0	90.0
Fuel	Natural gas, propane, butane, ethanol, biodiesel	Natural gas	Natural gas
Weight (kg)	–	–	125
L (mm)	600	–	516
H (mm)	1010	–	1667
D (mm)	660	–	480

cell-based MCHP systems available commercially. Table 15.4 describes the main FC-based micro-cogeneration units on the market (with electric output lower than 15 kW_{el}) in terms of electric output, thermal output, electric efficiency (ratio of electric power to fuel input power), overall efficiency (sum of electric and thermal efficiencies), fuel, weight, and dimensions (L = length, H = height, D = depth) based on manufacturers' nominal data.

15.2.4 Gas and Steam Micro-turbines (MT)

Micro-turbines extend combustion turbine technology to smaller scales. They are primarily fuelled with natural gas, but they can also operate with diesel, landfill gas, ethanol, gasoline, propane, hydrogen, and other bio-based liquid and gaseous fuels.

Table 15.5 Manufacturers' data on main FC-based micro-cogenerators

	FLOWGROUP (steam MT)	OTAG (steam MT)	MTT(steam MT)
Electric power (kW)	1.0	2.0	3.0
Thermal power (kW)	10.0	18.0	14.4
Electric efficiency (%)	10.0	10.4	15.0
Thermal efficiency (%)	80.0	83.6	72.0
Overall efficiency (%)	90.0	94.0	87.0
Fuel	–	Natural gas, LPG	Natural gas
Weight (kg)	–	195	225
L (mm)	–	62	610
H (mm)	–	126	970
D (mm)	–	83	1120
Noise (dB(A))	–	54	<58

In comparison to internal combustion engines, they offer a number of advantages such as more compact size, lower weight, shorter delivery time, smaller number of moving parts, and lower vibration and lower noise, with minimum maintenance requirements [maintenance costs are in the €0.006–0.01/kWh range (Onovwiona and Ugursal 2006)]. Additionally, micro-turbines have a significant advantage over reciprocating internal combustion engines in terms of emissions: current expectations for NO_x emissions from micro-turbines are already below those of ICEs. However, in the lower power ranges, micro-turbines have a lower overall efficiency (up to 80 %) when compared to reciprocating internal combustion engines. Their limitations are mainly due to high first-capital costs and a relatively short life. Other issues include relatively low electrical efficiency and sensitivity of efficiency to changes in ambient conditions. This technology has only recently been commercialized and is offered by a small number of suppliers. The electric capacity of micro-turbines currently on the market is usually 25 kW_{el} or above. Research is ongoing for systems with capacities less than 25 kW_{el}, which will be suitable for single-family residential buildings.

Table 15.5 describes the main MT-based micro-cogeneration units available on the market (with electric output lower than 15 kW_{el}) in terms of electric output, thermal output, electricity efficiency (ratio of electric power to fuel input power), overall efficiency (sum of electric and thermal efficiencies), fuel, weight, dimensions and noise level based on manufacturer nominal data.

15.2.5 Photovoltaic Thermal (PVT) Generators

Solar energy conversion to electricity and heat with a single device is obtained with the "photovoltaic thermal (PVT) collectors." A PVT collector is a module in which the photovoltaic (PV) system not only produces electricity, but also serves as a thermal absorber. PV cells utilize a fraction of the incident solar radiation to produce electricity and the remainder is turned mainly into waste heat in the cells and substrate raising the temperature of PV. The photovoltaic thermal (PVT) technology recovers part of this heat and uses it for practical applications (Chow 2010). In this way, both heat and power are produced simultaneously (Maghanki et al. 2013). The dual functions of the PVT result in a higher overall solar conversion rate (up to around 60–70 %) than that of solely PV or solar collector, and thus enable a more effective use of solar energy. Different types of PVT collectors are currently being used, such as PVT/air, PVT/water, and PVT concentrated collectors (Chow 2010). Currently, there are various PVT applications on the commercial level but it is still limited, due to product reliability and cost. Hence, significant research is required in the field of PVT, mainly in thermal absorber design and fabrication, material and coating selection, energy conversion and its effectiveness, cost minimization, performance testing, control, and the reliability of the system (Kumar et al. 2015).

In addition to the above-mentioned typology of PVT devices, an innovative system able to convert the radiant energy of combustion into electrical energy by using photovoltaic cells is under investigation. This technology (also known as thermal photovoltaic (TPV)) mainly consists of a heat source, an emitter, a filter, and an array of photovoltaic cells (Ferrari et al. 2014). The thermal production of the TPV is realized by recovering the heat from the cooling of the PV cells and the exhaust combustion products. The main advantages of TPV systems can be found in the (1) high fuel utilization factor (close to the unit, thanks to the recovery of most of the thermal losses); (2) low noise levels (due to the absence of moving parts); (3) easy maintenance (similar to a common domestic boiler); and (4) great fuel flexibility. According to the values reported in the literature associated with realized prototypes, the electric efficiency of TPV systems is low [from 0.6 to 11.0 % (Ferrari et al. 2014)], but the potential overall efficiency is always higher than 90 % (Ferrari et al. 2014). At present, the capital cost of thermophotovoltaic generators is high [around 6,000 €/m^2 (Bianchi et al. 2012)] and it does not ot appear very favorable for the development of this technology.

15.3 Operating Schemes

In this section the main operating schemes of MCHP applications are discussed. Figure 15.3 (Mohamed et al. 2014) presents a typical schematic diagram [thermal connections of the MCHP unit, and electrical connections of both the MCHP device and the photovoltaic (PV) modules (when used)] under the thermal load following control strategy, while Fig. 15.4 (Mohamed et al. 2014) presents a typical schematic diagram (electrical connections of the MCHP unit, and thermal connections of both the MCHP device and the solar thermal collectors (STC) modules (when used)) under the electric load, following control strategy. Both schemes can be used for domestic hot water (DHW) production, space heating, and electricity production.

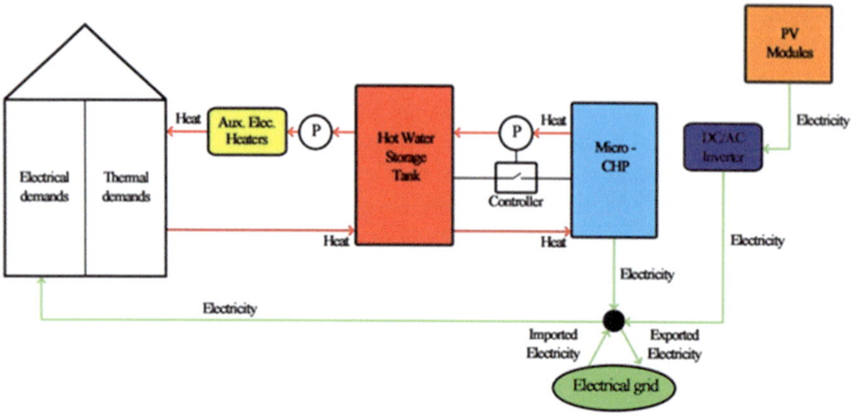

Fig. 15.3 The typical control principle of MCHP systems under thermal load following strategy (Mohamed et al. 2014)

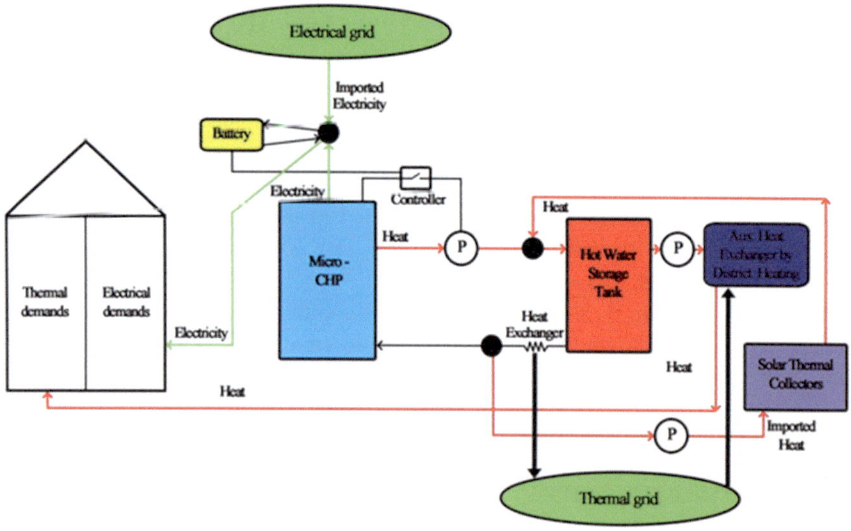

Fig. 15.4 The typical control principle of MCHP systems under electric load following strategy (Mohamed et al. 2014)

Some authors proposed micro-cogeneration plants using two separate tanks for space heating and domestic hot water (Fig. 15.5; Dorer and Weber 2009; González-Pino et al. 2014).

Additional operating schemes are suggested by the main manufacturers of the MCHP units, such as VAILLANT (Fig. 15.6), BOSCH (Fig. 15.7), and AISIN SEIKI (Fig. 15.8).

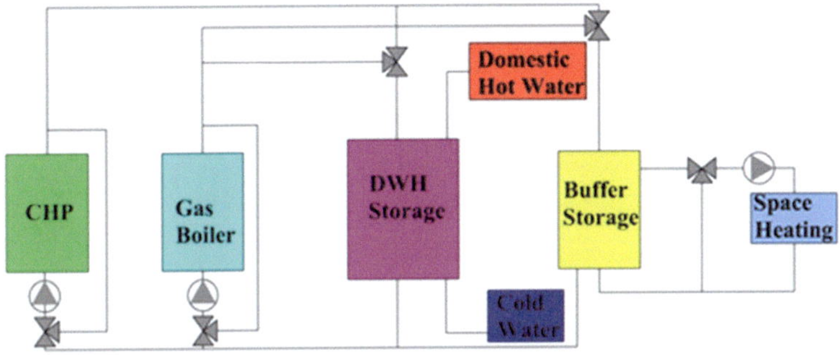

Fig. 15.5 Schematic of MCHP systems with two separate tanks (Dorer and Weber 2009)

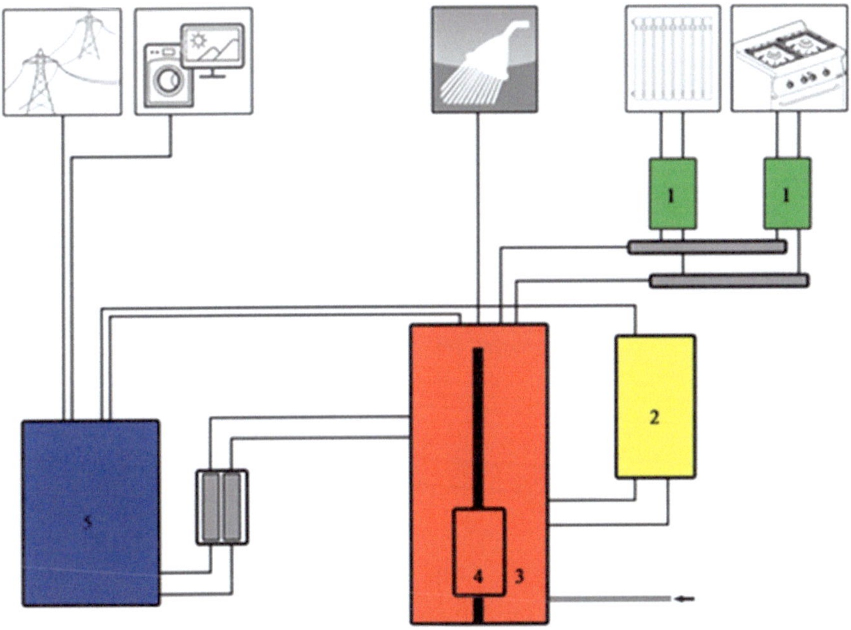

Fig. 15.6 Operating diagram of the MCHP scheme proposed by VAILLANT (*1* distribution and control system of the heating circuit; *2* boiler; *3* storage tank for space heating; *4* storage tank for DHW production; *5* MCHP unit)

Another interesting option is using the MCHP units under the so-called "load-sharing approach" (Cho et al. 2013). This approach mainly consists of combining opposing thermal load profiles in order to increase the operating hours of the micro-cogeneration devices and operate efficiently most of the time with improved part load conditions. This kind of operation is suggested in view of the fact that in residences, the main thermal loads are in the evening through the early morning, whereas the main thermal loads in offices are during the daytime. In this case, a

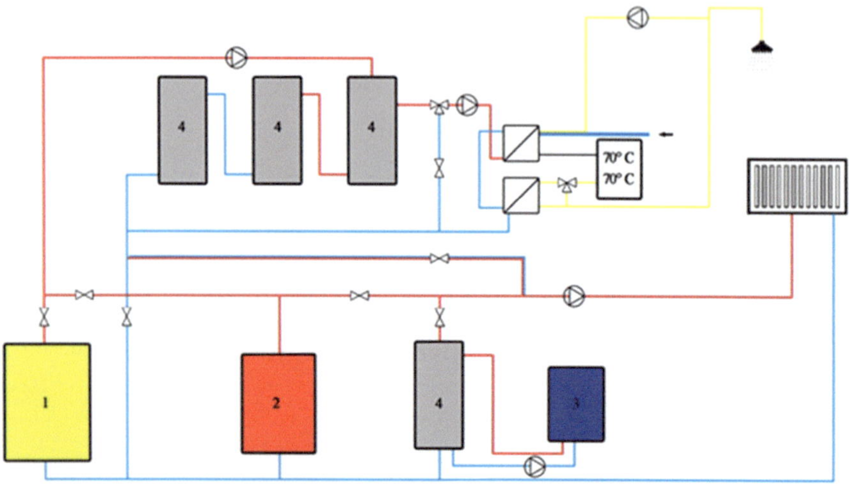

Fig. 15.7 Operating scheme of MCHP systems proposed by BOSCH (*1* MCHP unit; *2* condensing boiler; *3* air-to-water electric heat pump; *4* tanks)

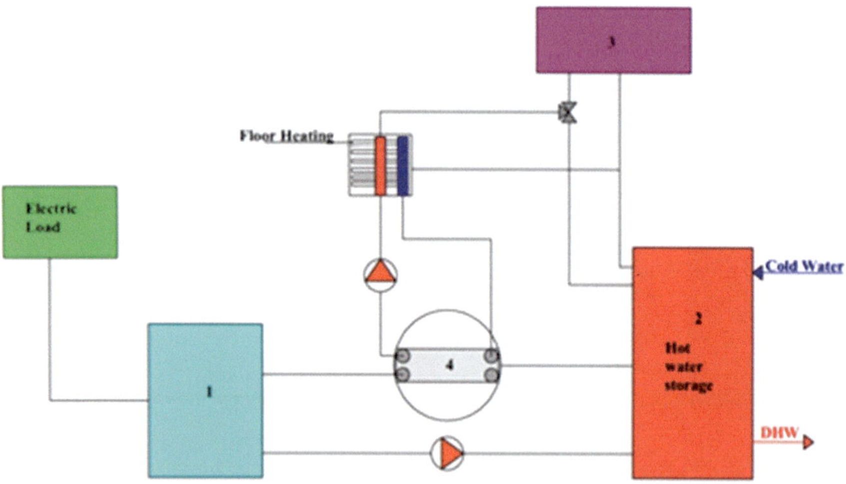

Fig. 15.8 Operating scheme of MCHP systems proposed by AISIN SEIKI (*1* MCHP unit; *2* combined tank; *3* air-to-water electric heat pump)

possible operating scheme for heating purposes is reported in Fig. 15.9 (Angrisani et al. 2014b).

Some authors (Canelli et al. 2015) investigated the "load-sharing approach," using a hybrid system composed of a fuel cell-based MCHP unit integrated with a ground source heat pump (GSHP) to satisfy the combined DHW, heating, and cooling demands of a residential application coupled with a typical office (Fig. 15.10).

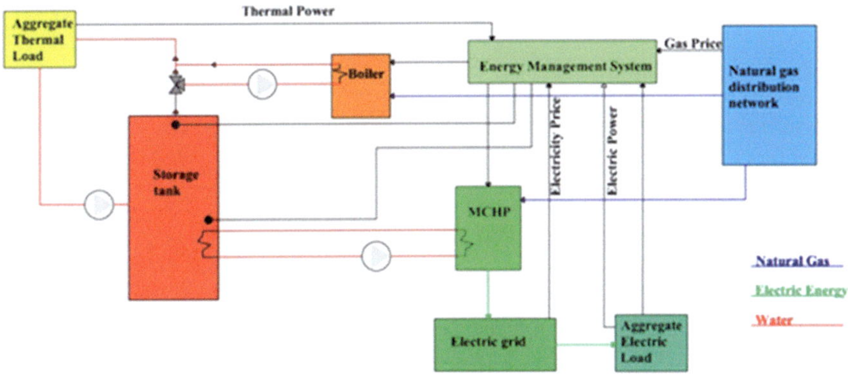

Fig. 15.9 Operating scheme of MCHP system for heating purposes in the case of a "load-sharing approach" (Angrisani et al. 2014b)

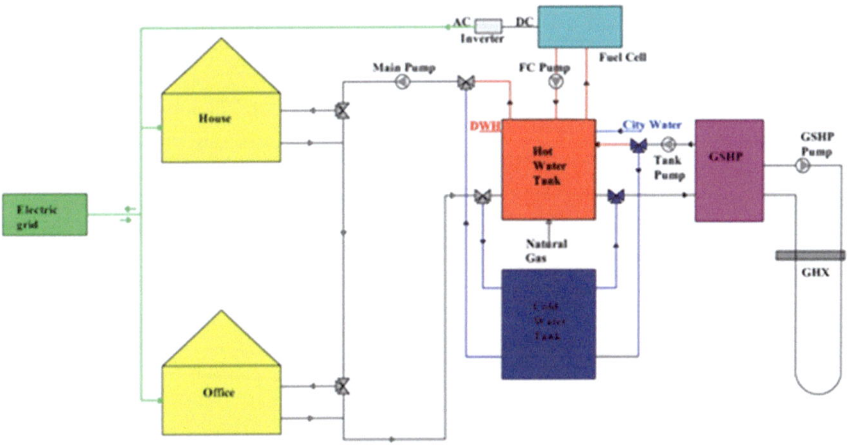

Fig. 15.10 Operating scheme using MCHP and GSHP for domestic hot water production, heating, and cooling demands in the case of a "load-sharing approach" (Canelli et al. 2015)

Integrated energy systems (known as "micro-grids") consisting of distributed generation systems (including micro-cogeneration technologies) and multiple electrical loads operating as a single, autonomous grid either in parallel to, or "islanded" from the existing utility power grid (Asmus 2010; Palizban et al. 2014; Bouzid et al. 2015) are emerging worldwide. Micro-cogeneration systems integrated into an ensemble of other distributed generation systems and load management technologies could also be centrally, remotely, and automatically controlled, forming so-called "virtual power plants" (Asmus 2010; Palizban et al. 2014; Bouzid et al. 2015), maximizing the performance for both end-users and distribution utilities.

15.4 Regulatory Framework

The economic suitability of micro-cogeneration projects are characterized by a high initial investment, the depreciation of which depends heavily on the following factors:

- Fuel price
- Price of electric energy purchased and sold
- Maintenance costs
- Utilization of cogenerated heat
- Yearly hours of operation
- Efficiency of the operating scheme
- Support schemes and other additional incentives

Within the EU member states, a wide variety of financial support mechanisms are in place—or in preparation—that are designed to improve the economics of cogeneration installations. The various European countries designed various support measures to support cogeneration. Table 15.6 summarizes the measures in

Table 15.6 Overview of support measures for CHP in the European Union in 2007 (Moya 2013)

Member state	Tax advantage	Feed-in tariff	Certificates	Grant	Other
Austria	0	1	0	0	1
Belgium	1	0	1	0	1
Bulgaria	0	1	0	0	1
Cyprus	0	0	0	0	1
Czech Republic	0	1	0	0	1
Denmark	0	0	0	0	0
Estonia	0	0	0	0	1
Finland	0	0	0	1	1
France	0	1	0	0	1
Germany	0	1	0	0	1
Greece	1	1	0	0	1
Hungary	0	1	0	0	1
Ireland	0	0	0	0	1
Italy	1	1	0	1	0
Latvia	0	1	0	0	1
Lithuania	0	1	0	0	1
Luxembourg	1	0	0	0	1
Malta	1	0	0	0	1
Netherlands	1	1	0	1	1
Poland	0	0	1	0	1
Portugal	0	0	0	1	1
Romania	0	1	0	1	0
Slovakia	0	1	0	0	0
Slovenia	0	1	0	0	1
Spain	1	1	0	0	1
Sweden	0	0	0	1	1
United Kingdom	1	1	0	1	0

1 measure in place; *0* measure not in place

operation in 2007 in the European Union (Moya 2013); they are divided into tax advantages, feed in tariffs, certificates, grants, or other kinds of additional support.

The most widely used support measure is the feed-in tariff. This is a special incentive for electricity supplied to the grid, mainly a generation bonus for total electricity generated in cogeneration mode or a fuel-related concession. Tax advantages are offered in seven countries, or capital grants for specific sizes of projects are offered in eight countries. Obviously, the effectiveness of support measures depends not only on their existence in the first place, but also on their intensity; a deep analysis of the change produced on the payback period—depending on the intensity of support measures—was carried out by Moya (2013).

15.4.1 Micro-cogeneration Testing Procedures

The support mechanisms for micro-cogeneration devices usually require the achievement of minimum energy performance, for example in terms of primary energy saving with respect to a benchmark case. Moreover, in some countries, MCHPs may be required to meet certain minimum standards to be marketable (Angrisani et al. 2014c). Therefore, it is useful to define a procedure for testing *ex-ante* the energy performance of a device, representative of a unit type, allowing for classification of the energy performance of the MCHP with experimental tests performed in a test facility, possibly certified by an independent third party. The diffusion of such standards procedures can also support the introduction of energy-labelling schemes for MCHP units, such as those already in place for various electric appliances, which could help potential users understand the achievable energy, and environmental and economic savings.

Standard testing procedures are available or at a discussion stage in many countries around the world. Following are examples of such standard procedures for small-scale cogeneration devices (Angrisani et al. 2014c):

- Italy: prUNI E0204A073: Draft of a proposed UNI standard: microcogeneration devices fuelled by gaseous or liquid fuels—*ex-ante* measurement of energy performance (in stand-by) (Bianchi et al. 2013).
- Germany: DIN 4709 (2011-11): Determination of the standard efficiency factor for micro-CHP-appliances of nominal heat input not exceeding 70 kW.
- UK: Publicly available specification 67 (PAS 67).
- USA: ASHRAE SPC 204—Method of testing for rating micro combined heat and power devices (in progress).
- Europe: prEN 50465: Gas appliances—Combined heat and power appliance of nominal heat input inferior or equal to 70 kW.
- Japan: Industrial standards for performance and safety testing of MCHP.

The principles of the main available national testing procedures are described and summarized in Angrisani et al. 2014c; they have many common general features:

- They require that the MCHP be heat-led.
- They refer to a control volume that includes the whole heating system (MCHP, integration boiler and storage tank).
- They require only a limited number of tests, both under nominal operating conditions and according to appropriate test cycles.
- They specify the equipment and instrumentation to be used, such as sensor accuracy.
- They define the reference testing conditions (supply and return water temperatures, ambient air temperature, etc.).

Nevertheless, some major differences can be detected among the national testing procedures; for example, the analyzed standards differ in terms of the limiting value of power (electric, thermal, or primary) for the applicability. A further dissimilarity can be found in the thermal load profile for testing: the Italian standard defines four day types (three for thermal and one for cooling loads); the German one defines a single profile, representative of an intermediate day; in the UK, the standard heat load profile is represented by the number of days per heating season at 13 part-load bands. A further major difference is that the Italian and German standards use energy-based performance indices to evaluate the stationary nominal and overall annual performance of the MCHP system, while the UK procedure is based on an environmental-based performance assessment parameter.

15.4.2 State of the Art: Experimental Results and Simulation Tools

The opportunity to use MCHP systems depends on factors such as heat and power demand variations, control modes, and the capacity and efficiency of the residential cogeneration system, as well as electricity import/export conditions and modes. Therefore the feasibility of a micro-cogeneration unit is a function of the design and size of the system as well as the building it is intended for. For these reasons, studying and evaluating the performance of various MCHP systems under different operating conditions is mandatory. Two main ways can be considered for this purpose:

- Running field/laboratory experiments.
- Developing accurate simulation models of micro-cogeneration devices.

Many laboratory and field tests of residential micro-cogeneration units have been performed worldwide (Entchev et al. 2004; Torrero and McClelland 2004; Van Herle et al. 2004; Thomas and Wyndorps 2004, 2005; Yagoub et al. 2006; Williams et al. 2006; DePaepe et al. 2006; Hubert et al. 2006; Thomas 2006; Kyocera 2006; Veitch and Mahkamov 2006; Possidente et al. 2006; Rosato and Sibilio 2013a, b) by individual manufacturers and/or by energy utility companies and/or by university researchers. Laboratory test results have been reported from all types of MCHP devices, although detailed results are not often given. Tests were conducted in steady-state mode for several load conditions, and for different supply and return temperature of the heat extraction circuit. Dynamic tests were carried out with MCHP systems including the storage components for typical space heating and domestic hot water load profiles. Measurements have also been taken in demonstration buildings with well-monitored boundary conditions and fully controlled internal loads. Many field trials with MCHP units were also conducted in a joint undertaking of MCHP manufacturers and energy service companies, but few results have been reported. National programs, such as the Carbon Trust in the UK, or the US DOE-DOD Residential PEM Fuel Cell Demonstration Program, promote the development, installation and field-testing of MCHP systems and publish their results.

A detailed review of the literature on the existing residential cogeneration performance assessment studies, on the methodologies and modelling techniques used, and on the assessment criteria and applied metrics, can be also found in Dorer (2007). This review indicated that, on the level of individual buildings, residential cogeneration systems are able to reduce the non-renewable primary energy demand compared to conventional gas boiler systems and grid electricity as the benchmark; the strong dependence of the achievable energy savings and, to an even greater extent, the resulting CO_2 emissions on the grid electricity generation mix is confirmed, as well as the strong dependence of cost savings on factors such as heat and power demand variations, control modes, the capacity and efficiency of the residential cogeneration system, and electricity import/export conditions and modes. The results of the performance assessment studies showed that big discrepancies may occur between the nominal efficiencies of the cogeneration device and the overall efficiency of the cogeneration system if the heat for starting up and cooling down the cogeneration device is not well recovered. In addition, the control mode was shown to have significant effects on the energy and environmental system performance: in many cases, heat load following modes showed the best energy efficiency, while electricity load following control modes reduced costs. In general, base-load control offered better energy savings compared to a peak-load oriented control.

The literature review (Dorer 2007) also emphasized that the potential design and operational combinations of factors affecting MCHP operation are almost limitless. These system integration issues, together considering that experimental analyses are both expensive and time-consuming, led to the need to use accurate and practical simulation models of micro-cogeneration devices as techno-economic analysis tools for studying and evaluating the performance of various systems under different load environments. In the past, it was common to model the performance of micro-cogeneration devices using performance-map methods, wherein the device's electrical and thermal efficiencies are treated as constant or as a parametric function of the device's loading. This approach essentially precludes an accurate treatment of the coupling between the building and MCHP system, and it neglects the inefficiencies associated with the transient operation of the system components. In response to this shortcoming, some authors (Voorspools and D'haeseleer 2002; Haeseldonckx et al. 2007; Onovwiona et al. 2007) recently proposed simple empirical models to simulate the performance of SE and ICE units in building-integrated cogeneration applications. All three of these models are parametric in nature, and all are closely based on empirical data collected for specific cogeneration devices. These models are well-suited for use in building simulations, and designed to predict system fuel use, power generation, and thermal output in response to part-load ratio. However, all three are directly derived from the performance data of specific systems, without the possibility of being readily recalibrated for other cogeneration products.

Recognizing the importance of investigating micro-cogeneration systems, two new generic models that characterize the performance of micro-cogeneration devices were developed (Ferguson et al. 2009; Kelly and Beausoleil-Morrison 2007): one for fuel-cell-based cogeneration systems (SOFC and PEMFC), and a second one for combustion-based systems (SE and ICE). These models rely extensively on parametric equations describing the relationships between key input and output parameters; each of these parametric equations requires empirical constants that characterize aspects of the performance of specific cogeneration devices. Like previous models (Voorspools and D'haeseleer 2002; Haeseldonckx et al. 2007; Onovwiona et al. 2007), the new models (Ferguson et al. 2009; Kelly and Beausoleil-Morrison 2007) are empirical in nature, but were also designed to support calibration with the measurements available during third party testing of the devices. Instances of new models were independently implemented into source code for four widely used building simulation tools [ESP-r (Clarke 2001), EnergyPlus (Crawley et al. 2001), TRNSYS, and IDA-ICE (Sahlin and Sowell 1989)]. A detailed calibration and validation exercise was undertaken for one fuel-cell-based cogeneration system [SOFC device (Beausoleil-Morrison and Lombardi 2006, Beausoleil-Morrison 2010)], and three combustion-based cogeneration systems [one SE device (Lombardi et al. 2010), two ICE devices (Beausoleil-Morrison 2007, Rosato and Sibilio 2012)].

At present, simulation and optimization tools are emerging as the best option to better understand and control micro-generation systems. The most popular tools used worldwide can be categorized as follows (Sonar et al. 2014):

- Economic tools (RETSCREEN, HOMER)
- Simulation tools (TRNSYS, EnergyPlus (Crawley et al. 2001))
- Optimization tools (EnergyPlan)
- Data-bases (CO2DB)
- Externalities and environment impact calculation tools (Extern E, ECOSENSE)

15.5 Conclusions/Discussion

The opportunity to use micro-cogeneration systems greatly depends on factors such as a building's thermal characteristics, prevailing weather, heat and power demand variations, characteristics and control logic of the MCHP device, unit costs of natural gas and electricity, etc. In order to obtain significant benefits when compared to the traditional energy supply systems, it is imperative that the thermal portion of the cogeneration device's output be well exploited and the utilization level of the units be typically more than 4,000 running hours per year.

Temperate climates are suitable for micro-cogeneration mainly in cases where the thermal or electric energy produced by the micro-cogeneration units is further utilized to provide space or process cooling throughout the year, thus allowing longer annual operating periods.

References

AISIN SEIKI. http://www.tecno-casa.com/EN/default.aspx?level0=prodotti&level1=mchp. Cited 23 April 2015

Asmus P (2010) Microgrids, virtual power plants and our distributed energy future. Electr J 23:72–82

Angrisani G, Roselli C, Sasso M (2012) Distributed microtrigeneration systems. Prog Energy Combust Sci 38:502–521

Angrisani G, Rosato A, Roselli C et al (2014a) Influence of climatic conditions and control logic on NOx and CO emissions of a micro-cogeneration unit serving an Italian residential building. Appl Therm Eng 71:858–871

Angrisani G, Canelli M, Rosato A et al (2014b) Load sharing with a local thermal network fed by a microcogenerator: thermo-economic optimization by means of dynamic simulations. Appl Therm Eng 71:628–635

Angrisani G, Marrasso E, Roselli C et al (2014c) A review on microcogeneration national testing procedures. Energy Procedia 45:1372–1381

ASHRAE Handbook (2000) HVAC systems and equipment. ASHRAE, Inc., USA

BAXI. http://www.baxi.co.uk/renewables/combined-heat-and-power/ecogen.htm. Cited 23 April 2015

Beausoleil-Morrison I (2007) Experimental investigation of residential cogeneration devices and calibration of annex 42 Models, IEA/ECBS Annex 42

Beausoleil-Morrison I (2010) The empirical validation of a model for simulating the thermal and electrical performance of fuel cell micro-cogeneration devices. Appl Energy 87:3271–3282

Beausoleil-Morrison I, Lombardi K (2006) The calibration of a model for simulating the thermal and electrical performance of a 2.8kWAC solid-oxide fuel cell micro-cogeneration device. J Power Sources 186:67–79

Bianchi M, De Pascale A, Melino F et al (2013) Micro-CHP system performance prediction using simple virtual operating cycles, In: Microgen III proceedings of the 3rd edition of the international conference on microgeneration and related technologies, Naples, Italy

Bianchi M, Ferrari C, Melino F et al (2012) Feasibility study of a Thermo-Photo-Voltaic system for CHP application in residential buildings. Appl Energy 97:704–713

BlueGEN. http://www.bluegen.info/What-is-bluegen/. Cited 23 April 2015

BOSCH. http://www.bosch-industrial.co.uk/products/combined-heat-and-power-modules/chp-modules.html. Cited 23 April 2015

Bouzid AM, Guerrero JM, Cheriti A et al (2015) A survey on control of electric power distributed generation systems for microgrid applications. Renew Sustain Energy Rev 44:751–766

Canelli M, Entchev E, Sasso M et al (2015) Dynamic simulations of hybrid energy systems in load sharing application. Appl Therm Eng 78:315–325

CARBON TRUST. http://www.carbontrust.com/resources/reports/technology/micro-chp-accelerator. Cited 23 April 2015

Chicco G, Mancarella P (2009) Distributed multi-generation: a comprehensive view. Renew Sustain Energy Rev 13:535–551
Cho S, Lee KH, Kang EC et al (2013) Energy simulation modeling and savings analysis of load sharing between house and office. Renew Energy 54:70–77
Chow TT (2010) A review on photovoltaic/thermal hybrid solar technology. Appl Energy 87:365–379
Clarke J (2001) Energy simulation in building design, 2nd edn. Butterworth-Heinemann, Oxford 2001
COGENGREEN. http://www.cogengreen.com/en/ecogen-12ag. Cited 23 April 2015
CO2DB. http://www.co2db.de/. Cited 23 April 2015
Crawley D, Lawrie L, Winkelmann F et al (2001) Energyplus: creating a new-generation building energy simulation program. Energy Build 33:319–331
DECENT project (2002) DECENTRALISED GENERATION: DEVELOPMENT OF EU POLICY. http://www.ecn.nl/docs/library/report/2002/c02075.pdf. Cited 23 April 2015
DePaepe M, D'Herdt P, Mertens D (2006) Micro-CHP systems for residential applications. Energy Convers Manage 47:3435–3446
Directive 2004/8/EC of the European Parliament and of the Council of the 11 February 2004 on the promotion of cogeneration based on the useful heat demand in the internal energy market and amending Directive 92/42/EEC. Official Journal of the European Union 2004
Dorer V (2007) Review of existing residential cogeneration systems performance assessments and evaluations. A report of Subtask C of FC + COGEN-SIM the simulation of building-integrated fuel cell and other cogeneration systems. http://www.ecbcs.org/docs/Annex_42_STC_Review_Cogen_Evaluations.pdf. Cited 23 April 2015
Dorer V, Weber A (2009) Energy and CO_2 emissions performance assessment of residential micro-cogeneration systems with dynamic whole-building simulation programs. Energy Convers Manag 50:648–657
EC POWER. Ec power. http://www.ecpower.eu/italiano/xrgir/dati-tecnici/xrgir-9.html. Cited 23 April 2015
ECOSENSE. http://www.dedicatedmicros.com/europe/products_details.php?product_assoc_id=228. Cited 23 April 2015
EnergyPlan. http://www.energyplan.eu/. Cited 23 April 2015
Entchev E, Gusdorf J, Swinton M et al (2004) Micro-generation technology assessment for housing technology. Energy Build 36:925–931
Evangelisti S, Lettieri P, Clift R et al (2015) Distributed generation by energy from waste technology: a life cycle perspective. Process Saf Environ Prot 93:161–172
Extern E. http://ecosenseweb.ier.uni-stuttgart.de/. Cited 23 April 2015
Ferguson A, Kelly N, Weber A et al (2009) Modelling residential-scale combustion-based cogeneration in building simulation. J Build Perform Simul 2:1–14
Ferrari C, Melino F, Pinelli M et al (2014) Overview and status of thermophotovoltaic systems. Energy Procedia 45:160–169
FLOWGROUP. http://flowgroup.uk.com/why-we-are-different/. Cited 23 April 2015
González-Pino I, Campos-Celador A, Pérez-Iribarren E et al (2014) Parametric study of the operational and economic feasibility of Stirling micro-cogeneration devices in Spain. Appl Therm Eng 71:821–829
Haeseldonckx D, Peeters L, Helsn L et al (2007) The impact of thermal storage on the operational behaviour of residential CHP facilities and the overall CO_2 emissions. Renew Sustain Energy Rev 11:1227–1243
HEXIS. http://www.hexis.com/en/system-data. Cited 23 April 2015
HONDA. http://world.honda.com/power/cogenerator/. Cited 23 April 2015
HOMER. http://www.homerenergy.com/software.html. Cited 23 April 2015
Hubert CE, Achard P, Metkemeijer R (2006) Study of a small heat and power PEM fuel cell system generator. J Power Sources 156:64–70
IEA/ECBCS Annex 42 (2007) The simulation of building-integrated fuel cell and other cogeneration systems (COGEN-SIM). http://www.ecbcs.org/annexes/annex42.htm. Cited 23 April 2015

IEA/ECBCS Annex 54 (2013) Integration of micro-generation and related energy technologies in buildings. http://iea-annex54.org/index.html. Cited 23 April 2015
Kelly N, Beausoleil-Morrison I (2007) Specifications for modelling fuel cell and combustion-based residential cogeneration devices within whole-building simulation programs, IEA/ECBCS Annex 42 report
Kyocera, Results of the First Domestic Trial Operations of Solid Oxide Fuel Cell (SOFC) Cogeneration System for Household Use, Kyocera Global News release, 16 May 2006
KYOCERA. http://global.kyocera.com/news/2012/0305_woec.html. Cited 23 April 2015
Kumar A, Baredar P, Qureshi U (2015) Historical and recent development of photovoltaic thermal (PVT) technologies. Renew Sustain Energy Rev 42:1428–1436
Lombardi K, Ugursal VI, Beausoleil-Morrison I (2010) Proposed improvements to a model for characterizing the electrical and thermal energy performance of Stirling engine micro-cogeneration devices based upon experimental observations. Appl Energy 87:3271–3282
Maghanki MM, Ghobadian B, Najafi G et al (2013) Micro combined heat and power (MCHP) technologies and applications. Renew Sustain Energy Rev 28:510–524
Mancarella P, Chicco G (2009) Global and local emission impact assessment of distributed cogeneration systems with partial-load models. Appl Energy 86:2096–2106
MICRO-MAP (2002) MICRO AND MINI CHP—MARKET ASSESSMENT AND DEVELOPMENT PLAN. http://www.microchap.info/MICROMAP%20publishable%20Report.pdf. Cited 23 April 2015
Mohamed A, Cao S, Hasan A et al (2014) Selection of micro-cogeneration for net zero energy buildings (NZEB) using weighted energy matching index. Energy Build 80:490–503
Moya JA (2013) Impact of support schemes and barriers in Europe on the evolution of cogeneration. Energy Policy 60:345–355
MTT. http://www.enertwin.com/cms/files/EnerTwin-specifications-2014-MR.pdf. Cited 23 April 2015
Onovwiona HI, Ugursal VI (2006) Residential cogeneration systems: review of current technology. Renew Sustain Energy Rev 10:389–431
Onovwiona HI, Ugursal VI, Fung AS (2007) Modelling of internal combustion engine based cogeneration systems for residential applications. Appl Therm Eng 27:948–961
OTAG. http://www.powerblock.eu/download-data/121106_Kurzinfo_lion-Powerblock.pdf. Cited 23 April 2015
Palizban O, Kauhaniemi K, Guerrero JM (2014) Microgrids in active network management—part I: Hierarchical control, energy storage, virtual power plants, and market participation. Renew Sustain Energy Rev 36:428–439
PANASONIC. http://panasonic.co.jp/ap/FC/en_doc03_00.html. Cited 23 April 2015
Possidente R, Roselli C, Sasso M et al (2006) Experimental analysis of microcogeneration units based on reciprocating internal combustion engine. Energy Build 38:1417–1422
Qnergy. http://www.qnergy.com/sites/Qnergy/UserContent/files/SAL-DataSheet-CHP-v15b-140700.pdf. Cited 23 April 2015
Residential PEM Fuel Cell Demonstration Program. www.eere.energy.gov/informationcenter. Cited 23 April 2015
RETSCREEN. http://www.retscreen.net/it/home.php. Cited 23 April 2015
Rosato A, Sibilio S (2012) Calibration and validation of a model for simulating thermal and electric performance of an internal combustion engine-based micro-cogeneration device. Appl Therm Eng 45–46:79–98
Rosato A, Sibilio S (2013a) Performance assessment of a micro-cogeneration system under realistic operating conditions. Energy Convers Manage 70:149–162
Rosato A, Sibilio S (2013b) Energy performance of a micro-cogeneration device during transient and steady-state operation: experiments and simulations. Appl Therm Eng 52:478–491
SENERTEC. http://www.senertec.de/en/derdachs.html. Cited 23 April 2015
Sahlin P, Sowell EF (1989) A neutral model format for building simulation models. In: Procedings of building simulation, IBPSA, Vancouver, Canada

San Martin JI, Zamora I, San Martin JJ et al (2010) Hybrid fuel cells technologies for electrical microgrids. Electr Power Syst Res 80:993–1005
SOLO 161. http://www.chp-goes-green.info/sites/default/files/SOLO_Stirling_161.pdf. Cited 23 April 2015
Sonar D, Soni SL, Sharma D (2014) Micro-trigeneration for energy sustainability: Technologies, tools and trends. Appl Therm Eng 71:790–796
SUNMACHINE. http://www.grunze-ht.de/Download/Sunmachine/Prospekt%20Sunmachine%20neu.pdf. Cited 23 April 2015
TEDOM. http://cogeneration.tedom.com/. Cited 23 April 2015
Thomas B (2006) Small SE CHP running on bio mine and sewage gas. In: Proceedings of the international stirling forum 2006, ECOS GmbH, Osnabrück
Thomas B, Wyndorps A (2004) Experimental Investigation of Micro-CHPs—focused on part load operation. In: Proceedings of the international stirling forum 2004, Osnabrück
Thomas B, Wyndorps A (2005) Experimental examination of micro-CHP's—stirling vs. IC engines. In: Proceedings of 12th international stirling engine conference (ISEC), Durham
Torrero E, McClelland R (2004) Evaluation of the field performance of residential fuel cells. In: Final report for the National Renewable Energy Laboratories, report no. NREL/SR-560-36229. NREL Golden, Colorado, USA
TRNSYS. http://www.trnsys.com/. Cited 23 April 2015
United Nations, Framework convention on climate change. http://unfccc.int/kyoto_protocol/items/2830.php. Cited 23 April 2015
VAILLANT. http://www.vaillant.co.uk/. Cited 23 April 2015
Van Herle J, Membrez Y, Bucheli O (2004) Biogas as a fuel source for SOFC co-generators. J Power Sources 127:300–312
Veitch DCG, Mahkamov K (2006) Experimental evaluation of performance of a WhisperGen Mk III Micro CHP unit. In: Proceedings of the international stirling forum 2006, ECOS GmbH. Osnabrück
VIESSMANN.http://www.viessmann.com/com/content/dam/internet-global/pdf_documents/com/brochures englisch/ppr fuel_cell_boiler.pdf. Cited 23 April 2015
Voorspools KR, D'haeseleer WD (2002) The evaluation of small cogeneration for residential heating. International J Energy Res 26:1175–1190
WHISPERGEN. http://www.whispergen-europe.com/productspec_en.php. Cited 23 April 2015
Williams MC, Strakey JP, Sudoval WA (2006) U.S. DOE fossil energy fuel cells program. J Power Sources 159:1241–1247
Wu DW, Wang RZ (2006) Combined cooling, heating and power: a review. Prog Energy Combust Sci 32:459–495
Yagoub W, Doherty P, Riffat SB (2006) Solar energy-gas driven micro-CHP system for an office building. Appl Therm Eng 26:1604–1610

Chapter 16
The State of the Art for Technologies Used to Decrease Demand in Buildings: Thermal Energy Storage

A. de Gracia, C. Barreneche, A.I. Fernández and L.F. Cabeza

Abstract The high energy consumption in the building sector, especially for heating and cooling, has promoted new and more restrictive energy policies around the world, such as the new European Directive 2010/31/EU on the energy consumption of buildings. Apart from enforcing stringent building codes that include minimum energy consumption for new and refurbished buildings, the IEA ETP 2012 highlights the necessity of using highly efficient technologies in the envelopes, equipment, and new strategies to address the high energy consumption of the sector. In this context, the use of appropriate thermal energy storage (TES) systems has a high potential to reduce the energy demand for both heating and cooling. The use of TES in the building sector not only leads to the rational use of thermal energy, which reduces the energy demand, but allows peak load shifting strategies, as well as manages the gap between possible renewable energy production and heating/cooling demands. In this chapter, various available technologies are analyzed where sensible, latent, or thermochemical storages are implemented in either active, passive, or hybrid building systems.

16.1 Introduction

Energy use in buildings means a large part of global energy demand becoming 30 % of energy-related CO_2 emissions (one-third of black carbon emissions) and 32 % of global energy used. Over 60 % of residential, and almost 50 % of

A. de Gracia · C. Barreneche · L.F. Cabeza (✉)
GREA Innovació Concurrent, Universitat de Lleida, Edifici CREA,
Pere de Cabrera s/n, 25001 Lleida, Spain
e-mail: lcabeza@diei.udl.cat

A.I. Fernández
Department of Materials Science and Metallurgical Engineering,
Universitat de Barcelona, Martí i Franqués 1-11, 08028 Barcelona, Spain
e-mail: ana_inesfernandez@ub.edu

S.-N. Boemi et al. (eds.), *Energy Performance of Buildings*,
DOI 10.1007/978-3-319-20831-2_16

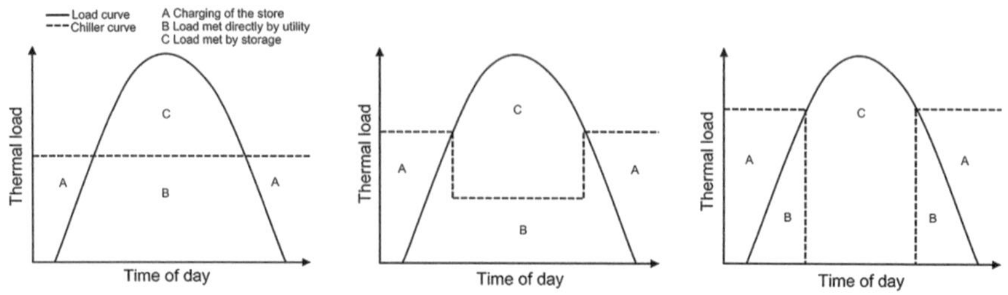

Fig. 16.1 Strategies for peak shaving when TES is used in buildings. From *left* to *right* load levelling, demand limiting, and full storage

commercial building energy, is expended for thermal application as water heating in residential buildings, and as a cooling supply in commercial buildings (IEA 2012; Ürge-Vorsatz et al. 2015).

Thermal energy storage (TES) is also known as heat and/or cold storage (Cabeza 2015; Mehling and Cabeza 2008). This technology allows storing thermal energy to be used at a later time. Implementing TES in buildings allows the use of renewable sources of heat and cold that otherwise would not be possible (Heier et al. 2015). The duration of the storage can be from short-term, daily storage, to long-term storage, usually between seasons and known as seasonal storage.

The application of storage in buildings can have various objectives. One of them is the reduction of the building energy demand, either by reusing the high internal gains or by storing exterior heat or cold to be used in the building. A second objective is "peak shaving" (also called "peak shifting"), as shown in Fig. 16.1; usually, according to this strategy, no energy is saved, but economic advantages are achieved. TES can also improve HVAC systems, increasing their efficiency or reducing emissions.

16.2 Materials Used for TES in Buildings

TES materials are becoming a key issue for the deployment of thermal heating and cooling technologies in the building sector. Meanwhile, the International Energy Agency (IEA) suggests energy storage as a possible answer to address the energy problems described in Sect. 16.1, focusing on materials and systems for building applications (IEA 2010). As previously stated, thermal energy can be stored following three processes: as sensible heat, as latent heat, or as thermochemical heat. These methods are described in the following sections as well as the desired properties for material selection, as are the main advantages and drawbacks of each technology.

16.2.1 Sensible Heat

This is the simplest method for storing thermal energy and consists of applying a temperature gradient to a media (solid or liquid); the heat that can be accumulated in this media is the sensible heat stored. Figure 16.2 shows the thermal profile of one substance or material when a temperature gradient is applied. The area under the curve is the total heat accumulated (Basecq et al. 2013).

The most common storage media used to store energy as sensible heat is water (see Fig. 16.2 as ice). Sensible heat storage has two main advantages: it is cheap and is without the risks derived from the use of toxic materials. Moreover, the material used to store energy is contained in vessels as bulk material, thereby facilitating the system design.

Heat capacity comes from either the content of the material used, volume, or mass. Sensible heat is often used with solids, such as stone or brick, or from liquids, such as water (high specific heat).

The materials mainly used to store sensible heat are based on common ceramics (cement, concrete, etc.), or natural stones, such as marble, granite, clay, sandstone, or polymers (PUR, PS, PVC). All of them are inexpensive materials, and sensible heat will depend on the mass and volume used. Therefore, waste materials or by-products from several industrial processes with proper thermophysical properties are becoming suitable candidates to be used for sensible heat TES (Fernandez et al. 2010; Navarro et al. 2011). Figure 16.3 shows the specific heat capacity versus density of all the materials mentioned before to be used as sensible heat media. Note that the waste materials are proper candidates because they have a high enough density and account for suitable specific heat capacity.

The more relevant properties for sensible heat TES are the ones collected in Table 16.1. The physical and technical requirements must be taken into account during the design step of the system to be implemented in buildings.

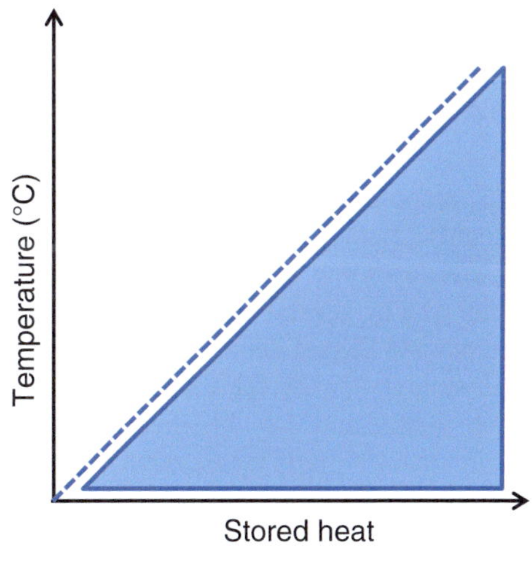

Fig. 16.2 Thermal energy storage profile versus temperature when heat is stored as sensible heat

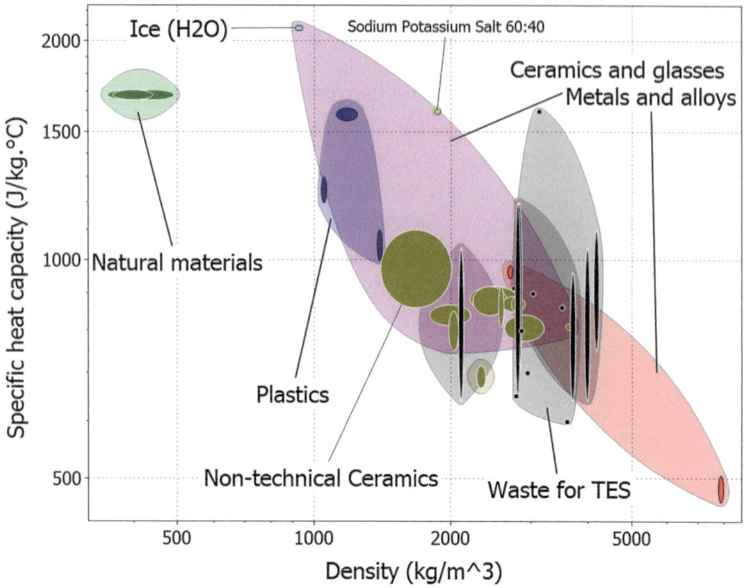

Fig. 16.3 Specific heat versus density of material candidates to be used for sensible heat thermal energy storage (SH-TES)

Materials to be used for sensible heat-TES can present two drawbacks. The first is related to corrosion incompatibility, but this drawback is easy to avoid if the material used to contain the storage media is well selected and/or the surface of the vessel material is coated with a material that is compatible with the sensible storage material. The second drawback is the thermal stability over thermal cycles. This behavior must be studied before the implementation under real thermal conditions fitted to the application.

16.2.2 *Latent Heat*

TES proposes phase change material (PCM) as materials able to store high amounts of energy as latent heat. PCMs increase the thermal mass of the building envelopes and building systems. Such materials have been extensively studied in the past by several researchers (Baetens et al. 2010; Khudhair and Farid 2004; Sharma and Sagara 2005; Zalba et al. 2003). Among all materials, those that have high storage density for small temperature ranges are considered PCM (Günther et al. 2006), and are classified as different groups, depending on the material nature (paraffin, fatty acids, salt hydrates, etc.).

Table 16.1 Requirement for materials to be used as TES media

	Sensible heat thermal energy storage	Latent heat thermal energy storage	Thermochemical storage
Physical and technical requirements	High density (ρ) temperature range fitted to the application High cyclic stability Large-scale production methods Non-corrosive low system complexity Low vapor pressure in the temperature range	Low density variation and small volume change High energy density Small or no subcooling No phase segregation Low vapor temperature Chemical and physical stability Compatible with other materials	Low density and small volume change High energy density No phase segregation Chemical and physical stability Compatible with other materials—non-corrosive
Thermal requirements	High energy density High thermal conductivity (κ) Low thermal diffusivity Good specific heat capacity (Cp) Thermal expansion coefficient (α)	Suitable phase change temperature fitted to application Large phase change enthalpy (ΔH) and specific heat (Cp) High thermal conductivity (except for passive cooling) Reproducible phase change Thermal stability Cycling stability	Reversible reaction Control of the kinetic model Control of the crystallographic structure changes Water stability within crystal structure Proper particle size Control of the impurities Solubility of the TCM in the working conditions (P, T) Suitable working temperature range Large energy involved in the reaction High thermal conductivity Thermal stability Cycling stability
Economical requirements	Low price Non-toxic	Low price Non-toxic	Low price Non-toxic
Recyclability	Recyclable or reusable Low CO_2 foot print Low embodied energy	Recyclable or reusable Low CO_2 footprint Low embodied energy	Recyclable or reusable Low CO_2 footprint Low embodied energy

A conventional profile—when one PCM is submitted to a temperature gradient—is shown in Fig. 16.4. The extremes correspond to sensible heat, and the central part to latent heat stored.

A temperature increment produces a temperature increase on the material evaluated. Then, the temperature remains constant during the phase change but the heat stored is increased. Further storage of heat results as sensible heat again when the phase change is finished.

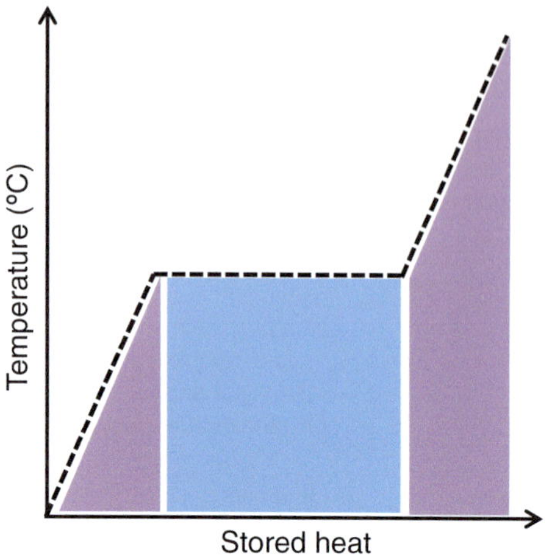

Fig. 16.4 Thermal energy storage profile versus temperature when heat is stored as sensible + latent heat

Buildings can use the latent heat from PCM by two different concept systems: passive latent heat energy storage systems (LHES), or using active LHES systems (Parameshwaran et al. 2012). However, independently of each system, PCM must be encapsulated or stabilized for technical use, and PCMs are introduced in building systems using several methods.

PCM encapsulation in structures is the direct inclusion in walls and ceilings. These building components offer large areas for passive heat transfer within every zone of the building (Zhang et al. 2010).

Impregnation (Khudhair and Farid 2004) is an extensively used method to introduce PCM in porous materials such as gypsum (Nomura et al. 2009), mortar, wood, etc. This technique has been widely studied lately (Barreneche et al. 2013a, b; De Gracia et al. 2011), and is based on introducing liquid PCM inside porous materials using vacuum systems.

A microencapsulation (Tyagi et al. 2011; Zhang et al. 2010) PCM process is projected as one way to incorporate PCM in building materials becoming a TES part of the building structure even for lightweight buildings. Microcapsules consist of small PCM containers using hard shells, although they allow volume changes.

A shape-stabilized PCM is the incorporation of PCM in a polymer as one shape stabilized material (Del Barrio et al. 2009; Li et al. 2012). PCM slurries consist of a microcapsuled PCM suspended on a thermal fluid to be used applying them in several active systems such as heat exchangers and thermal control systems (Delgado et al. 2012). Material candidates to be used as latent heat TES must undertake several required properties. Table 16.1 shows the properties required by PCM for building applications (Cabeza et al. 2011; Tyagi and Buddhi 2007).

The criteria that limited the storage process from the beginning are the segregation defined as the phase separation of the PCM when this material has more than one component, and the subcooling that appears when a PCM starts to solidify at a temperature below its congealing temperature. These two processes must be studied during the material selection step and must be avoided or minimized. In addition, there are other drawbacks that will limit the implementation of PCM but they will appear afterwards. Regarding corrosion problems between PCMs and containers, the compatibility will appear afterwards and it must be studied before the implementation; cycling stability at long term: one substance used as PCM must be stablized over cycles; and finally, the leakage is one of the most important drawbacks because if this effect occurs, the effectivity of the system will decrease drastically. For that reason, the techniques to encapsulate or stabilize PCM are a key issue in PCM implementation.

16.2.3 Thermochemical Reactions

Thermochemical materials (TCM) are materials that can store energy by a reversible endothermic/exothermic reaction/process and the resulting reaction products are easily separated (usually a gas-solid system or a liquid-solid system) (Cot-Gores et al. 2012). First of all, the charging process is performed: applying heat, the material reacts and is separated into two parts: A + B. Then, when the storing process takes place, the heat is stored because the products are easily separated and the storage could be complete at low temperature. Finally, the discharging process is performed—as a reversible reaction, when products A + B are placed together, and under the suitable pressure and suitable temperature conditions to react, the energy is released again. This process is shown in Fig. 16.5.

TCM systems are those that are more energy-efficient, but they are being developed (N'Tsoukpoe et al. 2009). There are almost no real applications implemented in the market, but researchers and companies are making a huge effort to bypass the barriers of this technology.

Building systems that include TCM are described for seasonal storage, in order to store the overheating of summers to be used to heat buildings during the winter.

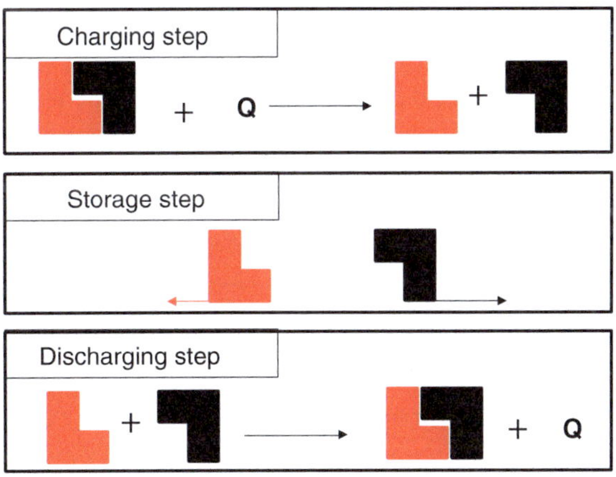

Fig. 16.5 Steps followed by TCM to store energy

Table 16.1 shows the properties required by TCM for building applications; these are the materials that present more technical requirements.

The main drawbacks of TCM storage are due to the corrosion, as with PCM and sensible storage, and due to the exhaustion of the solid-gas chemical reactions because they are poor regarding the heat and mass transfer performance in the reactive bed and present low thermodynamic efficiency of the basic cycle. In order to enhance the performance and well-operation of the storage system, the chemical reaction or the sorption process is the most important issue for storing energy through TCM. For that reason, precise control of the kinetics of the process/reaction as well as the boundary conditions of the system (pressure, temperature, water vapor flow, etc.) are priority issues. This challenge can be overcome by studying the details of the chemical reaction/sorption process.

16.3 Passive Technologies

16.3.1 Introduction

Buildings should be designed to provide thermal comfort for inhabitants with the minimal use of energy by the HVAC system (Soares et al. 2013). TES is a technology that can help achieve this objective, both with sensible and latent TES. The use of TES in the building components can significantly reduce the heating and cooling demands by absorbing solar radiation, internal heat gains, and/or free cooling strategies.

16.3.2 Sensible Passive Systems

16.3.2.1 Integration in Building Components

Passive storage with sensible heat storage is achieved mainly with high thermal mass materials, able to store bit amounts of energy, that give thermal stability and smooth indoor thermal fluctuations (Kosny et al. 2014). Examples of these materials are stone, rammed earth, concrete, stone, and brick. A comparison of rocks, clays, water and salts was carried out by Rempel and Rempel (2013); the authors compared the patterns of heat uptake of various materials and thicknesses (Fig. 16.6) to give an intuitive understanding of TES to architects and engineers (Fig. 16.7).

Solar walls can be used for energy savings in buildings (Saadatian et al. 2012). They can be classified as standard solar wall (Trombe wall), solar water wall, solar transwall, composite solar wall, and fluidized solar wall. Only the Trombe wall type and the solar water wall include storage in their operation.

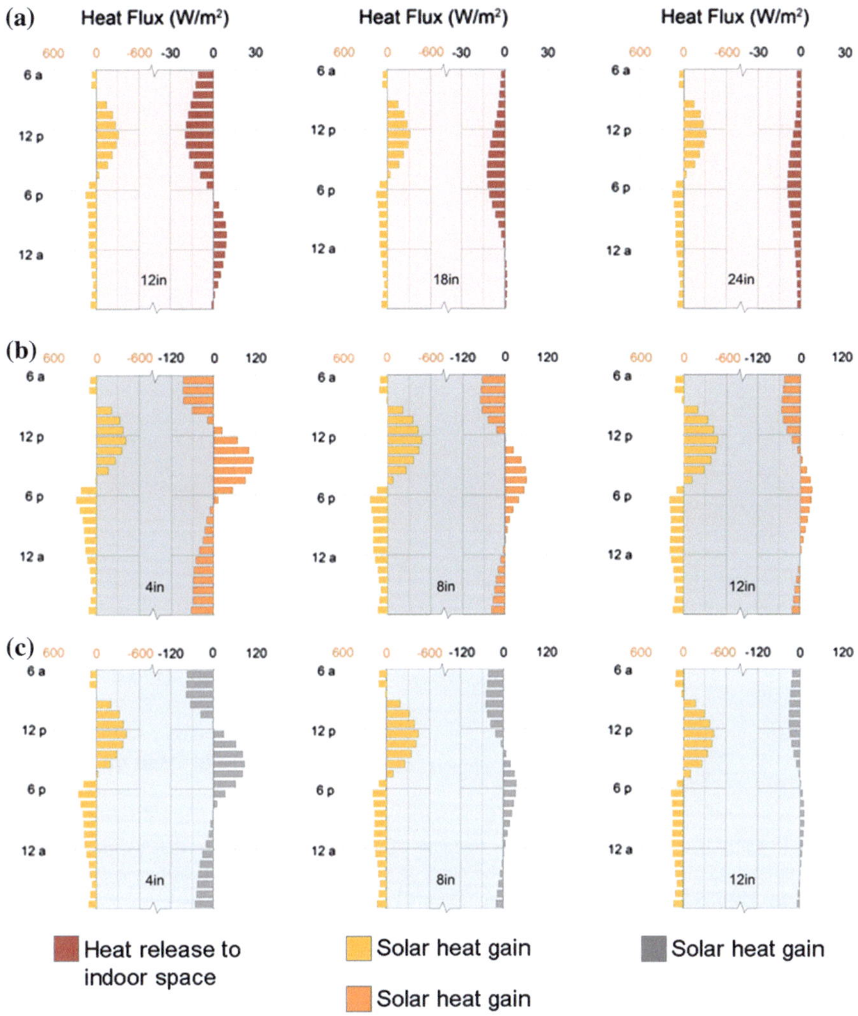

Fig. 16.6 Exterior mass walls. Heat fluxes across outdoor (*left axis*) and indoor (*right axis*) surfaces of exterior walls on a typically sunny January day in Denver, Colorado (Rempel and Rempel 2013)

A typical Trombe wall consists of a high thermal mass wall (concrete, stone, brick or earth wall) covered by glass, with an air channel in between (Stazi et al. 2012). Its operating principle can be seen in Fig. 16.8. The energy benefits of Trombe walls as a passive system can be found in the literature (Briga-Sá et al. 2014; Llovera et al. 2010). Solar water walls follow the same principle as the standard ones, but the massive part of it is replaced by water containers forming the wall (Adams et al. 2010).

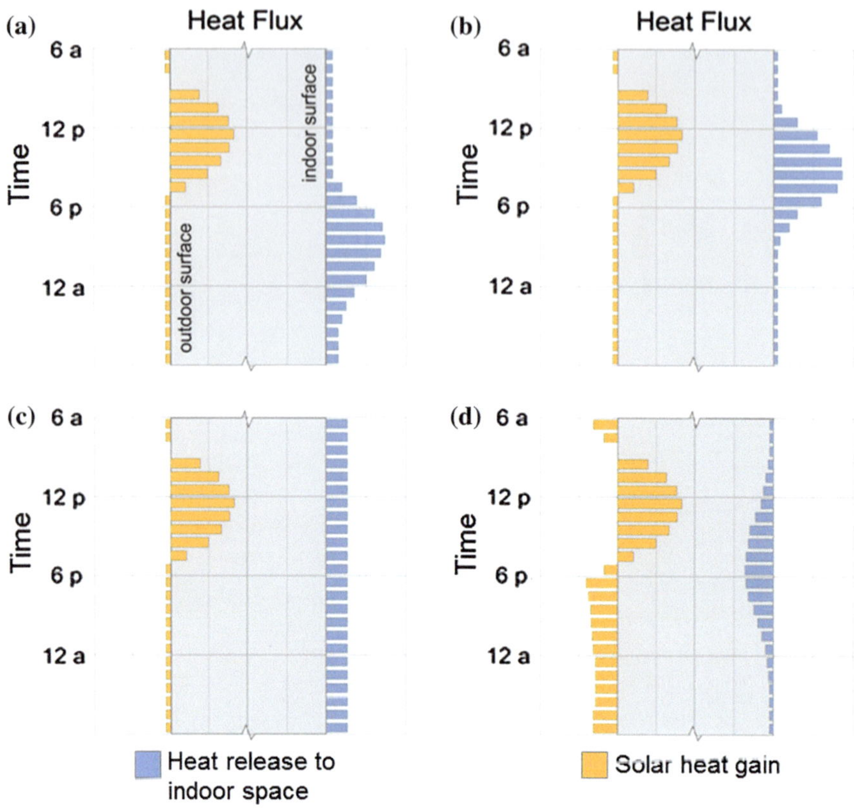

Fig. 16.7 Daily heat flux patterns across surfaces of thermal storage walls that would best meet design intents of **a** residential evening heating; **b** workplace daytime heating; **c** sunspace or greenhouse all-night heating; and **d** cooling (Rempel and Rempel 2013)

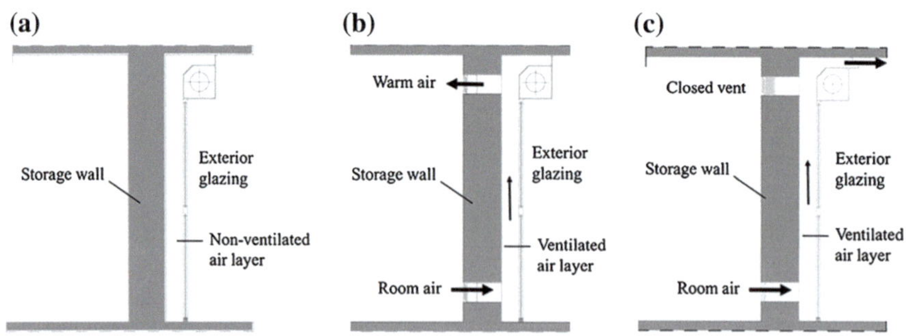

Fig. 16.8 Trombe wall configurations: **a** without ventilation; **b** winter mode with air circulation; and **c** summer mode with cross ventilation (Stazi 2012)

16.3.3 Latent Passive Systems

16.3.3.1 Integration in the Building

As previously stated, the most promising methods for incorporating PCM in construction materials are direct incorporation, immersion, encapsulation, microencapsulation, and shape-stabilized (Cabeza 2015; Hawes et al. 1993; Memon 2014).

A first way to incorporate PCM in building envelopes is through wallboards, as shown in many studies recently published, both experimental and numerical (Cabeza 2015; Heier et al. 2015; Kosny et al. 2014). The factors that will influence the final efficiency of PCM-containing wallboards are how the PCM is incorporated into the wallboard, the orientation of the wall, the climatic conditions, the direct solar gains, the internal gains, the color of the surface, the ventilation rate, the PCM chosen and its phase change temperature, the temperature range of the phase change, and the latent energy per unit area of the wall.

Recommended studies are those of Kuznik et al. (Kuznik et al. 2008a, b; Kuznik and Virgone 2009) (Fig. 16.9), with a wallboard with 60 % microencapsulated paraffin melting at 22 °C that reduced the air temperature fluctuations in a room and the overheating effect of solar energy (Fig. 16.9); Shilei et al. (2006), with a wallboard with 26 wt% of fatty acid tested in China; Diaconu and Cruceru (Diaconu and Cruceru 2010; Diaconu 2011), with a three-layer sandwich insulating panel, tested experimentally showing a contribution to the annual energy savings and a reduction of the peak heating and cooling loads (Borreguero et al. 2010), with the inclusion of microencapsulated commercial paraffin (RT27) to gypsum blocks; and, more recently (Lee et al. 2015a, b), with a PCM thermal board, tested experimentally achieving up to 27 % average heat flux reductions

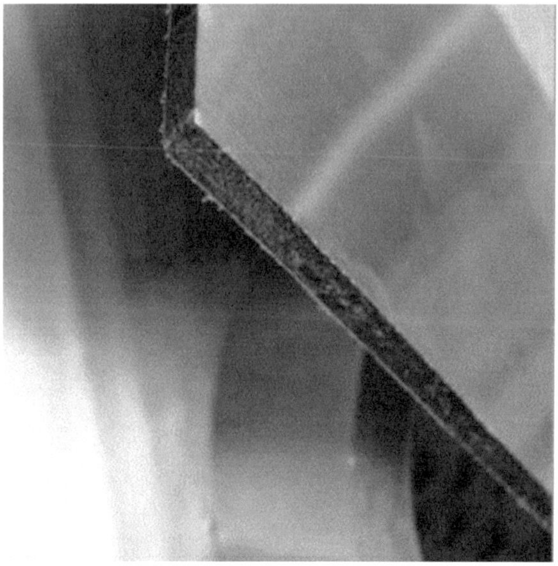

Fig. 16.9 Dupond PCM composite wallboard, studied by Kuznik et al. (Kosny et al. 2014)

and two to three hours of delayed peak heat transfer rate per unit of wall area (Biswas et al. 2014), with an innovative nano-PCM supported by expanded inter-connected nano-sheets of gypsum wallboard developed and tested (Lai and Hokoi 2014), with the addition of aluminium honeycomb in a PCM wallboard.

Zhou et al. (2014) evaluated the factors influencing both interior and exterior PCM wallboards under periodically changing environments, with solar radiation also being considered. Thermal properties, such as melting temperature, melting temperature range, latent heat, thermal conductivity and surface heat transfer coefficients, were qualitatively optimized in two design criteria: inner surface temperature history, and diurnal thermal storage.

Other ways to include PCM in passive walls are the addition of PCM to cellulose insulation (Evers et al. 2010) (with a reduction of 9.2 % of the average peak heat flux), or coupled with vacuum-insulating panels (VIP) (Ahmad et al. 2006), the addition of PCM to a structural insulated panel (Medina et al. 2008) (with a peak heat flux decrease of 62 %, and a daily heat transfer across the system reduction of 38 %, when 20 wt% PCM is added), or in rigid polyurethane foams (Castellón et al. 2010; Yang et al. 2015), or the addition of PCM as a new layer (Castell et al. 2010; Jin et al. 2014; Lee et al. 2015a, b) (with up to 20 % reduction of energy consumption), even with shape stabilized PCM (Barreneche et al. 2014, 2013a, b; Zhou et al. 2011). The inclusion of PCM in bricks (Fig. 16.10) showed similar results, reducing the thermal amplitude of the room and increasing the time delay of the peak temperature in about 3 h (Silva et al. 2012).

PCM has also been added to concrete and mortar. Cabeza et al. (2007) added 10 % microencapsulated paraffin in concrete panels and tested it in two real-size cubicles, showing up to 7 °C of indoor temperature oscillation decrease; Entrop et al. (2011) added microencapsulated PCM in concrete floors with a reduction

Fig. 16.10 Clay bricks including PCM macrocapsules (Kuznik et al. 2008a, b)

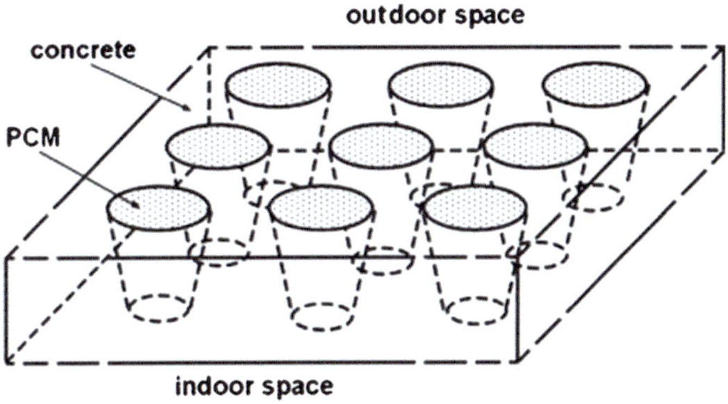

Fig. 16.11 Schematic representation of a concrete roof with frustum holes filled with PCM (Diaconu and Cruceru 2010)

Fig. 16.12 Hollow glass bricks filled with PCM (Biswas et al. 2014)

of the floor temperature of 7–16 %; Alawadhi and Alqallaf (2011) added PCM to a concrete roof with cone frustum holes (Fig. 16.11); and, more recently, Desai et al. (2014) added PCM to cementitious composites to be used as thin panels for façades; and Joulin et al. (2014) in cement-mortars.

PCM may also be added in shutters, window blinds and windows (which become the known translucent PCM walls). Soares et al. (2011) proposed a southward PCM shutter system to take advantage of the solar energy for winter nighttime heating, while Weinlaeder et al. (2011) used an interior sun protection system with vertical slats filled with PCM in office rooms. PCM was used together with a transparent insulation material (TIM) by Manz et al. (1997) to develop a selective optical transmittance solar radiation system. On the other hand, hollow glass bricks filled with PCM (Fig. 16.12) were studied for thermal management of an outdoor passive solar test room by Bontemps et al. (2011).

16.3.3.2 Environmental Impact

Life cycle assessment (LCA) is a good tool for the environmental evaluation of buildings and has been used in residential and non-residential buildings and in civil engineering constructions, as well as for construction product selection and for construction systems and processes evaluation (Cabeza et al. 2014). PCM inclusion in buildings was assessed with LCA by De Gracia et al. (2010) and Castell et al. (2013), who showed that the addition of PCM (paraffin and salt hydrate) in the building envelope, although decreasing the energy consumption during operation, did not significantly reduce the global impact throughout the lifetime of the building. Aranda-Usón et al. (2013) concluded that PCM in buildings can reduce the environmental impact, but that the reduction is strongly influenced by the climate conditions and the PCM introduced.

16.4 Active Systems

16.4.1 Introduction

The use of TES in building active systems is an attractive and versatile solution for several applications, such as the implementation of renewable energy sources in the HVAC of the building for space heating and/or cooling, the improvement in the performance of the current installations, or the possible application of peak load shifting strategies. In this section, the use of TES in active systems will be discussed, highlighting the performance of free cooling systems and building integrated active systems, such as thermal activated building system (TABS), and heat pumps with TES.

16.4.2 Free-Cooling Systems

In a free-cooling system, the TES medium is charged when outdoor temperature is lower in comparison to indoor, and this stored cold is discharged when required by the cooling demand. The advantage of free cooling using TES in comparison to night ventilated cooling is that the cold stored can be discharged whenever it is needed by circulating ambient or room air through the storage (Waqas and Ud Din 2013). Since there is a small thermal gap between the charging source (low outdoor temperature at night) and the desired indoor temperature (thermal comfort), the use of PCM as the storage medium is recommended because of its high energy density within the phase change. As previously stated, the free cooling sequence consists of two modes of operation: the charging process, when the PCM is solidified, and the discharging process, when cooling is supplied to the inner environment from the storage (Fig. 16.13).

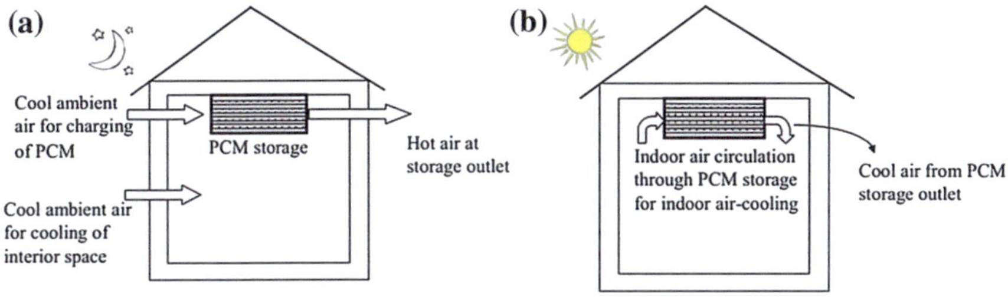

Fig. 16.13 Free-cooling working principle. **a** Charging process (working of PCM storage during nighttime), and **b** discharging process (working of PCM storage during daytime) (Hed and Bellander 2006)

This operating principle allows the storage to be placed and designed in various locations according to the application or requirements. Several authors have investigated the implementation of the storage in external heat exchangers (Turnpenny et al. 2001; Zalba et al. 2004), or packed bed systems integrated into the ventilation system (Arkar and Medved 2007; Arkar et al. 2007), while others integrated the storage in a building component, such as ceilings (Yanbing et al. 2003).

These free cooling systems consume around 9 % less electrical energy than conventional air conditioning units of the same power; however, the initial investment is 10 % higher, with a payback period of around 3–4 years (Arkar and Medved 2007). The main obstacle that this technology faces is the difficulty of ensuring the full solidification of the PCM in certain climates or summer periods, in which the night temperatures do not drop below the phase change range long enough. This problem can be overcome by enhancing the thermal conductivity of the PCM and using appropriate control strategies.

16.4.3 Building Integrated Active Systems

The active systems based on TES have been implemented using external storage systems or integrating the storage in the building elements. This last option avoids the occupation of space for the implementation of these technologies and makes use of large heat transfer surfaces between the storage and the application. The integration of the TES in the building can be done using the core of the building (core, floor, walls) in external solar façades, in suspended ceilings, ventilation systems, PV systems and water tanks, as shown in Fig. 16.14.

Regarding the use of the core of the building for the integration of the TES system, several diverse systems have been developed and tested. Navarro et al. (2014a) tested an innovative active slab with PCM in its hollows, in which air is the heat transfer fluid. On the one hand, the system stores solar energy during sunny hours in winter and provides a heating supply when required by the

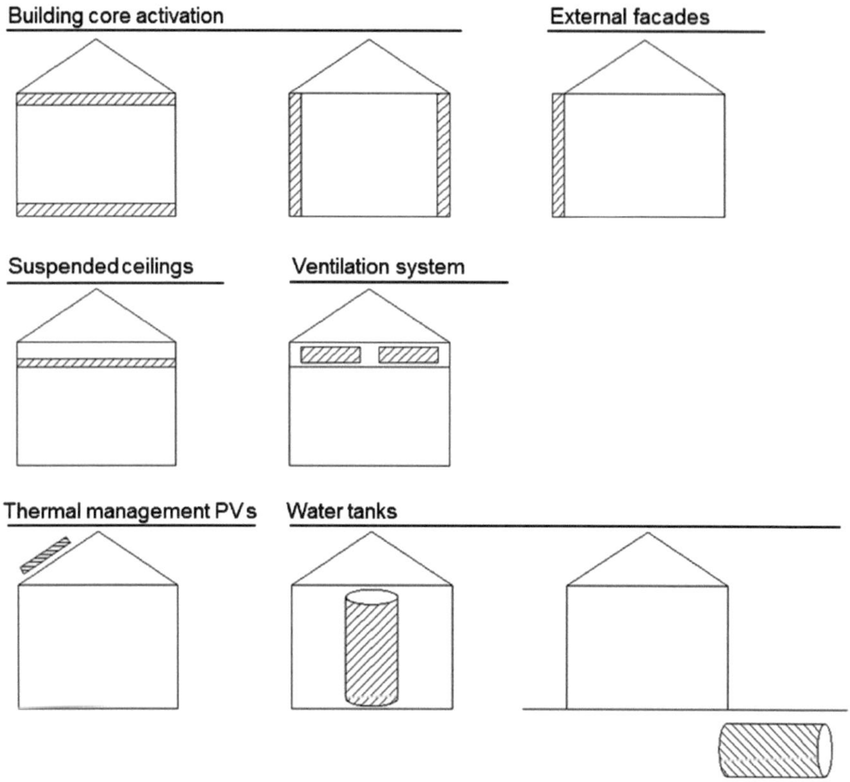

Fig. 16.14 Integration of the TES in buildings (Navarro et al. 2014b)

demand; on the other, during the summer it solidifies the PCM located inside its hollows, using the low night temperatures and supplies cooling in peak load hours (Fig. 16.15).

16.4.3.1 Integration of the TES Into the Core of the Building

Other storage systems that are integrated into the core of the building are the TABS, which makes use of a building element as a TES system, such as ceilings or envelopes, usually adding embedded pipes through water flows (Fig. 16.16). They are usually charged with the low temperatures at night (free cooling source) and designed to discharge passively the cooling store in the building structure (Saelens et al. 2011). The large area of these systems allows a high heat transfer rate between the space and the structure, and even the low thermal difference between them, which makes the system appropriate for managing internal heat gains (Lehmann et al. 2007).

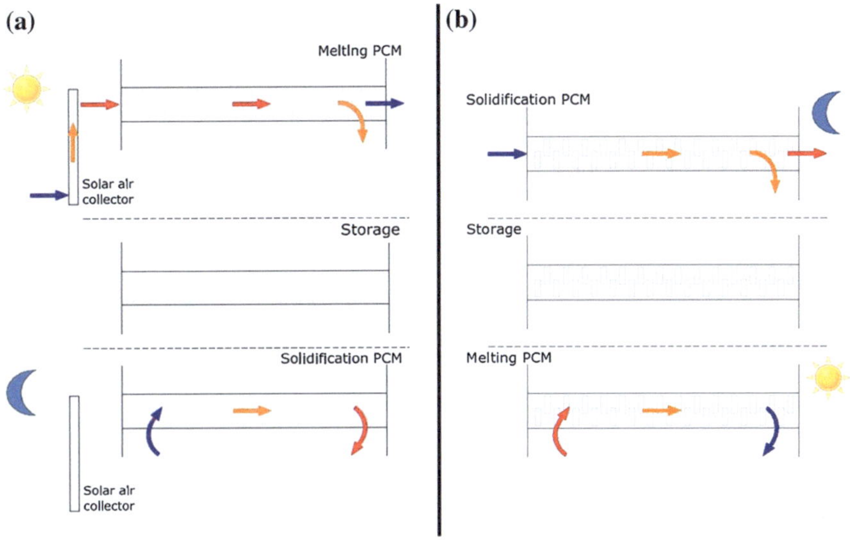

Fig. 16.15 Operating principle of the active slab with PCM: **a** heating, **b** cooling (Navarro et al. 2014a)

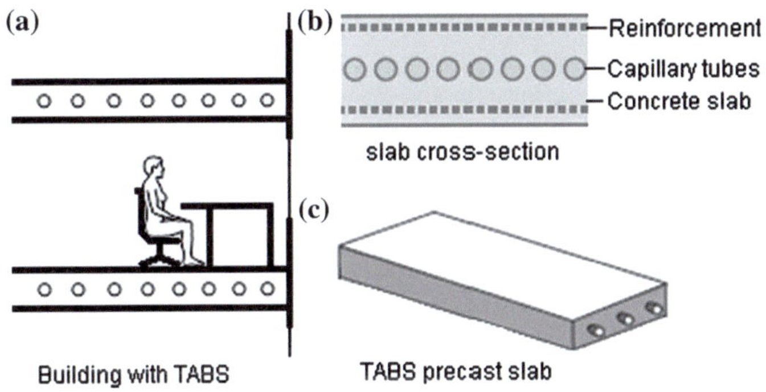

Fig. 16.16 TABS principle (Pavlov and Olesen 2011)

The typical daily cycle of the application of TABS is shown in Fig. 16.17, which is composed of the following steps:

- Initial temperature of the room air and surface (1)
- Heat sources are switched on, during the first interval of the heat sources releases its convection portion to the room air (1–2)
- Radiative portion is absorbed by the storage capacity of the surfaces (2–3)
- The heat sources are switched off at (3), and a similar pattern is found in the fast and slow heat discharge of the convective and radiative portion, respectively
- The system is charged by the water pipes for the following daily cycle.

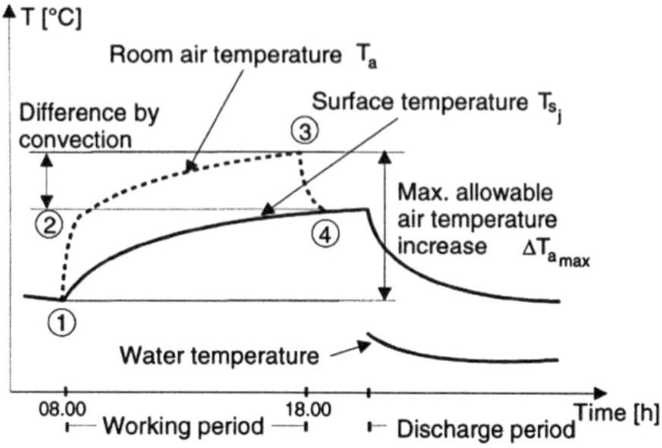

Fig. 16.17 Typical temperature development of room air, concrete slab surface and water during 24-h period (Koschenz and Dorer 1999)

The activation of the building mass using pipes embedded in the main concrete structure significantly reduces the cooling peak load, which drives to a reduction in the cooling capacity of the HVAC systems up to 50 % (Rijksen et al. 2010). These systems based on sensible heat storage (typically the use of concrete) have been used commercially because of low investment costs and because it provides favorable comfort conditions (Lehmann et al. 2011). Moreover, in order to increase the TES capacity of the system, the addition of microencapsulated PCM to a concrete deck (Pomianowski et al. 2012) has been explored. Furthermore, TABS with two different shape-stabilized PCMs have been investigated. The two layers of PCM have different melting temperatures, so one is used for heating purposes and the other for cooling (Jin and Zhang 2011). Despite the benefits that the use of latent heat storage can provide to the performance of TABS, its inclusion in commercial systems has been limited because of the cost.

16.4.3.2 Integration of the TES in External Façades

The implementation of storage in external façades, such as double-skin façade, has been discussed to reduce the risk of overheating and increase the system efficiency in both winter and summer (Fallahi et al. 2010). Moreover, De Gracia et al. (2012) tested the thermal performance of a ventilated façade with macro-encapsulated PCM in its air chamber. During the heating season, the façade acts as a solar collector during the solar absorption period (Fig. 16.18).

Once the PCM is melted and the solar energy is needed for heating, the heat discharge period begins. This discharge period is performed until no more thermal energy is needed or can be provided by the system. The authors registered a reduction of 20 % in the electrical energy consumption of the installed HVAC systems because of the use of this solar-ventilated façade.

Furthermore, the same system was tested for cooling purposes, using low temperatures at night to charge the PCM and/or provide free cooling. The authors developed a numerical model and evaluated the performance of the system under

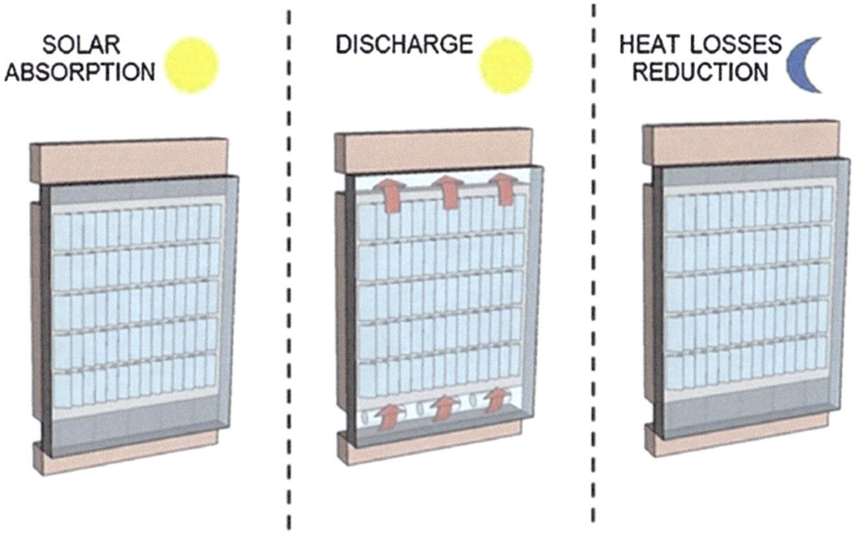

Fig. 16.18 Operational mode of the system during winter (De Gracia et al. 2012)

different weather conditions (De Gracia et al. 2015). It was shown that the use of this latent heat-activated system is suitable for temperate climates according to the Köppen Geiger climate classification (Kottek et al. 2006), while the potential of the system to provide benefits in arid and equatorial climates is very limited.

16.4.3.3 Integration of the TES in Suspended Ceilings and Ventilation Systems

A significant number of old buildings need serious, energetic retrofitting in order to accomplish the standards defined by the European Directives (Directive 2010). Hence, the integration of TES components in the suspended ceiling, such as actively charged radiant panels filled with water (Roulet et al. 1999) or PCM (Koschenz and Lehmann 2004), is a good solution as a cooling or heating system.

Moreover, the TES units can be located in the ventilation ducts behind the suspended ceilings as in the work proposed by Yanbing et al. (2003), in which PCM is solidified during the nighttime and discharged during the peak load hours as in Navarro et al. (2014a). The system was tested, and it was proved that its use improves the thermal comfort level of the indoor environment. Furthermore, Monodraught Ltd developed and commercialized a cooling and ventilation system called "Cool-Phase" (Cool Phase 2014), which consists of an air-handling unit and a storage device based on macro-encapsulated panels of PCM. This equipment, shown in Fig. 16.19, can provide space heating by a heat recovery cycle and cooling, making use of low night temperatures.

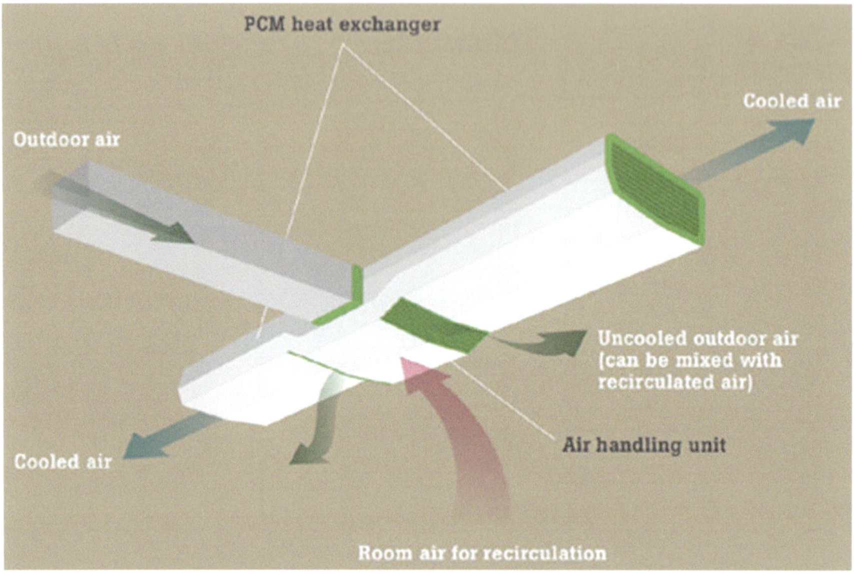

Fig. 16.19 TES integrated in the ventilation system. Cool-Phase system (Cool Phase 2014)

16.4.3.4 Integration of the TES in the PV System

The storage implemented in the PV system provides two main benefits. On the one hand, it can profit from the heat behind the PV panels for heating purposes, and on the other, the PV panels increase their performance when cooled. The conventional aluminium-fined PV panel with natural ventilation can reduce the temperature rise of the PV; however, the use of a PV/PCM system can greatly reduce the temperature rise of the PV. The PCM should have a flash point considerably higher than the maximum operating temperature of the PV system, and it should be non-flammable and non-explosive (Abhat 1983). The use of PCM was shown to provide significant PV temperature control and thus increased electrical conversion efficiency (Huang 2006).

16.4.3.5 Integration of the TES in Water Tanks

Water tanks are a widely used TES device for solar systems or domestic hot water facilities, but because of the high volume they occupy, their integration into the building has only recently been addressed. This integration was done by architecturally including the storage in the living areas (Fig. 16.20), or by using ground-integrated tanks. The use of underground storage is justified if seasonal TES strategies are considered. There are four main types of underground systems for seasonal TES (The European Project EINSTEIN 2014) (Fig. 16.21): tank thermal energy storage (TTES); pit thermal energy storage (PTES); borehole thermal energy storage (BTES); and aquifer thermal energy storage (ATES).

Fig. 16.20 Solar water tank integrated in a living room (Solaresbauen-Grundlagen 2014)

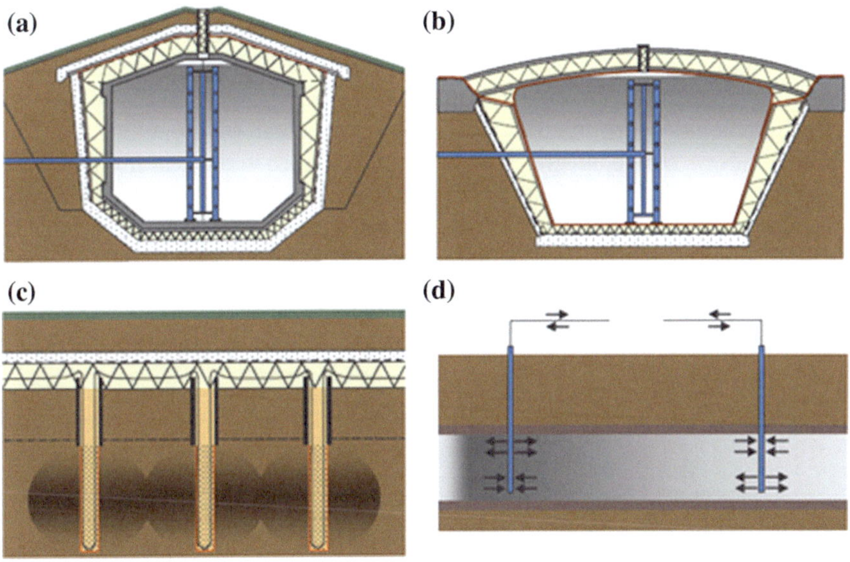

Fig. 16.21 Types of underground seasonal thermal energy storage (The European EINSTEIN Project 2014)

16.4.4 Use of TES in Heat Pumps

The use of heat pumps with TES systems are presented as a promising technology to shift electrical loads from high-peak to off-peak periods, thus serving as a powerful tool in demand-side management (DSM) (Arteconi et al. 2013). The benefits of this technology are focused on the energy cost savings by using thermal energy stored during low-cost electricity tariffs. The incorporation of these systems in the buildings has been investigated both for space heating and cooling. The TES is implemented in the heat pumps, not only for load shifting, but for achieving important energy savings on the defrosting system of the outdoor unit, or for heat recovery applications as well (Moreno et al. 2014a, b).

PCM thermal energy storage systems were coupled to conventional heat pumps for space heating in Agyenim and Hewitt (2010), which concluded that due to the low thermal conductivity of the PCM, the average storage tank size needed to cover the heating demand of a semi-detached house was 1116 l, which ruled out the use of PCM for this application, unless the thermal conductivity is enhanced. On the other hand, if the PCM storage tank is implemented in a modern low-temperature heating system, such as that studied by Leonhardt and Müller (Leonhardt 2009), the switching between the on and off of the heat pump is reduced, which represents cost savings. Moreover, the heat pumps with PCM were also studied in combination with other energy sources, such as solar (Kaygusuz 1999; Niu et al. 2013) or geothermal energy (Benli and Durmus 2009).

As previously written, a PCM module is implemented in air source heat pumps to solve the frosting problem in the outdoor coil, which limits the air passage area and dramatically decreases the performance of the unit. Currently, the most widely used defrosting method consists of a reverse-cycle defrost, which means that one defrosting is required; the outdoor unit acts as a condenser; and the indoor unit as an evaporator, which is costly and produces discomfort in the inner environment. The PCM can be integrated in the cycle to absorb waste heat from the compressor and release this heat during the defrosting period (IEA 1990) (Fig. 16.22). This system increases the COP of the system, reduces the energy consumption for defrosting (Dong et al. 2011), and increases the thermal comfort of the indoor environment as well (Minglu et al. 2010).

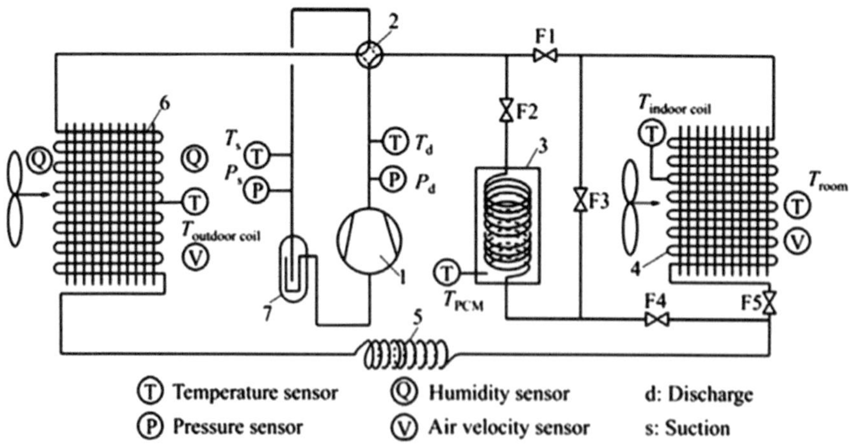

Fig. 16.22 Schematic diagram of operation for experimental ASHP unit (*left*) and physical view of PCM heat exchanger (*right*) evaluated by Jiankai et al. (Dong et al. 2011). *1* Compressor; *2* four-way reverse valve; *3* PCM heat exchanger; *4* indoor coil; *5* capillary tube; *6* outdoor coil; and *7* gas-liquid separator; *F1–F5* solenoid modulating valves

The use of PCM in heat pumps for space cooling is based on the cold storage from the evaporator, and the later use of this cooling during peak-load hours. Ice has been used as the storage media, implemented in heat storage tanks (Fang et al. 2009) or spherical capsules packed beds (Fang et al. 2010). In addition, Moreno et al. (2014a, b) implemented two PCM tanks in a heat pump cycle, one connected to the evaporator (cold TES tank) and the other to the condenser (hot TES tank). This allows the system to store cold and heat at once in order to cover the demand and apply peak load shifting strategies during hot or cold periods.

The use of waste heat from the compression cycle of the heat pump during the cooling season is based on the concept of storing the heat released by the condenser in a PCM tank and using it for domestic hot water (Gu et al. 2004). The ability to totally or partially fulfill the demand of domestic hot water might contribute in an important reduction of the CO_2 emissions in this field (Zhang et al. 2011).

16.5 Conclusions

This chapter provides the state of the art of the studied active and passive TES technologies integrated in the building sector, including sensible, latent, and thermochemical storage systems. Passive techniques are based on increasing the thermal inertia of envelopes and building elements, which provides thermal stability and hence reduces the energy consumption for space heating and cooling. The correct use of these passive systems requires temperate weather conditions, in which the storage can be passively discharged to the outer environment during nighttime in the cooling season, and enough solar radiation to be charged during the winter. Moreover, the inclusion of TES in active systems has been investigated for the implementation of renewable energies for space heating and/or cooling, for improving the performance of the current installations, and for using peak load-shifting strategies.

TES shows great potential for reducing the energy demand of buildings and/or improving the energy efficiency of their energy systems. Several groups and associations (i.e., IEA, RHC Platform, EASE) have identified this potential. However, some of challenges that have to be faced are reduction of the cost, increase of the compactness of the systems, increase of the energy density of the materials and systems, increase of the thermal conductivity of the materials, and development of new materials.

Acknowledgements This study was partially funded by the Spanish Government (ENE2011-22722, ENE2011-28269-C03-02, and ULLE10-4E-1305). The authors would like to thank the Catalan Government for the quality accreditation given to their research group GREA (2014 SGR 123) and DIOPMA (2014 SGR 1543). The research leading to these results has received funding from the European Union's Seventh Framework Program (FP7/2007-2013), under grant agreement No. PIRSES-GA-2013-610692 (INNOSTORAGE).

References

Abhat A (1983) Low temperature latent heat thermal energy storage: heat storage materials. Sol Energy 30:313–332

Adams S, Becker M, Krauss D, Gilman C (2010) Not a dry subject: optimizing water Trombe wall. In: ASE S (ed) SOLAR 2010 conference

Agyenim F, Hewitt N (2010) The development of a finned phase change material (PCM) storage system to take advantage of off-peak electricity tariff for improvement in cost of heat pump operation. Energy Build 42(9):1552–1560. doi:10.1016/j.enbuild.2010.03.027

Ahmad M, Bontemps A, Sallée H, Quenard D (2006) Thermal testing and numerical simulation of a prototype cell using light wallboards coupling vacuum isolation panels and phase change material. Energy Build 38:673–681

Alawadhi EM, Alqallaf HJ (2011) Building roof with conical holes containing PCM to reduce the cooling load: numerical study. Energy Convers Manag 52(8–9):2958–2964

Aranda-Usón A, Ferreira G, López-Sabirón AM, Mainar-Toledo MD, Zabalza Bribián I (2013) Phase change material applications in buildings: an environmental assessment for some Spanish climate severities. Sci Total Environ 444:16–25. doi:10.1016/j.scitotenv.2012.11.012

Arkar C, Medved S (2007) Free cooling of a building using PCM heat storage integrated into the ventilation system. Sol Energy 81(9):1078–1087. doi:10.1016/j.solener.2007.01.010

Arkar C, Vidrih B, Medved S (2007) Efficiency of free cooling using latent heat storage integrated into the ventilation system of a low energy building. Int J Refrig 30(1):134–143. doi:10.1016/j.ijrefrig.2006.03.009

Arteconi A, Hewitt NJ, Polonara F (2013) Domestic demand-side management (DSM): role of heat pumps and thermal energy storage (TES) systems. Appl Therm Eng 51(1–2):155–165. doi:10.1016/j.applthermaleng.2012.09.023

Baetens R, Jelle BP, Gustavsen A (2010) Phase change materials for building applications: a state-of-the-art review. Energy Build 42:1361–1368

Barreneche C, De Gracia A, Serrano S, Navarro ME, Borreguero AM, Fernández AI, Cabeza LF (2013) Comparison of three different devices available in Spain to test thermal properties of building materials including phase change materials. Appl Energy 109:544–552. Retrieved from http://www.scopus.com/inward/record.url?eid=2-s2.0-84879270070&partnerID=40&md5=b9a5b9836401f86c9f005c65f1cb7569

Barreneche C, Fernández AI, Niubó M, Chimenos JM, Espiell F, Segarra M, Cabeza LF (2013) Development and characterization of new shape-stabilized phase change material (PCM)—polymer including electrical arc furnace dust (EAFD), for acoustic and thermal comfort in buildings. Energy Build 61:210–214. Retrieved from http://www.scopus.com/inward/record.url?eid=2-s2.0-84875178352&partnerID=40&md5=86e72298677da691830b12cdcfb601f5

Barreneche C, Navarro ME, Niubó M, Cabeza LF, Fernández AI (2014) Use of PCM-polymer composite dense sheet including EAFD in constructive systems. Energy Build 68(PARTA):1–6. Retrieved from http://www.scopus.com/inward/record.url?eid=2-s2.0-84886433203&partnerID=40&md5=fb540fb7b5cc6dd73a2a6cb22a664d64

Basecq V, Michaux G, Inard C, Blondeau P (2013) Short-term storage systems of thermal energy for buildings: a review. Adv Build Energy Res 7(1):66–119

Benli H, Durmuş A (2009) Evaluation of ground-source heat pump combined latent heat storage system performance in greenhouse heating. Energy Build 41(2):220–228. doi:10.1016/j.enbuild.2008.09.004

Biswas K, Lu J, Soroushian P, Shrestha S (2014) Combined experimental and numerical evaluation of a prototype nano-PCM enhanced wallboard. Appl Energy 131:517–529

Bontemps A, Ahmad M, Johannès K, Sallée H (2011) Experimental and modelling study of twin cells with latent heat storage walls. Energy Build 43(9):2456–2461

Borreguero A, Carmona M, Sanchez M, Valverde J, Rodriguez J (2010) Improvement of the thermal behavior of gypsum blocks by the incorporation of microcapsules containing PCMS obtained by suspension polymerization with an optimal core/coating mass ratio. Appl Therm Eng 30:1164–1169

Briga-Sá A, Martins A, Boaventura-Cunha J, Lanzinha J, Paiva A (2014) Energy performance of Trombe Walls. Adaptation of ISO 13790:2008(E) to the Portuguese reality. Energy Build 74:111–119

Cabeza L (2015) Advances in thermal energy storage systems. In: Cabeza J (ed) Methods and applications. Woodhead Publishing, Cambridge

Cabeza LF, Castellón C, Nogués M, Medrano M, Leppers R, Zubillaga O (2007) Use of microencapsulated PCM in concrete walls for energy savings. Energy Build 39(2):113–119. Retrieved from http://www.scopus.com/inward/record.url?eid=2-s2.0-33845749671&partnerID=40&md5=9ea560a5ff7136bf555816b3b0c5c873

Cabeza LF, Castell A, Barreneche C, De Gracia A, Fernández AI (2011) Materials used as PCM in thermal energy storage in buildings: a review. Renew Sustain Energy Rev 15(3):1675–1695

Cabeza LF, Rincón L, Vilariño V, Pérez G, Castell A (2014) Life cycle assessment (LCA) and life cycle energy analysis (LCEA) of buildings and the building sector: a review. Renew Sustain Energy Rev 29:394–416. Retrieved from http://www.scopus.com/inward/record.url?eid=2-s2.0-84884633626&partnerID=40&md5=39e37dbc12cc2071136d75807458b360

Castell A, Martorell I, Medrano M, Pérez G, Cabeza LF (2010) Experimental study of using PCM in brick constructive solutions for passive cooling. Energy Build 42(4):534–540. Retrieved from http://www.scopus.com/inward/record.url?eid=2-s2.0-77649183812&partnerID=40&md5=892b975b0f63d8318c264e2d1eeabf63

Castell A, Menoufi K, de Gracia A, Rincón L, Boer D, Cabeza LF (2013) Life cycle assessment of alveolar brick construction system incorporating phase change materials (PCMs). Appl Energy 101:600–608. Retrieved from http://www.scopus.com/inward/record.url?eid=2-s2.0-84869869833&partnerID=40&md5=4fa882f68ee30f00e51a1520030acdb1

Castellón C, Medrano M, Roca J, Cabeza LF, Navarro ME, Fernández AI, Zalba B (2010) Effect of microencapsulatedphase change material in sandwich panels. Renew Energy 35(10):2370–2374

Cool Phase (2014) Natural cooling and low energy ventilation system. Available from http://www.cool-phase.net/

Cot-Gores J, Castell A, Cabeza LF (2012) Thermochemical energy storage and conversion: a-state-of-the-art review of the experimental research under practical conditions. Renew Sustain Energy Rev 16(7):5207–5224. Retrieved from http://www.scopus.com/inward/record.url?eid=2-s2.0-84862739473&partnerID=40&md5=8b344aa8d631967b13fac135a9660879

De Gracia A, Rincón L, Castell A, Jiménez M, Boer D, Medrano M, Cabeza LF (2010) Life cycle assessment of the inclusion of phase change materials (PCM) in experimental buildings. Energy Build 42(9):1517–1523. Retrieved from http://www.scopus.com/inward/record.url?eid=2-s2.0-78651454787&partnerID=40&md5=bd9ae4dc9db35879be7f15c2086a992f

De Gracia A, Barreneche C, Farid MM, Cabeza LF (2011) New equipment for testing steady and transient thermal performance of multilayered building envelopes with PCM. Energy Build 43(12):3704–3709

De Gracia A, Navarro L, Castell A, Ruiz-Pardo A, Alvárez S, Cabeza LF (2012) Experimental study of a ventilated facade with PCM during winter period. Energy Build. Retrieved from http://www.scopus.com/inward/record.url?eid=2-s2.0-84869487163&partnerID=40&md5=74059d06d245d5c1a6b78902d38c784c

De Gracia A, Navarro L, Castell A, Cabeza LF (2015) Energy performance of a ventilated double skin facade with PCM under different climates. Energy Build 91:37–42

Del Barrio EP, Dauvergne JL, Morisson V (2009) A simple experimental method for thermal characterization of shape-stabilized phase change materials. Sol Energy Eng. Trans ASME 131(4):0410101–0410108

Delgado M, Lázaro A, Mazo J, Zalba B (2012) Review on phase change material emulsions and microencapsulated phase change material slurries: Materials, heat transfer studies and applications. Renew Sustain Energy Rev 16:253–273

Desai D, Miller M, Lynch JP, Li VC (2014) Development of thermally adaptive engineered cementitious composite for passive heat storage. Constr Build Mater 67:366–372

Diaconu BM (2011) Thermal energy savings in buildings with PCM-enhanced envelope: influence of occupancy pattern and ventilation. Energy Build 43(1):101–107

Diaconu BM, Cruceru M (2010) Novel concept of composite phase change material wall system for year-round thermal energy savings. Energy Build 42(10):1759–1772

Directive 2010/31/EU (2010) 2010/31/EU: Directive of the European parliament and of the council of 19 May 2010 on the energy performance of buildings. Available from 30 Oct 2012: http://www.epbd-ca.eu

Dong J, Jiang Y, Yao Y, Zhang X (2011) Operating performance of novel reverse-cycle defrosting method based on thermal energy storage for air source heat pump. J Cent South Univ Technol 18(6):2163–2169

Entrop AG, Brouwers HJH, Reinders AHME (2011) Experimental research on the use of micro-encapsulated phase change materials to store solar energy in concrete floors and to save energy in Dutch houses. Sol Energy 85(5):1007–1020

Evers AC, Medina MA, Fang Y (2010) Evaluation of the thermal performance of frame walls enhanced with paraffin and hydrated salt phase change materials using a dynamic wall simulator. Build Environ 45(8):1762–1768

Fallahi A, Haghighat F, Elsadi H (2010) Energy performance assessment of double-skin façade with thermal mass. Energy Build 42(9):1499–1509. doi:10.1016/j.enbuild.2010.03.020

Fang G, Liu X, Wu S (2009) Experimental investigation on performance of ice storage air-conditioning system with separate heat pipe. Exp Thermal Fluid Sci 33(8):1149–1155. doi:10.1016/j.expthermflusci.2009.07.004

Fang G, Wu S, Liu X (2010) Experimental study on cool storage air-conditioning system with spherical capsules packed bed. Energy Build 42(7):1056–1062. doi:10.1016/j.enbuild.2010.01.018

Fernandez AI, Martinez M, Segarra M, Martorell I, Cabeza LF (2010) Selection of materials with potential in sensible thermal energy storage. Sol Energy Mater Sol Cells 94(10):1723–1729. Retrieved from http://www.scopus.com/inward/record.url?eid=2-s2.0-77955427486&partnerID=40&md5=57c2ce552a5eaba44202c57daf33dfc0

Gu Z, Liu H, Li Y (2004) Thermal energy recovery of air conditioning system—heat recovery system calculation and phase change materials development. Appl Therm Eng 24(17–18):2511–2526. doi:10.1016/j.applthermaleng.2004.03.017

Günther E, Hiebler S, Mehling H (2006) Determination of the heat storage capacity of PCM and PCM-objects as a function of temperature. Proc ECOSTOCK, 10th Int. Conference on Thermal Energy Storage

Hawes DW, Feldman D, Banu D (1993) Latent heat storage in building materials. Energy Build 20(1):77–86

Hed G, Bellander R (2006) Mathematical modelling of PCM air heat exchanger. Energy Build 38(2):82–89. doi:10.1016/j.enbuild.2005.04.002

Heier J, Bales C, Martin V (2015) Combining thermal energy storage with buildings—a review. Renew Sustain Energy Rev 42:1305–1325

Huang M, Eames P, Norton B (2006) Experimental performance of phase change materials for limiting temperature rise building integrated photovoltaics. Sol Energy 80:1121–1130. Available from http://www.einstein-project.eu/fckeditor_files/D_9_2_EINSTEIN_leaflet_English.pdf

IEA (1990) Heat Pump Centre Newsletter

IEA (2010) Technology Roadmap. Solar Heating and Cooling
IEA (2012) Energy Technology Perspectives (ETP) 2012. International Energy Agency
Jin X, Zhang X (2011) Thermal analysis of a double layer phase change material floor. Appl Therm Eng 31(10):1576–1581. doi:10.1016/j.applthermaleng.2011.01.023
Jin X, Medina M, Zhang X (2014) On the placement of a phase change material thermal shield within the cavity of buildings walls for heat transfer rate reduction. Energy 73:780–786
Joulin A, Zalewski L, Lassue S, Naji H (2014) Experimental investigation of thermal characteristics of a mortar with or without a micro-encapsulated phase change material. Appl Therm Eng 66:171–180
Kaygusuz K (1999) Investigation of a combined solar–heat pump system for residential heating. Part 1: experimental results. Int J Energy Res 23:1213–1223
Khudhair AM, Farid MM (2004) A review on energy conservation in building applications with thermal storage by latent heat using phase change materials. Energy Convers Manag 45(2):263–275. Retrieved from http://www.scopus.com/inward/record.url?eid=2-s2.0-0141510020&partnerID=40&md5=afb9a447d9699dcde0d44119197faa4a
Koschenz M, Dorer V (1999) Interaction of an air system with concrete core conditioning. Energy Build 30(2):139–145. doi:10.1016/S0378-7788(98)00081-4
Koschenz M, Lehmann B (2004) Development of a thermally activated ceiling panel with PCM for application in lightweight and retrofitted buildings. Energy Build 36(6):567–578. doi:10.1016/j.enbuild.2004.01.029
Kosny J, Petrie T, Gawin D, Childs P, Desjarlais A, Christian J (2014) Energy savings potential in residential buildings
Kottek MJ, Grieser J, Beck C, Rudolf B, Rubel F (2006) World map of Köppen-Geiger climate classification updated. Mereorol Z 15:259–263
Kuznik F, Virgone J (2009) Experimental investigation of wallboard containing phase change material: data for validation of numerical modeling. Energy Build 41(5):561–570
Kuznik F, Virgone J, Noel J (2008) Optimization of a phase change material wallboard for building use. Appl Therm Eng 28(11–12):1291–1298
Kuznik F, Virgone J, Roux JJ (2008) Energetic efficiency of room wall containing PCM wallboard: a full-scale experimental investigation. Energy Build 40(2):148–156
Lai C, Hokoi S (2014) Thermal performance of an aluminum honeycomb wallboard incorporating microencapsulated PCM. Energy Build 73:37–47
Lee K, Medina M, Raith E, Sun X (2015a) Assessing the integration of a thin phase change material (PCM) layer in a residential building wall for heat transfer reduction and management. Appl Energy 137:699–706
Lee K, Medina M, Sun X (2015b) On the use of plug-and-play walls (PPW) for evaluating thermal enhancement technologies for building enclosures: Evaluation of a thin phase change material (PCM) layer. Energy Build 86:86–92
Lehmann B, Dorer V, Koschenz M (2007) Application range of thermally activated building systems tabs. Energy Build 39(5):593–598. doi:10.1016/j.enbuild.2006.09.009
Lehmann B, Dorer V, Gwerder M, Renggli F, Tödtli J (2011) Thermally activated building systems (TABS): energy efficiency as a function of control strategy, hydronic circuit topology and (cold) generation system. Appl Energy 88(1):180–191. doi:10.1016/j.apenergy.2010.08.010
Leonhardt C, Müller D (2009) Modelling of residential heating system using a phase change material storage system. In: Proceedings of the 7th Modelica Conference, pp 20–22. Como, Italy
Li L, Yan Q, Jin L, Yue L (2012) Preparation method of shape-stabilized PCM wall and experimental research of thermal performance. Taiyangneng Xuebao/Acta Energiae Solaris Sinica 33(12):2135–2139
Llovera J, Potau X, Medrano M, Cabeza L (2010) Design and performance of energy-efficient solar residential house in Andorra. Appl Energy 88:1343–1353

Manz H, Egolf PW, Suter P, Goetzberger A (1997) TIM-PCM, external wall system for solar space heating and daylight. Sol Energy 61(6):369–379

Medina M, King J, Zhang M (2008) On the heat transfer rate reduction of structural insulated panels (SIPs) outfitted with phase change materials (PCMs). Energy 33(4):667–678

Mehling H, Cabeza L (2008) Heat and cold storage with PCM. Springer, Berlin. Retrieved from ISBN-13: 9783540685562

Memon S (2014) Phase change materials integrated in building walls: A state of the art review. Renew Sustain Energy Rev 31(2014):870–906

Minglu Q, Liang X, Deng S, Yiqiang J (2010) Improved indoor thermal comfort during defrost with a novel reverse-cycle defrosting method for air source heat pumps. Build Environ 45:2354–2361

Moreno P, Castell A, Solé C, Zsembinszki G, Cabeza LF (2014) PCM thermal energy storage tanks in heat pump system for space cooling. Energy Build 82:399–405. Retrieved from http://www.scopus.com/inward/record.url?eid=2-s2.0-84905819417&partnerID=40&md5=772d19eccde216de15ebe2f2d1aed4dc

Moreno P, Solé C, Castell A, Cabeza LF (2014) The use of phase change materials in domestic heat pump and air-conditioning systems for short term storage: a review. Renew Sustain Energy Rev 39:1–13. Retrieved from http://www.scopus.com/inward/record.url?eid=2-s2.0-84905005142&partnerID=40&md5=32ea981f4780f4d5948103a4718b9ba4

N'Tsoukpoe KE, Liu H, Le Pierrès N, Luo L (2009) A review on long-term sorption solar energy storage. Renew Sustain Energy Rev 13(9):2385–2396. Retrieved from http://www.scopus.com/inward/record.url?eid=2-s2.0-68749105582&partnerID=40&md5=ac8244725fd2851c4b71bd2c45dfb995

Navarro ME, Martínez M, Gil A, Fernández AI, Cabeza LF, Py X (2011) Selection and characterization of recycled materials for sensible thermal energy storage. In: 30th ISES Biennial Solar World Congress 2011, SWC 2011, vol 6, pp 4875–4881. Retrieved from http://www.scopus.com/inward/record.url?eid=2-s2.0-84873840715&partnerID=40&md5=6c1dae385096f833c6631c7483d6019f

Navarro L, de Gracia A, Castell A, Alvarez S, Cabeza L (2014a) Experimental study of a prefabricated concrete slab with PCM. In: Eurotherm Seminar #99, Advances in Thermal Energy Storage

Navarro L, de Gracia A, Castell A, Alvarez S, Cabeza L (2014b) Experimental study of a prefabricated concrete slab with PCM. In: Proceedings of Eurotherm Seminar #99, Advances in Thermal Energy Storage

Niu F, Ni L, Yao Y, Yu Y, Li H (2013) Performance and thermal charging/discharging features of a phase change material assisted heat pump system in heating mode. Appl Therm Eng 58(1–2):536–541. doi:10.1016/j.applthermaleng.2013.04.042

Nomura T, Okinaka N, Akiyama T (2009) Impregnation of porous material with phase change material for thermal energy storage. Mater Chem Phys 115(2–3):846–850

Parameshwaran R, Kalaiselvam S, Harikrishnan S, Elayaperumal A (2012) Sustainable thermal energy storage technologies for buildings: a review. Renew Sustain Energy Rev 16(5):2394–2433. Retrieved from http://www.scopus.com/inward/record.url?eid=2-s2.0-84858401061&partnerID=40&md5=c6d943a3bab67da38f3d55073ffaa4d2

Pavlov G, Olesen B (2011) Building thermal energy storage—concepts and applications. In: Roomvent Proceedings, 12th International conference on air distribution in rooms. Norway

Pomianowski M, Heiselberg P, Jensen RL (2012) Dynamic heat storage and cooling capacity of a concrete deck with PCM and thermally activated building system. Energy Build 53:96–107. doi:10.1016/j.enbuild.2012.07.007

Rempel AR, Rempel AW (2013) Rocks, clays, water, and salts: highly durable, infinitely rechargeable, eminently controllable thermal batteries for buildings. Geosciences 3:63–101

Rijksen DO, Wisse CJ, van Schijndel AWM (2010) Reducing peak requirements for cooling by using thermally activated building systems. Energy Build 42(3):298–304. doi:10.1016/j.enbuild.2009.09.007

Roulet CA, Rossy JP, Roulet Y (1999) Using large radiant panels for indoor climate conditioning. Energy Build 30(2):121–126. doi:10.1016/S0378-7788(98)00079-6

Saadatian O, Sopian K, Lim C, Asim N, Sulaiman M (2012) Trombe walls: a review of opportunities and challenges in research and development. Renew Sustain Energy Rev 16(8):6340–6351

Saelens D, Parys W, Baetens R (2011) Energy and comfort performance of thermally activated building systems including occupant behavior. Build Environ 46(4):835–848. doi:10.1016/j.buildenv.2010.10.012

Sharma SD, Sagara K (2005) Latent heat storage materials and systems: a review. Int J Green Energy 2(1):1–56

Shilei L, Neng Z, Guohui F (2006) Impact of phase change wall room on indoor thermal environment in winter. Energy Build 38(1):18–24

Silva T, Vicente R, Soares N, Ferreira V (2012) Experimental testing and numerical modelling of masonry wall solution with PCM incorporation: a passive construction solution. Energy Build 49:235–245

Soares N, Samagaio A, Vicente R, Costa J (2011) Numerical simulation of a PCM shutter for buildings space heating during the winter. In: World renewable energy congress, pp 8–13. Linköping, Sweden

Soares N, Costa JJ, Gaspar AR, Santos P (2013) Review of passive PCM latent heat thermal energy storage systems towards buildings' energy efficiency. Energy Build 59:82–103

Solaresbauen-Grundlagen. Sonnenhaus-Institu e.V. Available from: www.sonnenhaus-institut.de (2014)

Stazi F, Mastrucci A, di Perna C (2012) The behaviour of solar walls in residential buildings with different insulation levels: an experimental and numerical study. Energy Build 47:217–229

The European Project EINSTEIN (2014) Effective Integration of Seasonal Thermal Energy Storage Systems IN existing buildings. Available from http://www.einstein-project.eu/fckeditor_files/D_9_2_EINSTEIN_leaflet_English.pdf

Turnpenny JR, Etheridge DW, Reay DA (2001) Novel ventilation system for reducing air conditioning in buildings. Part II: testing of prototype. Appl Therm Eng 21(12):1203–1217. doi:10.1016/S1359-4311(01)00003-5

Tyagi VV, Buddhi D (2007) PCM thermal storage in buildings: a state of the art. Renew Sustain Energy Rev 11:1146–1166

Tyagi VV, Kaushik SC, Tyagi SK, Akiyama T (2011) Development of phase change materials based microencapsulated technology for buildings: a review. Renew Sustain Energy Rev 15:1373–1391

Ürge-Vorsatz D, Cabeza LF, Serrano S, Barreneche C, Petrichenko K (2015) Heating and cooling energy trends and drivers in buildings. Renew Sustain Energy Rev 41(0):85–98. doi:http://dx.doi.org/10.1016/j.rser.2014.08.039

Waqas A, Ud Din Z (2013) Phase change material (PCM) storage for free cooling of buildings—a review. Renew Sustain Energy Rev 18:607–625. doi:10.1016/j.rser.2012.10.034

Weinlaeder H, Koerner W, Heidenfelder M (2011) Monitoring results of an interior sun protection system with integrated latent heat storage. Energy Build 43(9):2468–2475

Yanbing K, Yi J, Yinping Z (2003) Modeling and experimental study on an innovative passive cooling system—NVP system. Energy Build 35(4):417–425. doi:10.1016/S0378-7788(02)00141-X

Yang C, Fischer L, Maranda S, Worlitschek J (2015) Rigid polyurethane foams incorporated with phase change materials: A state-of-the-art review and future research pathways. Energy Build 87:25–36

Zalba B, Marín JM, Cabeza LF, Mehling H (2003) Review on thermal energy storage with phase change: materials, heat transfer analysis and applications. Appl Therm Eng 23(3):251–283.

Retrieved from http://www.scopus.com/inward/record.url?eid=2-s2.0-0037289573&partnerID=40&md5=7a2e981c4d5bec884b0a0e58a47ba49f

Zalba B, Marín JM, Cabeza LF, Mehling H (2004) Free-cooling of buildings with phase change materials. Int J Refrig 27(8):839–849. Retrieved from http://www.scopus.com/inward/record.url?eid=2-s2.0-10044257361&partnerID=40&md5=2e6493def785d26d2ce79d3182a819bb

Zhang P, Ma ZW, Wang RZ (2010) An overview of phase change material slurries: MPCS and CHS. Renew Sustain Energy Rev 14:598–614

Zhang X, Yu S, Yu M, Lin Y (2011) Experimental research on condensing heat recovery using phase change material. Appl Therm Eng 31(17–18):3736–3740. doi:10.1016/j.applthermaleng.2011.03.040

Zhou G, Yang Y, Xu H (2011) Performance of shape-stabilized phase change material wallboard with periodical outside heat flux waves. Appl Energy 88:2113–2121

Zhou D, Shire G, Tian Y (2014) Parametric analysis of influencing factors in phase change material wallboard (PCMW). Appl Energy 119:33–42

Chapter 17
Solar Thermal Systems

L.M. Ayompe

Abstract This chapter presents an overview of solar thermal systems used to supply energy for domestic hot water provision as well as space heating and cooling of buildings. Solar energy collectors are the main component of solar thermal systems; they play a vital role in converting solar radiation to heat. Water and air heating collectors are the most widely used in either glazed or unglazed configurations, with solar water heating systems the most popular means of utilizing solar energy. Solar water heating systems consist of forced circulation and thermosyphon systems, and solar thermal cooling systems use heat from the sun to drive absorption and adsorption chillers, desiccant and ejector systems to provide cooling in buildings. Solar air heating systems heat ventilation air for buildings and can also provide domestic hot water when connected to a suitable heat exchanger. Solar collectors can be integrated into elements of building envelopes, such as roofs, façades, balconies, and walls.

17.1 Introduction

Most of the energy used in buildings in developed countries is for heating, cooling, ventilation, and domestic hot water provision. In two-thirds of households, heating alone accounts for more than 50 % of all energy use (CEREN 2011). The sun is a free and abundant source of heat that can be used in buildings to provide most of their heating and cooling needs. Our reliance on fossil fuels can be substantially reduced if we harness and use heat from the sun; this can be achieved by transforming inefficient buildings to efficient or passive, or even positive energy buildings. The refurbishment of energy systems in existing buildings offers

L.M. Ayompe (✉)
International Energy Research Center, Tyndall National Institute, Lee Malting, Dyke Parade, Cork, Ireland
e-mail: lacourmody@yahoo.com

S.-N. Boemi et al. (eds.), *Energy Performance of Buildings*,
DOI 10.1007/978-3-319-20831-2_17

a valuable opportunity to introduce alternative technology options, such as solar thermal systems, to provide heating and cooling.

Solar thermal systems is now a proven technology in terms of reliability, cost-benefit, and low environmental impact. The integration of solar thermal systems and installations into the design of buildings can provide a clean, efficient, and sustainable low-energy solution for heating and cooling while, taken in a wider context, contribute to climate protection.

Solar thermal energy is adequate for providing the heating and cooling needs of buildings. Relevant solar energy technologies are those that require low-temperature heat, such as domestic water heating and space heating. Building cooling needs can also be met by solar applications, since cooling needs match with the availability of solar energy during sunny days. Peak loads in buildings often occur in summer, when electrically driven compressor chillers are operated to provide cooling. Solar thermal cooling technologies have low electricity consumption and environmental footprints, so they therefore have enormous unexploited market potential.

17.1.1 Solar Energy Collectors

Solar energy collectors are the major component of solar thermal systems. They are devices that absorb incoming solar radiation, convert it to heat, and transfer this heat to a fluid (usually air, water, oil, or a mixture) flowing through the collector. The solar energy thus collected is carried from the circulating fluid, either directly to the hot water or space-conditioning equipment, or to a thermal energy storage tank from which heat can be drawn for use at night and/or on cloudy days.

Two types of solar energy collectors commonly used to supply energy to buildings include solar water heating collectors and solar air heating collectors. Water heating collectors are the most commonly used type; both water heating and air heating collectors are either unglazed or glazed. A glazed solar collector traps heat, using a glazing surface on top of the absorber. The glazing of solar collectors is a critical issue, considering the weather (wind speed, rain, and dust) of the location, which may have negative effects on the performance of the collector, such as deterioration of the absorbing surface and dust deposition.

Solar energy collectors are either non-concentrating or concentrating. A non-concentrating collector has the same area for intercepting and absorbing solar radiation, whereas a concentrating one usually has concave reflecting surfaces to intercept and focus the sun's beam radiation to a smaller receiving area, thereby increasing the radiation flux.

Solar fraction (SF) is the most widely used technical indicator to evaluate the performance of solar thermal systems. SF measures the ratio of thermal energy produced by solar collectors to the total energy required to drive the heating or cooling system. It depends on several factors that include solar collector area, thermal storage size, available solar radiation, and load. An auxiliary heating system is used to deliver the required energy when there is insufficient heat supplied by the solar collector. SF is expressed as

$$\mathrm{SF} = \frac{Q_c}{Q_c + Q_{aux}} \tag{17.1}$$

where Q_c is the thermal energy produced by solar collectors in kW, and Q_{aux} is the thermal energy produced by the auxiliary heating system in kW.

Solar energy collectors contribute significantly to hot water production in many countries, and increasingly to space heating and cooling. In 2012, the world added 55.4 GW_{th} (more than 79,000,000 m^2) of solar heat capacity. This increased the cumulative installed capacity of all collector types in operation to 283.4 GW_{th} (REN21 2014). The breakdown of newly installed capacity in 2012 by collector type is 81.0 % evacuated tube collectors, 15.9 % glazed flat-plate collectors, 3.0 % unglazed water collectors, and 0.2 % glazed and unglazed air collectors. The breakdown of the cumulative capacity in operation in 2012 by collector type is 64.6 % evacuated tube collectors, 26.4 % glazed flat-plate collectors, 8.4 % unglazed water collectors, and 0.6 % glazed and unglazed air collectors (Mauthner and Weiss 2014).

17.1.1.1 Solar Water Heating Collectors

Solar collectors are the main components of solar water heating systems, with flat plate and evacuated tube collectors being the most widely deployed in buildings. Collector efficiencies are typically greater than 30 % for flat plates and greater than 45 % for evacuated tubes. Solar water heating collectors should be oriented directly towards the equator, facing south in the northern hemisphere and north in the southern. The optimum tilt angle of the collector is equal to the latitude of the location, with angle variations of 10–15°, more or less depending on the application (Kalogirou 2003).

17.1.1.2 Flat-Plate Collector

Figure 17.1 shows a schematic of a typical flat-plate solar water heating collector. It consists of a transparent glazing material, side and bottom insulation, absorber plate, risers, header tubes, casing and sealing gaskets. When solar radiation passes through the transparent cover and impinges on the absorber surface of high absorptivity, a large portion of this energy is absorbed by the plate and then transferred to the transport medium in the fluid tubes to be carried away for storage or use. The underside of the absorber plate and the side of the casing are well insulated to reduce conduction losses. The liquid tubes can be welded to the absorber plate, or they can be integrated to the plate. The liquid tubes are connected at both ends by header tubes.

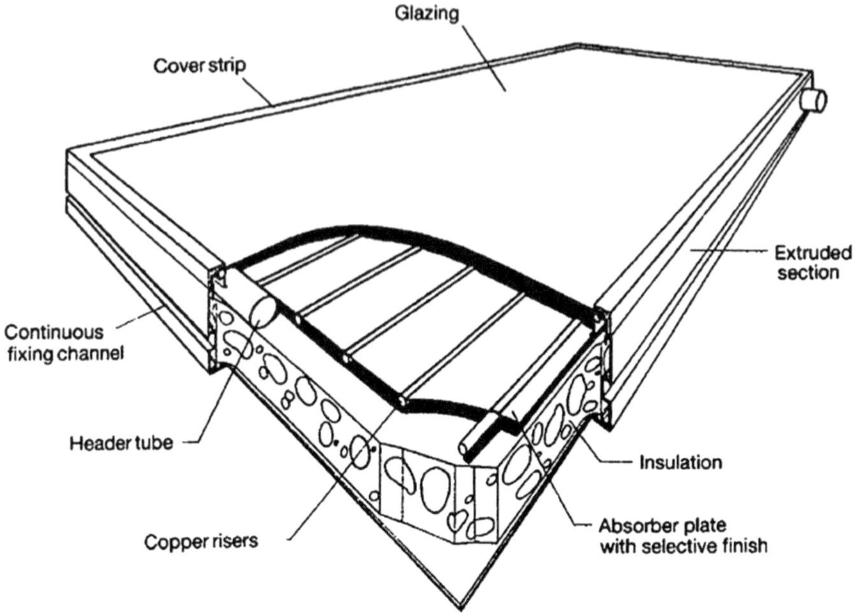

Fig. 17.1 Schematic of a flat-plate collector (Kalogirou 2004)

Flat-plate collectors (FPCs) are either glazed or unglazed; unglazed ones are mainly used for heating swimming pool water. In a glazed FPC, the transparent cover is used to reduce convection losses from the absorber plate through the restraint of the stagnant air layer between the absorber plate and the glass. It also reduces radiation losses from the collector as the glass is transparent to short wave radiation received by the sun but it is nearly opaque to long-wave thermal radiation emitted by the absorber plate.

Glass has been widely used to glaze solar collectors because it can transmit as much as 90 % of the incoming shortwave solar irradiation while transmitting virtually none of the longwave radiation emitted outward by the absorber plate. Glass with low iron content has a relatively high transmittance for solar radiation (approximately 0.85–0.90 at normal incidence), but its transmittance is essentially zero for longwave thermal radiation (5.0–50 mm) emitted by sun-heated surfaces (Kalogirou 2004).

The collector plate absorbs as much of the irradiation as possible through the glazing, while losing as little heat as possible upward to the atmosphere and downward through the back of the casing. The collector plates transfer the retained heat to the transport fluid. The absorptance of the collector surface for shortwave solar radiation depends on the nature and color of the coating and on the incident angle. Usually black is used although various color coatings have been proposed by Tripanagnostopoulos et al. (2000), mainly for esthetic reasons.

17.1.1.3 Heat Pipe Evacuated Tube Collector

Heat pipes are structures of very high thermal conductance. They permit the transport of heat with a temperature drop, which are several orders of magnitude smaller than that for any solid conductor of the same size. Heat pipes consist of a sealed container with a small amount of working fluid. The heat is transferred as latent heat energy by evaporating the working fluid in a heating zone and condensing the vapor in a cooling zone; the circulation is completed by return flow of the condensate to the heating zone through the capillary structure that lines the inner wall of the container (Faghri 1995).

A heat pipe-evacuated tube collector (HP-ETC) consists of a series of heat pipes inside vacuum-sealed tubes mounted into a heat exchanger (manifold), as shown in Fig. 17.2. Water or water/glycol mixture flows through the manifold and picks up the heat from the tubes. The vacuum envelope reduces convection and conduction losses, so the collector can operate at higher temperatures than FPC. Like FPC, they collect both direct and diffuse radiation, but their efficiency is higher at low incidence angles. This effect tends to give HP-ETC an advantage over FPC in day-long performance.

17.1.1.4 Water-in-Glass Evacuated Tube Collector

Figure 17.3 shows a schematic of a water-in-glass evacuated tube collector connected to a horizontal water tank. A water-in-glass evacuated tube collector typically consists of about 15–40 flooded, single-ended tubes directly connected to a

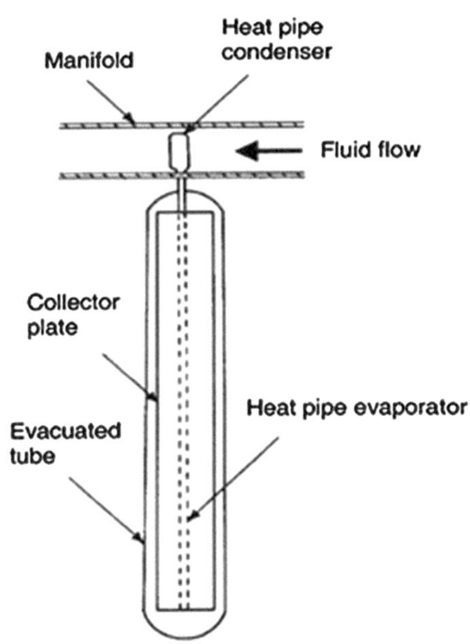

Fig. 17.2 Schematic of a heat pipe-evacuated tube collector (Kalogirou 2004)

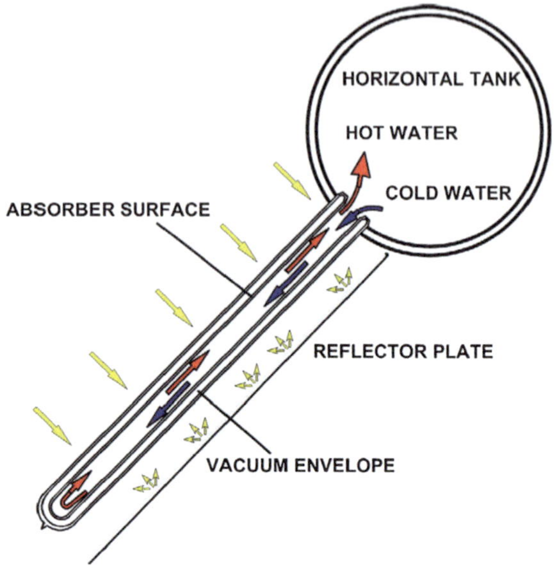

Fig. 17.3 Water-in-glass evacuated tube collector (Budihardjo et al. 2007)

horizontal water tank. The tubes consist of two concentric glass tubes sealed at one end with a vacuum in the annular space between the tubes, and a selective surface coated on the outer surface of the inner tube. Heat is driven by natural circulation of water through the single-ended opening into the tank. Water in the tubes is heated by solar radiation, which then rises along the top of the tube to the storage tank and is replaced by colder water from the tank entering the bottom of the tube opening. These collectors are widely used in China because of their high thermal efficiency, simple design and low manufacturing cost.

17.1.2 Solar Air Heating Collectors

Solar air heating collectors (SAHCs) convert solar energy to heat air. They can be used to support space heating through the ventilation system of buildings. Unglazed solar air heating collectors (USAHC) and glazed solar air heating collectors (GSAHC) are two types of commercially available SAHCs. The basic difference between USAHC and GSAHC is a glazing cover and design of the absorber surface.

17.1.2.1 Unglazed Solar Air Heating Collector

USAHC generally consists of an absorber plate with a parallel back plate. The space between the absorber and the back plate forms a plenum, as shown in

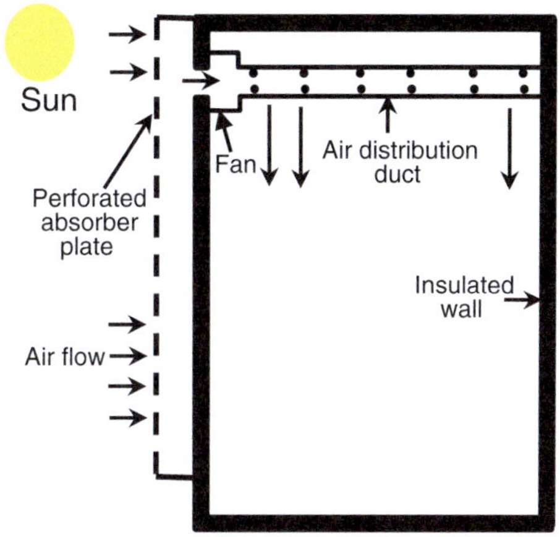

Fig. 17.4 Schematic of an unglazed transpired solar air heating collector

Fig. 17.4. Outside air to be heated is drawn through the perforation and the plenum, using an extraction fan. There are two types of USAHC: transpired absorber, and non-transpired absorber (back pass); the difference between the two systems is the absorber perforation. The most popular type of USAHC is the unglazed transpired collector (UTC), known as the "solar wall," which was invented and patented in the 1990s by John Hollick (Shukla et al. 2012).

UTCs use solar energy to heat the absorber surface, which transmits energy to the ambient air. The absorber surface is generally a metallic sheet which can be integrated to the building's façade. The contact surface between the metal skin and air is increased by drawing air through the multiple small perforations into the cavity between the skin and façade. Heated air is drawn into the building to provide the heating of the space.

Continuous improvements to UTC have focused on the material of the transpired absorber (Gawlik et al. 2005), diameter and pitch of the perforation and thickness of the absorber (Van Decker et al. 2001). This technology is an effective, low-cost option to meet the heating demands in the buildings. However, the temperature rise of the air is comparatively much lower than that for glazed solar collectors at maximum efficiency. UTC loses heat from the exposed absorber perforation due to reverse flow and radiation losses from unglazed absorbers cannot be avoided.

17.1.2.2 Glazed Solar Air Heating Collector

A glazed solar air heating collector (GSAHC) consists of an absorber plate with a parallel plate below, forming a passage through which the air to be heated flows. Figure 17.5a–c show various designs of GSAHCs, notably finned absorber, glass absorber, and corrugated absorber. Roughness inducing wires and wavy passage has been used to boost collector heat transfer (Karim and Hawlader 2006). Many

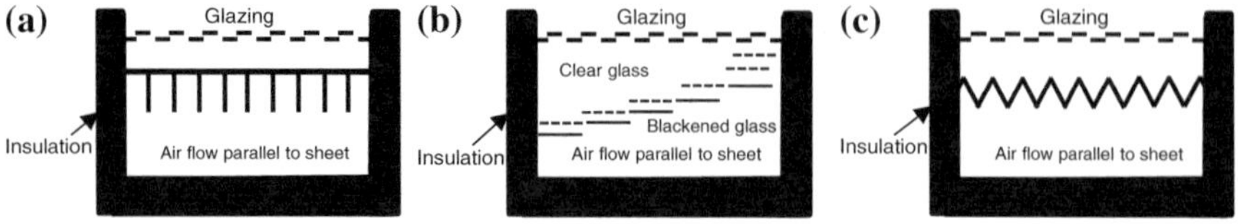

Fig. 17.5 Glazed solar air heating collector with, **a** finned absorber; **b** glass absorber; **c** corrugated absorber

kinds of fins, such as offset strip fins (Karsli 2007) and continuous fins (Moummi et al. 2004) have been studied extensively. Rough surface and porous media were often used to enhance heat transfer mechanism (Yousef and Adam 2008).

Conventional glazed SAHCs have low efficiencies due to the low thermal capacity of air, low absorber to air heat transfer coefficient and inadequately addressed design. These systems inherit poor convective heat transfer coefficients between the absorber plate and the air flow, resulting in the absorber temperature being much higher than the air stream temperature. This causes greater heat losses to the surroundings. The heavy metallic absorber also makes air heating systems difficult to integrate in buildings. In order to overcome the limitations inherent in conventional glazed SAHCs, Shams (2013) designed and demonstrated the performance of a stationary novel air heating system that integrates a "transpired air-heating collector" (TAC) and an "asymmetric compound parabolic concentrator" (ACPC), as shown in Fig. 17.6.

The perforated absorber faces downward to reduce radiation loss. It allows air to flow through the absorber to enhance heat transfer between the absorber and flowing air. The concentrator increases the concentration ratio of the incident solar radiation onto the inverted perforated absorber, and the parallel reflector below the inverted absorber improves stratification and enhances heat transfer. The glazing acts as a heat trap for emitted radiation from the absorber surface.

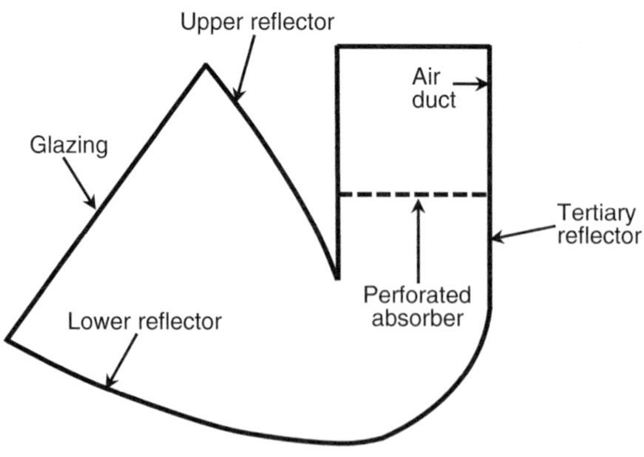

Fig. 17.6 Schematic of the CTAH (Shams 2013)

17.1.3 Solar Water Heating Systems

Water for domestic use in urban areas is generally heated by burning fuels such as firewood, kerosene, oil, liquid petroleum gas (LPG), coal, or electricity (either by geysers or immersion heaters). The utilization of solar energy through solar water heating systems (SWHSs) can play an important role in offsetting the quantity of conventional energy used in domestic hot water provision. These systems therefore have significant potential in reducing environmental emissions arising from the use of fossil fuels.

SWHSs are the most popular means of utilizing solar energy, since they are a relatively well-established technology and are perceived as economically attractive compared with other kinds of solar energy technologies. Solar water heater technology has been well developed and can be easily implemented at low cost (Xiaowu and Ben 2005). In a solar domestic hot water system, the solar collector is the main component of the system, hence its optimal performance is important (Luminosu and Fara 2005). Flat-plate and evacuated tube collectors are the main types of collectors used in SWHSs for domestic application; these absorb both diffuse and direct solar radiation and would therefore function even under overcast skies. The ease of operation and low cost of solar water heating collectors makes them suitable for low temperature applications below 80 °C.

There are several main types of SWHSs for use in buildings (The German Solar Energy Society 2007):

1. Thermosyphon and forced circulation systems: Thermosyphon systems do not require pumps (as in this case, gravity is used for liquid transport), whereas systems with forced circulation require circulating pumps for this purpose.
2. Open and closed systems: Open systems have an open container at the highest point of the solar circuit, which absorbs the volumetric expansion of the liquid caused by the temperature changes. The pressure in open systems thereby corresponds at its maximum to the static pressure of the liquid column. Closed (sealed) systems operate at a higher pressure (1.5–10 bar), which influences the physical properties (such as the evaporation temperature) of the solar liquid. Closed systems require special safety devices.
3. Single-circuit (direct) and twin-circuit (indirect) systems: In direct systems, the domestic water circulates from the storage vessel to the collector and back again. Indirect systems have two separate circuits—the solar circuit and domestic water circuit. The solar circuit includes the collectors, the ascending pipes, the solar pump with safety equipment, and a heat exchanger. A mixture of water and antifreeze agent can be used as the heat transfer fluid. The domestic water circuit includes the storage vessel as well as the cold water and hot water installations of the house.
4. Filled and drainback systems: Indirect systems can have a collector circuit that is either completely filled or only partially filled. In the latter systems, which are called "drainback systems," the collector drains completely when the collector pump is switched off.

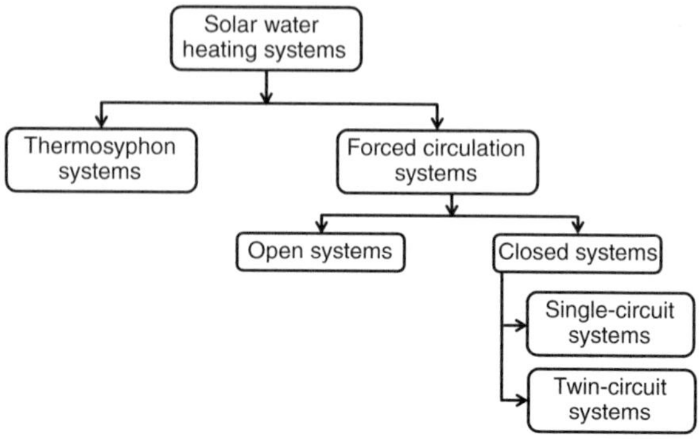

Fig. 17.7 Classification of solar water heating systems

Systems with forced circulation are predominant in temperate climates, such as in central and northern Europe. In southern Europe, Australia, Israel, and other mostly sunny countries, thermosiphon systems are the most common type. Figure 17.7 shows a classification of SWHSs.

SWHSs are an established technology. Over the years, SWHSs have witnessed significant increases in performance and production volumes as well as installed cost reductions due to the economies of scale, learning, and technological developments. For example, improved methods of production and the use of surfaces with increased absorptivities and lower emissivities have increased efficiencies. Modularization has facilitated production optimization while improved pump designs have been adapted to suit various flow regimes. Development in control devices have also contributed to system improvements; furthermore, reductions in collector and hot water tank heat losses have been achieved due to improvements in insulating materials. Various auxiliary heating systems have also been integrated into storage tanks, while new tank designs with enhanced stratification mechanisms have been developed, thus reducing system losses (Duffie and Beckman 1991).

Most solar thermal systems are used for domestic water heating and typically meet 40–80 % of annual hot water demand; solar thermal systems have also been used for space heating, where they provide 15–30 % of the demand. Combi-systems (for water and space heating) account for about 4 % of the global solar thermal heat market.

The installed cost of SWHS varies from 1100–2200 $/kW_{th}$ for single-family homes, and from 950–2050 for multi-family homes. The relationship between collector area and capacity is $1\ m^2 = 0.7\ kW_{th}$ (kilowatt-thermal). The system's useful life is typically 20 years. Table 17.1 details the characteristics of thermosyphon and forced circulation SWHSs.

SWHSs increasingly contribute to domestic hot water provision and space heating in many countries. Global cumulative installations of glazed SWH collectors have risen steadily from 44 GW_{th} in 2000 to 277 GW_{th} in 2012. The global market for glazed SWH collectors grew by 24.1 % between 2011 and 2012. Figure 17.8 shows the evolution of the global cumulative installed capacity of glazed SWHSs between 2000 and 2012.

Table 17.1 Characteristics of thermosyphon and forced circulation SWHSs (REN21 2014)

Parameter	Value	
	Domestic hot water	Domestic heat and hot water
Service life (years)	20	20
Learning rate (%)	10	10
Collector types	Flat-plate, evacuated tube (thermosyphon and pumped systems)	Flat-plate, evacuated tube (thermosyphon and pumped systems
System efficiency (%)		
FPC	>30	>30
ETC	>45	>45
Plant size (kW_{th})		
Single family	2.1–4.2	7–10
Multi-family	35	70–130
Capital cost (\$/kW_{th})		
Single family	1100–2140 (OECD, new build) 1300–2200 (OECD, retrofit)	1100–2140 (OECD, new build) 1300–2200 (OECD, retrofit)
Multi-family	950–1850 (OECD, new build) 1140–2050 (OECD, retrofit)	950–1850 (OECD, new build) 1140–2050 (OECD, retrofit)

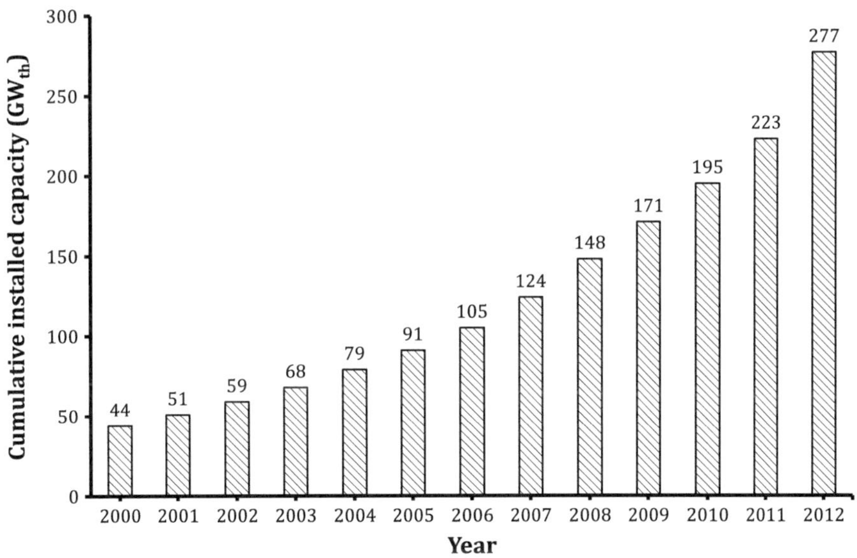

Fig. 17.8 Global cumulative installed capacity of glazed SWH collectors (REN21 2014)

17.1.4 Forced Circulation SWHS

In a forced circulation SWHS, water is heated in the collectors, and a pump is used to circulate a water glycol mixture used as the heat transfer fluid. A solar

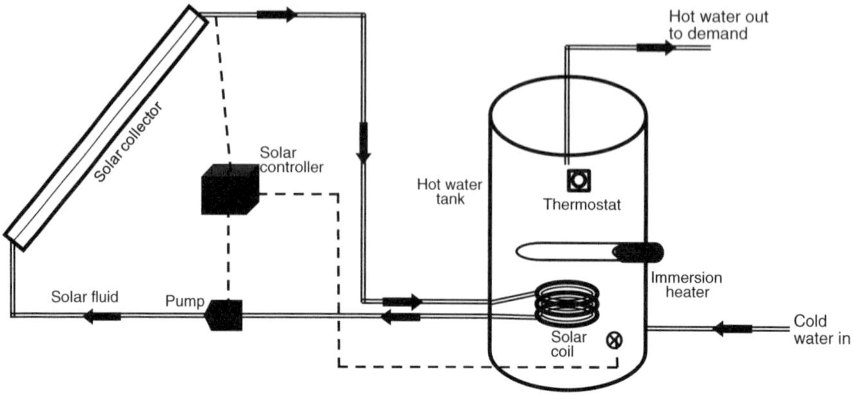

Fig. 17.9 Schematic diagram of an electricity-boosted SWHS

controller triggers the pump when the solar fluid outlet temperature from the collector is above a set value of the water temperature at the bottom of the storage tank. A solar coil at the bottom of the hot water tank is used to heat water. The solar fluid should have some desirable properties, such as low freezing and high boiling points. An auxiliary heating system is used to raise the water temperature during periods when there is less heat available from the solar collector. Figure 17.9 shows the basic components of an electricity-boosted SWH system used in Ireland.

Promising new designs include"combi-systems" that combine water heating and space heating. This extends the operation period, thus improving economic performance. Active solar space and water heating systems usually need a back-up system that uses electricity, bioenergy, or fossil fuels. Other designs integrate a solar-assisted heating system with a heat pump resulting in ultra-high efficiencies of 125 % to 145 %, compared to a condensing boiler at around 107 % (OECD/IEA 2008).

17.1.5 Thermosyphon SWHS

In a thermosyphon SWHS, the heat absorbed by the collector heats up water and causes its density to decrease. Denser cold water fed through the bottom of the collector then forces the warmer water to rise through natural circulation (thermosyphon effect) into an overhead storage tank. Thermosyphon SWHSs are suitable for sunny climates. Figure 17.10 shows a schematic of a thermosyphon SWHS; these are cheaper than their forced circulation counterparts and are easier to maintain, since they do not have any moving parts.

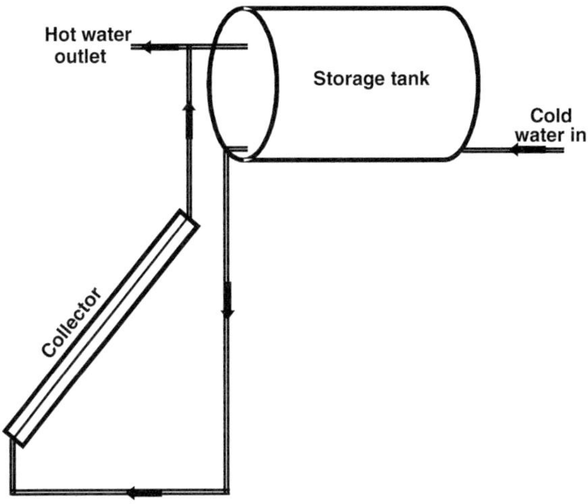

Fig. 17.10 Schematic of a thermosyphon SWHS

17.2 Solar Thermal Cooling Systems

In solar thermal cooling systems, heat from the sun is required to drive the cooling process. This is achieved by collecting solar radiation and converting it to heat, using solar thermal collectors. The heat is then used to drive cooling technologies involved in closed-cycle (absorption and adsorption) and open-cycle (desiccant) systems. These systems consist of several components: the heat-driven system, the air conditioning systems, heat-driven cooling devices, solar collectors, a heat buffer storage, a cold storage and auxiliary subsystem (Kalkan et al. 2012), and pumps to regulate the flow rate and controllers for the automatic operation of the complete system (Ghafoor and Munir 2015). Four types of commercially available solar thermal-driven cooling systems include absorption, adsorption, desiccant, and ejector cooling systems.

In addition to replacing primary energy or fossil fuels, solar thermal cooling systems can reduce electric grid load at peak cooling demand periods by partially or fully replacing the electricity needed for conventional vapor compression chillers or room air conditioners. Solar cooling technologies are particularly beneficial due to the strong correlation between the supply of the solar resource and demand for cooling during the daytime, while efficient heat storage techniques can also fully or partially cover cooling demands at night (IEA 2012).

The efficiency of solar thermal refrigeration systems is expressed by the coefficient of performance (COP), which is defined as the ratio of energy required from the refrigerated space to the work input. The COP is expressed mathematically as

$$\mathrm{COP} = \frac{Q_{\mathrm{i}}}{W_{\mathrm{c}}} \tag{17.2}$$

where Q is the energy required from the refrigerated space (kW) and W_c is the work input (kW).

The global market for solar cooling systems grew at an average annual rate exceeding 40 % between 2004 and 2012. About 1050 solar cooling systems of all technology types and sizes were installed by 2013. One important market driver for solar cooling systems is the potential to reduce peak electricity demand, particularly in countries with significant cooling needs. The cost of solar cooling kits declined by 45–55 % between 2007 and 2013 (REN21 2014).

17.2.1 Solar Absorption Cooling System

The basic physical process of an absorption chiller consists of two chemical components—one serving as refrigerant, and the other as sorbent or absorbent. A fluid pump is used to circulate both the refrigerant and the sorbent. In a solar-driven absorption chiller, the mechanical compressor is replaced by a thermal compressor that consists of an absorber, a generator, a pump, a condenser, an evaporator, and a circulating valve as shown in Fig. 17.11.

The absorption cooling cycle starts in the evaporator where the refrigerant evaporates in a low partial pressure environment to the absorber, and the evaporation of the refrigerant extracts from the surroundings and cools down the chilled

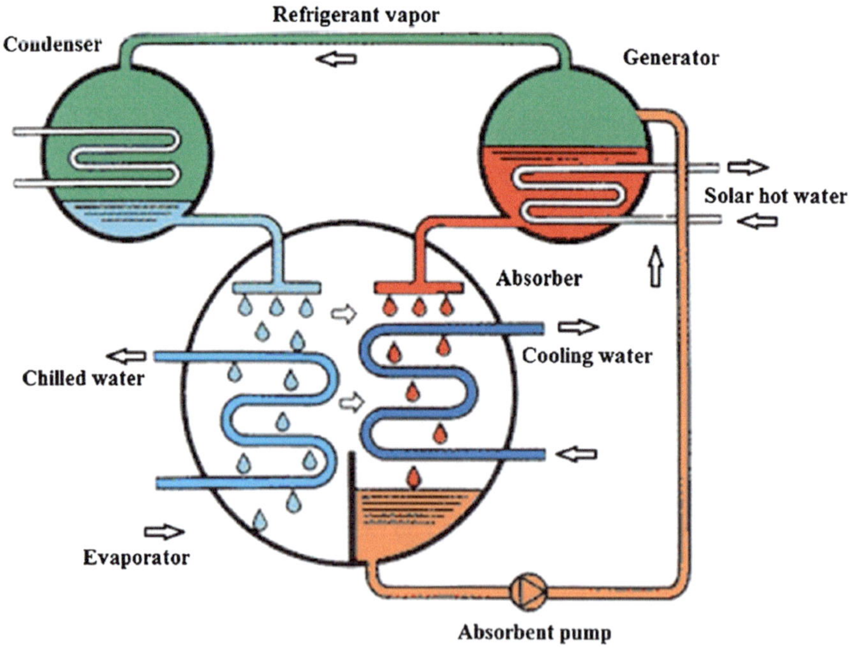

Fig. 17.11 A schematic of absorption chiller (Coast DoEG 2012)

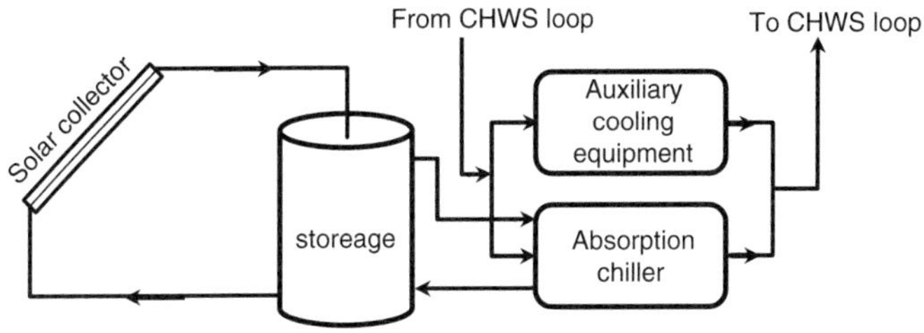

Fig. 17.12 Solar cooling integration with auxiliary cooling equipment (Tamasauskas et al. 2014)

water. The gaseous refrigerant is then absorbed into the absorbent, causing its partial pressure to be reduced in the evaporator, thereby allowing more liquid to evaporate. The diluted liquid of the refrigerant and the absorbent material are pumped into the generator, where the liquid mixture is heated, causing the refrigerant to evaporate and then condense on the condenser, which refills the supply of liquid refrigerant in the evaporator through a circulation valve. The energy used to run the pump is negligible compared to the energy required by a conventional compressor (Tsoutsos et al. 2010).

In a solar-driven absorption chiller operating in parallel with auxiliary cooling equipment, heat from the sun is used to meet a portion of the cooling load. When applied to a building, it reduces its peak electrical demand during the summer months. Heat obtained from solar collectors is stored in a storage tank, between 70 and 95 °C. This low temperature heat can easily be achieved using flat-plate or evacuated tube collectors. The stored energy is then used at the generator of an absorption chiller sized to meet the peak cooling demand of the building. The absorption chiller does not operate if there is insufficient solar energy supplied from the storage tank. Auxiliary cooling equipment is used to provide additional cooling requirements. Figure 17.12 is a schematic of a hybrid solar-driven absorption chiller operating in parallel with auxiliary cooling equipment.

The COP of le is defined as

$$\mathrm{COP} = \frac{Q_{\mathrm{L}}}{Q_{\mathrm{gen}} + W_{\mathrm{pump}}} \frac{Q_{\mathrm{L}}}{Q_{\mathrm{gen}}} \tag{17.3}$$

where Q_{L} is the heat transferred from the refrigerated space (kW), Q_{gen} is the heat input to the generator (kW), and W_{pump} is the work input to the pump which is negligible.

Absorption chillers are the dominant technology for solar cooling systems. Most market-available absorption chillers based on H_2O–Br working pairs use water (H_2O) as refrigerant and lithium-bromide (LiBr) as sorbent. Small- and medium-sized absorption chillers available on the market range between 4.5 and 20 kW; they work on H_2O–LiBr, H_2O–LiCl, and NH_3–H_2O. The driving temperatures range between 65 and 90 °C; cooling temperatures are between 24 and 35 °C; and chilled temperatures between 7 and 18 °C (Jakob et al. 2008; Jakob and Pink 2007; Jakob and Mittelbach 2008).

Commercially available single-effect absorption refrigeration cycle systems have COPs around 0.6, while double-effect systems have COPs between 0.8 and 1.2. Triple-effect and quadruple-effect absorption refrigeration cycles are still under development and will provide COPs up to 2 (Kalkan et al. 2012).

17.2.2 Solar Adsorption Cooling System

Adsorption is the process of bonding a gas or other material on a solid surface. The adsorption refrigeration system uses a solution containing water and lithium bromide salt to absorb heat from the surroundings, and is driven by hot water, steam, or combustion. The substance is transferred from one phase to another and penetrates the second substance to form a solution (Kalkan et al. 2012). The common type of adsorbents used in adsorption chillers are silica-gel, zeolites, and activated carbons (IEA 2009). The main difference in adsorption systems is that the sorbent/adsorbent is a solid and not a liquid, as in the case of an absorption chiller.

Basic adsorption refrigeration systems do not require a pump and a rectifier. The pressure difference arising inside the system is the result of the transfer of a substance from one phase (i.e., vapor) followed by condensation on the surface. Electricity consumption during the adsorption process is minimal. Adsorption chillers consist of two compartments (1 and 2), an evaporator, and a condenser, as shown in Fig. 17.13.

Typical COP of existing adsorption chillers are in the range of 0.2–0.6. However, adsorption chillers can work at lower generator inlet temperatures in the range of 45–65 °C (Ghafoor and Munir 2015). The absence of pumps and noiseless operation are other advantages of adsorption chillers; they are currently available in the market range between 5.5 and 15 kW.

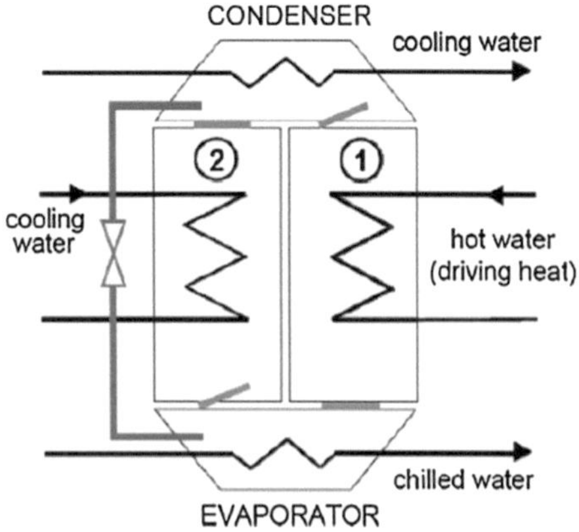

Fig. 17.13 Schematic of a typical adsorption chiller (ESTIF 2012)

17.2.3 Solar Desiccant Cooling Systems

Solar desiccant cooling uses low-grade heat delivered by solar collectors and can help eliminate the use of refrigerant gases. Solar desiccant cooling technology consists of three sub-systems: solar thermal system, desiccant dehumidifier, and evaporative cooler. In the desiccant cooling process, the main concept depends on the desiccant material's ability to reduce the air moisture content for the air cooling process and dehumidification. As a result, there is constant air drying, and the dehumidified air is heated above ambient temperature. Evaporative cooling or heat exchangers are used to cool the heated dry air to near-ambient temperature.

Solar desiccant cooling is more effective when used in hot and humid climates, since the prime feature of desiccant cooling systems is that they can treat sensible and latent heat loads separately. Figure 17.14 shows a schematic of a solar desiccant cooling system.

The COP of the desiccant cooling system is defined by speculation on no demand for a vapor compression system

$$\mathrm{COP} = \frac{Q_{\mathrm{L}}}{Q_{\mathrm{H}}} \tag{17.4}$$

where Q_{L} is the heat extracted from the conditioned space (kW), and Q_{H} is the heat added to the heater (kW).

17.2.4 Solar Ejector Cooling System

The solar ejector refrigeration system is a thermally driven technology that has the ability to produce cooling using low-grade heat from the sun at temperatures

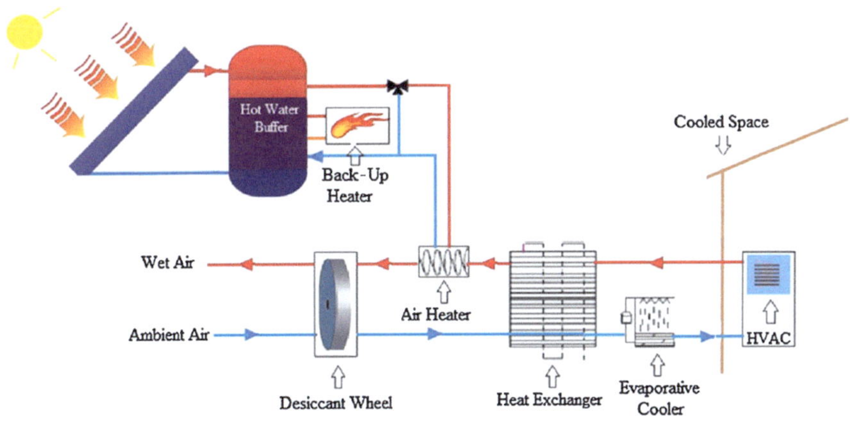

Fig. 17.14 Schematic of a solar desiccant cooling system (Baniyounes et al. 2013)

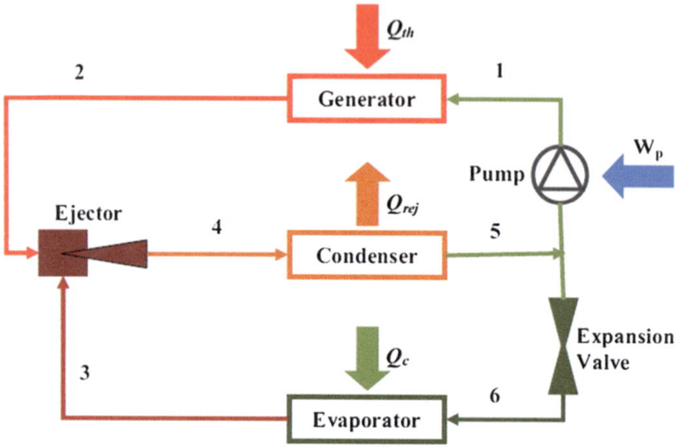

Fig. 17.15 Schematic of an ejector refrigeration system (Al-Zubaydi 2011)

as low as 65 °C. The compressor in a conventional vapor compression cycle is replaced with a boiler, an ejector, and a pump. The ejector system shown in Fig. 17.15 consists of two loops: power and refrigeration. In the first loop, from process 1 to 2, high pressure vapor (primary fluid) is generated, using heat supplied by solar collectors or an auxiliary unit (e.g., a boiler) Q_{th}. They have a much lower COP, usually under 0.3, and condenser temperatures between 85 and 95 °C, and 28 and 32 °C, respectively (Al-Zubaydi 2011). However, they offer the advantages of simplicity, no moving parts, and low operating and installation costs.

The COP of an ejector refrigeration system is defined as

$$\mathrm{COP} = \frac{Q_L}{Q_H} \tag{17.5}$$

where Q_L is the heat extracted from the evaporator (kW), and Q_H is the heat added into the boiler (kW).

17.2.5 Advantages and Disadvantages

The main advantages and disadvantages of solar cooling systems used in buildings are listed in Table 17.2.

17.2.6 Overview of Solar Cooling Systems

Table 17.3 details the characteristics of market-available solar cooling systems. The refrigerant cycles used are either closed or open cycles, with liquid or solid

Table 17.2 Advantages and disadvantages of solar cooling systems (Ghafoor and Munir 2015; IEA 2009)

Technology	Advantages	Disadvantages
H_2O-LiBr absorption chillers	• Low maintenance and operating cost • Longer service life • Non-toxic and operate at low pressure • Cheaper and have relatively higher COP compared to adsorption and DEC chillers • Absorbent is non-volatile • Low electricity requirement • Easier to implement • Silent operation and high reliability	• Larger and heavier than electric chillers • Require large cooling tower • Water as refrigerant limits the chilled water temperature to values above 0 °C • Supports low pressure because of water as refrigerant • Li-Br is corrosive
NH_3–H_2O absorption chillers	• Pressurized system avoiding complex vacuum vessels • Chilled temperatures below 0 °C are possible • More flexible in the chilled water and heat rejection temperatures compared to Li-Br systems	• Have a slightly lower COP compared to Li-Br chillers • Typical COP (0.5–0.6)
Adsorption chillers	• Robust • No danger of damage due to high temperatures • Use environmentally friendly materials • Requires very little amount of electricity • Low maintenance cost because of few moving parts • Low manufacturing cost	• Requires high vacuum tightness of the container • Slightly lower COP compared to absorption systems • Few suppliers in the market • Requires careful design of external hydraulic circuits due to cyclic temperature variation in the hydraulic circuit
Desiccant evaporative cooling (DEC) system	• Systems with liquid desiccants have higher air dehumidification • Possibility of high energy storage using concentrated hygroscopic material solutions • Continuously pass a large volume of air through an enclosed structure as a result of more fresh air introduced into an enclosed structure • COP slightly higher than absorption and adsorption chillers	• The application of this cycle is limited to temperate climates • Relatively larger and heavier as compared to absorption, adsorption and electric chillers • Air leaks reduce the system efficiency

(continued)

Table 17.2 (continued)

Technology	Advantages	Disadvantages
Ejector cooling cycle	• Simplicity in design and smaller in size • Fewer moving parts resulting in silent operation of the system • Low electricity consumption compared to other chillers • High reliability and lower running and maintenance cost • Lower initial cost and longer life	• Very low COP compared to other cooling techniques • Higher generator temperatures between 85 and 95 °C produce COP in the range of 0.2–0.33

Table 17.3 Overview of market-available solar cooling technologies (SOLAIR 2009)

Refrigerant cycle principle	Closed cycle Chilled water production		Open cycle Direct treatment of air	
Type of sorbent	Liquid	Solid	Solid	Liquid
Market technology	Absorption	Adsorption	Solid desiccant	Liquid desiccant market introduction stage
Working pair	i. H_2O–LiBr ii. NH_3–H_2O	i. H_2O–Silicagel ii. H_2O–Zeolites	i. H_2O–Sillcagel ii. H_2O–LiCl	i. H_2O–LiCl ii. H_2O–CaCl
Cooling capacity	4.5–5000 kW	50–430 kW	20–350 kW	–
COP	0.6–0.8 (SE) 1.1–1.2 (DE)	0.5–0.7	0.5–>1	>1
Driving temperature	60–90 °C (SE) 140–160 °C (DE)	55–90 °C	45–90 °C	45–70 °C
Collectors	ETC or CPC	FPC or ETC	FPC or solar air collectors	FPC or solar air collectors

SE Single effect, *DE* Double effect

sorbents. Absorption, adsorption, and solid desiccant are market-ready technologies, while liquid desiccant systems are in their infancy stages.

17.2.7 Solar Cooling System Costs

The costs of solar thermal cooling systems have decreased over the years due to a decrease in the prices of FPC and ETC along with sorption chillers. The specific cost of FPC, with its accessories for a heating loop, is in the range of €150–200/m^2, while the specific cost of ETC lies in the range of €250–300/m^2. The introduction and rapid expansion of the Chinese market in Asia and Europe has significantly contributed to reducing the cost of solar thermal systems. Currently, the

Table 17.4 Cost of different types of complete solar cooling systems (Ghafoor and Munir 2015)

Type	Cost of complete solar cooling kit (€/kW_c)	Remarks
H_2O–LiBr absorption cooling system with FPC	1500–2000	• FPC: €150–200/m^2 • Chiller cost: €300–350/kW_c • Backup boiler: €150–200/kW • Storage tank: €500–600/m^3 • Cooling tower: €80–100/kW • Mountings and accessories: €200–300/kW (fan coils, pipe work, pumps, sensors and controllers etc.) • Installation cost: €150–200/kW_c (10 % of capital investment)
NH_3–H_2O absorption cooling system with ETC	2500–3000	• ETC: €250–300/m^2 • Chiller cost: €500–600/kW_c (rest of components same as above)
Adsorption chiller using FPC	2000–2500	• Chiller cost: €400–450/kW_c (rest of components same as above)
Desiccant evaporative cooling system with FPC	3000–4000	• Complete DEC system cost: €2000–2500 coupled with FPC

implementation of solar thermal based cooling technologies seems to be a technically as well as economically feasible option for medium- to large-scale applications. Table 17.4 shows estimates of the average specific costs of complete solar cooling technologies.

17.3 Solar Air Heating System

The use of solar energy to heat air for various purposes in buildings has attracted much interest. Solar air heating systems (SAHSs) are well established, with applications in space heating and space ventilation in buildings. SAHSs are of greater benefit in airtight buildings with low energy demands. Figure 17.16 shows the schematic of an SAHS where hot air is supplied by the solar collector. The system has two fans; the first is the collector's fan, while the second is the heating fan. The fans ensure that air flows through the solar collector to the space being heated. An air-to-water heat exchanger can be installed at the collector's exit to provide hot water for use in conventional radiators, radiant floors or walls, or to a domestic hot water tank, and excess heat is stored in the thermal storage room when the solar collector supplies more heat than is required.

Research on system modifications approved by the committee of the International Energy Agency and funded by nine countries under "Task 19"

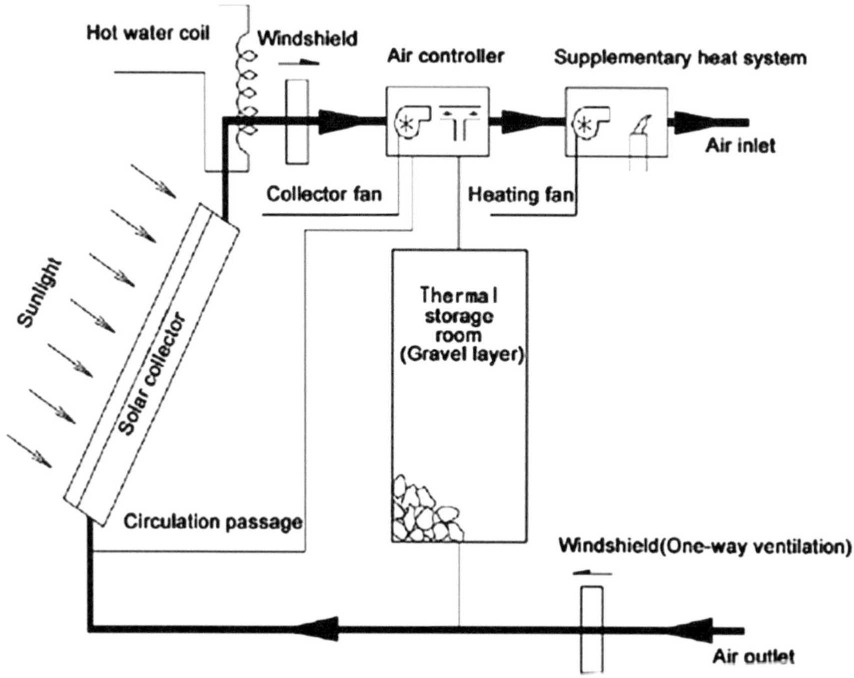

Fig. 17.16 Solar air heating system (Yang et al. 2014)

proposed six types of SAHSs. Details of the six models are shown in Fig. 17.17a–f. The models include: (a) solar heating of ventilation air in an open loop system; (b) closed loop heating system where air is circulated into the collector where it is heated; (c) collector-heated air is circulated through a cavity in the building envelope; (d) closed loop collector/storage system where warm air is circulated through channels in a floor/wall that then radiates heat into the building; (e) open single loop collector to the building spaces; and (f) collector heated air transferred to water via an air/water heat exchanger.

17.3.1 Solar Absorption Heat Pump System

Figure 17.18 shows the proposed integration of solar thermal technologies with an absorption heat pump. This concept uses solar energy to complement the operations of an indirect fired absorption heat pump, with solar energy operating in priority. The solar-driven heat pump is sized to meet the building's loads, and acts as the primary source of heating and cooling energy.

Whenever possible, energy from the solar heat store is used at the heat pump generator to drive operations. When the heat store's temperature drops below the

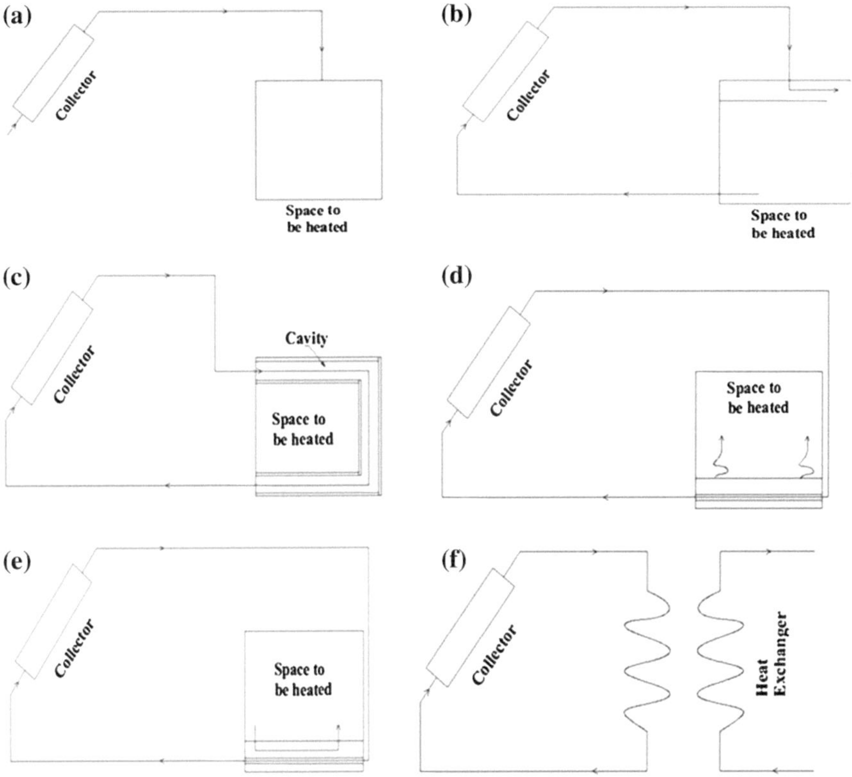

Fig. 17.17 System modifications of SAHSs (Hastings and Morck 2000)

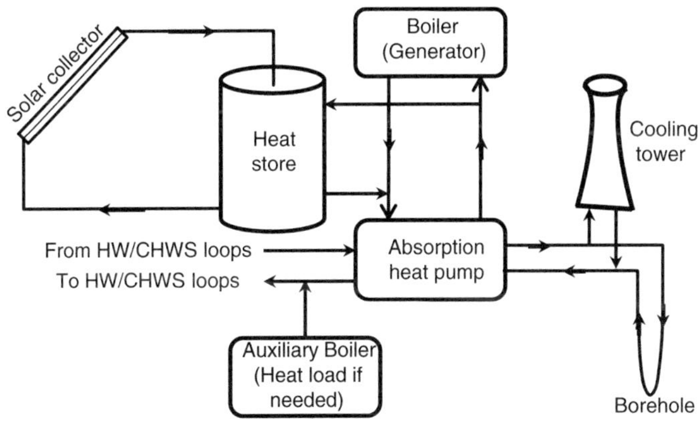

Fig. 17.18 Solar-driven absorption heat pump system (Tamasauskas et al. 2014)

defined lower limit (85 °C in winter or 70 °C in summer), an auxiliary boiler is used to boost the temperature to the level required for operations. A series of vertical boreholes acts as the thermal source in winter and thermal sink in summer, while a cooling tower is used to balance the ground loads. A second auxiliary boiler is also installed on the building side of the system to ensure that sufficiently heated water is supplied to the low temperature radiators.

17.4 Building Integrated Solar Thermal Systems

Building integrated solar thermal (BIST) systems are integrated into elements of building envelopes, such as the roof or the façade. These systems are very important because they serve the dual function of building skin, replacing conventional materials, and energy generator, and they modify the architectural appearance of the construction. Before their application, the site and the orientation of the building must be carefully considered because the impact of shading on the solar collector would reduce the amount of energy collected (Moschella et al. 2013).

Most often, solar collectors are considered only as technical elements and are confined to the rooftop, whether flat or pitched, where a bad integration is less visible and the architectural impact is minor (Munari et al. 2007). The main requirement for integration of solar collectors with buildings has two aspects: the installation modes of solar collectors, in which solar collectors should be integrated with roofs, walls, and balconies without destroying building façades, and the design of solar thermal systems. Solar collectors and heat storage tanks should be separated, which means that heat storage tanks should be installed in attics or equipment rooms.

17.4.1 Façade Integrations

Façade integrations are still very rare since they are, in fact, much more delicate than their roof counterparts because of the higher visibility of the collectors. As the façade is the public face of the architecture, the collectors cannot simply be used as added technical elements; their architectural integration needs to be satisfactory and the design controlled (Munari Probst et al. 2004).

Solar collectors are installed to occupy the whole parapet area and are used as parapet external finishing. They appear as an integral part of the building. The size and shape of the modules fit the grid and match the rhythm of the façade. The collector glazing and the black color of the absorber match the color and material of the window openings above and are dimensioned to completely cover their area. For a hangar, an unglazed solar system works both as a solar collector and façade cladding. Solar air-collectors can be integrated into façades and used as façade cladding. The collectors have color freedom that can be exploited.

17.4.2 Roof Integrated Systems

Unglazed roof integrated collectors are well appreciated. Collectors occupy the whole roof area and have the dual function of solar absorbers and upper layer of the covering system. Generally speaking, the integration of solar collectors into roofs of buildings has the disadvantage of serious thermal loss occasioned by long pipe lengths. Furthermore, the inspections that may have to be carried out on the roof are sometimes difficult and dangerous.

17.4.3 Balconies and Walls

The integration of solar collectors with balconies and walls makes it convenient for subsequent repairs and inspection. They also require shorter pipelines and less civil engineering work. However, the proper solar collector type must be chosen to suit the building façade. In addition, the building has to be designed to prevent the shadow of the building itself from blocking the sunlight. At the same time, while designing the building to integrate and effectively utilize solar energy systems, the primary function of providing usable space should also be ensured (Wang and Zhai 2010). For integration into glass curtain walls of buildings, solar collectors that have colors that match the glass curtain wall can serve the dual purpose of supplying hot water and becoming components of the building.

References

Al-Zubaydi AYT (2011) Solar air conditioning and refrigeration with absorption chillers technology in Australia—an overview on researches and applications. J Adv Sci Eng Res 1:23–41

Baniyounes AM, Liu G, Rasul MG et al (2013) Comparison study of solar cooling technologies for an institutional building in subtropical Queensland, Australia. Renew Sustain Energy Rev 23:421–430

Budihardjo I, Morrison GL, Behnia M (2007) Natural circulation flow through water-in-glass evacuated tube solar collectors. Sol Energy 81(12):1460–1472

CEREN (Centre d'Etudes et de Recherches Economiques sur l'Energie) (2011) Consommation totale d'énergie des secteurs résidentiel et tertiaire, par usage, de 1990 à 2009, http://www.statistiques.developpementdurable.gouv.fr June 2011

Coast DoEG (2012) CHP thermal technologies. Available via DIALOG. http://gulfcoastcleanenergy.org/CLEANENERGY/CombinedHeatandPower/Thermaltechnologies/tabid/1789/Default.aspx. Cited 10 Mar 2015

Duffie JA, Beckman WA (1991) Solar engineering of thermal processes. Wiley, New York

European Solar Thermal Industry Federation (ESTIF) (2010) Solar thermal markets in Europe: trends and market statistics 2009. Available via DIALOG http://www.estif.org/fileadmin/estif/content/market_data/downloads/2009%20solar_thermal_markets.pdf. Cited 10 Mar 2015

European Solar Thermal Industry Federation (ESTIF) (2012). Solar thermal markets in Europe. Trends and Market Statistics 2011, Brussels, June

Faghri A (1995) Heat pipe science. Taylor & Francis, London

Ghafoor A, Munir A (2015) Worldwide overview of solar thermal cooling technologies. Renew Sustain Energy Rev 43:763–774
Gawlik K, Christensen C, Kutscher C (2005) A numerical and experimental investigation of low-conductivity unglazed, transpired solar air heaters. J Sol Energy Eng Trans ASME 127:153–155
Hastings SR, Morck O (2000) Solar air system: a design handbook. UK: James and James Science
IEA (2012) Technology roadmap: solar heating and cooling. International Energy Agency, Draft report
IEA (2009) Solar air-conditioning and refrigeration. International Energy Agency, Technical report Task-38
Jakob U, Spiegel K, Pink W (2008) Development and experimental investigation of a novel 10 kW ammonia/water absorption chiller–chillii® PSC for air-conditioning and refrigeration systems. In: Proceedings of the 9th internal IEA heat pump conference, Zurich, Switzerland, 20–22 May
Jakob U, Pink W (2007) Development of an ammonia/water absorption chiller–chilli PSC–for a solar cooling system. In: Proceedings of the international conference solar air conditioning, OTTI, Tarragona, Spain, 18–19 Oct, pp 440–445
Jakob U, Mittelbach W (2008) Development and investigation of a compact silica-gel/water adsorption chiller integrated in solar cooling systems. VII Minsk international seminar "Heat pipes, heat pumps, refrigerators, power sources", Minsk, Belarus, 8–11 Sept
Kalkan N, Young EA, Celiktas A (2012) Solar thermal air conditioning technology reducing the footprint of solar thermal air conditioning. Renew Sustain Energy Rev 16:6352–6383
Kalogirou SA (2004) Solar thermal collectors and applications. Prog Energy Combust Sci 30:231–295
Kalogirou S (2003) The potential of solar industrial process heat applications. Appl Energy 76(4):337–361
Karim MA, Hawlader MNA (2006) Performance investigation of flat plate, v-corrugated and finned air collectors. Energy 31:452–470
Karsli S (2007) Performance analysis of new-design solar air collectors for drying applications. Renew Energy 32:1645–1660
Luminosu I, Fara L (2005) Determination of the optimal operation mode of a flat solar collector by exergetic analysis and numerical simulation. Energy 30(5):731–747
Mauthner F, Weiss W (2014) Solar heat worldwide: markets and contribution to the energy supply 2012. IEA solar heating and cooling programme, June
Moschella A, Salemi A, Lo Faro A, Sanfilippo G et al (2013) Historic buildings in mediterranean area and solar thermal technologies: Architectural Integration vs Preservation Criteria. Energy Procedia, vol 42 pp 416–425
Moummi N, Youcef-Ali S, Moummi A et al (2004) Energy analysis of a solar air collector with rows of fins. Renew Energy 29:2053–2064
Munari Probst MC, Roecker C, Schueler A (2004) Impact of new developments on the integration into facades of solar thermal collectors. In: Proceedings EUROSUN, Freiburg im Breisgau, Germany
Munari Probst MC, Roecker C (2007) Towards an improved architectural quality of building integrated solar thermal systems (BIST). Solar Energy 81(9):1104–1116
OECD/IEA (2008) Energy technology perspectives: Scenarios and strategies to 2050, Paris
REN21 (2014) Renewables 2014 global status report. Available via DIALOG http://www.ren21.net/portals/0/documents/resources/gsr/2014/gsr2014_full%20report_low%20res.pdf. Cited 10 Mar 2015
Shams NSM (2013) Design of a transpired air heating solar collector with an inverted perforated absorber and asymmetric compound parabolic concentrator. Thesis submitted in fulfilment of the requirements for the degree of Doctor of Philosophy to the Dublin Institute of Technology, March 2013

Shukla A, Nkwetta DN, Cho YJ et al (2012) A state of art review on the performance of transpired solar collector. Renew Sustain Energy Rev 16:3975–3985

SOLAIR (2009) Market report for small and medium-sized solar air-conditioning appliances. Analysis of market potential. Available via DIALOG. http://www.solair-project.eu. Cited 10 Mar 2015

Tamasauskas J, Kegel M, Sunye R (2014) An analysis of solar thermal technologies integrated into a canadian office building. Energy Procedia 48:1017–1026

The German Solar Energy Society (2007) Planning and installing solar thermal systems: a guide for installers, architects and engineers. James and James, UK

Tripanagnostopoulos Y, Souliotis M, Nousia T (2000) Solar collectors with colored absorbers. Sol Energy 68:343–356

Tsoutsos T, Aloumpi E, Gkouskos Z et al (2010) Design of a solar absorption cooling system in a Greek hospital. Energy Build 42:265–272

Van Decker GWE, Hollands KGT, Brunger AP (2001) Heat-exchange relations for unglazed transpired solar collectors with circular holes on a square or triangular pitch. Sol Energy 71:33–45

Wang RZ, Zhai XQ (2010) Development of solar thermal technologies in China. Energy 35(11):4407–4416

Xiaowu W, Ben H (2005) Exergy analysis of domestic-scale solar water heaters. Renew Sustain Energy Rev 9:638–645

Yang L, He B, Ye M (2014) The application of solar technologies in building energy efficiency: BISE design in solar-powered residential buildings. Technol Soc 38:111–118

Yousef BAA, Adam NM (2008) Performance analysis for flat plate collector with and without porous media. J Energy Southern Africa 19:32–42

Chapter 18
Solar Energy for Building Supply

Theocharis Tsoutsos, Eleni Farmaki and Maria Mandalaki

Abstract Photovoltaics (PVs) is an established technology, but there are still potential applications that require further development, and undoubtedly appropriate architectonic integration. This chapter is a short guide for architects and engineers, providing an overview of the main types of PVs and of the alternative ways that PV modules can be integrated into buildings by using various technical arrangements to replace existing construction elements by PV modules in roofs, façades, and other parts of the buildings. Presented are the main types of PV cells—depending on the semiconductor material (single-crystal, multicrystalline silicon, amorphous, thin film), on the type of junction (homojunction and heterojunction), on the method of manufacture (thin film, stacked, vertical multifunction), and on the devices of the system that utilize solar radiation. The main types of PV building integration (roof, façade, sun screening) are discussed at the end of the chapter.

18.1 Introduction

The enormous development of photovoltaic (PV) technology in the past few years is due to—among other factors—certain special characteristics of PV systems that distinguish them from other renewable energy sources, such as (Roman et al. 2008a, b):

- They do not produce greenhouse gases during their operation and they are noiseless, making their installation even in urban areas (houses, public buildings, etc.) attractive. There is also the possibility of being placed on the roofs of buildings (building attached PV panels—BAPV) or even integrated into the structure of buildings as a façade or shading elements (building integrated

T. Tsoutsos (✉) · E. Farmaki · M. Mandalaki
Renewable and Sustainable Energy Lab, Technical University of Crete, Chania, Crete
e-mail: theocharis.tsoutsos@enveng.tuc.gr

S.-N. Boemi et al. (eds.), *Energy Performance of Buildings*,
DOI 10.1007/978-3-319-20831-2_18

PV—BIPV), resulting in the reduction (or elimination) of the potential aesthetic nuisance. They require minimal maintenance and have a long lifespan (20–30 years).
- They are able to operate as stand-alone, autonomous, or grid-connected.
- They can be combined together in a modular way, generating various kinds of power from a few tens of mW to MW (which increases their unlimited range of applications).
- They can be combined with other energy forms (particularly renewable, such as winds) to operate as hybrid systems.

The environmental impact and the esthetics of the integration of PVs are important factors. Tsoutsos et al. (2005) point out the potential of PVs as a method for reducing the building's environmental impact (visual, noise, pollution, waste management, economic impact), making them more cost-effective.

PV is an established technology, but there are still potential applications that require further development, and their architectonic integration as BIPVs is undoubtedly one of them. Additionally, BIPV requires an additional effort to encourage the designers, the engineers, and the stakeholders to increase the number of successful examples, and to solve legislative and normative issues, etc. BIPVs can provide a new approach and uses for solar energy, such as façades, roofs, and sun screening or sun shading components can be replaced altogether by integrating solar modules. Obviously the energy production by using the envelope of the building is an eco-friendly action.

The main question is if the integration of PVs in the building creates new esthetics and if it can be adapted to the architectural language of the building. What are the alternatives to the esthetic result? Can these types of technologies be "adapted" to an existing urban environment?

This chapter is a short guide for architects and engineers, providing an overview of the main types of PVs and of the alternative ways that PV modules can be integrated into buildings, by using various technical arrangements to replace existing construction elements by PV modules in roofs, façades, and other parts of the building.

18.2 PV Modules and Cells

18.2.1 Electricity Production

In order to produce electricity from silicon (Si) cells, current must flow from the positive to the negative terminal (the same way as in batteries). For this purpose, PV cells are composed of two layers—a positively and a negatively "doped" one. Light shining on the cell generates a voltage between the two layers that appears at the terminals. A single cell generates only a limited amount of electrical power (Fig. 18.1).

Each cell has an estimated output of just 2.5–4.0 Wp, so they are connected together to form modules that are interconnected to form a complete PV generator

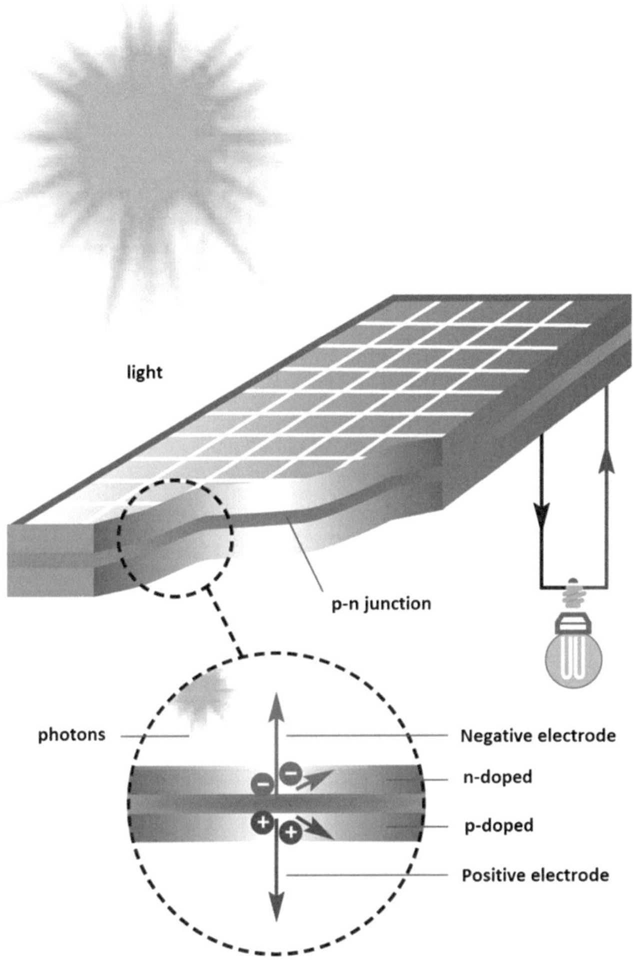

Fig. 18.1 The operation of the PV cell (commons.wikimedia.org)

(Fig. 18.2). The solar modules are contained in a frame and covered by a glass plate to protect them against external effects. Before the electricity produced by the solar module is sent to the local supply grid, it has to be converted from direct current (DC) to alternating current (AC) by an inverter.

18.2.2 The Components

The *solar generator* consists of a specified number of PV modules, depending on its manufacturing specifications. The support structure of the module can be

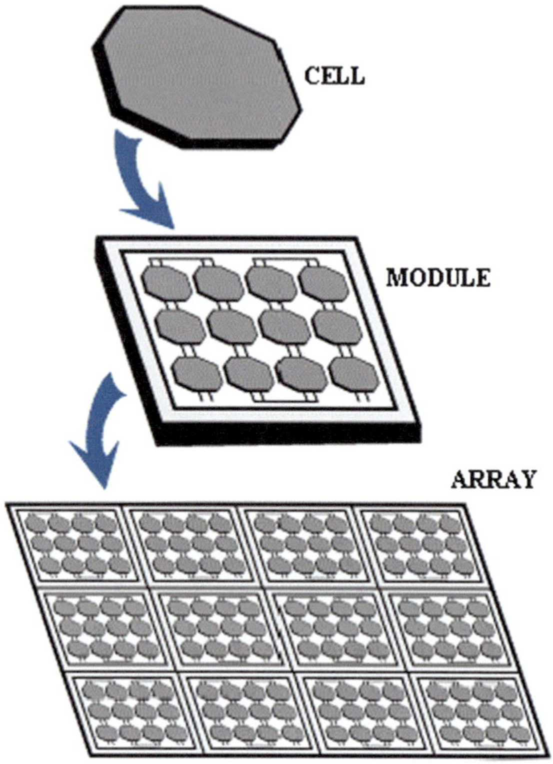

Fig. 18.2 PV cell, module and array (commons.wikimedia.org)

directly secured to the building so that no major change of the existing roofing is required. Modules mounted on flat roofs are usually fitted on stands for optimum positioning, but this is by no means a risk for potential damage to the roof covering. Obviously, by using different techniques, there is an optimum solution for any form of roof structure.

The *connection cables* for the solar modules are protected by a weather- and UV-resistant sheath, also fitted with pluggable connectors. This is very practical because it simplifies installation and prevents an accidental reverse of polarity in the connections.

The *inverter* converts the DC voltage produced by the solar cells into AC voltage that can be fed into the existing supply grid. The inverter operation is completely automatic, i.e., it switches on at dawn as soon as electrical energy is generated and switches off at dusk. After the inverter, the generated electricity passes through a feed meter that is used to calculate revenues and/or netmetering credits (Fig. 18.3).

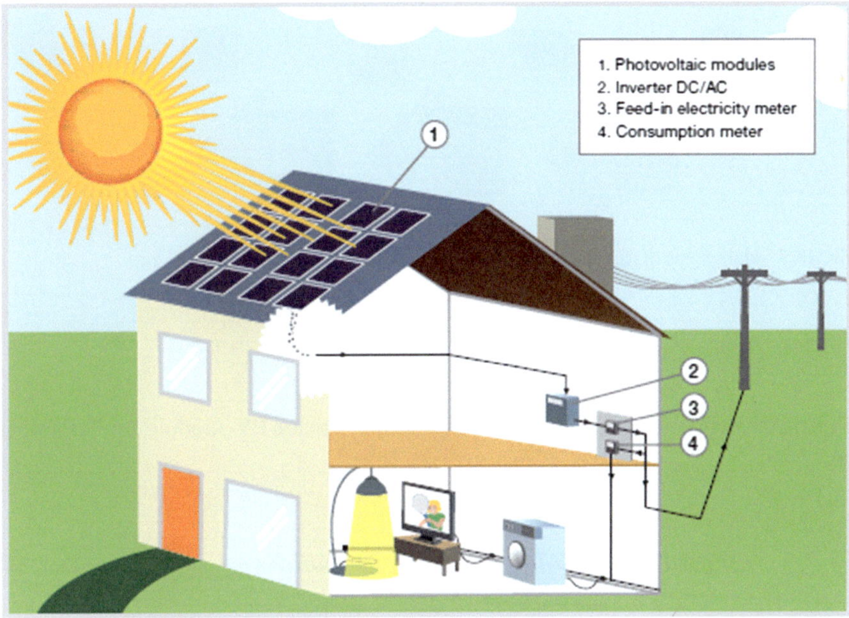

Fig. 18.3 Grid connected system (EPIA)

18.2.3 Dependency of Energy Generated on System Installation

The quantity of electricity produced depends on the geographical district, alignment, and tilt angle. A yield of approximately annually 700–1500 kWh/kWp a year can be expected with a space requirement of approximately 10 m^2 (Fig. 18.4).

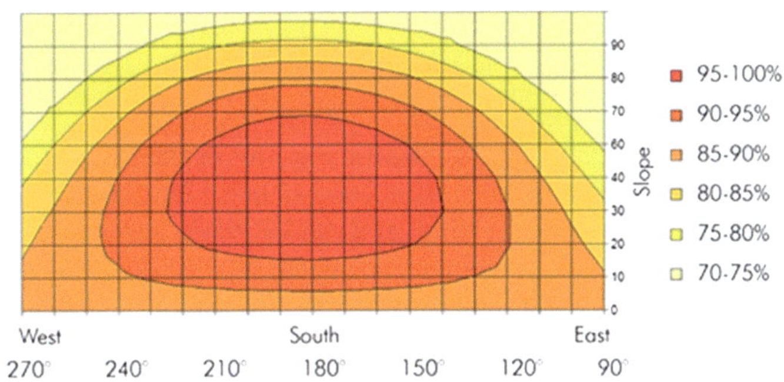

Fig. 18.4 Change in performance due to panel orientation (www.heatshine.com)

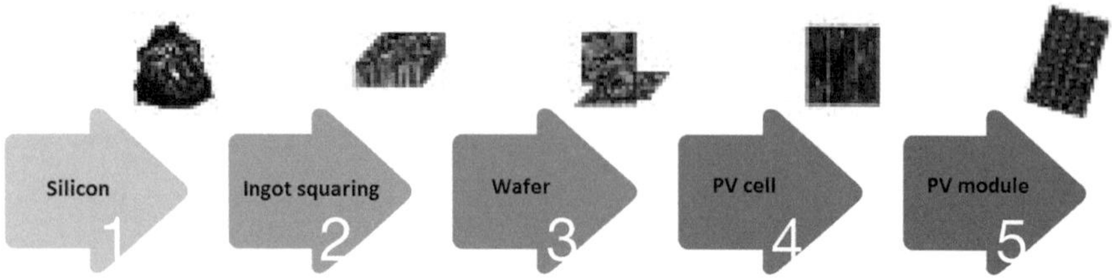

Fig. 18.5 Main steps of manufacturing a crystalline silicon PV module (EPIA)

18.2.4 Production

The main steps of manufacturing a crystalline silicon PV module are shown in Fig. 18.5.

18.2.5 Integration of Solar Modules in Buildings

Although there has been a major increase in interest in BIPVs, the number of their installations is still limited in contrast to the number of common PV plants. The main reasons are:

- The high cost of integrating PV into façades and roofs, although façades are often built from marble and other high-cost materials
- The lack of knowhow of the new technology
- The effort involved in planning and configuring a BIPV system is no different than that required to build a "normal" glass façade/roof or a standard PV plant. It can almost be planned and built like a normal glass façade or roof, and electrically connected like a conventional solar plant. However, BIPVs offer a chance for double exploitation of building envelopes: protect the climate change, and produce ecofriendly energy.

18.3 Types of PV Cells

Although silicon PV cells comprise the vast majority of commercial PV frames, innovative and different PV types are continuously being developed as regards the semiconductor material, its processing, and the way of construction. This is an ongoing process that aims to reduce the manufacturing and operating costs of PV cells and increase their performance. The various types of PV cells are briefly presented below, although the number of types is not limited, as changes and developments in this area are constant. For the best classification of the different types of PV cells, they can be distinguished according to:

- The basic semiconductor materials that are manufactured today
- The type of junction
- The method of manufacture
- The devices that utilize solar radiation

18.3.1 Types of PV Cells Depending on the Semiconductor Material

Following are the most common PV modules, which are those made from silicon:

Single-crystal silicon PVs

The monocrystalline silicon at high purity levels has been used from the 1950s in the semiconductor industry (for electronic applications); this manufacturing method was inherited by the manufacturing PV industry. This method is expensive, which is why the share of monocrystalline silicon PV cells is very small in the PV market (since polycrystalline or amorphous silicon PV are more beneficial in economic terms). The material's thickness in the wafer is relatively large (~300 mm). Their performance under laboratory conditions is ~20–24 %, while in commercial applications it is ~14–18 %. The color of the PV cells is usually dark blue.

Polycrystalline or multicrystalline silicon

In these PV cells, the semiconductor material is polycrystalline silicon, so that the manufacturing process of wafers is different and cheaper than the manufacturing of monocrystalline PV cells. But their performance is lower, as the discontinuities in the crystal lattice due to their structure increases the rate of electrons-holes reconnection. Lately, however, results of research has enabled increasing their performance, and in combination with the relatively low manufacturing cost they have taken over a very large market share. Their performance under laboratory conditions is ~17–20 %, whereas in commercial applications it is ~10–14 %. The color of the PV cells is typically light blue.

Amorphous silicon

The amorphous silicon, used in these PV modules, is manufactured so that the raw material significantly loses its crystalline structure (thus it is named "amorphous" rather than "crystalline"), but it sufficiently maintains the semiconductor properties. Of course, the structure favors, to a great extent, the hole-electron reconnections resulting in a lower performance of the PV modules. But what is lost in yield can be gained in cost and ease of processing, as the material (amorphous silicon) is pliable and can be used in manufacturing thin-film solar cells, which have a very small thickness and have other advantages (such as flexibility because it is elastic) and have many commercial applications in roofs, etc.). Also, the manufacturing technology of amorphous silicon PV cells is used to fabricate composite PV cells (e.g., Tandem) consisting of two or three layers of different semiconductor compositions, each having a different energy gap in order to be able to utilize a larger range of the solar spectrum. For example, the three contact PV modules from silicon, carbon and germanium a-SiGe (Eg ~1.4 eV), a-SiC (Eg = 1.85 eV) have a stable efficiency ~13 % (Fragkiadakis 2004). An additional advantage is that their performance is slightly affected by temperature.

PV thin film cells of cuprum, indium, gallium, selenium (thin film Cu(InGa)Se₂ [CIGS])

The semiconductor material composed of the above-listed chemical elements is selected based on its ability to utilize solar radiation because of the value of the energy gap that these materials have and that has led to yields of 20 % in the laboratory (greater than the respective thin-film PV cells that are made of silicon and have a yield of ~12 % in laboratory conditions).

PV thin film cells CdTe

The semiconductor that is created from these two chemical elements has an energy gap Eg equal to 1.5 eV that is very close to the ideal energy gap for the absorption of solar radiation in the air mass spectrum. The PV thin film cells manufactured have a laboratory efficiency that reaches 16 %. The advantage of such PV cells (apart from their performance) is that to manufacture them, much less semiconductor material is used—compared to silicon (film thickness is much shorter), with obvious economic and technical advantages. But they have two major disadvantages: their unstable performance and the environmental risks associated with cadmium-tellurium.

18.3.2 Types of PV Cells According to the Type of Junction

The PV modules can be separated as homojunction and heterojunction cells; the term "junction" means the p-n junction.

Homojunction PV cells

In these PV cells, the semiconductor base material remains the same throughout the whole p-n junction and they may be changes only in the concentration of impurity. Therefore, the energy gap is stable over the entire length of the p-n junction. These kinds of PV modules are the PV silicon modules presented above.

Heterojunction PV cells

In these PV cells, the semiconductor base material is changed within the extent of the p-n junction and therefore the energy gap is different at different (spatially) areas of p-n junctions; this means that the absorption of radiation (photons) may occur at two different frequencies (corresponding to different energy gaps). Thus, the share of photons (of solar radiation) that can (theoretically) be absorbed is greater than the share of a similar junction. Usually the material with the wider energy gap is (structurally) on top of the PV cell and then follows the one with the narrowest energy gap. Initially, heterojunction PV cells were constructed with two different energy gaps (such as PV cells Cd /CdTe or $Ga_{1-x}Al_xAs/GaAs$), but recently heterojunctions with three different energy gaps were announced.

18.3.3 Types of PV Cells According to the Method of Manufacture

The most common types of PV cells are the thin film, stacked, and vertical multijunction:

Thin film PV cells

One type of such PV cell was mentioned above (e.g., PV amorphous silicon). The semiconductor material of such PV cells may be amorphous silicon (such PV cells constitute the majority of commercial applications), or other semiconductors such as CdTe, $CuInSe_2$ (CIS), etc. Their main characteristics are the lower weight (due to the smaller amount of semiconductor material needed), the range of applications they can be used at (because of the flexibility they can show), and the increased difficulty of the various construction techniques (especially when the material is not silicon).

PV stacked cells

These are heterojunctions PV cells, which means that along the p-n junction, semiconductors are arranged in order of two (or more), with different energy gaps (the one with the wider energy gap is positioned externally, and then the one (or more) with the narrowest).

PV vertical multifunction cells

These are PV cells in which hundreds of p-n microjunctions are created and placed next to each other.

18.3.4 Types of PV Cells According to the Devices of the System that Utilizes Solar Radiation

PV cells with a concentrating radiation device

The use of PV concentrators is a method of increasing the power generation of integrated PVs. In these modules, the incidental solar radiation is concentrated through the collecting system of mirrors and prisms. The concentrated (condensed) solar radiation increases the efficiency of PV cells. To maximize the performance of these systems, the solar radiation should incident vertically on the surface of the PV cells; to achieve this, the presence of solar trackers is necessary, as well as a cooling system to prevent the yield from decreasing by the temperature increase resulting from the concentrated solar beam. The presence of trackers and cooling systems increases the overall construction and maintenance costs of these systems and reduces their reliability. Comparing these disadvantages with the main advantage of the increased performance is the main criterion for their successful dissemination in the future.

Thermophotovoltaic PV cells

These cells are designed with devices that can utilize the low frequency radiation emitted by a surface when heated by solar radiation. The solar radiation is concentrated and does not incident on the PV cells but on a highly absorbent surface that absorbs solar radiation and gets heated. The surface (depending on the temperature) emits a low frequency radiation (having a maximum in the infrared section of the spectrum). This radiation incident on the PV cell, which uses as photoconductive material a semiconductor with a small energy gap that matches with the energy of the photons of infrared radiation (incident upon the PV cell), is thus strongly absorbed. The performance that has been reported in laboratory applications is ~40 % (Twidell and Weir 2006). These systems could be used in order to exploit the excess heat in industrial applications.

Organic or dye PV sensitive cells

The essence of the operation of these PV cells lies in the fact that they utilize the conversion of solar radiation into electrical energy through the process of photosynthesis, just as plant organisms do (e.g., dye-sensitive cells) (Tsoutsos and Kanakis 2013).

Nano-PV cells

Nano-PV cells are often called "third-generation PV cells." According to this approach, first-generation PV cells are considered to be those of crystalline silicon, while the second generation includes the thin film PVs. In nano-PV cells, there is no separation between the layers, which is characteristic of the first two generations of PV cells. On the contrary, the material of nano-PV is uniform and is the result of mixing printable and flexible polymeric materials with conductive nano-materials. This type of PV cell is expected to become commercially available in a few years and dramatically reduce the high cost of PV cells.

18.3.5 Semitransparent Modules (Crystalline Glass-Glass Module)

The glass-glass modules are more popular for architect integration, due to their design and the fact that they can be manufactured as insulated glass. The front

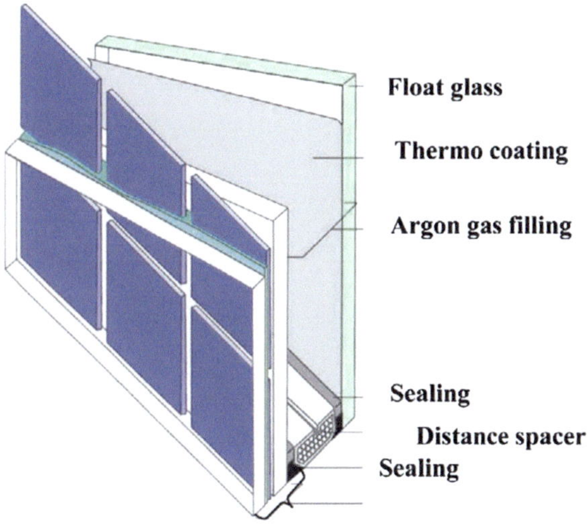

Fig. 18.6 Structure of an insulated PV module for a PV system integrated into the building (Scheuten Solar) In the figure itself, is it "thermo coating" or "thermal coating"?

position of an insulated glass-glass module consists of an extra white pane and a float glass. Between these two glass panels is a special resin with the solar cells embedded. The additional insulation part consists of a distance support with a seal on either side, and another pane with a thermal coating (Fig. 18.6).

18.4 *I–V* Curve and Losses

18.4.1 Characteristic I–V Curve of a PV Cell—Power Curve

Figure 18.7 shows the dependence of the current (I) flowing through the circuit of a PV cell as a function of the voltage (V) generated at its ends. This graph is a typically characteristic I–V curve of an illuminated PV cell and represents its identity.

As seen from the graph below, the current I starts from a maximum value (I_{sc}), remains nearly constant (a small decrease) for increasing values of voltage up to a point (which is about the maximum power point-MPP-) (V_{mp}, I_{mp}), and then starts decreasing vigorously until the voltage reaches the maximum value of Voc where the current is zero.

The maximum value of the current is called short-circuit current (I_{sc}) and it is the value that the current has in the ideal case that the value of the resistors is zero (then the diode current ID $= 0$ and I is equal to I_{ph}, which means that $I_{sc} = I_{ph}$). The maximum voltage is called "open circuit voltage V_{oc}" and is equal to the voltage present at the ends of the PV cell when the photocurrent I_{ph} is equal to the diode current ID. Then the current is zeroed $I = 0$ and the voltage present in the circuit is the V_{oc}. This is the voltage at the PV cell's ends when it is not illuminated (for example, during the night).

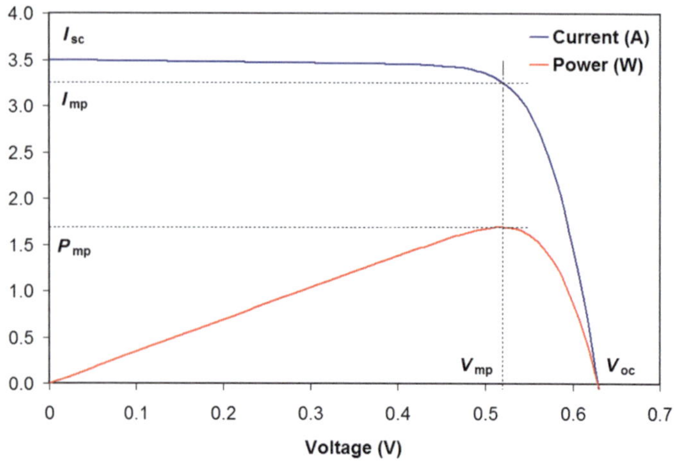

Fig. 18.7 Characteristic *I*–*V* curve and power curve of a PV cell (commons.wikimedia.org)

The electrical power that a PV element can produce (as any other electrical source) depends on the product of voltage *V* and intensity *I* of the current:

$$P = V \cdot I$$

Therefore, at some point of the characteristic *I*–*V* curve, the values of voltage and current are such that the power generated (as the product $V * I$) will take the maximum power (P_m). This characteristic operating point of the PV circuit is called Maximum Power Point (MPP), while the *V* and *I* values corresponding to P_m are called maximum power voltage (V_m or V_{mp}) and maximum power current (I_m or I_{mp}) respectively. The graph that gives the dependence of the power produced by a PV cell to the voltage that is at its ends is shown in Fig. 18.7.

According to the rates of characteristic quantities, I_{sc}, V_{oc}, I_m, V_m of a PV cell, and the Fill Factor (FF) is equal to

$$\mathrm{FF} = \frac{V_m \cdot I_m}{V_{oc} \cdot I_{sc}}$$

The numerator of the fraction represents the maximum power P_m produced by a particular PV cell (in certain solar radiation and temperature values), while the denominator is the maximum theoretical power that this PV cell could produce if the voltage and intensity are maximum. The fill factor is a quality indicator of a PV cell, as high values of the fill factor (definitely greater than 0.7) indicate that the PV is energy-efficient.

Efficiency of PV cell and losses

The efficiency of a PV cell is an indicator associated with the percentage of incident solar radiation that the PV element can convert into electricity.

Table 18.1 The efficiency of specific PV cells in the laboratory (Markvart and Castaner 2003)

	Efficiency (%)	I_{sc} (mA/cm^2)	V_{oc} (V)	FF (%)
Single-crystal silicon PV cell (c-Si)	24.7	42.2	0.706	82.8
GaAs	25.1	28.2	1.022	87.1
InP	21.9	29.3	0.878	85.4
Crystal silicon multijunction GaInP/GaAs/Ge tandem	31.0	14.11	2.548	86.2
Thin film PV cells single junction				
GdTe	16.5	25.9	0.845	75.5
CIGS	18.9	34.8	0.696	78.0
Thin film PV cell multijunction				
a-Si/a-SiGe tandem	13.5	7.72	2.375	74.4
Photoelectrochemical				
dye sensitive PV cells TiO_2	11.0	19.4	0.795	71.0

Usually the performance of a PV cell (n) is defined as the ratio of the maximum power provided by the cell to the incident solar power

$$n = \frac{I_m \cdot V_m}{P_m} = \frac{(FF) \cdot I_{sc} \cdot V_{oc}}{P_{in}}$$

where

P_{in}: the incident solar power on the PV
I_m V_m: maximum electric power
and the remaining quantities are defined above.

The efficiency of PV cells continuously grows with the development of technology and research. Today the efficiency of commercial PV cells made of silicon reaches about 18 %, while for those made of other semiconductor materials it exceeds 20 %. Table 18.1 shows the performance of specific PV cells, both in the laboratory and in commercial applications.

The efficiency of a PV cell depends on several parameters. The main ones are the semiconductor material of which the PV cell is made, the density of the light (solar) radiation and the temperature of the PV cell (which is closely related to the ambient temperature).

18.5 Types of Building Integration

There are many alternatives to integrating PV into buildings. Generally speaking, there are three areas of a building where PV-modules can easily be integrated:

- the roof
- the façade
- the sun screening components

Figure 18.8 shows these different potential integration methods:

The roof

There are three alternatives for installing solar-modules on a roof (Fig. 18.9).

The most common way is not to integrate them into the building, but to add them to the surface of the roof (Fig. 18.10).

Another possibility is to integrate them directly into the roof (Fig. 18.11).

The third—and most fully integrated solution—involves making the PV-modules act as the roof itself (Fig. 18.12).

The façade

Ecofriendly solutions have rarely looked this good in practice: the shapes and colors of façade elements can be manufactured in a number of different ways in order to adapt perfectly to the appearance of the façade. The design in which solar cells are embedded in resin between two glass panes means that solar elements can be significantly larger than conventional components for the same production capacity. In that way they provide a wide range of design opportunities allowing the architect to develop his creativity.

Modern façades have various functions, such as heat protection, insulation glass, sun protection, and noise protection. Like roofs, there are three options for integrating solar modules into façades (Fig. 18.13). The solar cells can be

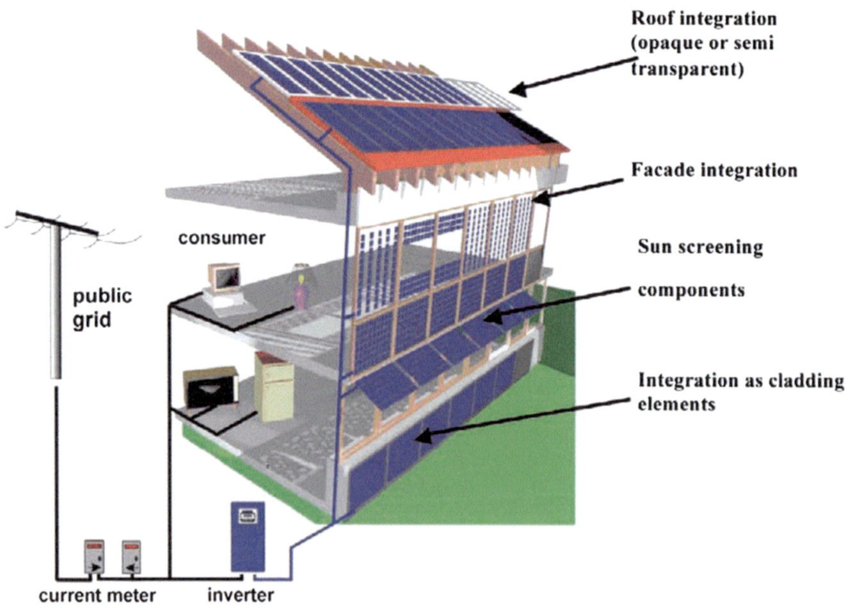

Fig. 18.8 Alternative integration methods (Roman et al. 2008a)

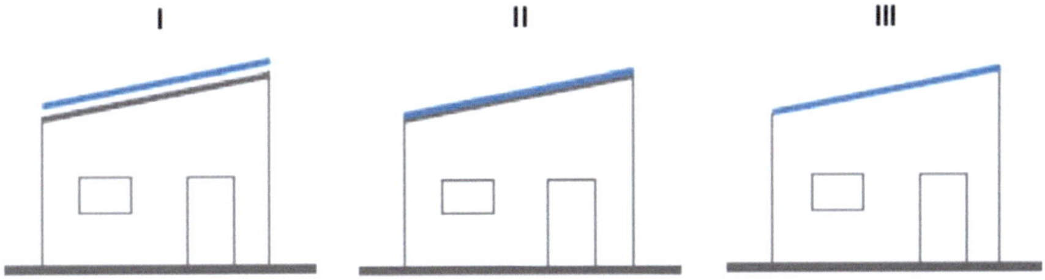

Fig. 18.9 Alternative types of PV on a roof (Landesgewerbeamt Baden Württemberg)

Fig. 18.10 A residential grid-tied PV installation in Glastonbury, Connecticut (NREL)

integrated in a cold façade, such as a curtain wall façade, or in warm façades (Fig. 18.14).

PV modules are fully integrated into the roof and the façade. The modules replace the insulated façade or roof, saving the cost of these structures in a new building. In Fig. 18.15, an existing façade was renovated and replaced with a very modern energy-producing system.

An important parameter of the PV modules integrated into façades is the function and the operating schedule of the building on which they are being installed. As PVs produce electricity only during the day, for this reason high energy loads of building should be limited during the daytime in order to reduce energy demand

Fig. 18.11 A roof-integrated PV system in San Jose, California (NREL)

Fig. 18.12 A PV system replacing the roof of the building (commons.wikimedia.org)

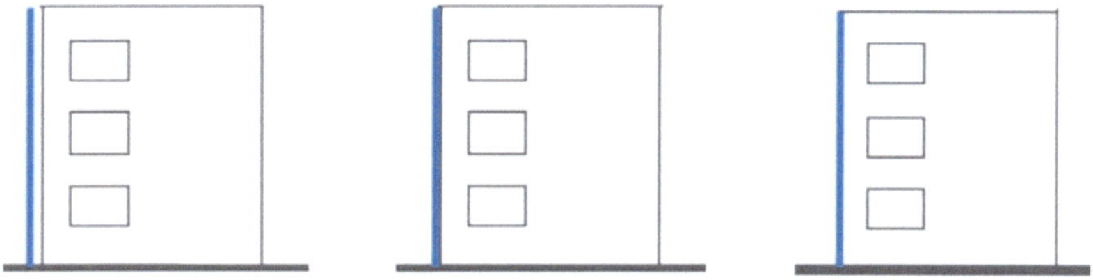

Fig. 18.13 Alternative types of PV into façades (Landesgewerbeamt Baden Württemberg)

Fig. 18.14 A curtain wall system integrated in the Future Business Center (FBC) in Cambridge (www.polysolar.co.uk)

from the grid and maximize energy produced by the PV. The schedule of organizations occupying office buildings is generally suitable for the function of the PVs, due to the fact that office buildings are mostly operational in the daytime, when energy production from the PVs is high.

In 2007, Yun et al. introduced the term "effectiveness of a PV façade (PVEF)" which is used to evaluate the overall energy performance of a PV façade. PVEF takes into account the energy produced from PV, the reduction in electric lighting needs due to daylight control, and the heating and cooling consumption. The formula they used is

$$\mathrm{PVEF} = \frac{L_{\mathrm{saving}} + E_{\mathrm{output}}}{H_{\mathrm{energy}} + C_{\mathrm{energy}}}$$

Fig. 18.15 Ökotec building in Berlin (Scheuten)

where

L_{saving} = the energy saving in lighting
E_{output} = the output energy of PV modules
H_{energy} = the energy spent for heating, and
C_{energy} = the energy spent for cooling

Sun screening components

Using PV for sun screening has two benefits (Mandalaki et al. 2012): on the one hand, you can save otherwise essential sun screens because the solar cells in the glass-glass module provide sufficient shade, so you can choose the see-through rate, depending on how much shade is needed; on the other hand, the PV modules produce electricity, which means a significant investment in the future. The big advantage to using PV modules as sun protection is that the best inclination for producing the most energy is the same angle that provides the most shade (see Fig. 18.16).

The integration of PVs in shading devices (SDs) is an intermediate solution falling between the BIPV and BAPV (Peng et al. 2011). This type of PV integration has the advantages of BIPV, is architecturally clean and attractive, and

Fig. 18.16 Example of PV for sun screening (PURE project)

offsets the cost of roofing, façade or glazing materials, but also keeps the advantage of BAPVs; in case they are damaged, the buildings' internal functioning is not affected.

The potential of replacing conventional building materials with PV structural materials (especially SDs) has been researched for Finland, Austria, Denmark, Switzerland, and Germany, as an esthetically appealing and energy-saving solution (Hestnes 1999). For Norwegian office buildings, some PV systems are analyzed from the esthetic point of view, according to PV types, color, and final surface. The thin film technology as a competitor to the crystalline silicon technology is being proposed as a lower-cost solution (Hermstad 2006).

At the State University of New York, Albany, PV modules have been used as sunshades providing 15 kWp of energy, simultaneously reducing cooling loads since 1996 (Eiffert and Kiss 2000). In order to make the shading devices more competitive in the market, the glass content of the PV louvers was minimized. Weight reduction was achieved by replacing glass components of PV modules (at least in part) with flexible membranes (Zentrum für Sonnenenenergie 2007). The only resulting disadvantage was that these types of flexible PV modules have a lower efficiency factor due to the type of material of the PV used. Amorphous silicon was used to substitute the glass PV components in order to make them flexible. Due to their disadvantage of low efficiency, the progress in market penetration of these types of systems was not the one anticipated—as can be seen, for example, in Korea, according to Hwang et al. (2012).

A simulation analysis of an office is presented by Bloem (2008) with 99 PV modules mounted on a horizontal spandrel enclosure on the south façade. The system works as a window-shading system producing 36 Wp in Standard Test Conditions. Natural ventilation was assumed in the module enclosure via vents in the upper and lower surfaces. He claimed that although PV modules cover only part of the energy demand, their performance can be improved by changing their inclination according to the season of the year. The combination of produced electricity with the improvements in the indoor quality conditions makes the use of BIPV on shading systems a very promising application of technology.

Another application that has been proposed on the market is the integration of PV louvers between two sheets of glass. The main advantage is that PV blinds are fixed between two sheets of glass and therefore cannot get dirty. The PV slats consist of tandem amorphous silicon cells deposited on glass, and have, according to company specifications, an efficiency of 6 %. Since amorphous silicon cells operate more efficiently at higher temperatures than crystalline silicon cells, the placement of the PV slats between the insulating glass sheets, where temperatures of 70 °C are easily reached, leads to very good performance (www.photon-international.com/products/products_01-03_syglas.htm).

Fig. 18.17 South-facing conditions for an office unit (Mandalaki et al. 2014)

The use of PV integration in shading systems has been promoted by various researchers. In Brazil, Cronemberger et al. (2012) argued that "for non-vertical façades (40º $\leq \beta \leq$ 90ºS) the solar potential represents between 60 and 90 % of the maximum global solar irradiation, even when facing south, indicating that the use of sloped building envelope surfaces, such as atriums and shading elements on façades and windows, should be promoted."

Various types of PV shading systems composed of glass and Si polycrystalline technology (the use of the most efficient technology of PVs) have been examined by Mandalaki et al. (2014), according to their efficiency in supplying a typical office unit and according to the most visual comfort conditions. Shading systems, such as Brise Soleil and Canopy, inclined, when duplicated in front of the window with integrated PV; they were proved to be efficient enough, compared with louver systems in terms of electricity supply for artificial light when needed for an office unit and provide the most comfortable visual conditions when facing south (Fig. 18.17).

18.6 Conclusion

In this chapter, we have shown various built examples and experimental works that use the integration of PV in buildings. The choice of technology should take into account not only the energy efficiency parameters, but also the appropriate and suitable solution that fits the esthetics of the building and the surrounding built environment. The integration of PV in buildings creates a specific architectural and esthetic result that should be integrated into the hole and should upgrade the shading of the building and its surroundings, especially in case of temperate climates; in these cases, the additional reflections of solar beams that PV can create should be carefully and thoroughly examined so that it does not increase the surrounding temperatures and the interior-exterior visual comfort.

References

Bloem JJ (2008) Evaluation of a PV-integrated building application in a well-controlled outdoor test environment. Build Environ 43(2):205–216

Cronemberger J, Caamano-Martın E, Vega Sanchez S (2012) Assessing the solar irradiation potential for solar photovoltaic applications in buildings at low latitudes–making the case of Brasil. Energy Build 55:264–272

Eiffert P, Kiss JG (2000) Building—integrated photovoltaic, design for commercial and institutional structures, a source book for architects. US Department of Commerce, Springfield

Fragkiadakis I (2004) Photovoltaic systems. Ziti, Greece (in Greek), pp 444

Hermstad K (2006) Architectural integration of PV in Norwegian Office Buildings. SINTEF Building and Infrastracture, Architecture and Building Technology, Trondheim

Hestnes A (1999) Building integration of solar energy systems. Sol Energy 67:181–187

Hwang T, Kang S, Kim JT (2012) Optimization of the building integrated photovoltaic system in office buildings-Focus on the orientation, inclined angle and installed area. Energy Build 46:92–104

NREL, images. http://www.nrel.gov. Accessed 01 Apr 2015
Mandalaki M, Zervas K, Tsoutsos T, Vazakas A (2012) Assessment of shading devices with integrated PV for efficient energy use. Sol Energy 86(9):2561–2575
Mandalaki M, Tsoutsos T, Papamanolis N (2014) Integrated PV in shading systems for Mediterranean countries: balance between energy production and visual comfort. Energy Build 77:445–456
Markvart T, Castaner L (2003) Practical Handbook of photovoltaics. Elsevier, Amsterdam
Peng Ch, Huan Y, Wu Z (2011) Building-integrated photovoltaics (BIPV) in architectural design in China. Energy Build 43:3592–3598
Roman E, Lopez JR, Alves L, Eisenschmid I, Melo P, Rousek J, Tsoutsos T (2008a), BIPV Technical Solutions and best practices. European Commission, DG Energy and Transport
Roman E, Lopez JR, Alves L, Eisenschmid I, Melo P, Rousek J, Tsoutsos T (2008b) Potential and benefits of BIPV. European Commission, DG Energy and Transport
Tsoutsos T, Kanakis I (2013) Renewable Energy Sources. Technologies and Environment, Papasotiriou (in Greek)
Tsoutsos T, Frantzeskaki N, Gekas V (2005) Environmental impacts from the solar energy technologies. Energy Policy 33:289–296
Twidell T, Weir AD (2006) Renewable energy resources. Taylor & Francis, UK
Various. www.photon-international.com/products/products_01-03_syglas.htm.s Accessed 20 Dec 2012
Yun GY, McEvoy M, Steemers K (2007) Design and overall energy performance of a ventilated photovoltaic facade. Sol Energy 81:383–394
Zentrum Fur Sonnenenenergie- Und Wasserstoff-Forschung Baden-Wurttemerg (ZSW) (2007) Lightweight PV louvres for multi-functional solar control and daylighting systems with improved building integration. Final report

Chapter 19
The State of the Art for Technologies Used to Decrease Demand in Buildings: Thermal Insulation

Stella Chadiarakou

Abstract Existing building stock in European countries accounts over 40 % of final energy consumption in the European Union (EU) member states, of which residential use represents 63 % (Hellenic Statistical Authority 2015) of total energy consumption in the building sector. Taking into consideration the significantly low rate of new construction and the implementation of the recasting of the Energy Performance Buildings Directive, the need for energy conservation measures in the existing building stock is of great importance. Consequently, the amelioration of a building's energy performance can constitute an important instrument towards the alleviation of the EU energy import dependency and comply with the Kyoto Protocol to reduce carbon dioxide emissions. Thermal insulation materials are the appropriate tool towards the mitigation of energy loss. There is a variety of materials based on their technical characteristics that fulfill the requirements of any construction; furthermore, there is great progress in innovative materials (thermo chromic, change face material, cooling). Moreover, the EU enacted legislation ensuring adequate quality for any construction. In this chapter we will investigate all available types of insulation materials, their technical parameters, and the best practices for the most well-known construction elements.

19.1 Thermal Insulation Materials

There are several types of insulation materials that are used in construction (Fig. 19.1); these are used for thermal, sound insulation, and fire protection. The most common ones are foamed materials (mainly expanded and extruded polystyrene) and mineral materials (stonewool and glass wool). Research on new, advanced materials is based on improving its thermal transmittance and the solar

---

S. Chadiarakou (✉)
Antisthenous 13, 54250 Thessaloniki, Greece
e-mail: shadiarakou@gmail.com

S.-N. Boemi et al. (eds.), *Energy Performance of Buildings*,
DOI 10.1007/978-3-319-20831-2_19

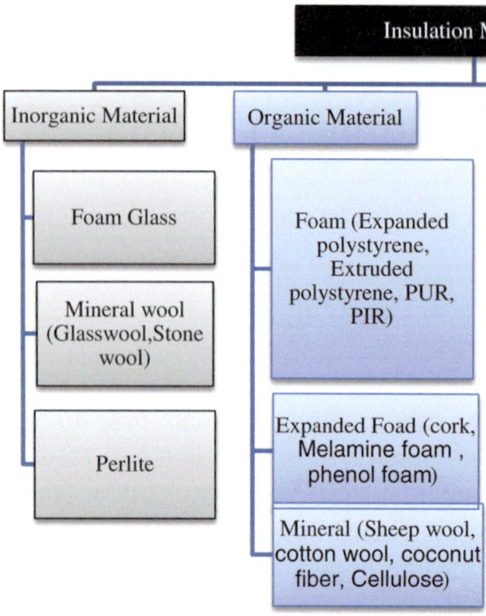

Fig. 19.1 Types of insulation materials

reflectance. Furthermore, the need for greater thickness promotes advanced production materials.

In order to understand its type of material we will study the most common ones in the next paragraphs.

19.2 Foamed Materials

We begin with foamed materials, and extruded and expanded polystyrene. Extruded polystyrene production is based on the extrusion of the mixture of raw material, with the appropriate blowing agents and fire retardants. It is important to note that since 2000, the European Union (EU) has banned the use of HFC. In addition, during the last few years, the implementation of the REACH directive has stricken more production of the extruded polystyrene in the scope of environment friendly production—products (Fig. 19.2). Extruded polystyrene products are in the shape of boards, with thickness ranging from 20 to 250 mm (Fig. 19.3). The extrusion gives the molecular structure of the XPS almost 97 % of closed shells, which is why XPS material has an extremely high resistance to water. Furthermore, the coherence of the structure provides a board with very high compressive strength.

The basic difference in the expanded polystyrene production is that the raw material is expanded and not extruded. Moreover, the blowing agent that EPS production uses (Fig. 19.4) is air, and not chemical gases. Moreover, with expansion we have the creation of large open shells that can be distinguished. The production

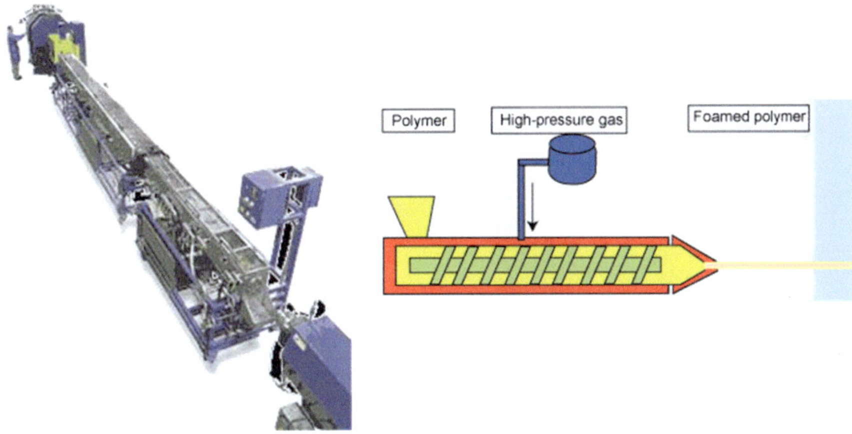

Fig. 19.2 XPS production

Fig. 19.3 XPS board

gives a cube that is formed in any shape requested (Fig. 19.5). The most important disadvantage of expanded polystyrene boards is the open structure shells, because in the presence of high humidity or a wet environment, it absorbs water, replacing the air of the shells. Furthermore, the water cannot get out of the shell and the material cannot be drawn off. The result is to destroy the good initial thermal conductivity factor and the mechanical characteristics.

19.3 Fibrous Materials

On the contrary, the production of the fibrous material is more complicated. The production of the glasswool is based on the melting of sand and other physical raw materials. The final product is a result of fiberizing (Fig. 19.6), except the minerals

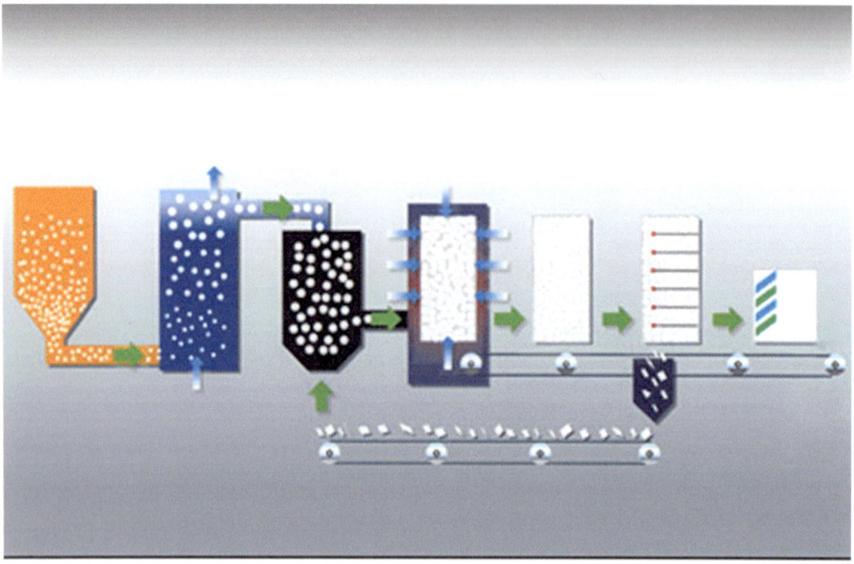

Fig. 19.4 EPS production

Fig. 19.5 EPS board

the addition of binder is added in order to acquire coherence and create a board in the shape of a roll (Fig. 19.7). The main characteristic of glasswool is the high usage temperature of 300 °C.

Furthermore, fibrous material, and glasswool in particular, behave well towards sound reduction; however, the disadvantage of glasswool is its sensitivity to water. Glasswool can absorb water because of its open structure, which can reduce the thermal conductivity factor. Another disadvantage of glasswool is the low mechanical properties; it should be noted that glasswool is mainly used in industry and not in building in Greece.

Stonewool production has similarities and dissimilarities like glasswool. The basic idea is melting of minerals in an electric (mostly) furnace (Fig. 19.8). Lava

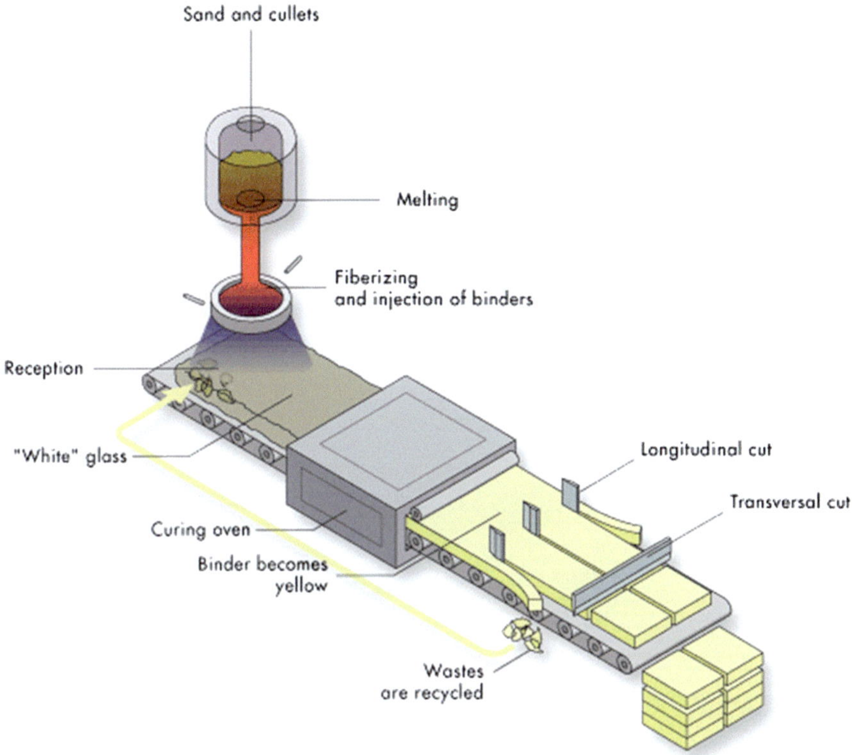

Fig. 19.6 Glasswool production

Fig. 19.7 Board and roll of glasswool

creates fibers through a spinning device, while adding binder the basic glue, oil for a smoother surface, and silicone for water repellency. After curing, stonewool has its final thickness and dimensions. It can be produced on boards or rolls

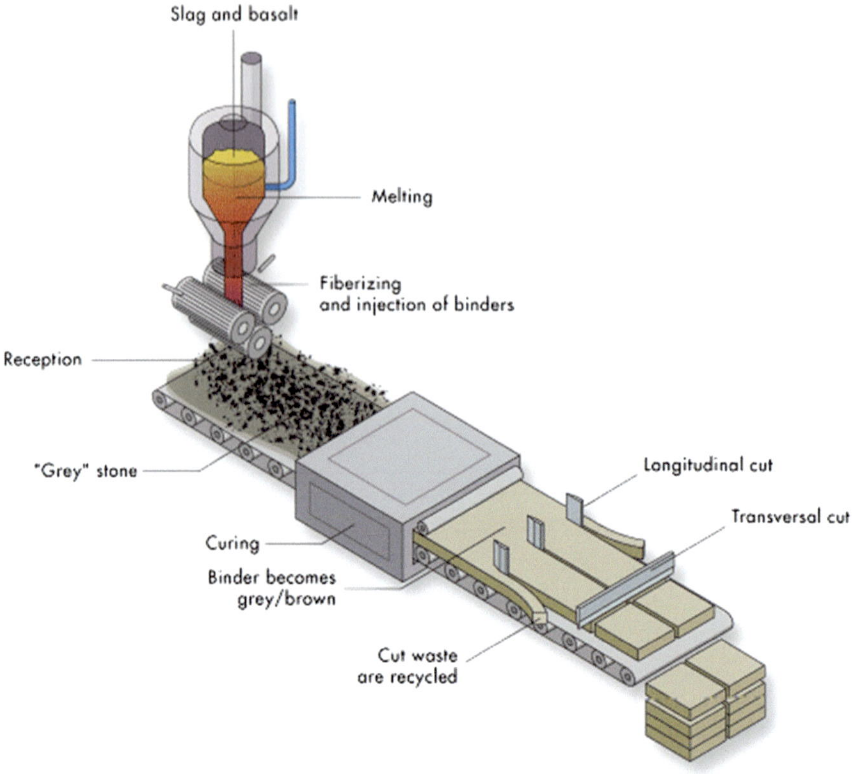

Fig. 19.8 Stonewool production

Fig. 19.9 Board of stonewool

(Fig. 19.9). As for the characteristics, stonewool has a very good thermal conductivity factor; moreover, certain types of it acquire very high (for stonewool) compressive strength. One of the most significant advantages of stonewool is its fire behavior. Stonewool is rated as A1 class (according to European Norm 13501-1),

Table 19.1 Technical characteristics of the products

Material	Organic			Inorganic		
	Expanded polystyrene	Extruded polystyrene	Polyurethane	Stone wool	Glass wool	Wood wool
Thermal conductivity (W/mK)	0.040–0.048	0.033–0.038	0.040	0.033–0.035	0.035–0.040	0.040
Fire reaction	F	E	F	A1	A1	A2
Compressive stress	Good	Very good	N.A.	Good	Poor	Good
Water absorption	Poor	Very good	N.A.	Good	Poor	Good
Sound absorption	None	None	None	Very good	Very good	Good
Mechanical properties	Good	Very good	N.A.	Good	Poor	Good
Passive ventilation	None	None	N.A.	Very good	Good	None

which corresponds to lighting a fire or contributing to the spreading of fire. Furthermore, stonewool does not release dangerous gases for humans. In addition, the service temperature of stonewool exceeds 700 °C and its melting point is more than 1000 °C. Another characteristic is the sound absorption and, consequently, sound reduction. Stonewool can absorb both airborne and impact noises, making it the proper solution for sound insulation. It faces problems with water, but not humidity because it has an open structure that allows the humidity to pass through, controlling the humidity in the building element. However, Stonewool can absorb water when it is put in place; thus it is not recommended for use when water can affect the insulation layer.

In conclusion, Table 19.1 collects and compares the most important technical characteristics of the products discussed before. It must be stressed that our market has all the possible insulation materials for any possible building solution.

19.4 Construction Solutions

We will investigate the typical construction applications that we come across in most buildings. It must be stressed that because of the level of seismicity in southern Europe, all buildings consist of an armed concrete structure that is usually about 25 cm thick. We have divided the buildings' elements into vertical and horizontal ones. Verticals are the walls (external, internal), beams, and columns, while horizontals are the roof (flat or sloping), floors, pilotis (floor or ceiling) and balcony. Furthermore, we will investigate the appearance of thermal bridges in any element.

19.4.1 Vertical Building Elements

External walls are usually constructed as a cavity with two brick layers, at least 6 cm (internal) and 9 cm thick (external—see Fig. 19.10). Between them, a layer of expanded or extruded polystyrene or stonewool is put in for thermal insulation.

After 2000, another façade construction solution was introduced (Fig. 19.11); external composite insulation systems, known as External Thermal Insulation Composite Systems (ETICS), appear in some constructions when retrofitting

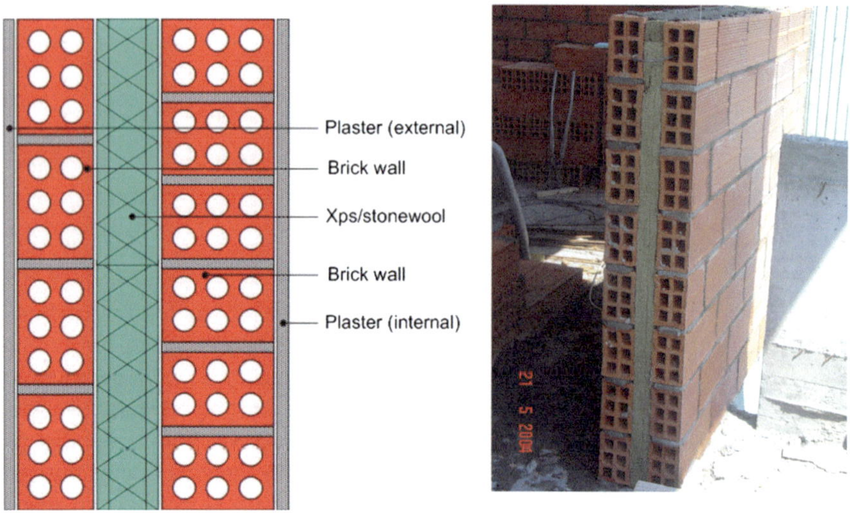

Fig. 19.10 Typical double brick façade

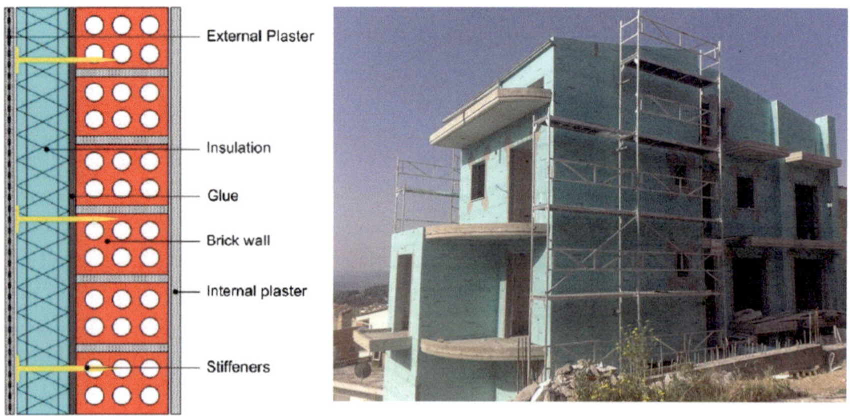

Fig. 19.11 Typical ETICS system

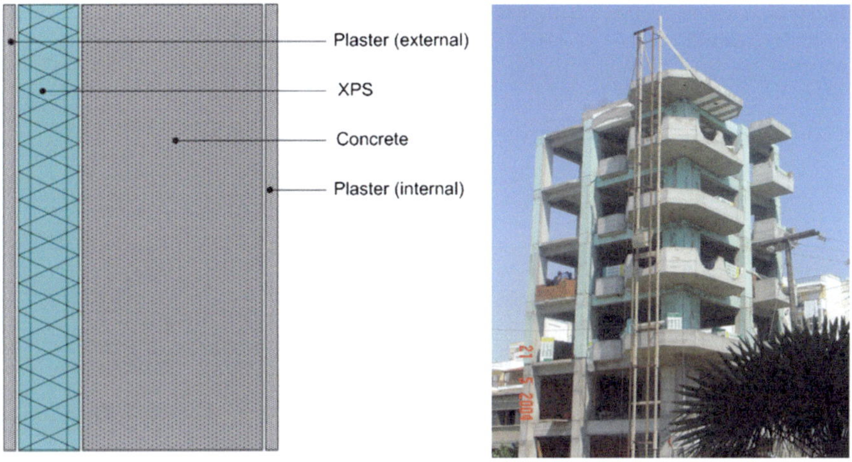

Fig. 19.12 Typical armed concrete insulation

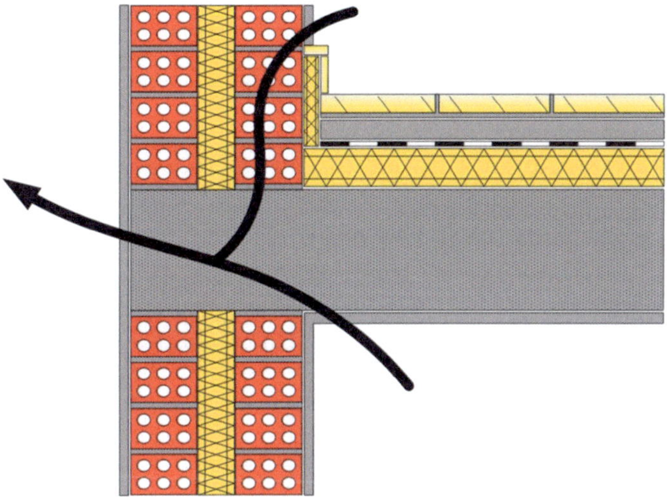

Fig. 19.13 Thermal bridge of double brick wall with pillar

insulations is mandatory. In the market, there are several certified systems—organic and inorganic—with extruded polystyrene and stonewool.

Pillars, piers, and slabs, with a minimum thickness of 25 cm, are insulated during the construction. Insulation is commonly placed inside the wooden frame, so when they remove it the insulation is stuck on the armed concrete (Fig. 19.12).

For example, from a thermal transmission point of view, we can find thermal bridges in many parts of the typical Greek constructions. In the double brick façade technique, there are linear thermal bridges in the joints between walls and piers, or pillars and "senaz," as can be seen in Figs. 19.13 and 19.14.

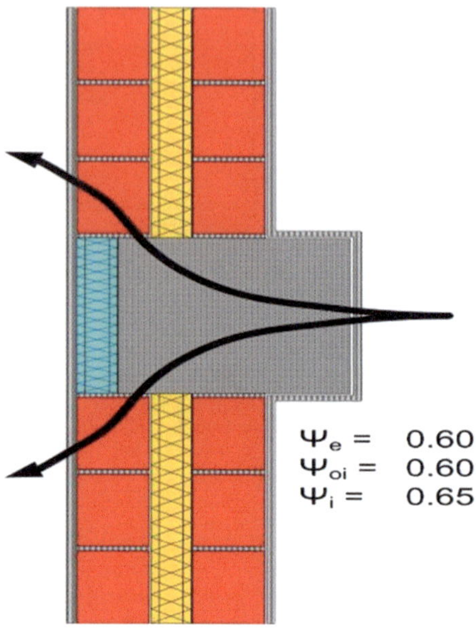

Fig. 19.14 Thermal bridge of double brick wall with "senaz"

19.4.2 Horizontal Building Elements

Almost all apartment buildings have accessible flat roofs, used as terraces. They are constructed as conventional flat roofs or as upside-down ones (inverted) (Figs. 19.15 and 19.16). The final layer is usually gravel or pavement plates.

The conventional roof used to be the only solution up until 1985, when the inverted roof was first introduced. The basic difference between the two solutions is the displacement of the waterproofing in relation to the insulation. Conventional roof insulation protects thermal insulation, putting all the stress on the waterproofing membrane; this unfortunately has a maximum lifespan of only 10–15 years. Inverted roofs consider the waterproofing membrane to be more sensitive and put all the stress on the thermal insulation. Studies have shown that inverted solutions give better fluctuation of the surface temperature and the apartment below. Rarely, in colder regions in northern Greece, apartment buildings with tiled roofs resting on inclined concrete slabs may be found.

Floors do not need to have any thermal protection if they separate heated apartments. The most common materials covering the floors are ceramic tiles (Fig. 19.17), marbles, or wood. Unfortunately, the lack of thermal protection in floors causes it to be lacking from the floor or ceiling of balconies (Fig. 19.18).

When the floor is characterized as a piloti floor, thermal insulation is mandatory. There are two different insulation solutions: the first (Fig. 19.19) is to insulate the pilots floor and the second to insulate the piloti ceiling (Fig. 19.20).

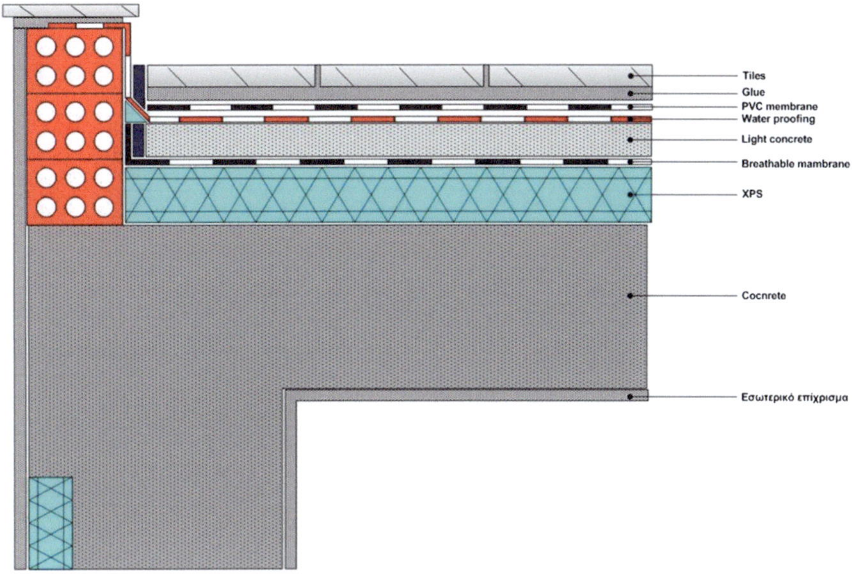

Fig. 19.15 Conventional roof application

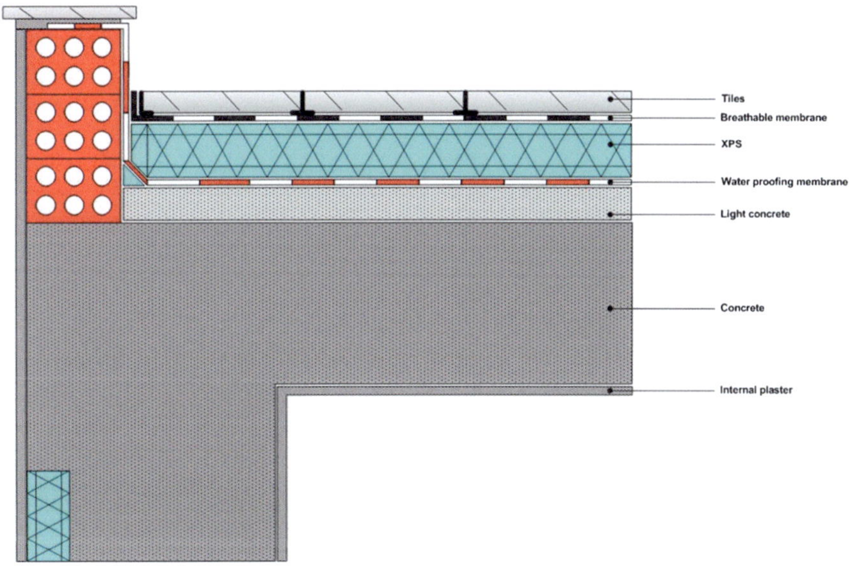

Fig. 19.16 Inverted roof application

The last building element that we characterize as horizontal is a sloping roof. This type of construction appeared after 1990, usually in stand-alone houses. Sloping roofs are rarely used in MFH. Insulation is usually put underneath the tiles and protected with a membrane (Fig. 19.21).

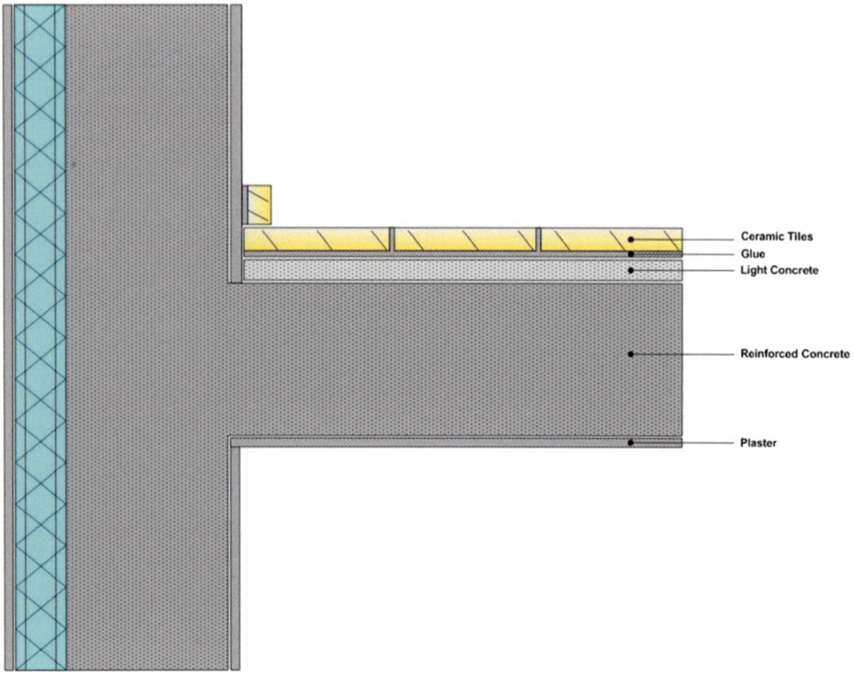

Fig. 19.17 Floor construction

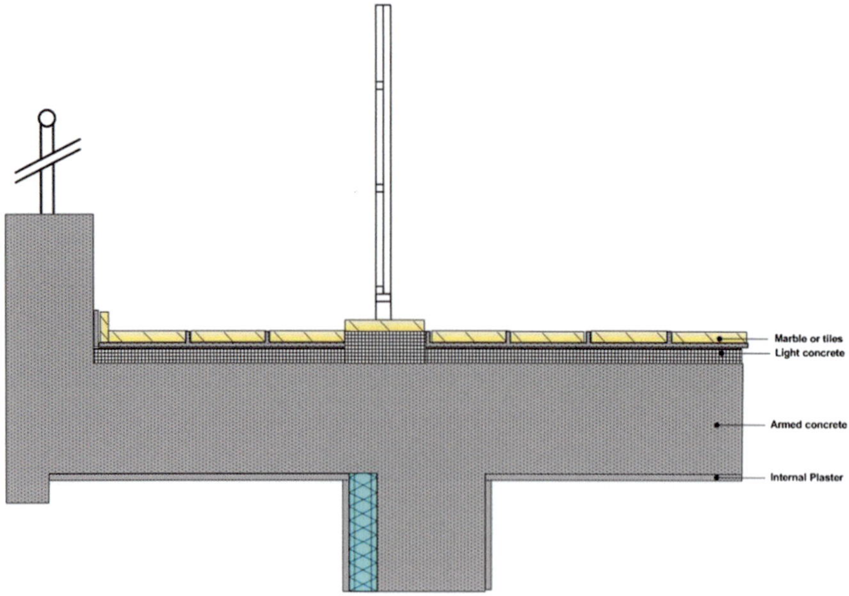

Fig. 19.18 Cross section of a balcony

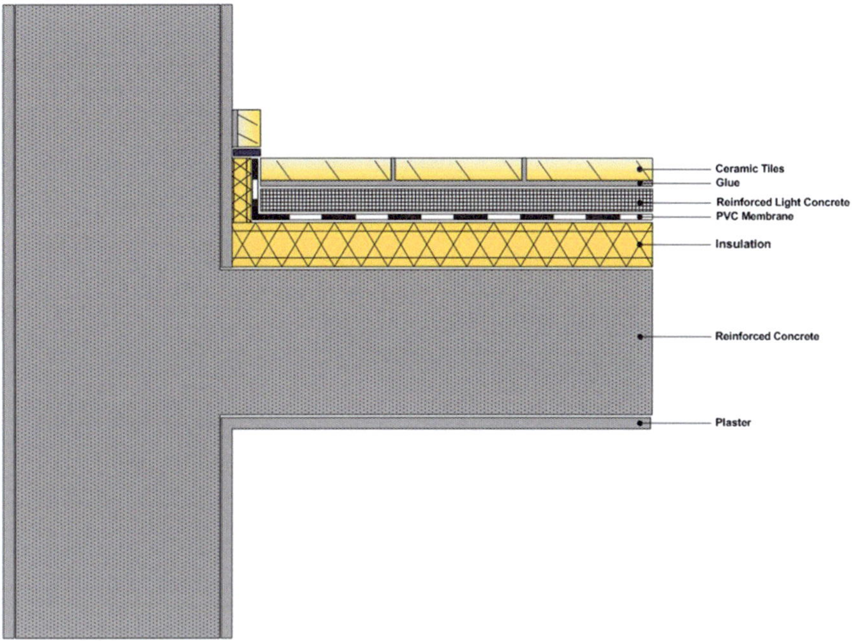

Fig. 19.19 Piloti floor insulation

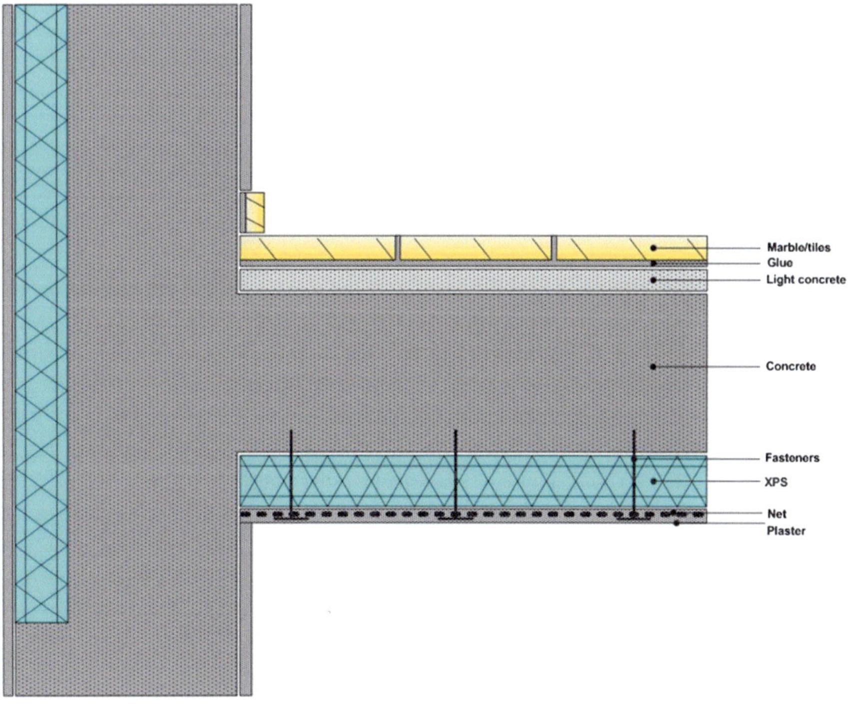

Fig. 19.20 Piloti ceiling insulation

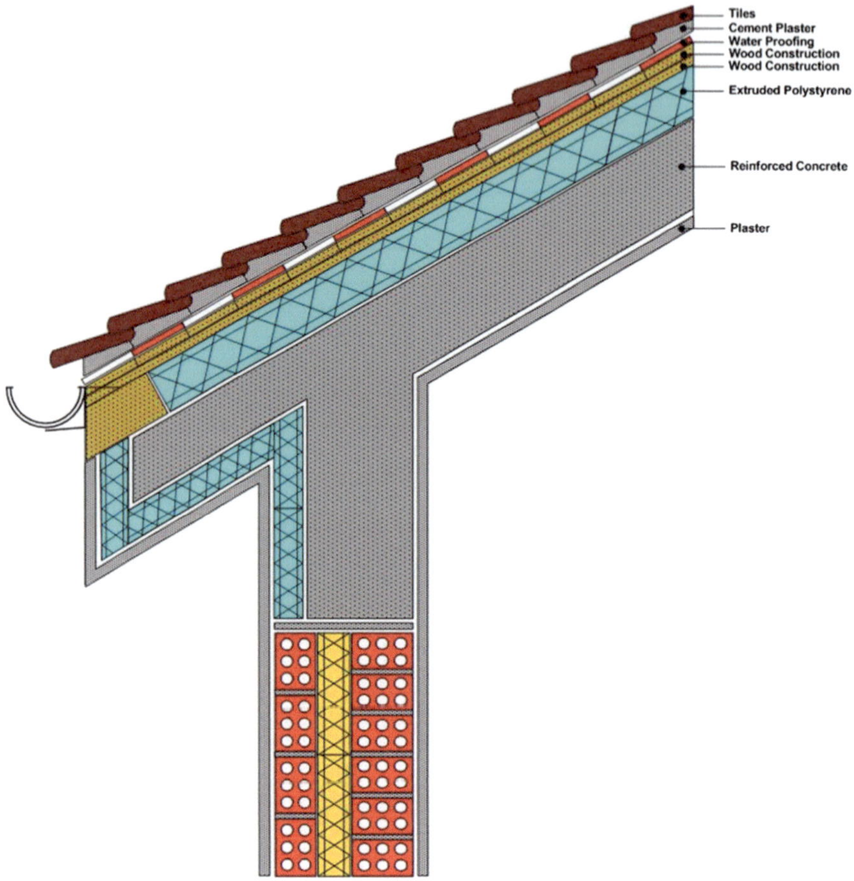

Fig. 19.21 Pitch roof insulation

As we have mentioned, in several cases vertical elements also appear in thermal bridges. First of all, there is the connection of the balcony with the wall element (Fig. 19.22). This type of thermal bridge is very important because the length of the balcony is usually the same as that of the total floor area. Thus, energy losses from this type of thermal bridge cannot be negligible.

A thermal bridge of equal importance is that appearing at the conjunction of a piloti floor and the vertical beams (Fig. 19.23). This usually happens because the beams of the open parking space are not insulated; the insulation begins on the first floor level.

One other type of thermal bridge that we can recognize is that of the flat roof with vertical beams. Insulation is stopped on the ceiling level of the last floor and does not continue on the low wall of the flat roof (Fig. 19.24).

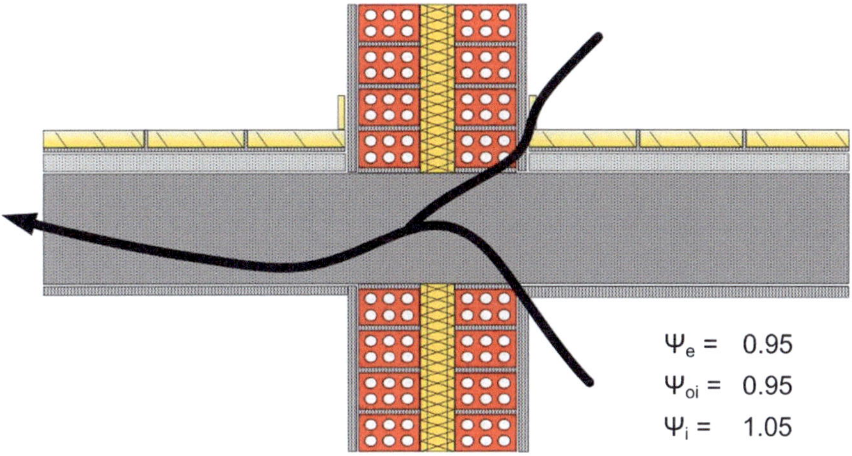

Fig. 19.22 Thermal bridge of balcony

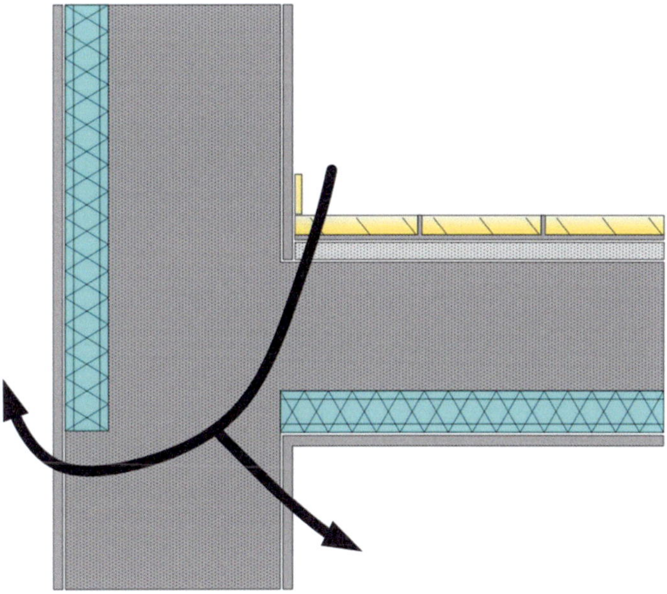

Fig. 19.23 Thermal bridge of piloti floor/ceiling

In conclusion, both horizontal and vertical construction solutions can offer sufficient thermal protection, although the creation of several types of thermal bridges can significantly increase the thermal losses.

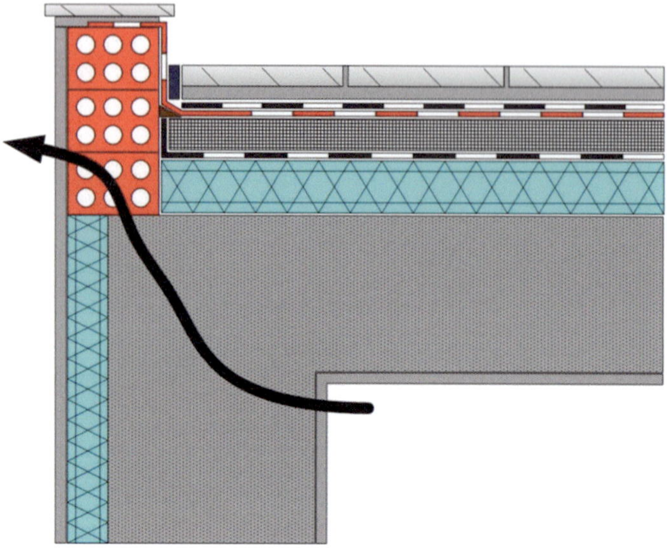

Fig. 19.24 Thermal bridge of flat roof

19.5 Conclusions

Thermal insulation material plays an important role in the improvement of the energy performance of buildings. Each type of material can contribute to thermal, sound, or passive fire insulation. The "correct" application of each material can improve the thermal resistance of the construction solution.

Reference

Hellenic Statistical Authority (EL. Stat.) (2015) Statistics, Population, Population Census. Available via DIALOG. http://www.statistics.gr/portal/page/portal/ESYE/PAGE-themes?p_param=A1602&r_param=SAM01&y_param=2001_00&mytabs=0. Cited 25 Mar 2015

Chapter 20
Cool Materials

Michele Zinzi

Abstract Vernacular architecture in the Mediterranean area has some peculiarities; one of them is the use of light colors for the building envelope to reduce solar gains during the hot summer period. Cool materials merge ancient concepts and new technological solutions in modern buildings. Cool materials can be used for the building skins and for urban applications. They are characterized by high thermal emissivity and higher solar reflectance than that of conventional construction and building materials, thus reducing the heat released to the ambient air by convection and to the built environment through conduction. The various technologies available on the market, as well as innovative technologies still in the research phase, are described. Potentialities and limits of the technology for building and urban applications are presented via numerical analyses; case studies of cool roof application are also presented.

20.1 Introduction

Traditional Mediterranean architecture was characterized by strategies and technologies able to ensure the thermal comfort in highly variable climatic conditions throughout the year, especially during summer. Solutions included heavy structures with high thermal mass to modulate the heat conduction and enhance the potentialities of the natural ventilation; bioclimatic strategies, design and materials to optimize solar gains and control through transparent elements in the different seasons; and the use of highly reflective materials to reduce the solar gains of opaque envelope components.

The latter was, in particular, a trademark of the vernacular architecture in the Mediterranean basin for a century, in particular along the coastal areas. Shining

M. Zinzi (✉)
ENEA, Via Anguillarese 301, 00123 Rome, Italy
e-mail: Michele.zinzi@enea.it

S.-N. Boemi et al. (eds.), *Energy Performance of Buildings*,
DOI 10.1007/978-3-319-20831-2_20

Fig. 20.1 View of a village on Santorni Island, Greece

examples still exist on the southern and northern shores to remind us how thermal issues shaped buildings and urban settlements in past centuries. Figures 20.1 and 20.2 offer landscapes of, respectively, Santorini, a world-famous Greek island, and Ostuni, known as Città Bianca (White City), located in Puglia in southeast Italy.

Recent environmental and energy issues recalled building solutions put aside for most of the twentieth century, among them reflective materials for the building envelope. The growing interest in these materials is related to global warming, the heat island effect, and the higher expectations of building occupants for increased cooling—worldwide and, in particular, in Mediterranean countries during the last years.

Fig. 20.2 View of Ostuni, the "White City," in Puglia, Italy

Modern materials and components overcame the limitations of traditional materials, mainly connected with durability and construction technologies, which might be not acceptable for the contemporary building construction sector. The combination of ancient concepts and new technologies becomes, consequently, an opportunity to upgrade thermal and energy performance of buildings in the Mediterranean region.

20.2 Construction Materials Under Solar Radiation

The thermal response of construction materials under solar radiation depends on the solar radiation absorbed by the sample (being $\rho_e + \alpha_e = 1$); the long wave radiation emitted by the material to the sky vault; convective heat transfer to ambient air; and the conduction of heat transfer through the construction material(s) to/from the built environment or the soil. The equation can hence be written as follows:

$$\alpha_e H = \delta\varepsilon(T_S^4 - T_{SKY}^4) + h_C(T_s - T_a) + \lambda\frac{dT}{dz} \tag{20.1}$$

when

α_e	solar absorptance of the material surface (-);
ε	infrared emissivity of the material surface (-);
λ	thermal conductivity of the layer construction layer (W/mK);
ρ_e	the solar reflectance of the material surface (-);
σ	the constant of Stefan–Boltzmann (5.67×10^{-8} W/m^2 K^4);
H_0	the global solar radiation (W/m^2);
h_c	the convection coefficient (W/m^{2}K);
T_s	the material surface temperature (K);
T_{SKY}	the sky radiant temperature (K);
T_a	the air temperature (K);
$\frac{dT}{dx}$	gradient temperature through the construction layer (Km^{-1})

It must be noted that for opaque materials, the following basic equation applies: $\rho_e + \alpha_e = 1$, with ρ_e the solar reflectance of the material surface.

It is the surface temperature that drives the heat transfer to the built environment by conduction and the heat released to the ambient air by convection and, due to the thermal capacity and the eventual thermal insulation of the structure, it is mainly the function of the solar reflectance and the emissivity of the materials. These properties can be defined as follows:

- The solar reflectance (ρ_e, also indicated as SR), is a measure of the ability of a surface to reflect the incident solar radiation. It is defined as the ratio of the total hemispherical radiation reflected by the material (including the specular and diffuse components) on the incident radiation, integrated over the solar spectrum. It is measured on a scale of 0–1 (or 0–100 %).

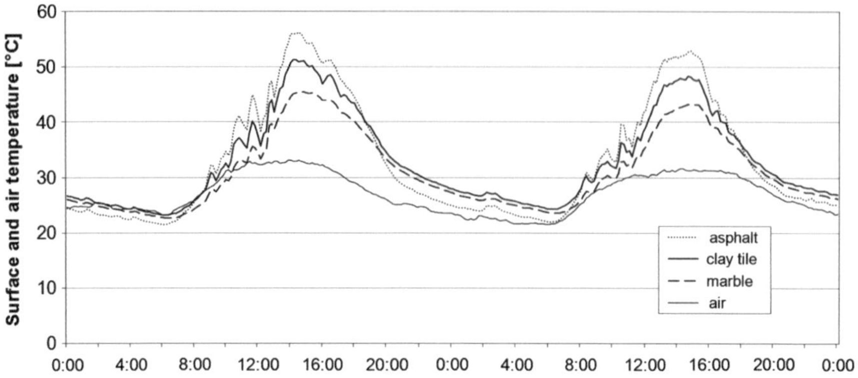

Fig. 20.3 Air and materials surface temperature profiles measured in Rome during summer

- The thermal infrared emissivity (ε, also indicated as IE), also emittance according to some sources, is a measure of the ability of a surface to release the absorbed heat at long wave radiation. It is defined as the ratio between the heat flow radiated away by the material surface at a certain temperature and the heat flux radiated by a black body at the same temperature. The infrared emissivity is measured on a scale from 0 to 1 (or 0–100 %).

Usual construction materials, except metals, have high emissivity (typically in the 0.85–0.94 range) and low solar reflectance values (or alternatively high solar absorptance), with the result that they can reach high surface temperatures—black materials as asphalts can get up to 40 °C hotter than the air temperature. Figure 20.3, as an example, shows the temperature profiles of construction materials measured in Rome in August. The response of different materials is observed; the aged asphalt (estimated reflectance 15 %) gets up to 23 °C hotter than the air temperature during the daytime; a lower, but relevant temperature increase can be observed for the other materials. Surface temperatures get very close to the ambient air temperature at night, and within a few hours they get even cooler; this happens to the high emissivity that allows the materials to cool down, radiating away the absorbed heat to the sky vault.

The solar reflectance of materials depends on the intrinsic characteristics of the materials themselves, as well as the color used to finish the construction element, which implies a huge variety of possible solutions. Some examples of solar reflectance for widely used materials are reported in Table 20.1, all of them characterized by a high infrared emissivity.

Reflective materials can be used to reduce the surface temperature of the construction components reducing the thermal stress to the built and urban environments. These materials are known as cool materials; products dedicated to the building roof are known as cool roofs. Cool materials can be defined as materials with a high solar reflectance (white and light-colored), or with a solar reflectance higher than that of a conventional material of the same color (cool-colored). Cool materials are also characterized by a high thermal emissivity; low emissivity materials may exhibit a cool behavior only in case of high solar reflectance.

Table 20.1 Solar reflectance of typical construction materials

Material	ρ_e (−)
Concrete tile	0.2–0.35
Terra cotta red clay tile	0.25–0.35
Marble tile	0.35–0.55
New asphalt	0.05
Aged asphalt	0.1–0.15
Dark membranes	0.1–0.2
Pre-painted metal roofs	0.1–0. 75

A crucial issue emerges from the above definitions; the response of the material depends on the reflectance across the whole solar range, between 200 and 2500 nm, and not only on its color, which is a photometric parameter and takes into account the solar radiation in the 380–780 nm range. Taking the solar irradiation spectral distribution defined in ISO 9050, it can be observed that only 53 % of the solar radiation falls in the visible range, a small part (2 %) in the UV range, and the remaining 45 % is near infrared radiation (NIR). This implies that the color plays a relevant role in the *cool* behavior of the construction material, but the performances need to be assessed across the whole solar range, due to the significant amount of infrared radiation.

It is important to note that even if solar reflectance and emissivity are the main physical properties of the surface material, there is another parameter that is widely used for cool materials: the solar reflectance index (SRI). It represents a measure of the roof capacity of rejecting the solar gains as a combined function of the two main properties—solar reflectance and infrared emissivity. The method is based on Formula 1, neglecting the conduction heat transfer and setting fixed reference radiation and temperature conditions. SRI is defined so that for a standard black body (ρ_e 0.05, ε 0.90) is 0 and for a standard white body (ρ_e 0.80, ε 0.90) is 100; the method allows calculating the SRI of any material of known radiative properties by an interpolation technique. This indicator is often used in product catalogues.

A crucial issue to mention is that solar reflective materials for building application reduce the cooling demand in summer but increase the heating demand in winter because of the reduction of solar gains. It is therefore important to assess the potentialities of the technology, taking into account the benefit/penalty balance throughout the year.

20.2.1 Construction and Building Solutions for Cool Applications

There has been a growing interest in the technology over the past 20 years, mainly for roofing system applications, and today many cool roof solutions are available, as a function of the roof geometry. The latter can be categorized as follows:

- Flat roof. Only a small slope is given, so that the water runs off to the drainage system, without ponding.
- Low-slope roof. The surface has a maximum slope of a 5 cm rise for a 30 cm run, corresponding to less than a 10-degree inclination.
- Steep-slope (sloped) roof. The surface has a minimum slope of a 5 cm rise for a 30 cm run, corresponding to more than a 10-degree inclination.

Cool materials are today available for many roofing technologies. Some of the most frequently used are:

- Elastomeric Coating. They have elastic properties, and can stretch under the summertime heat and then return to their original shape without being damaged. These products overcome traditional paints, plasters, and sealants bound to dry hard and become brittle, with a lifespan of a very few years. Elastomeric coatings are rubber-like and flexible, and are also UV-resistant and do not blister, crack, or flake under severe climatic conditions. Roof coatings are available in many colors, thus spreading the architectural solutions. They can be applied on several substrates (mineral, asphalts, membranes, and eventually on metals).
- Single-ply membrane. A flexible or semi-flexible pre-manufactured roofing system, generally made of rubber or plastic materials. Single-ply membranes are categorized into two main groups: thermosets and thermoplastics. According to the different techniques, the membranes can be loosely laid over the substrate or bound to the substrate by adhesives or glues.
- Shingles. Technology consisting of individual overlapping elements. These elements are normally flat, rectangular shapes that are laid in rows. Shingles are made of various materials, such as wood shingle, slate shingle, asbestos-cement, bitumen-soaked paper covered with aggregate (asphalt shingle), or ceramic.
- Tiles. Technology consisting of individual elements, laid in rows or, in some cases, overlapping. They can be ceramic (typically clay tile), based on cement concrete, or manufactured using other (mixed) stones. The outer layer finishing is manufactured according to the product's requirements.
- Metal roof. A roofing system made from metal pieces or tiles, usually coated and based on corrugated galvanized iron, stainless steel, zinc and aluminium, and/or silicon materials. The product has several applications: agricultural, residential and commercial.

Recently, the application of cool materials for building façades was explored as well. Several technologies were tested for possible applications:

- Water- and acrylic-based (elastomeric) paints
- Mineral-based coatings, renders, and mortars.

The application of cool materials on building facades is, in principle, less effective than on roofs, being the solar irradiation on the horizontal greater than on vertical surfaces during the summer season; however, this application can be of interest to mitigate the urban temperature under the canopy, improving the outdoor thermal comfort and reducing the building cooling demand by lowering the air temperature in proximity of the building itself.

The last sector that is gaining interest is the application of cool materials for urban applications, such as road paving, sidewalks, and pedestrian areas. Interesting potentialities can be found for the various products' categories:

- Thin elastomeric coatings to be placed on an existing substrate (asphalt, concrete, etc.) and characterized by higher solar reflectance than the original construction.
- Asphalt and concrete pavements, produced using different percentages of binder and aggregate (gravel, stone, etc.).
- Porous and permeable structures (mainly concrete-based), which can be cooled by the evaporation of the water stored in the substrate. This solution is obviously of interest in wet climatic conditions.

It should be noted that the application of cool materials for the urban environment involves glare problems, which may occur in cases of high light reflectance with relevant specular component.

20.3 Cool Materials for Building Applications

Today, several solutions are available for cool applications, and much research and technology development is taking place that can exploit the full potentials of the technology. They mainly focus on roof products; however, there are some applications for building façades, solar protection devices, and air-exposed air conditioning machines. Descriptions of some older, or promising technologies for building applications, follow.

20.3.1 White and Light-Colored Materials

White is, obviously, the most reflective of colors; however, it has already been noted that the solar radiation is dependent on the response in the near-infrared range and not only in the visible range. The roughness of the material is another parameter that can have an impact on the reflectance. For the above reasons, not all whites have the same response in terms of cooling performances. Figure 20.4 shows the spectral curves of several white products (acrylic and water paints, membrane, enamel), and the variability in the near range (780–2500 nm) can be inferred. While the light reflectance of white products is about 0.88–0.90, the solar reflectance presents a wider variation, typically in the 0.75–0.90 range. The best commercial products slightly exceed 90 % solar reflectance; an experimental ceramic tile reached 94 % in laboratory measurement. Solar reflectance up to 70 % is measured for white tiles, where the color is achieved by surface coating or coloring in mass. A 45 % solar reflectance was obtained for white-coated asphalt shingles, and higher values are obtained for built-up roof systems, but the increase

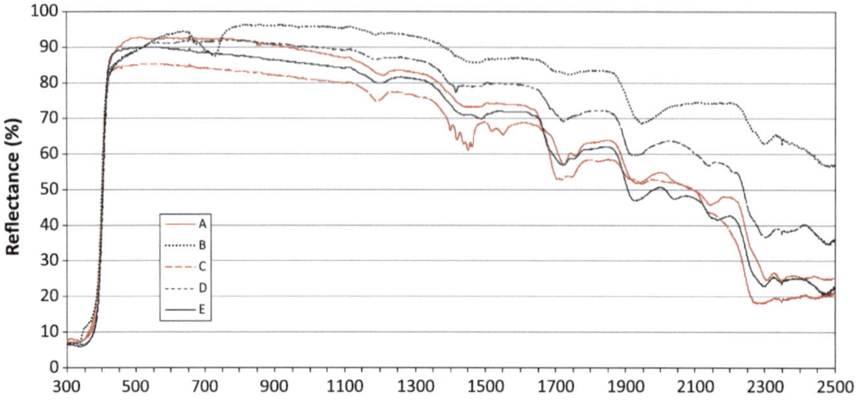

Fig. 20.4 Spectral reflectance of white materials

in these cases is due to the coating deposit on the product more than the improvement of the product itself. It must be noted that solar reflectance is not necessarily linked to a specific material, but is strongly related to the design and optimization process carried out for the specific product.

It is important to note that the reflectance of the materials is affected by the exposure and tends to decrease during the product's life cycle. Keeping the reflectance as close to the initial values as possible is a challenging task, depending on the material characteristics and the environmental conditions. It is generally expected that a white product experiences a 0.15–0.25 solar reflectance reduction during the first three years of its life.

Light-colored materials are a valid opportunity whenever white products cannot be used. Light colors still have a high reflectance in the visible range, while the near infrared response can be adequately optimized. In principle, the same application listed for white materials applies also for light-colored materials. The graphs in Fig. 20.5 refer to an ecological organic paint based on milk and vinegar, used to develop light-colored solutions. The examples in the figure have the following solar reflectance values: 79 % green, 79 % red, 74 % brown, 83 % yellow.

20.3.2 Cool Colored Materials

It is not possible to use light reflective products in many cases, hence whites and light colors cannot be applied—typically for architectural integration and for eventual glare problems. It is still possible to take advantage of solutions by using cool-colored materials, which are characterized by a solar reflectance higher than that of conventional materials of the same color. An example of cool-colored materials and conventional color comparison is shown in Fig. 20.6.

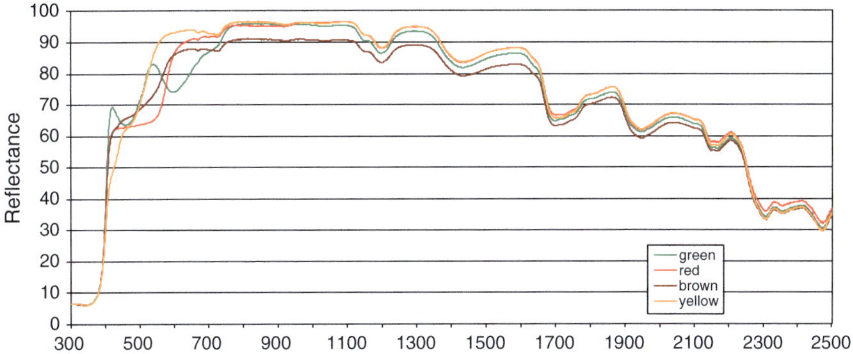

Fig. 20.5 Spectral reflectance of light-colored materials

Fig. 20.6 Visual comparison of conventional and cool-colored colors (*below*)

These products are already on the market, mainly in the form of coatings. As an example, Table 20.2 reports the solar reflectance values for a set of colors obtained using conventional and cool-colored materials. The solar reflectance of dark coloured samples increase by factor 2 to 4, as expected the black samples exhibits the

Table 20.2 Solar reflectance of conventional and cool-colored coatings

Sample	ρ_e cool (−)	ρ_e conv. (−)	$\Delta\rho_e$ (−)	$\Delta\rho_e$ (%)
Orange	0.63	0.53	0.1	19
Light blue	0.42	0.4	0.02	5
Blue	0.33	0.18	0.15	83
Green	0.27	0.2	0.07	35
Anthracite	0.26	0.07	0.19	271
Brown	0.27	0.09	0.18	200
Black	0.27	0.05	0.22	440

best performance. Smaller increases of solar reflectance are obtained for lighter colors. In absolute figures, the solar reflectance increase is below 0.1 for light colors and up to 0.22 for dark colors.

20.3.3 Advanced Materials

Innovative solutions for building integrated cool materials have been on the way for several years, addressing various technologies and strategies. Of particular interest are the thermal responsive technologies that are able to change the response of the building envelope as a function of the outdoor climatic conditions.

Thermochromic materials change their colors and, consequently, their optical response as the materials' temperatures rise. This change is reversible, which means that the materials return to the original color as the temperature decreases below the switching value. The basic technology relies on organic leuco-dye mixtures based on three main components: the color former, the color developer (usually a weak acid that causes the reversible color switch and color intensity of the final product), and the solvent. The components concur to determine the magnitude of the color transition and the melting (switching) temperature. Thermochromic pigments were developed and incorporated into conventional white coatings, eventually with the addition of titanium dioxide (TiO_2). Table 20.3 shows the solar reflectance in the colored and bleached states of an experimental product prepared in various colors.

The results show a limited benefit in terms of solar reflectance increase for the thermochromic materials coupled to a conventional white binder; the green sample has a 0.12 increase, while it is always lower than 0.07 for the other color. Better results are achieved with the addition of titanium dioxide, with solar reflectance increasing between 0.13 and 0.22. These materials exhibit peak surface temperatures under the solar radiation up to 10°C lower than conventional materials of the same color. The technology is not yet fully developed and most of the work needs to be addressed due to durability and stability problems.

Phase change materials (PCM) are able to store the heat in the latent form, with more relevant heat storage capacity per unit volume than that of conventional building materials. The chemical bonds of the materials are used to store and release heat as a consequence of the ambient temperature rises and drops. The technology has been investigated for building applications for decades, with

Table 20.3 Solar reflectance of conventional and cool-colored coatings

Material	TiO_2 ρ_e col. (−)	TiO_2 ρ_e bleach (−)	ρ_e col. (−)	ρ_e bleach (−)
Green with TiO_2	0.51	0.73	0.33	0.45
Yellow with TiO_2	0.78	0.81	0.70	0.73
Blue with TiO_2	0.59	0.71	0.41	0.54
Gray with TiO_2	0.53	0.78	0.34	0.40

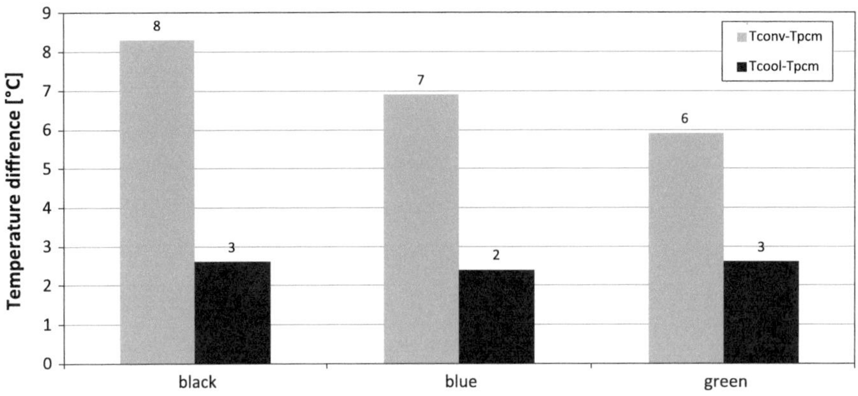

Fig. 20.7 Visual comparison of conventional and cool-colored colors (*below*)

the main objective to modulate the indoor temperature and reduce the cooling demand. Coatings containing paraffin as PCM are produced and investigated; several melting temperatures and PCM concentrations were used in order to exploit the potentialities of the technology.

An extract of the results is in Fig. 20.7, where the maximum surface temperature reduction is plotted for three samples (black, blue, and green) with respect to conventional and cool samples of the same color. The results show a significant improvement of the performances with respect to the conventional materials, but the temperature reductions do not exceed the 3 °C when PCM coatings are compared to the cool coatings. Also, this technology is not yet marketable.

20.4 Cool Materials for Urban Applications

Urban materials do not directly affect the energy performances of buildings, but they do it in an indirect way, affecting the ambient temperature, which in turn establishes the heat transfer with the built environment. Urban materials include pavements for roads, parking lots, and sidewalks and pedestrian areas, which represent a relevant portion of the city footprint. A basic requirement is, in this case, to avoid glare problems for the people; this implies that white or other too-light colors should be avoided.

A solution is to increase the solar reflectance of the asphalt by using light-colored aggregates (like off-white gravel and stones) or pigments in the asphalt mixture, instead of only black tar. It is thus possible to achieve a 30 % solar reflectance, higher than the conventional asphalt that ranges from 5 to 15 % for,new and older pavements, respectively.

An alternative solution to increase the solar reflectance of urban pavements and areas is the application of thin coatings to be laid, using the existing pavement

Fig. 20.8 Different chromatic versions of a commercial coating for road pavement. Noteworthy is the effect of the cool pigment, which gives an NIR reflectance higher than the luminous reflectance up to a factor 4 (for the *green* and the *red* samples); increments are limited for the lighter colors. One of the best performing coating for pavements has an 83 % NIR reflectance and 23 % light reflectance

Table 20.4 Solar, light and near-infrared reflectance of a thin coating for asphalts

Sample	ρ_e (−)	ρ_v (−)	ρ_{NIR} (−)
Off-white	0.55	0.45	0.63
Green	0.27	0.10	0.39
Red	0.27	0.11	0.40
Yellow	0.44	0.26	0.51

as the coating substrate. The optical requirements are similar to those for building applications: keeping moderate to low light reflectance values, and increasing the response in the near infrared region. Higher performances are of course required in terms of resistance of the coating to the mechanical stress, due to the road traffic. Figure 20.8 presents different colorations of a commercial coating (Table 20.4).

Another solution is using another material, more reflective than tar, for road pavements. Concrete is a feasible solution, since it is already used in some applications (airports) and has a solar reflectance of between 0.25 and 0.40; lighter materials used as binders could even increase such values. Figure 20.9 show a porous concrete pavement with various solutions in terms of chromatic response and size of the aggregates. The solar reflectance of these materials is: 0.45 for 1; 0.56 for 2; 0.30 for 3; and 0.22 for 4. Such values are higher than those of new and older asphalts; however, the products do not present an NIR selectivy, since the reflectance curves flat across the whole solar spectrum.

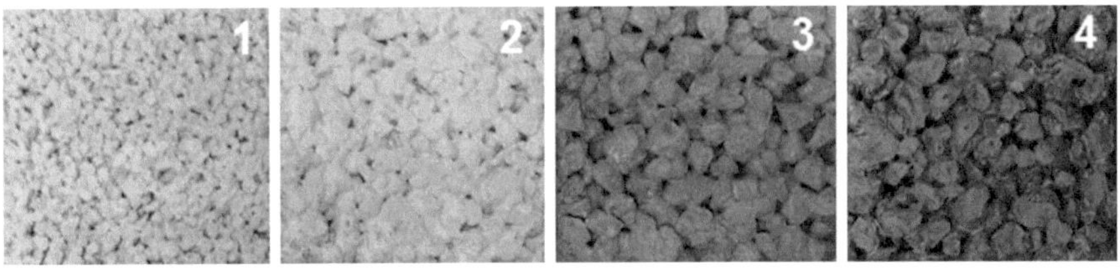

Fig. 20.9 Concrete pavements in different colors and sizes of the aggregate

20.5 Potentialities of Cool Materials Applications

Potentialities of cool roof and cool materials application were widely explored over the past few years, focusing on various building uses, climatic conditions, roofing technologies, and urban layout and construction characteristics. Following are some examples related to the Mediterranean region.

20.5.1 Saving Energy with Cool Roofs

Potential energy savings that are achievable using cool roofs are explored for several cities in the Mediterranean region, including on European, Asian and African shores. The calculations are carried out for a rowhouse, whose relevant geometric data are reported in Table 20.5.

Two insulation configurations are selected, and the related data are summarized in Table 20.6. It must be noted that the *U* values of the insulated configuration are to be considered for exercise purposes and not taken by country-dependent building codes.

The calculations are carried out, taking into account the net energy only, which means that the efficiencies of the energy systems are not taken into account (in

Table 20.5 Geometric data of the reference building

Geometric data	Unit	Row
Net area	m^2	115
Gross volume	m^3	427
Gross external surface	m^2	211
Roof area	m^2	68
Total window area	m^2	18
Roof-to-volume ratio	m^{-1}	0.16
Roof-to-external surface ratio	–	0.32
External surface-to-volume ratio	m^{-1}	0.49

Table 20.6 Geometric data of the reference building

Envelope component	U_{ins} (W/m^{2}K)	U_{not_ins} (W/m^{2}K)
Wall	0.7	1.4
Ground floor	0.8	1.4
Roof	0.6	1.4
Window	2.5	4.5

other words, they are 100 % efficient). Set points are 20 °C air temperature and no humidity control in winter, and 26 °C air temperature and 60 % relative humidity in summer. It is assumed that the building is continuously cooled or heated. The calculations are carried out to compare the performances achieved by the reference building equipped with the roof having 20 and 80 % as solar reflectance values.

The results for the not-insulated configuration are presented in Fig. 20.10 and confirm that the cool roofing is a net energy-saving technology for this analysis. In the northern zones (Marseille, Barcelona, Rome) the energy savings are between 1 and 3 kWh/m^2 per year. The southern European Mediterranean cities (Athens, Palermo, Seville) have savings in the 6–10 kWh/m^2 range. Energy savings in the southeast rim cities (Cairo, Tel Aviv) are in the 12–16 kWh/m^2 range. Net cooling energy savings range from 8 kWh/m^2 in Marseille to 19 kWh/m^2 in Cairo.

The results for the insulated configuration show that net energy savings improve for the northern European cities, thanks to the reduction of the heating penalties, which are below 1 kWh/m^2 for all the selected cities. Net cooling and energy savings are lower, yet significant, for the other cities, ranging between 6 kWh/m^2 in Athens and 10 kWh/m^2 in Cairo (Fig. 20.11).

It is important to note that these results show the potentiality of the technology on the one hand, but that on the other, they cannot be considered to be generally

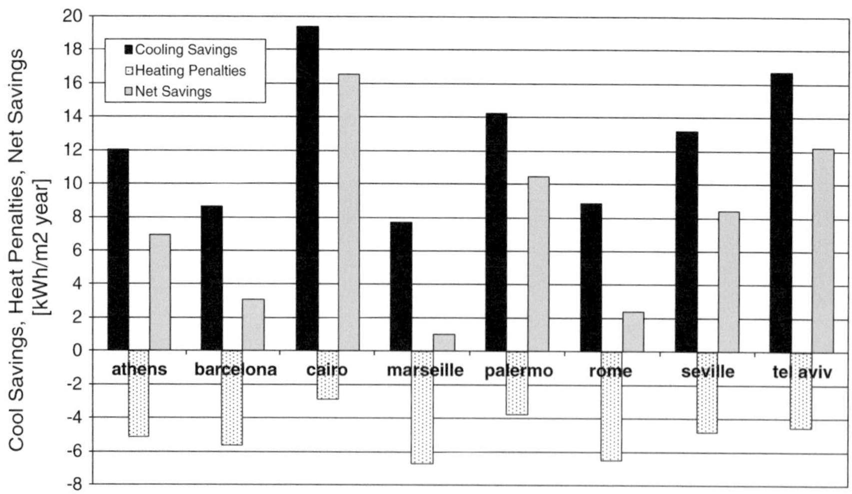

Fig. 20.10 Net energy demands in Mediterranean cities—non-insulated building

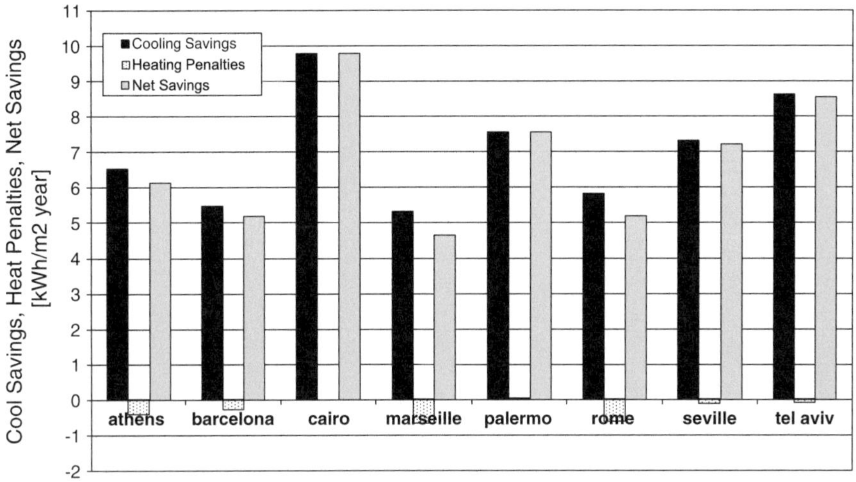

Fig. 20.11 Net energy demands in Mediterranean cities—insulated building

valid; other building geometries, thermal characteristics, and uses may lead to different results. Effective energy savings potentials should be analyzed for each specific case.

20.5.2 Mitigating the Urban Temperatures with Cool Materials

Using cool materials instead of conventional ones reduces the ambient temperature in urban areas, with improvement of thermal conditions for people and potential cooling energy savings in buildings. The impact of cool asphalt (solar reflectance increasing from 0.1 to 0.4) is investigated for a densely built area in Rome by a calculation using EnviMet software. Geometries, vegetation and construction data are inputted and the simulation is carried out for a mid-July day. The calculation shows a reduction of about 3 °C under the canopy between the reference asphalt and the cool materials during mid-day (12:00–18:00). The impact on the building's cooling performances is calculated in terms of peak power reduction of a reference apartment block. The reduction is 10 % for a non-insulated building and 8 % for the same building in an insulated configuration (Fig. 20.12).

Another calculation exercise was carried out in a densely built suburb of Athens. The simulations were carried out using the Phoenics CFD calculation package. The air temperature evolution along a main commercial road during the summer period was investigated, with considering replacing the existing asphalt (solar reflectance 0.04) with an off-white coating (ρ_e 0.55). The average temperature difference that was calculated at a height of 1.5 m is 5 °C. Both exercises

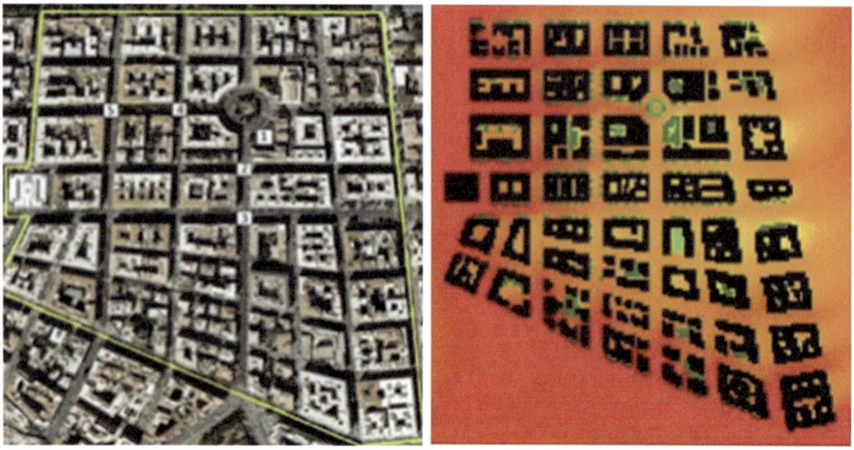

Fig. 20.12 Net energy demands in Mediterranean cities—non-insulated building

confirm the potentialities of cool technologies in reducing the urban ambient temperatures, improving the thermal quality of outdoor spaces, and improving the thermal and energy performances of the built environment.

20.6 Cool Roofs Case Studies

Performances of cool roofs in real applications are carried out under several climatic conditions in different geographic zones and documented in publications. However, the attention here is on case studies carried out in the Mediterranean region.

20.6.1 Senior Recreation Building in Rome, Italy

This case study was performed in Rome during the summer of 2010. The single-floor building belongs to the Municipality of Rome and is used as recreation center for the elderly. The building is parallel shaped, with the large façades facing north and south, and it hasa surface of 275 m^2. The building envelope is not thermally insulated, windows are single-glazed with aluminum frames and without thermal break; some of the windows have internal solar protection, and the roof finishing layer consists of a dark bitumen membrane. These features made the building an overheating hazard during the summer, with the additional trouble of the cooling system often having maintenance problems. The implementation of the cool roof was in this case a chance to improve the quality of the built environment and the quality of life for visitors.

Fig. 20.13 Implementation of a cool roof system in a recreation center in Rome

The technology used for this building was a white membrane, finished with an ecological coating base of mill and vinegar. The cool materials have a solar reflectance of 86 % and thermal emissivity of 0.88, while the original membrane had 14 % solar reflectance and 0.9 emissivity. The cool roof was applied on a 130 m^2 roof portion, corresponding to the main room of the building. The membrane was applied on July 19th and 20th; photos of the building during the work and after completion are shown in Fig. 20.13. The building was monitored before and after the roofing work in order to compare the performances during the summer. The quantities that were measured on site were ambient air temperature and relative humidity; horizontal global solar radiation; indoor and outdoor surface temperatures; and indoor air temperature and relative humidity in different rooms of the building. The monitoring was carried out under free-floating conditions with the cooling system not working during the period.

Concerning the results, a reduction of the peak temperatures between 2.5 and 3 °C were found. The cooling energy savings were calculated, under the hypothesis of a continuously working cooling system. The effect of the cool roof is a reduction of 9 kWh/m^2 of the cooling demand, corresponding to 36 % of the initial value.

20.6.2 Office/School Building in Trapani, Sicily, Italy

This block is part of a secondary school, and it hosts the school administration offices and some laboratories. The building is used in winter as well as in summer, for the recruitment of teachers in the Province of Trapani. The building is single-story and has a flat roof of 760 m^2. It dates back to the 1970s and, according to the construction technology of the period, has no thermal insulation on the envelope components; windows are single-glazed with internal shading and some are partially shaded by concrete fins. The flat roof has concrete tiles as a finishing layer (ρ_e 25 %). The building experienced severe overheating problems due to the hot Sicilian summer and is densely occupied during the recruitment period.

Fig. 20.14 Implementation of a cool roof system in a school in Trapani, Italy

The cool technology adopted for the roof is the same as that in 20.6.2, even if the product was prepared as a roller coating and not as a membrane. Figure 20.14 shows the cool roof on the office block, to be compared with the original roof visible in the foreground of the picture. Monitoring in free-floating conditions included an external microclimatic station; surface temperature on a cool and an original roof; internal air temperature, humidity, and air speed; and mean radiant temperature.

Concerning the results, in one laboratory the average temperature during the observation period decreased from 27.9 to 24.6 °C. In another monitored room, the average temperature dropped by 1.8 °C during the observation period. The surface temperature was monitored during the entire summer, and Fig. 20.15 shows the fluctuations together with the ambient temperature. The graph shows the signficant reduction of the surface temperature profile after the installation of the cool roof; not monitoring was possible during the working phase, as documented by the interruption of the red curve. The difference between the two values is a few degrees, while a difference of up to 25 °C was measured before.

Predicted energy performances estimate a 56 % reduction of the cooling demand due to the cool roof application, which is the most effective among the passive strategies and technologies.

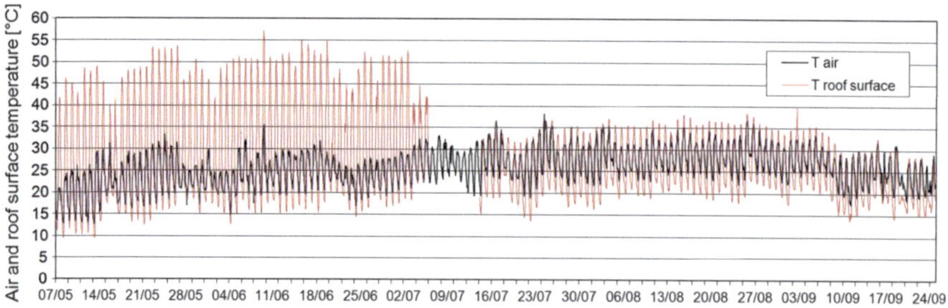

Fig. 20.15 Surface and ambient temperature evolution in the Trapani school

20.6.3 School Building in Athens, Greece

This case study was implemented in Athens in the summer of 2009. The building, dating to 1980, has two floors and is part of a larger complex hosting a secondary school. The building hosts offices and classrooms. The structure is not thermally insulated, apart from the windows that were replaced with double glazing units in 2008; it has a heating system but not a cooling system, as it is closed from the end of June until the end of August. The roof, over 400 m^2 in size, had cement and gravel screed as a finishing layer, with ρ_e 20 %, replaced by a white elastomeric waterproof coating (ρ_e 89 %). Roofing work and the completed installation are shown in Fig. 20.16. Ambient air relative humidity and temperature were measured, as well as roof surface temperature, which was then combined with a nearby meteorological station. Surface temperature measurements carried out with an infrared thermal camera showed a 12 °C reduction of the cool roof respect to the original finishing layer, nevertheless meaurements were performed during a not intense irradiation period.

Calculations carried out on a building-calibrated model resulted in an average temperature reduction of 1.8 °C in a classroom with the roof exposed to the sun. The prediction of cooling energy savings is 40 % for the actual building; it decreases to 35 % under the hypothesis of the building thermal insulation; and the peak cooling demand is reduced by 20 %. Another relevant result is the reduction of the daily surface temperature fluctuation, decreasing from an average of 22–8 °C after the cool roof application, as inferred from Fig. 20.16. This is an added benefit of cool roofs—in fact, reducing the thermal stress and fatigue is crucial to lengthening the lifespan of the product.

20.6.4 School Building in Heraklion, Crete, Greece

This case study is the only one to have been carried out on an already energy-efficient building. It is located in Heraklion, Crete, and serves as an administrative

Fig. 20.16 Implementation and completed installation of the cool roof on the Athens school

Fig. 20.17 The cool roof installation in Heraklion, Crete

office on a university campus and is used throughout the year. The building has thermal insulation on the walls and roof, which is partly flat and partly tilted; only the flat part was used for the cool roof application. Windows are equipped with coated double-glazing units, to have thermal insulation in winter and solar control in summer. The same coating as 20.6.3 was applied to the flat roof area of the building.

Since the building is part of a campus, a fully instrumented meteorological station was available on site. The building is equipped with energy metering, and the indoor air temperature and relative humidity are measured. The case study was carried out with the support of a calibrated model to predict the thermal and energy response of the building. The effect of the cool roof is a 1.5 °C reduction of the indoor air temperature with respect to the original configuration. In terms of energy performance, the building, after the installation, registers a 27 % cooling energy saving. To be noted that the green roof was taken into as alternative technology to cool the roof in the numerical analysis. The green roof provides a 31% cooling energy saving, but the cool roof is the most performing technology on annual basis (Fig. 20.17).

Bibliography

Akbari H, Konopacki S (2005) Calculating energy-saving potentials of heat-island reduction strategies. Energy Policy 33:721–756

Akbari H, Bretz S, Kurn D, Hartford H (1997) Peak power and cooling energy savings of high albedo roofs. Energy Build 25:117–126

Butera F (1994) Energy and buildings in Mediterranean countries: present and future. Renewable Energy 5:942–949

Carnielo E, Zinzi M (2013) Optical and thermal characterisation of cool asphalts to mitigate urban temperatures and building cooling demand. Build Environ 60:56–65

Cheng V, Givoni B (2005) Effect of envelope color and thermal mass on indoor temperatures in hot humid climate. Sol Energy 78:528–534
Doulos L, Santamouris M, Livada I (2004) Passive cooling of outdoor urban spaces. The role of materials. Sol Energy 77(2):231–249
Fanchiotti A, Carnielo E, Zinzi M (2010) Cool roofs: monitoring a retrofit project in Rome, Italy. In: Proceedings of 40th annual conference of the American Solar Energy Society
SOLAR (2010) Raleigh. NC, USA
Hernández-Pérez I, Álvarez G, Xamán J, Zavala-Guillén I, Arce E, Simá J (2014) Thermal performance of reflective materials applied to exterior building components—a review. Energy Build 80:81–105
Karlessi T, Santamouris M, Apostolakis K, Synnefa A, Livada I (2009) Development and testing of thermochromic coatings for buildings and urban structures. Sol Energy 83:538–551
Karlessi T, Santamouris M, Synnefa A, Assimakopoulos D, Didaskalopoulos P, Apostolakis K (2011) Development and testing of PCM doped cool colored coatings to mitigate urban heat island and cool buildings. Build Environ 46(3):570–576
Kolokotsa D, Maravelaki-Kalaitzaki P, Papantoniou S, Vangeloglou E, Saliari M, Karlessi T et al (2012) Development and analysis of mineral based coatingsfor buildings and urban structures. Sol Energy 86:1648–1659
Kolokotsa D, Diakaki C, Papanoniou S, Numerical Vlissidis (2012) 'Experimental analysis of cool roofs application on laboratory building in Iraklion, Crete, Greece. Energy Build 55:85–93
Levinson R, Berdahl P, Akbari H (2005a) Spectral solar optical properties of pigments Part I: model for deriving scattering and absorption coefficients from transmittance and reflectance measurements. Sol Energy Mater Sol Cells 89:319–349
Levinson R, Berdahl P, Akbari H (2005b) Spectral solar optical properties of pigments. Part II: survey of common colorants. Sol Energy Mater Sol Cells 89:351–389
Libbra A, Tarozzi L, Muscio A, Corticelli A (2011) Spectral response data for development of cool coloured tile coverings. Opt Laser Technol 43(2):394–400
Ma Y, Zhang X, Zhu B, Wu K (2002) Research on reversible effects and mechanism between the energy-absorbing and energy-reflecting states of chameleon-type building coatings. J Solar Energy 72:511–520
Pisello AL, Cotana F (2014) The thermal effect of an innovative cool roof on residential buildings in Italy: Results from two years of continuous monitoring. Energy Build 69:154–164
Romeo C, Zinzi M (2013) 'Impact of a cool roof application on the energy and comfort performance in an existing non-residential building. A Sicilian case study. Energy Build 67:647–657
Santamouris M (2007) Heat island research in Europe—the state of the art. Adv Build Energy Res 1(1):123–150
Santamouris M, Synnefa A, Karlessi T (2011) Using advanced cool materials in the urban built environment to mitigate heat islands and improve thermal comfort conditions. Sol Energy 85(12):3085–3102
Synnefa A, Santamouris M, Akbari H (2007a) Estimating the effect of using cool coatings on energy loads and thermal comfort in residential buildings in various climatic conditions. Energy Build 39(11):1167–1174
Synnefa A, Santamouris M, Apostolakis K (2007b) On the development, optical properties and thermal performance of cool coloured coatings for the urban environment. Sol Energy 81:488–497
Synnefa A, Santamouris M, Livada I (2006) A study of the thermal performance and of reflective coatings for the urban environment. Sol Energy 80:968–981
Synnefa A, Karlessi T, Gaitani N, Santamouris M, Papakatsikas C (2011) Experimental testing of cool colored thin layer asphalt and estimation of its potential to improve the urban microclimate. Build Environ 46(1):38–44
Synnefa A, Saliari M, Santamouris M (2012) Experimental and numerical assessment of the impact of increased roof reflectance on a school building in Athens. Energy Build 55:7–15

Taha H, Sailor D, Akbari H (1992) High albedo materials for reducing cooling energy use. Lawrence Berkley Laboratory Report 31721, UC-350, Berkley CA
Zinzi M (2010) Cool materials and cool roofs: potentialities in Mediterranean buildings. Adv Build Energy Res 4:201–266
Zinzi M, Carnielo E, Agnoli S (2012) Characterization and assessment of cool coloured solar protection devices for Mediterranean residential buildings application. Energy Build 50:111–119

Chapter 21
Shading and Daylight Systems

Aris Tsangrassoulis

Abstract Shading can have an impact on both building's energy balance affecting not only annual energy consumption but peak loads as well. In addition influences users' visual and thermal comfort. Proper selection of shading system is a key factor for reducing unwanted solar gains satisfying simultaneously daylight adequacy and aesthetic desires. With today's pressing problem of global warming, optimization of building façade in terms of daylighting/solar radiation control can be considered as a corner stone strategy for low-energy building design. This chapter present basic information on the parameters that affect the selection of shading & daylighting systems. This includes a review of metrics used to characterize these systems according various European norms and energy codes. Antagonistic phenomena are discussed in detail together with synergies with glazing selection. In addition a number of selected systems are presented along with their operational principles.

21.1 Introduction

Control of solar radiation is a fundamental parameter of any building's energy balance. Although, in principle, the intent of the shading strategy is relatively simple in the early stages, to reduce solar gain during the summer, the interaction with other inter-related and mutually antagonistic parameters has to be taken into account. This turn the whole design process into a more complicated task, trying to use solar radiation at the time needed (winter) and reducing unwanted solar gains during the summer. Glazing selection, the opening position on the façade and its size together with the shading system adopted have a decisive effect on both the building's peak loads and total energy consumption. The afore mentioned

A. Tsangrassoulis (✉)
Department of Architecture, University of Thessaly, Pedion Areos, 38334 Volos, Greece
e-mail: atsagras@uth.gr

S.-N. Boemi et al. (eds.), *Energy Performance of Buildings*,
DOI 10.1007/978-3-319-20831-2_21

parameters are complementary and should be selected in such a way as to achieve the design goal. Therefore, the shading system's role is to improve thermal and visual comfort by reducing the likelihood of overheating and glare. In addition, shading affects the visual contact with the external environment, privacy and natural ventilation. The results of the ESCORP EU-25 (2005) project, which analyzed building simulation results for a number of European climates, indicate that the use of blinds and shutters can decrease cooling energy demand, up to about 40 kWh/m^2 for southern and eastern European regions. Apart from energy savings, it seems that users prefer the proximity to windows, which offers higher levels of activity and wellness (Dubois 2001). This preference is based mainly on the visual connection to the external environment (information on the time of day, weather conditions) and stops when discomfort phenomena or lack of privacy are experienced. Daylight's dynamic character contributes to that, since variability in both intensity and spectrum creates a visually interesting internal environment. The importance of daylighting within energy codes was demonstrated in detail in 2013, when a proposal to be adopted by ASHRAE 189.1 (ASHRAE 2014) standard was withdrawn. The proposal was related to a 25 % in glazing area (from 40 % WWR (Window to Wall Ratio) to 30 %) for buildings less than 25,000 ft^2 Called "the battle for the wall", during its course, it was proved that the proposed reduction is not necessarily linked to a possible reduction in the building's energy consumption. It seemed that large size windows equipped with high performance glazing together with shading systems are a much more preferable choice against small size windows. Although highly efficient glazings have been available for decades, 44 % of the windows in EU are still single glazed (TNO 2011). Any strategy to increase the percentage of solar control glazing either in existing or future buildings might contribute to the EU target for CO_2 reduction for buildings in 2020. This contribution is between 1.5–27 % depending on the scenario examined with the south Europe share 50–75 % (TNO 2007).

Thus, according to the above, a shading system should:

- Regulate (or block when necessary) direct solar radiation.
- Regulate diffuse radiation (either from the sky or reflected by external obstacles). However, a balance should be achieved between solar gains and daylighting.
- Reduce the luminance of the opening, limiting any likelihood of visual contact with the solar disk.
- Enhance daylight uniformity together with good color rendering.
- Allow visual contact with the external environment.
- Provide privacy.
- Be esthetically integrated with the building façade.
- Not significantly limit the ability for ventilation when needed.

The first classification of shading systems is based on their angular selectivity. Systems (such as louvers, blinds, fins, overhangs, egg-crates etc.), can block direct radiation from certain areas of the sky, while diffuse radiation can be used for space illumination. One of their characteristic parameters is the cut-off angle

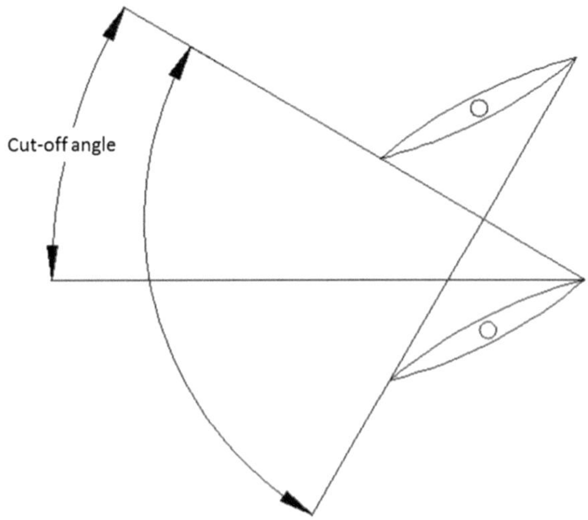

Fig. 21.1 Schematic representation of the cut-off angle for louvers

which is used during the initial design approach to determine the projected-on-the-façade height of the sun, above which there is no direct penetration. Its definition is presented schematically in Fig. 21.1.

Systems presenting no (reduced to be precise) angular selectivity (i.e., performance almost independed form sun's angle of incidence) are screens, curtains, rollers and diffusing opal glass which reduce similarly both direct and diffuse radiation. Depending on their material, the luminance of the opening might increase when the sunlight is present on it.

A second classification is based on the position of shading systems in relation to the glazing. The shading systems can be external, in between double glazing or internal. The mounting position clearly affects the transmitted energy with external shading systems exhibiting lower values of solar heat gain coefficients than internal. Typical external systems are overhangs, vertical fins, eggcrates, louvers, blinds, rollers or perforated metal sheets. These can be static or movable. If static, orientation plays a crucial role to their performance. In south oriented openings, horizontal shading is preferable, while in east/west orientation, vertical shading is more suitable. The latter orientations are quite difficult in terms of shading due to the lower elevation of the sun which requires a balance between the reduction of solar gain and daylight/view/glare. Therefore, a combination of horizontal and vertical shading can contribute to these orientations (Fig. 21.2).

Internal shading systems include fabric rollers, curtains, and blinds. Their main purpose is the reduction of glare and the preservation of privacy. Solar gains are larger when interior shading is used, in comparison to the external ones, since solar radiation has already entered the space and can be even higher when a dark color is used. In some cases (heat reflective shades) a highly reflective aluminum layer is used on their outside surface, facing glazing, in an effort to reflect back a large portion of solar radiation. Such a solution can increase energy savings during winter as well, reducing heat loss. Since they are operated by the users, their inefficient use

Fig. 21.2 Non-opaque overhangs used for shading. Such systems control solar radiation not only because of their form (overhang) but also due to their steel gratings (flat metal bars)

is rather frequent. Internal systems, in most cases, have a lower cost compared to the external ones. The mid pane systems combine easy handling with simultaneous weather protection but with larger solar heat gains coefficients than the external ones. An additional benefit of mid pane systems is that highly reflective materials can be used without the weather deterioration of their reflectance properties. Such systems can redirect solar radiation offering better distribution of daylight in the space.

Shading systems can be classified as movable or static as well. The most common movable systems are motorized louvers or blinds. The environmental parameters most commonly used to define the control strategy—besides manual override—for such systems might be vertical irradiance, indoor/outdoor temperature, and interior/exterior illuminance.

The introduction of dynamic movement aims to address performance requirements for heating, lighting, cooling. and energy generation (Kostantoglou et al. 2013). The majority of the dynamic systems examined in the research literature are mostly venetian blinds. Work by Bülow-Hübe (2007) showed that energy savings for motorized blinds reach up to 50 % on a yearly basis. Lee et al. (1998) used an automated blind system together with a lighting control system. The energy savings achieved were from 7–15 % for cooling and 19–52 % for lighting respectively. Concerning dynamic daylighting systems, research is even more limited. Konstantoglou et al. (2013) examined the performance of a fully dynamic system with daylighting and shading components. The results demonstrated that dynamic light-shelves increase daylighting levels in non-daylight areas and that Daylight Autonomy values (DA) increase up to 50 %. DA.represents the percentage of occupied hours on a yearly basis during which illuminance values are greater than a predefined threshold (i.e., 300 lux). Automated blinds increase energy savings by up to a factor of 1.53 compared to static ones. Meek and Breshears (2010) conducted a study on the facade system for the new Health Science Research Laboratory of the University of California San Diego. Their results have shown that the dynamic exterior shading system in the view window provides the highest indoor quality and energy efficiency with the minimum building ventilation rates.

It is therefore evident that the selection of proper shading system can be a complicated task, since it has to be integrated in the building façade an affect postively building's energy balance. As a first step to this approach, the statistical analysis of solar radiation and daylighting data in a region, can be used as a design tool to support design options relating e.g., to the orientation or size of the openings. Solar radiation data are systematically measured in various locations, but not quite the same is happening for daylighting data. A number of daylight measurement stations were deployed in various parts of the world during the International Daylight Measuring Program of the International Commission on Illumination (CIE). For this data analysis, existing climate records are used to form weather files for building simulation. Various formats exist such as the Typical Meteorological Year or ASHRAE's International Weather for Energy Calculations (IWEC). In addition, software might be used to generate weather data either statistically or by interpolating existing data. The chart below presents the monthly values of global radiation on four main vertical orientations for Athens, Greece using METEONORM software (Remund and Kunz 2004) (Fig. 21.3).

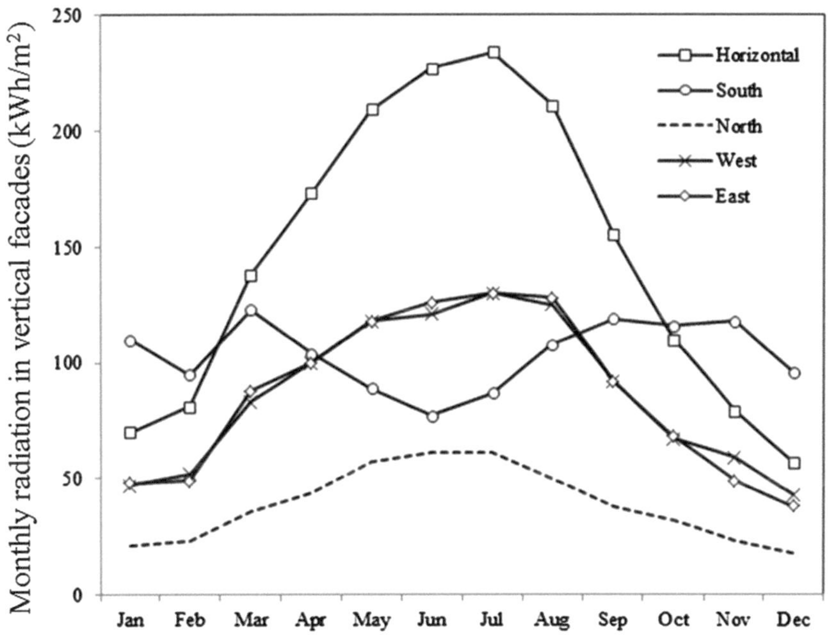

Fig. 21.3 Monthly distribution of global radiation in four cardinal orientations for vertical surfaces using Athens, Greece data as produced by METEONORM software

Global horizontal monthly radiation values are quite increased in comparison with other orientations, especially in summer months. This puts a restriction in the size of horizontal openings as a large one can increase considerably cooling loads. Thus a shading system have to be used, reducing opening daylight efficiency. A more preferable solution is the adoption of a north facing saw tooth structure, if possible. This can act as diffuse daylight provider, avoiding direct solar radiation. South oriented surfaces receive more radiation than horizontal during winter months. In summer, east/west vertical radiation increases, making the shading of the west oriented façades necessary especially when the operation schedule of the building is extended till late evening.

During the initial design phase there are two basic questions that need to be answered. The time period when shading is necessary and the geometry of the system. An interesting chart that can be used for the estimation of the shading period is based on the hourly time series of ambient temperature together with the estimation of the thermal comfort zone (Szokolay S. 2007). This zone is defined by two (upper and lower) temperature limits. The period when the average monthly temperature exceeds the lower limit temperature, determines the shading period. In the figure below, the shading period is from May to September. Today due to the widespread use of simulation algorithms, the shading period can be estimated more accurately since internal operative temperature can be used, which in turn is affected by many other parameters (thermal mass, orientation, internal gains, etc.).

For the estimation of shading systems's geometry a sun path diagram together with a shading mask can be used. It is actually a simple approach and can be used to estimate the effectiveness of a shading system by determining the time period that direct solar radiation is blocked (100% shading). Although percentage of the opening that is in shade for a certain period is an indication of the shading system's effectiveness, what is really needed for the estimation of a building's energy balance is the calculation of direct and diffuse radiation at the opening. An example of using a shadow mask in combination with the sun path diagram is presented in Figs 21.4 and 21.5.

By superimposing the shading mask to a sun path diagram, the time period with 100% shading can be estimated. The figure below presents this for a southern oriented window. Fig. 21.5

Solar elevation at the end of the shading period can be used a design guideline for horizontal static systems such as overhangs (Lechner 2001) Fig. 21.6. The angle that is formed between the line which connects overhang edge with window sill and horizontal, should be equal to the sun elevation at the end of the shading period. For Athens, Greece for example, this angle is approximately 500. A similar approach can be applied for vertical fins placed in west oriented facades. Slant angle is related to the latitude to provide shade from sunlight annualy between 9:00–17:00.

Belia et al (2013) examined the effects of solar shading on building's energy balance for Italian climates. Two shadings systems have been used for a reference office building, overhang to south oriented facades and a louver system for east/

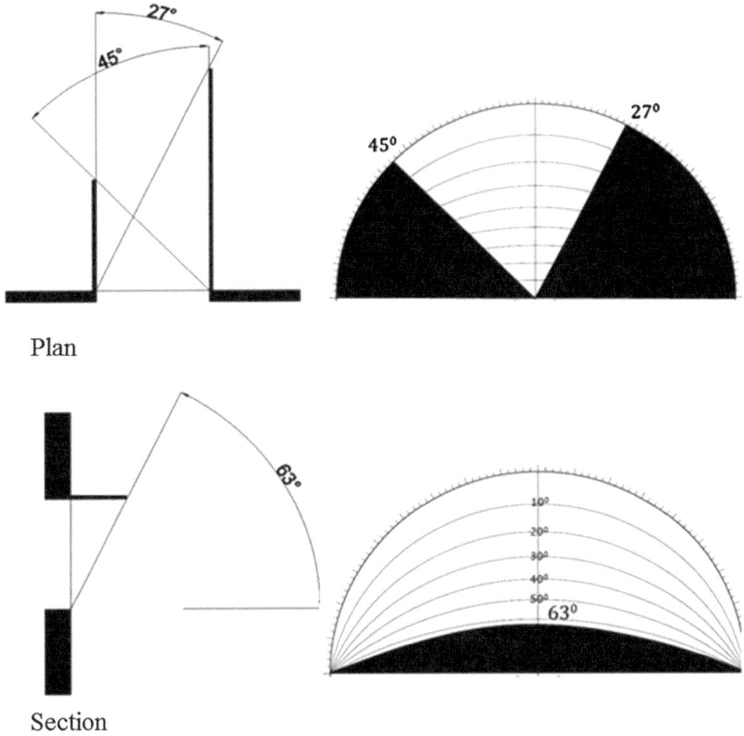

Fig. 21.4 Shading mask creation (full shading) (http://www.jaloxa.eu/resources/daylighting/sun path.shtml)

west orientations. Best energy savings achieved with the 1 m depth overhang with these savings reduced when latitude increased (20 % in Palermo, 8 % for Milan). The optimal depth depends on the window height which in this case is 1.5 m. Energy savings due to shading are increased when the building is well insulated. One interesting issue raised by the study was the fact that the global energy requirements of the building having Window to Wall Ratio 30 % without shadings are about equal to those of the building with WWR ¼ 60 % and the shadings. When the building is oriented to west/east the adoption of louvers (cut-off angle < 1 degree) energy saving decrease considerably (5 % for Palermo) and even negative for Milan.

Mandalaki et al. (2012) tested thirdteen static shading systems for two Greek cities (Chania and Athens). For Chania the best shading systems in terms of energy savings for heating, cooling lighting was the surrounding shade followed by the brise-soleil full façade while for Athens surrounding shade was again the best option followed by canopy inclined double.

Daylight illuminance data can be presented using similar graphs to those used for hourly radiation values. An interesting presentation is the temporal map plots as the following one Fig. 21.7 :

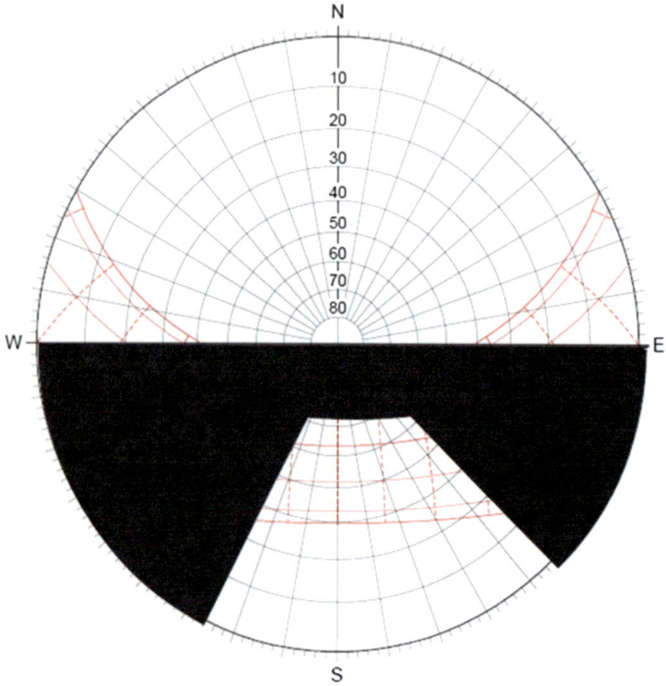

Fig. 21.5 Estimation of time period with 100 % shading by superimposing a shading mask on a sun path diagram (http://www.jaloxa.eu/resources/daylighting/sunpath.shtml)

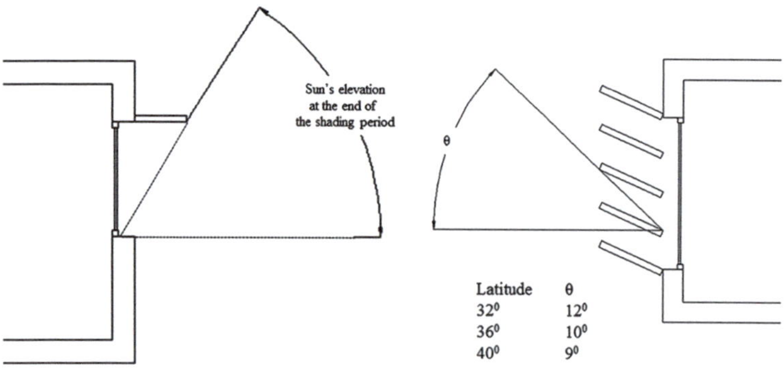

Fig. 21.6 Design recommendation for overhangs and side fins (Lechner 2001)

The amount of information presented in these graphs has to be reduced in order for it to be more easily understood by designers. Among various ways of presenting a time series analysis of illuminance is the cumulative diffuse illuminance distribution for either horizontal or vertical surface. Using this graph, a designer can

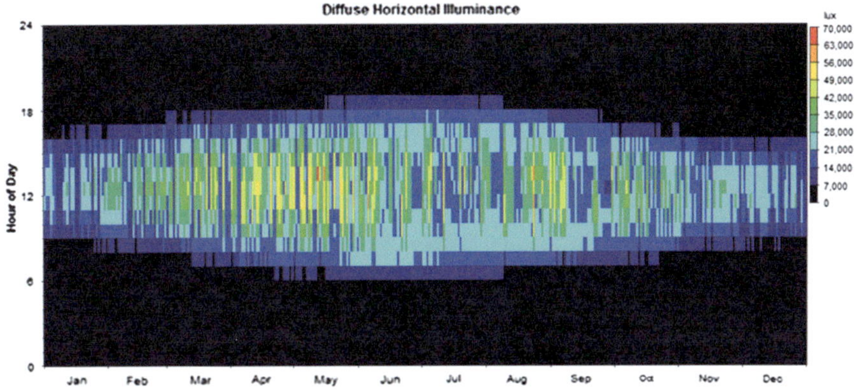

Fig. 21.7 Map of horizontal diffuse illuminance. Graphs created with Dview 2.0 NREL 2015 (https://beopt.nrel.gov/downloadDView)

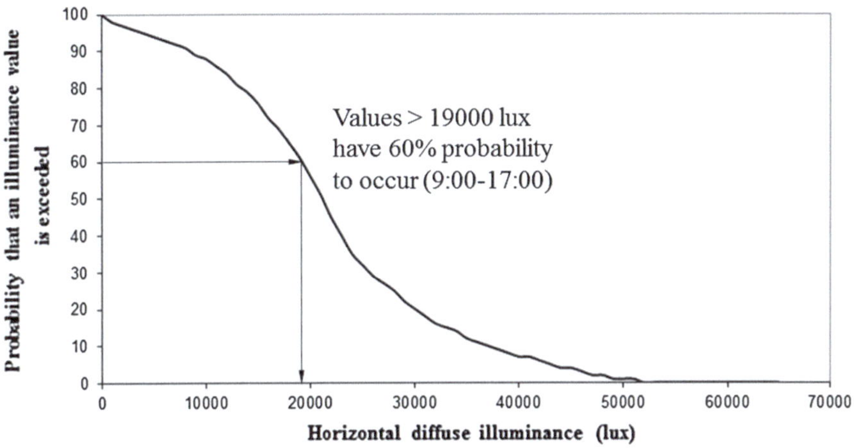

Fig. 21.8 Cumulative probability distribution for diffuse illuminance (8:00–17:00), Athens

estimate the probability of exceeding a specific illuminance value, for a specific operational schedule (i.e., 8:00–18:00). The graph together with a Daylight Factor (ratio between indoor and outdoor illuminance under overcast sky) value can be used to estimate possible lighting energy savings Fig. 21.8.

21.2 Shading

As mentioned above, the control of the solar radiation can be achieved with either orientation in conjunction with urban geometry, or by selecting the appropriate opening size and glazing properties together with a shading system. The shading requirements may vary according to the climate, but a general guideline for static systems

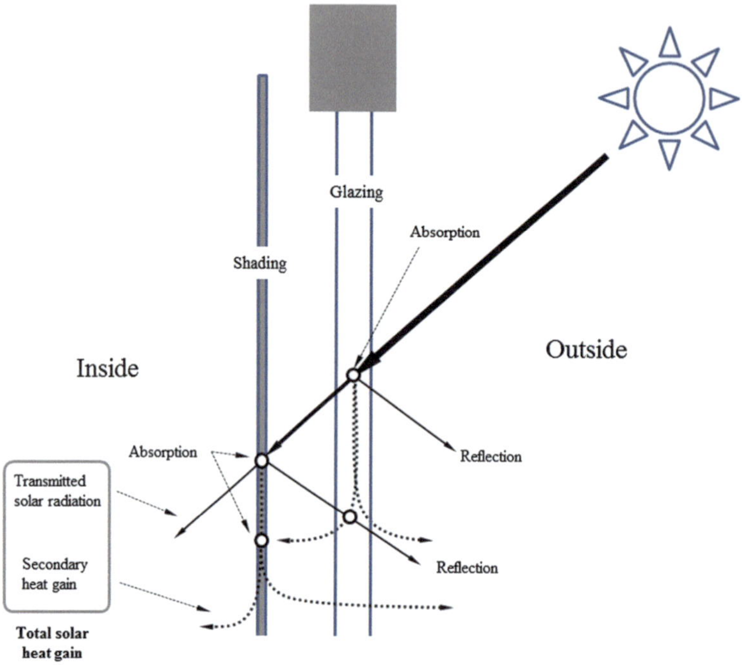

Fig. 21.9 Schematic representation of overall solar heat gain

is the use of horizontal shading in south orientation and vertical for east/west orientations. Movable horizontal shading can be used in east-/west-oriented openings as well, which may adversely affect the design objectives for glare-free daylighting.

Luminous and thermal characteristics for glazing are the thermal transmittance (U-value), the solar heat gain coefficient (g-value), and the visible/solar transmittance and reflectance. The standards applied for the measurement of these characteristics are the EN 410-2011 and EN 673:2011. The solar heat gain coefficient is the fraction of incident solar radiation that actually enters a building through the entire based on the estimation of a characteristic window assembly as heat gain. It replaces the shading coefficient, which is calculated as the ratio of examined glazing g-value to the g-value of a 3 mm clear float glass (0.87) Fig. 21.9.

Typical values for glazing properties are presented in Table 21.1.

Tsikaloudaki et al (2015), using simulation of a reference building unit for various Mediterranean cities concluded that for offices, when glazing with low g-value (0.3) is used, window's U-value and cooling loads are inversely related. Thus, when such a glazing is selected, it should combined with a relatively high U-value fenestration system, allowing dissipation of heat to the external environment. Luminous characteristics used for a shading system are no different, in principle, from those used for glazing.

The effect of shading on transmitted solar gains can be examined using the total solar gain coefficient (GTOT), which refers to the shading system combined with

Table 21.1 Typical values of U, g, and Tvis for three types of glazing (according to EN 13363-1:2003 + A1)

Glazing	U value (W/m^{2}K)	g-value	T_{visible}
Single glazing	5.7	0.85	0.90
Double glazing	3	0.75	0.82
Double clear low-e	1.6	0.70	0.75

glazing. European standard EN 13363-1: 2003 + A1: 2007 and EN 13363-2: 2005 describe the methodology for calculating the g-value and Tvis for combinations of shading and glazing. For example, the simplified method (when installation conditions are unknown) uses the glazing's U- and g-values together with solar transmittance and reflectance of the shading system to estimate the total solar heat gain coefficient (shading-glazing). The calculation used to estimate g_{tot} for an exterior shading system is as follows:

$$g_{\text{tot}} = t_{\text{solar-shading}} * g + (1 - t_{\text{solar-shading}} - \rho_{\text{solar-shading}}) * G/10 + t_{\text{solar-shading}} * (1 - g) * G/5$$

where $G = \sqrt{\frac{1}{U} + 0.3}$, g is the solar gain coefficient for the glazing, U is the center of glazing U-value, t $t_{\text{solar-shading}}$ and $\rho_{\text{solar-shading}}$ is the solar transmittance and the solar reflectance of the shading system (facing incident radiation).

Another way to characterize a shading system is through the use of the "shading factor" (Fc), which is defined as the ratio of the g-value of the combined glazing and solar protection device (g_{tot}) to that of the glazing alone (g). In general, interior shading systems have higher values of shading factors in relation to the exterior ones. The shading factor's typical values are 0.09–0.2 for exterior systems, 0.2–0.3 for midpane systems, and 0.3–0.6 for the interior ones. Table 21.2 shows values of gtot and FC from different shading systems.

Due to the sun's movement on the sky vault, the angle of incidence on the shading constantly changes, and this leads to hourly varied solar gain coefficients. Figure 21.10 shows the change of gtot for a double glazing 6/12/6 and direct-direct transmittance for blinds positioned horizontally with a cut-off angle equal to 45°.

An interesting approach to the estimation of the shading system's efficiency (Littlefair 2005) is the introduction of an "effective g-value" for the shading (g_{eff}).

Table 21.2 Typical values of g_{tot} and F_c (KEEPCOOL, 2015)

	g_{tot}	Fc
Exterior blinds	0.11	0.15
Exterior roller shutters	0.14	0.19
Exterior louvers, balconies	0.39	0.6 for south orientation
Exterior side fins	0.42	0.65 for east/west orientation
Exterior roller blinds	0.17	
Indoor solar shadings (blinds, fabrics, curtains, foils)	0.4	0.56

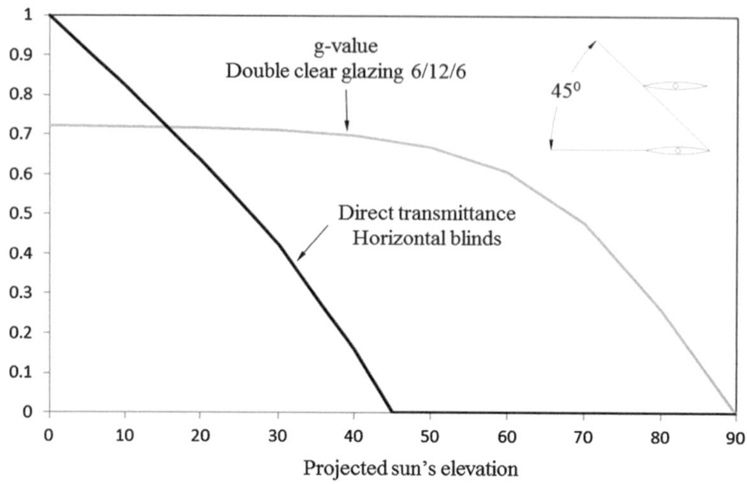

Fig. 21.10 Double clear glazing *g*-value and solar direct-direct transmittance of a blind system with 45° cut-off angle versus projected sun's elevation

This is defined as the ratio of solar gains through the window and shading system for a period, with a possibility of overheating, to the solar gains—for the same period—through unshaded and unglazed apertures. To perform a conversion from glazing *g*-value to the *g*-eff, a multiplication with a precalculated factor is needed. This factor can be used for direct comparison of different shading systems.

In EN 13790-2008, the impact of shading on the estimation of monthly energy consumption is performed through the use of a shading reduction factor for two periods (heating and cooling). This factor depends on climate, orientation and the type of shading system used and its calculation is based on the estimation of a characteristic angle (horizontal angle for external obstructions or ovehang/fin angle for overhangs and fins respectively). This procedure is presented schematically in the following figure for external obstructions Fig. 21.11.

According to the Greek Regulation for the Energy Efficiency of Buildings, the reference building should have the following values for the shading reduction factor. During the summer period at least 0.70 for the south oriented facades, 0.75 for west and east facades. For intermediate orientations the reduction factor is 0,80 for northeastern and northwestern, 0,73 for southeastern and southwestern and ofcourse 1,00 for north oriented facades.

A similar to the above described method is used in ASHRAE 90.1-2010. Fenestration g-value is modified according to a multiplier (0-1) which depends on the orientation and a parameter called Projection Factor. The latter is the ratio of the horizontal depth of the exterior shading system to its vertical distance from the fenestration sill.

EN 14501:2005 describes a number of properties that have to be taken into account when comparing shading systems. It applies to shutters, awnings and blinds. The Properties are related to thermal and visual comfort. For the thermal

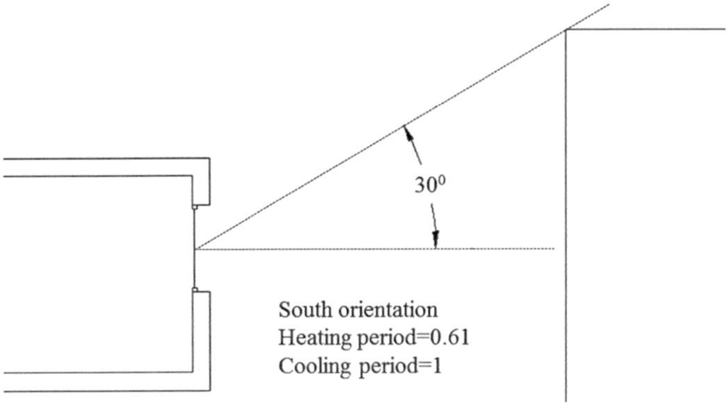

Fig. 21.11 Estimation of shading reduction factor (*1* no shading, *0* 100 % shading). South orientation, Athens

comfort, g_{tot}, a secondary heat transfer factor and direct solar transmittance are used while for the visual comfort, opacity control, glare control, night privacy, visual contact with the outside, daylight utilization and rendering of colours. The parameters used to estimate all properties in relation to visual comfort are the following visible transmittances: a) normal/normal b) normal/diffuse c) diffuse/hemispherical and d) spectral normal/hemispherical. For the classification of a shading system, a scale from 0–4 is used according to the effect that a specific parameter has on a specific property. For example for daylight utilization, diffuse-hemispherical visible transmittance (τdif-h) is used. The classification is presented in the Table 21.3. Most often it is difficult for a single shading system to meet all requirements and this is why an "onion approach" might be adopted. According to that, external shading, glazing and interior shading can work collaboratively achieving the design objectives for energy consumption and comfort. Shading form -per se- affects the way we perceive the external environment. A characteristic case is the use of a perforated metal sheet as a shading system. The same perforation ratio (percentage of holes against the total surface) can be maintained with either a large number of small diameter holes or vice versa. Although both cases can handle direct solar radiation similarly, the "quality" of visual contact with the external environment is different Fig. 21.12.

According to EN 14501:2005, visual contact is defined as the ability of a person standing on the inside, 1 m away from the fully extended shading device, to recognize a person or an object which is outside 5 m from the façade.

Table 21.3 Classification of a shading system according EN 14501:2005

Class	0	1	2	3	4
$\tau_{dif\text{-}h}$	$\tau_{dif\text{-}h} < 2\,\%$	$2\,\% \leq \tau_{dif\text{-}h} < 10\,\%$	$10\,\% \leq \tau_{dif\text{-}h} < 25\,\%$	$25\,\% \leq \tau_{dif\text{-}h} < 40\,\%$	$\tau_{dif\text{-}h} \geq 40\,\%$

Fig. 21.12 Visual contact with the exterior environment through a perforated metal sheet

21.3 Daylight Systems

Besides the reduction of solar gains, an additional function that a shading system should have is the ability to redistribute daylighting in a way that improves uniformity, reduces glare, and retains visual contact with the external environment. This can be achieved by proper handling of both sunlight and skylight as well. Visible transmittance and reflectance are key parameters affecting the system's performance. The spectral distribution of daylight is related to the glazing spectral properties and spectral reflectance of the material of the shading systems. Figure 21.13 presents spectral transmittance for three types of glazing:

The accurate calculation of daylight distribution in a space equipped with a complex fenestration system (CFS) requires knowledge of its luminous intensity spatial distribution. This can be determined by the bi-directional transmission distribution function (BTDF) of the shading/daylight system that can be either measured or simulated (depending on the system's geometrical and optical complexity). Thus, the redirection ability of the system can be demonstrated with the help of a photometric diagram under a predefined sun position; Such a graph is presented in the Fig. 21.14 for the Lightlouver system (www.lightlouver.com):

Today in a number of cases, BSDF can be created (in.xml format) using, for example, Windows 7.3, and then it can be used in radiance, either for visualization or for hourly based calculation of illuminances. Most daylighting systems

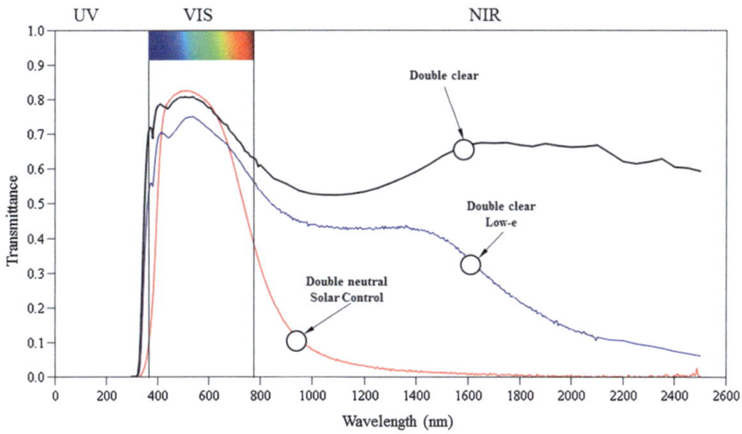

Fig. 21.13 Spectral visible transmittance for three types of double glazing

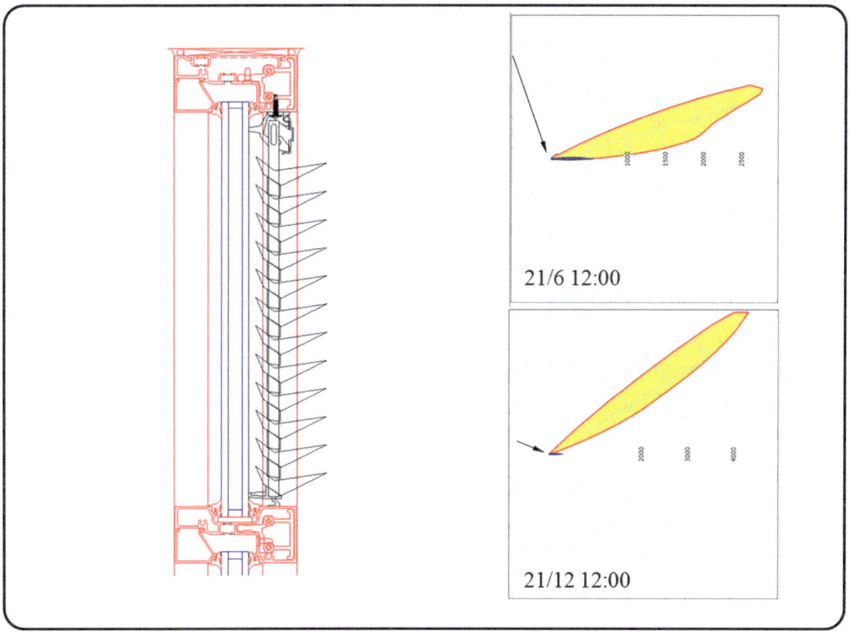

Fig. 21.14 Seasonal redirection properties of the Lightlouver ® daylighting system. Latitude 40 degrees, southern orientation, 0.5 sq m opening

have been tested during research programs, but their market penetration is rather limited, due mainly to their high cost, which is a result of the precision required during the manufacturing process; these systems help achieve an improvement in lighting quality and better occupant health, comfort, and productivity (Andersen et al. 2012). During the last decade, daylight metrics are moved from static to dynamic ones. Some building energy rating/certification schemes like LEED v.4[www.usgbc.org/v4] moved from simplistic metrics (Window to Floor Ratio) to the new dynamic ones, spatial daylight autonomy (sDA) and annual sunlight exposure (ASE). The criteria used are the percentage of measurement points achieving more than 300 lux for 50 % of the time (8:00–18:00) presented as sDA300/50 % and the percentage of measurement points having more than 1000 lux only from direct sunlight for more than 250 hours annually, presented as ASE1000,250h.

The exploitation of daylight can considerably reduce lighting energy consumption and peak electric loads as well. There are many studies that present possible energy savings, both via simulations and measurements in office buildings, due to the use of a photo sensor-controlled lighting system; literature (Doulos et al. 2008) reports lighting energy savings up to 77 %. It must be pointed out that in a number of cases, these systems do not operate satisfactorily (Papamichael et al. 2006). Conventional fenestration systems provide sufficient daylight in areas near windows called "perimeter zones." Although maximizing the areas that benefit from daylight can reduce the building's electric lighting consumption, solar heat gains and glare issues should be carefully balanced.

The building area benefited from daylight can be increased either by an increase of the perimeter zones (narrow plan) or by using daylighting systems which offer shading and sunlight redirection at the same time. Using a high reflectance ceiling, increased values of daylighting levels far away from the perimeter zone can be achieved, together with an improvement in uniformity. Various energy codes around the world propose a method for the determination of daylight zones close to façade apertures as a first step to an effective energy saving strategy. Depending on illuminance levels achieved, a daylight harvesting control system can increase energy savings considerably for both lighting and cooling. ASHRAE 90.1-2010 establishes primary and secondary daylight zones while EN 15193-2007 define the daylight zones in a slightly different way as presented in the following figure Fig. 21.15:

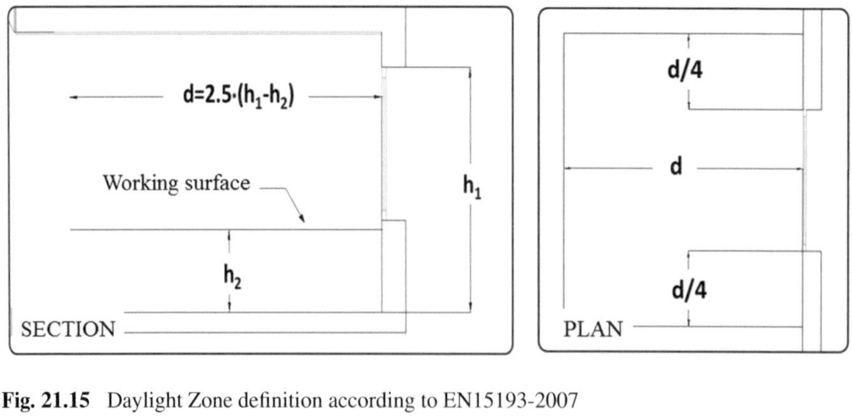

Fig. 21.15 Daylight Zone definition according to EN15193-2007

One of the design challenges is to increase the size of these zones by transferring daylight to the building core and in some cases, this can be achieved through the use of a daylight system. A very good classification of these systems together with a short description of the operation performed by Kischkoweit-Lopin (2002) and Mayhoub (2014). Depending on their installation, daylighting systems can be splitted into a) envelope systems located on the building's façade/roof capable of offering shading and b) core systems which are capable of transferring daylight to the core of the building. A number of daylight systems are presented in the following paragraphs.

21.3.1 Lightshelf

A review of the various designs that have been proposed over the years was presented by Tsangrassoulis et al. (2014). These designs varied from simple stationary projecting mirror arrays (Stiles and McCluney 1994) to optically complicated anidolic ones (Courret et al. 1998), in an effort to supplant artificial lighting with glare-free daylight (if possible). Lightshelves divide the window into two parts, a lower-view window, and a clerestory that acts as a daylight provider. Despite its form, a common characteristic is the use of highly reflective material (specular or diffuse) on its upper surface. Since sunlight reflection on this surface can cause glare, it is advisable that it not be seen by the occupants of the space. Exterior lightshelves offer better shading, while interior ones offer better glare protection. Parameters that have an impact on the proper selection of a lightshelf are:

- Tall windows, which are preferable
- High ceilings are recommended. Lightshelves work best with ceilings higher than 3 m (Littlefair 1995). The best ceiling shape, in an effort to increase uniformity, is one that is curved in the front and in the rear of the room (Freewan et al. 2008).
- Exterior lightshelf length can be up to 1.5 times the height of the clerestory window (Selkowitz et al. 1983), or smaller than its distance from the working surface (Littlefair 1995). For interior ones, the depth is roughly equal to the height of the clerestory, and for the exterior (Freewan et al. 2008), no more than its distance from the working plane. Lightshelf length should be optimized mainly according to ceiling and clerestory height.
- Overall, lightshelves can perform best when there are external obstructions increasing core illuminance by 15 %.
- Lightshelf tilt can affect the shading and—depending on the upper surface material—enhance sunlight penetration during the summer. Moore (1991) suggests shelves painted white for the south, a tilt equal to 40 degrees—(latitude/2).
- Research on movable lightshelves is limited, although in 1990 Place and Howard (1990) tested a sun-tracking mirror shelf, increasing daylight to 14 m away from the window (Fig. 21.16).

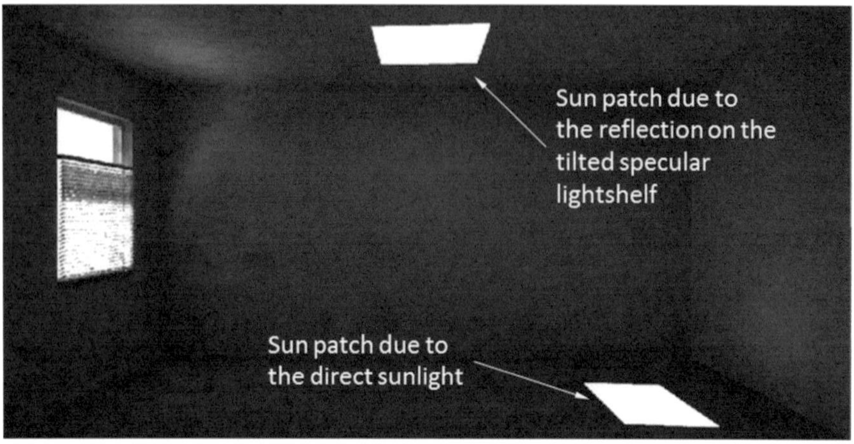

Fig. 21.16 Sunpatches due to direct and reflected sunlight

21.3.2 Blinds

Blinds are the most common system and can be installed either externally or internally, or can be sealed in a double-glazing unit. The use of highly reflective materials on the upper surface may redistribute daylight onto the roof; their tilt is the main mechanism for controlling solar gains, and in many cases that can be antagonistic towards the visual contact with the external environment. Two solutions have been proposed in order to improve this problem:

- A part of the slats' surface is perforated.
- The use of a static slat with tailor made profile presenting seasonal selectivity to solar radiation.

The use of highly reflective materials helps redirect the direct sunlight deeper into a space, but this can cause glare problems. An interesting approach is the separation of a blinds system into two parts, with the upper part acting as daylight provider and the lower one used for viewing (Jia and Svetlana 2011) (Fig. 21.17).

The change in the tilt angle has an impact on the luminance distribution, especially when there is direct sunlight in the space; this is shown in Fig. 21.18. For the creation of the images, a fish-eye lens (Sigma 8 mm f4 EX Circular Fisheye) was used, together with the "high dynamic range" technique and Hdroscope software[1] for the analysis.

Obviously, the tilt control can be done automatically using a controller, having as input either a radiation incident on the façade and/or daylight/temperature within the room. Although existing technology allows for continuous activation or modification of the tilt angle, this ability may turn into a discomfort factor. Hence

[1]http://courses.washington.edu/hdrscope/download.html.

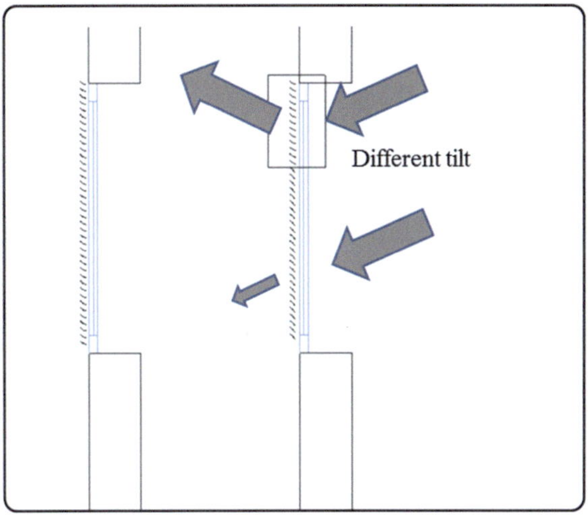

Fig. 21.17 Optimizing solar gains-daylighting by splitting a blind system into two parts with different tilts

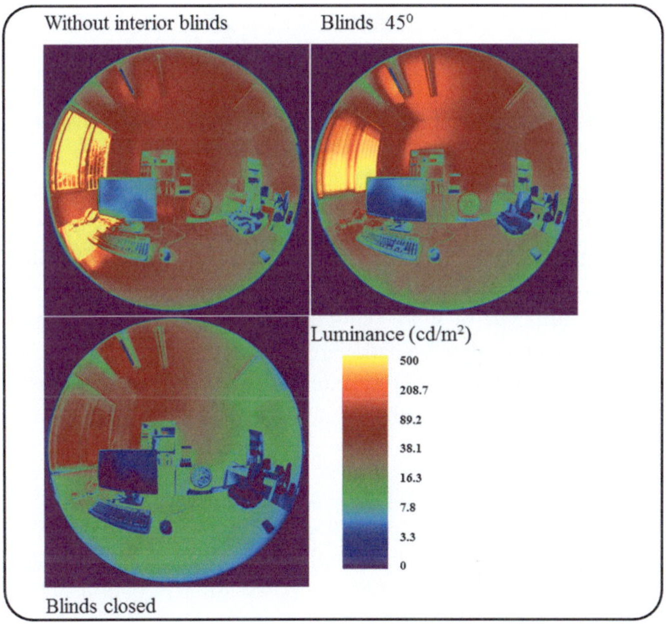

Fig. 21.18 Luminance distribution change due to interior blinds tilt angle. High dynamic range photographs

Fig. 21.19 Retrolux system

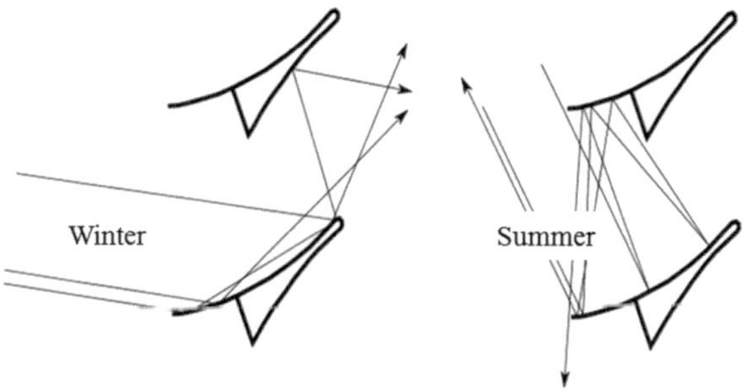

Fig. 21.20 Operating principle of Okasolar system

the activation frequency plays an important role in both energy consumption and visual comfort. Increasing tilt angle (0° represents a horizontal position) substantially reduces daylighting levels, while for angles greater than 45°, there is no additional benefit in terms of glare reduction. In addition to conventional blinds, there are systems that have been developed mainly in order to satisfy requirements for shading, visual contact with the external environment, and direct sunlight redistribution Figs. 21.19 and 21.20.

Examples of such systems are:

- Retrolux system (Koester Lichtplanung 2015), which is a toothed, two-part blind system that is capable of presenting a seasonal selectivity to the radiation of the sun together with redirection, maintaining a constant position. The outer part of the blind is used for retro-reflection of solar radiation, while the inner part is for redirection (Fig. 21.19).
- Okasolar-W (2015). The system per se is located in the double-glazing cavity. It is static and highly reflective, thus the cavity offers weather protection. Due to various existing geometrical variations, the visible transmittance of the system,

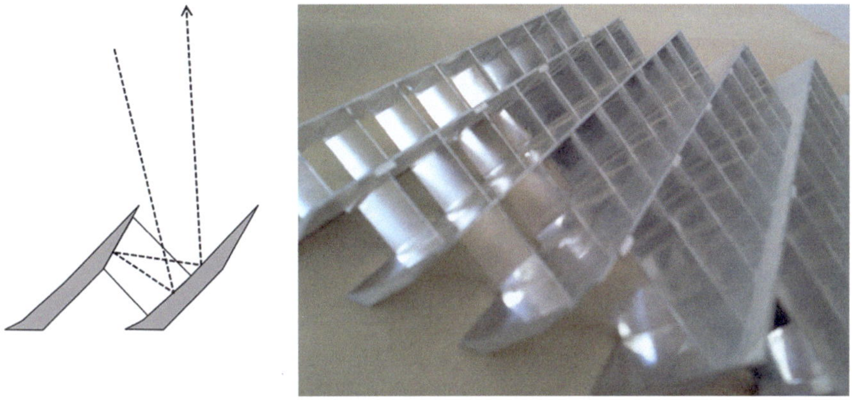

Fig. 21.21 Operation principle of Micro Sun Shielding Louver system

when low-e glazing is used, is between 5 and 57 % and *g*-value between 17 and 45 %. Similarly, cut-off angles vary from 25–30° while the percentage of horizontal vision blockage is between 38 and 45 % (Fig. 21.20).

- Micro Sun Shielding Louver.[2] A mid-pane system that is placed in horizontal openings. Highly reflective material is used, while its shape is designed to use sky light from the northern part of the sky and reject solar radiation. Proper design of its acceptance angle determines the shading performance. Its *g*-value is 14 %; visible direct-direct transmittance is between 0 and 55 %, while visible diffuse transmittance is from 12 to 38 % (Fig. 21.21).

All the aforementioned systems use reflection to achieve redirection. However, similar results can be achieved using transparent materials such as PMMA. By modifying the surface geometry, affecting reflectance and refraction, the incident solar radiation can either be rejected or redirected, simultaneously allowing the entrance of diffuse skylight. The view is distorted, so for their installation, openings located far from the line of sight are used. These systems can be movable, with their most common installation spot the cavity between two glazing panes. Examples of such systems are:

- **Prismatic Panels**

  One surface of the panel consists of small prisms, one side of which might be coated with highly reflective aluminum. Any differentiation of the prism's geometry can modify its optical properties. It can be used as either sunlight excluding or redirecting systems. Their operation principle is shown in Fig. 21.22.

The manufacturing process can affect their performance and should be carried out accurately. Edge inaccuracies can create the dispersion of the sunlight into its component colors.

[2]http://www.siteco.com/en/products/daylight-systems/micro-sun-shielding-louvre.html.

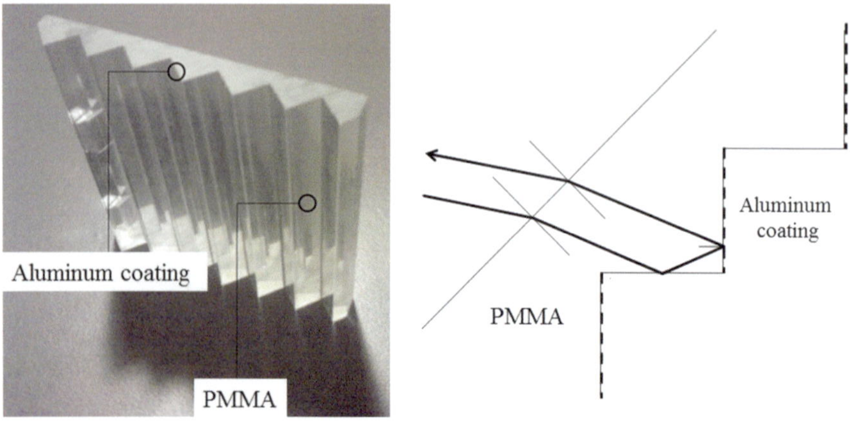

Fig. 21.22 Picture and operating principle of a prismatic panel

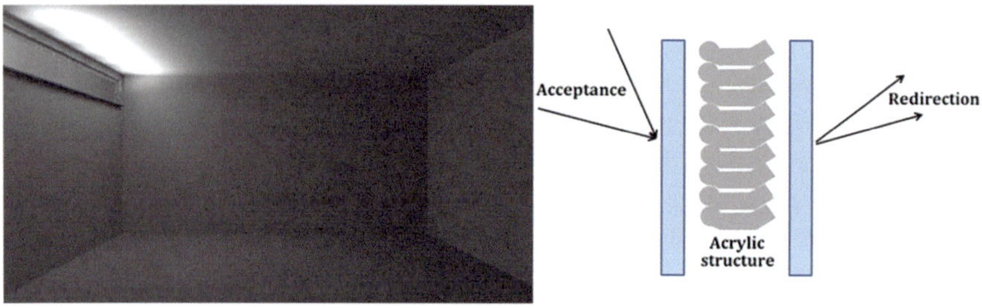

Fig. 21.23 Schematic representation of Lumitop SGG system (*right image*) and simulation (*left image*) (*g*-value = 30 %)

- **Acrylic structures**
  This acrylic structure[3] is placed in the double-glazed cavity, usually in clerestory windows offering redirection of sunlight towards the ceiling (Fig. 21.23).
- **Laser-cut panels**
  Their construction is relatively simple and carried out with the help of small incisions on a sheet of acrylic. The incisions (~75 % of the sheet's depth for reasons of structural stability) cause total reflection of incident radiation with no distortion of the visual contact with the outside environment (Fig. 21.24).

Results that are similar to the above-mentioned systems can be achieved if there is a reduction in the dimensions of the systems to make films. Two typical products of this kind are:

[3]http://merint.com/site/ourbusiness/SGG%20LUMITOP.pdf.

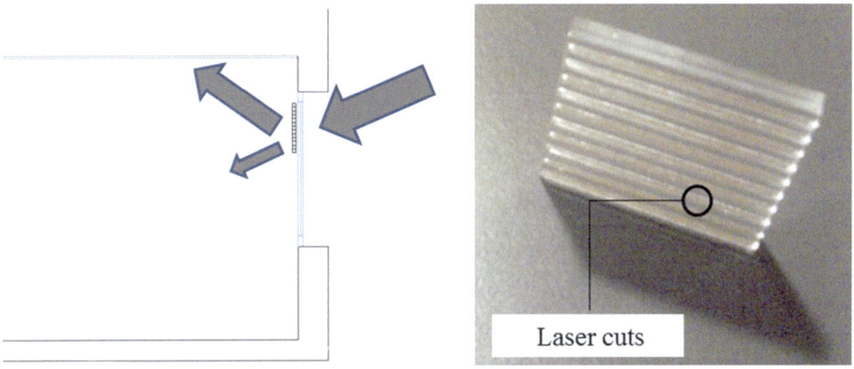

Fig. 21.24 Laser-cut panel picture (*right*) with indicative installation

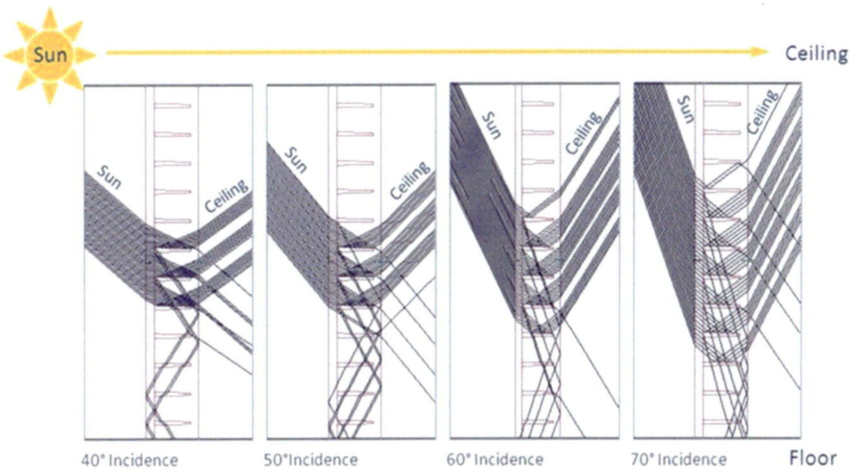

Fig. 21.25 Sunlight redirection for various sun's elevation angle using SerraGlaze ®

- **SerraGlaze,**[4] which is an acrylic film containing a multitude of microscopic internal elements that redirect sunlight to the ceiling through total internal reflection while permitting a near normal view outwards (Fig. 21.25).
- **Holographic Optical Elements (HOE)**

Holographic optical elements are films that are laminated between two panes of glazing. Modifying their density, selective sunlight redirection can be achieved through diffraction as presented schematically in Fig. 21.26.

[4]http://serraluxinc.com/serraglaze-daylight-redirecting-film/serraglaze-behavior/.

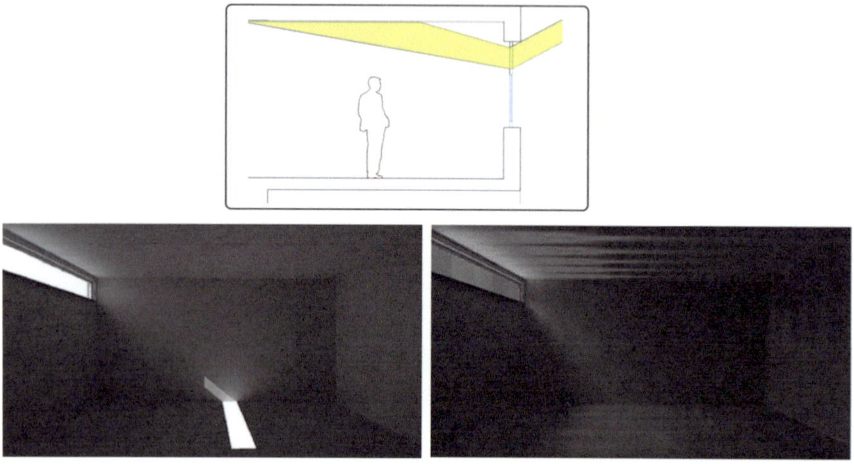

Fig. 21.26 Typical installation of HOE in a vertical window. Simulation of two different cases. Clear glazing (*left image*), HOE (*right image*)

21.3.3 Daylight Transporting Systems

Although daylight transporting technology is considered a cutting-edge technology today, the first prototype system was realized in 1881 when William Wheeler designed a light transmission system very similar to a current optical fiber system. These systems consist of (1) a collector; (2) a transmission system, and (3) an emitter.

Collection is achieved using sunlight concentration either through the use of a mirrored optical system or a lens (Fresnel). Since the sun's angular dimension is small, any tracking system has to be quite accurate. Excessive concentration can lead to significant overheating problems with the transport material. Low concentration can probably make transportation uneconomical when compared with the cost of conventional light sources.

The simplest transmission system is a highly reflective guide (solar tube), although transmission can be achieved with hollow prismatic guides or, rarely, a solid core. It is a passive system, in the sense that there are no moving parts to track the sun's position.

Solar tube ends have to be covered with a transparent material (or translucent for the emitter), since the ingress of dust reduces system performance. Other factors that affect performance are the aspect ratio (ratio of length to diameter), the curvature, and the number of bends. Any increase in the number of reflections that occurred inside the guide reduces the emitted light flux. Thus, for a vertical solar tube, low sun altitude angle during the winter reduces efficiency. A number of manufacturers, in an effort to solve the problem, adopt the idea of changing the direction of the sun's rays by using redirection systems such as transparent prismatic elements or appropriate modification of the outer transparent cover, creating a kind of lens.

A hollow prismatic guide can be realized using a thin film incorporating microscopic prisms (Optical Light Film, 3 M) http://starlight-sl.de/mb/3M_Lighting_Films/OLF2301_neu.pdf. Its thickness is small enough to be configured to form a square or circular cross-section tube, while its operation is based on achieving total reflection for rays that strike the smooth side of the film at an angle greater than the critical angle. For the plastic material used in OLF, light must enter the guide at an angle of 27.6° or less from the axis of the guide in order for total reflection to be achieved. If the angle is greater than the light's escape from the guide, the reflectivity of this material reaches up to 99 % (depending on the angle of incidence). The total reflection condition can be interrupted either by placing a diffuser inside the guide, or creating holes or modifying the surface. The majority of these systems are placed vertically in an attempt to transfer natural light to areas of the building with little or no access to it. But there are also designs in which this transfer is made horizontally, usually through the space above the false ceiling. Urban geometry can be a problem due to the reduction of available light flux to the systems' entrance.

In the following graph, an anidolic ceiling (Scartezzini and Courret 2002) is presented. It consists of an anidolic zenithal collector, a light duct, and a diffuser at the end. The overall efficiency of the system is 37 %, depending on the reflectance of its surfaces. The system can increase daylight factors by 1.7 times at a distance of 4.5 m from the window (Fig. 21.27).

Light transmission can be achieved with optical fibers as well. These have a core of a transparent material (plastic or glass) that is surrounded by a material with a lower refractive index so that the total reflection conditions can be achieved,and the whole structure is surrounded by an outer cover for protection. Depending on their diameter, a large number of fibers may be bundled together, offering greater flexibility in designing a lighting system.

Fibers that are typically used for illumination transfer have a diameter between 0.25 and 5 mm. The diameter determines the flexibility of the fiber, which in turn

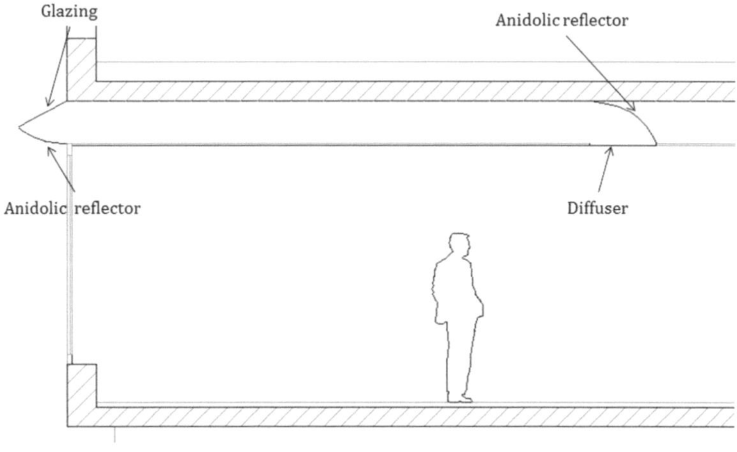

Fig. 21.27 Schematic representation of an anidolic ceiling

defines the radius of curvature. Various transparent materials are used to manufacture fibers, but for daylight transfer applications, acrylic (PMMA) is prevalent, due to good transmission and heat resistance properties (−55–70 °C in low humidity environments) with a PVC housing. Glass fibers are produced by the extrusion of borosilicate glass. The use of silica glass in the core of the fiber can significantly reduce light attenuation. Acrylic (PMMA) fibers are more transparent at short wavelengths, while the opposite occurs with the glass ones (due to Rayleigh scattering). The fiber's acceptance angle represents a cone of light that can be guided by its core. Its sine is multiplied by the refractive index of air called a "numerical" aperture, and it is a measure of how much light can be collected. There are various systems available, such as the PARANS system presented in Fig. 21.28.

Fig. 21.28 Operational principle of the PARANS system (http://www.parans.com/eng/)

Fig. 21.29 Solux system tested during Universal Fiber Optics project

The output of a fiber-optic system can increase by using a concentrator; such a system is the Solux system, created by Bomin Solar Research. It uses a 1 m diameter fresnel lens to concentrate sunlight into a liquid fiber-optic, which is in turn connected to a flat Prismex emitter (Tsangrassoulis, 2005) (Fig. 21.29).

21.3.4 Heliostat

Sunlight transmission can be achieved by using mirrors (heliostats) as well, as shown in Fig. 21.30.

The operating principle of a heliostat is simple. Usually, a flat, movable, automatically controlled mirror is used, sending reflected rays to a static target—which can be another mirror. In most cases, sunlight rays are considered parallel, without the accuracy of shading calculations being significantly affected.

However, when sunlight is transferred over long distances using reflections, the finite dimensions of the source (sun) should be taken into account. Thus, the error in the calculations of transmitted flux due to the size of the solar disk is

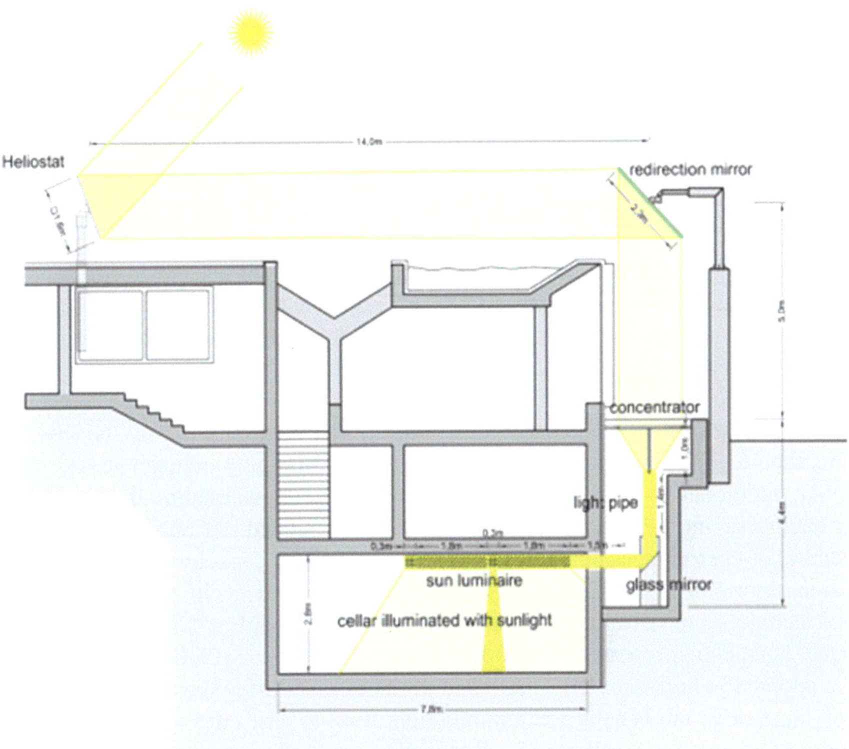

Fig. 21.30 Sunlight transfer in a building's basement (Bartenbach 2015)

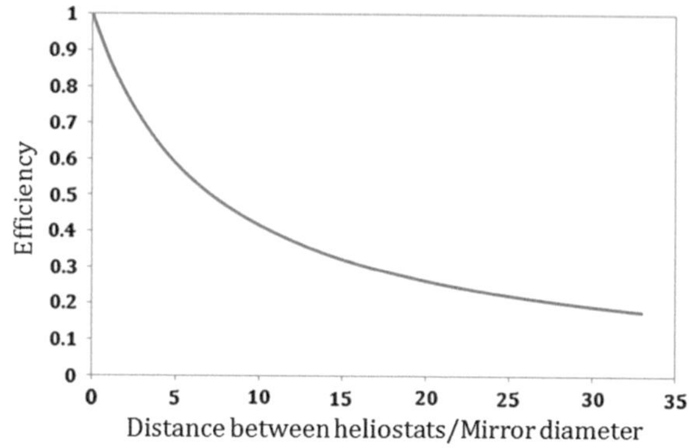

Fig. 21.31 Sunlight transmission efficiency when a heliostat and a same-size static mirror are used

proportional to its angular diameter, which is the angle that the sun subtends to an observer on the earth. This is equal to $2 \times e$ ($e = 0.260$). Consequently, each time sunlight is reflected onto a mirror, flux is reduced—due to the mirror's surface irregularities ($2 \times \delta$), curvature errors ($2 \times \beta$), and alignment errors either between heliostat and sun or between mirrors ($2 \times 2\alpha$). The sum of all the above errors determines flux transmission efficiency. Adding up these errors, and assuming an overall divergence of $2 \times 4°$, the efficiency of sunlight transmission from a heliostat to another static mirror of the same size, in relation to the ratio distance/diameter, is shown in Fig. 21.31.

The use of natural lighting in the core of the building can contribute significantly to lighting energy savings. With proper management of solar gains, total energy consumption and peak loads can be reduced as well. However, the design and installation of such systems presents a series of drawbacks that slow down their widespread adoption. It has to be realized that there isn't a solution that satisfies every demand. Climatic conditions play an important role since—depending on prevailing conditions (i.e., cloudiness)—a different system has to be selected. Some provisions during a building's initial design phase should be taken when daylight transportation systems are planned to be used in order to transfer flux in the most efficient way (small distances, a small number of redirections). Their disadvantages are actually a combination of a limited flexibility during the installation, their high cost for the light flux delivered, and difficulty in maintenance.

An additional problem is the difficulty in accurately estimating the possibility of energy saving during the design phase. This is related not only to the lack of goniophotometric measurements, but also to the limited capacity of existing simulation algorithms to manage this type of data. Nevertheless, in recent years tools and techniques have begun to emerge (five-phase method, E + CFS that can effectively handle the properties of complex fenestration systems. Although the systems are adopted in high-profile projects and in many cases offer superior light quality, any increase in the benefit (i.e., emitted light flux) to cost ratio would be valuable for their adoption in smaller projects as well.

References

Andersen M, Mardaljevic J, Lockley SW (2012) A framework for predicting the non-visual effects of daylight—part i: photobiology-based model, lighting. Res Technol 44:37–53

ASHRAE 90.1-2010 (Energy Standard for Buildings Except Low-Rise Residential Buildings

ASHRAE Standard 189.1-2014, Standard for the Design of High-Performance Green Buildings

Bartenbach (2015) Available via DIALOD www.bartenbach.com. Cited 10 Jan 2015

Bellia L, de Falco F, Minichiello F (2013) Effects of solar shading devices on energy requirements of standalone office buildings for Italian climates. Appl Therm Eng 54:190–201

Bülow-Hübe (2007) Solar shading and daylight redirection. Demonstration project for a system of motorized daylight redirecting venetian blinds and light controlled luminaire, Energy and Building Design, Lund University

Courret G, Scartezzini J-L, Francioli D, Meyer J-J (1998) Design and assessment of an anidolic light-duct. Energy Build 28:79–98

Doulos L, Tsangrassoulis A, Topalis F (2008) Quantifying energy savings in daylight responsive systems: the role of dimming electronic ballasts. Energy Build 40:36–50

Dubois M-C (2001) Impact of shading devices on daylight quality in offices. Simulations with radiance, Report No TABK -01/3062, Lund 2001, ISSN 1103-4467

EN 13363-1:2003 + A1:2007 Solar protection devices combined with glazing—calculation of solar and light transmittance—part 1: simplified method

EN 13363-2:2005 Solar protection devices combined with glazing—Calculation of total solar energy transmittance and light transmittance—part 2: Detailed calculation method

EN 13790-2008 (energy performance of buildings—calculation of energy use for space heating and cooling)

EN 14501:2005 Blinds and shutters. Thermal and visual comfort. Performance characteristics and classification

EN 410-2011 Glass in building. Determination of luminous and solar characteristics of glazing

EN 673:2011 Glass in building. Determination of thermal transmittance (U value). Calculation

ESCORP—EU25 (2005). Energy savings and CO_2 reduction potential from solar shading. Available via DAILOG www.es-so.eu. Cited 15 Jan 2015

Freewan A, Shao L, Riffat S (2008) Optimizing performance of the lightshelf by modifying ceiling geometry in highly luminous climates. Sol Energy 22:287–384

Jia H, Svetlana O (2011) Illuminance-based slat angle selection model for automated control of split blinds. Build Environ 46:786–796

Kischkoweit-Lopin Martin (2002) An overview of daylighting systems". Sol Energy 73(77–82):2002

Koester-lichtplanung (2015) Available via DIALOD http://www.koester-lichtplanung.de/pages_gb/product_01.html Cited 10 Mar 2015

Konstantoglou M, Tsangrassoulis A (2012) Dynamic building envelope system: a control strategy for enhancing daylighting quality and reducing energy consumption. In: Energy forum conference, 06–07 December 2012, Bressanone, Italy.

Kostantoglou M, Kontadakis A, Tsangrassoulis A (2013) Dynamic building skins: performance criteria integration. In: PLEA2013—29th conference, sustainable architecture for a renewable future, Munich, Germany, 10–12 Sept 2013

Lechner N (2001) Heating, cooling, lighting design methods for architects. Wiley Publ. ISBN 0-471-24143-1

Lee ES, DiBartolomeo DL, Vine EL, Selkowitz ES (1998) Integrated performance of an automated venetian blind/electric lighting system in a full-scale private office. In: ASHRAE/DOE/BTECC conference, thermal performance of the exterior envelopes of buildings VII

Littlefair PJ (2005) Summertime solar performance of windows with shading devices, BRE Trust Report FB9 (Garston: BRE)

Littlefair P (1995) Light shelves: Computer assessment of daylight performance. Light Res Technol 27:79–91

Mandalaki M, Zervas K, Tsoutsos T, Vazakas A (2012) Assessment of fixed shading devices with integrated PV for efficient energy use. Solar Energy 86:2561–2575
Mayhoub MS (2014) Innovative daylighting systems challenges: a critical study. Energy Build 80:394–405
Meek C, Breshears J (2010) Dynamic solar shading and glare control for human comfort and energy efficiency at UCSD. In: Sol. 2010 conference, Phoenix, USA, 17–22 May 2010
Meteonorm (2015) Remund J, Kunz S (2004) METEONORM handbook. METEOTEST
Moore F (1991) Concepts and practice in architectural daylighting. Van Nostrand Reinhold, New York
NREL 2015. https://beopt.nrel.gov/file/DView125.exe
Okalux (2015) Available via DIALOD http://www.okalux.de/fileadmin/Downloads/Downloads_englisch/Infotexte/i_okasolar_w_e.pdf Cited 10 Mar 2015
Papamichael K, Page E, Graeber K (2006) Cost effective simplified controls for daylight harvesting". ACEE Summer Study Energy Effi Build 3:208–218
Place W, Howard TC (1990) Daylighting multistorey office buildings. North Carolina Alternative Energy Corporation
Report on Available Products and Passive Cooling Solutions, KEEPCOOL project, EIE/07/070/SI2.466264 Keep Cool II
Scartezzini JL, Courret G (2002) Anidolic Daylighting systems. Sol Energy 73:123–135
Selkowitz S, Navvab M, Mathews S (1983) Design and performance of light shelves. In: International daylighting conference, Phoenix, AZ
Stiles MR, McCluney R (1994) Daylighting commercial and educational rooms to 750 lux with stationary projecting reflector arrays (SPRA): a simulation. In: ACEEE summer study proceedings
Szokolay S (2007) Solar geometry: passive and low energy architecture international in association with Department of Architecture, The University of Queensland Brisbane, 2007, ISBN 0 86766 634 4
TNO Built Environment and Geosciences—Glazing type distribution in the EU building stock—TNO Report TNO-60-DTM-2011-00338 –February 2011
TNO report 2007-D-R0576/B (2007) Impact of solar control glazing on energy and CO_2 savings in Europe
Tsangrassoulis A, Kostantoglou M, Kontadakis A (2014) Lighting energy savings due to the use of a sun tracking mirrored lightself in office buildings. In: ASHRAE hellenic chapter 3rd international conference "ENERGY in BUILDINGS", Athens
Tsangrassoulis A, Doulos L, Santamouris M, Fontoynont M, Maamari F, Wilson M, Jacobs A, Solomon J, Zimmerman A, Pohl W, Mihalakakou G (2005) On the energy efficiency of a prototype hybrid daylighting system. Sol Energy 79:56–64

Chapter 22
The State of the Art for Technologies Used to Decrease Demand in Buildings: Electric Lighting

Wilfried Pohl

Abstract In temperate climates, lighting represents a significant part of the consumed end-energy of a building; considering the primary factor, it is very often the biggest part. Modern lighting techniques, especially the groundbreaking LED technology, offer a great variety of energy-conscious solutions. Replacing antiquated and inefficient lighting installations by energy-efficient techniques is a huge energy-saving potential, and, in parallel, the lighting quality could be significantly improved. By doing so, we should not lose sight of the primary goal of the system: to lighten a space according to architectural, visual, and biological requirements. LEDs as a digital light source offer completely new technical possibilities for fulfilling these special needs.

22.1 Energy Consumption by Electric Lighting

According to the statistics, 19 % of the world's electric energy consumption is used for lighting, causing 2.4 % of the primary energy consumption, and we are facing an increase of approximately 80 % by 2030. Lighting is a substantial energy consumer and a major component of the service costs in many buildings.

For temperate climates, lighting is a major consumer of building energy. The exemplary evaluation of a standard office room in Innsbruck, Austria (Central Europe) shows this kind of lighting and energy performance:

The south-oriented façade is modeled with a parapet and a clear glazing in the upper part. In front of the clear glazing, standard external Venetian blinds are at a 45° tilted position whenever the luminance experienced at the façade from inside exceeds 1000 cd/m^2 or—during the cooling period only—the irradiation at the façade exceeds 150 W/m^2. Assuming a new, efficient building standard, one

W. Pohl (✉)
Bartenbach GmbH, Rinner Strasse 14, 6071 Aldrans, Austria
e-mail: Wilfried.Pohl@bartenbach.com

S.-N. Boemi et al. (eds.), *Energy Performance of Buildings*,
DOI 10.1007/978-3-319-20831-2_22

can see that in that particular moderate climate, hardly any cooling is needed. It also shows that the lighting demand is an essential part of the overall annual energy need—in this example around one-third, about half as much as the heating demand. If we take into account the primary factor, it is most likely the biggest part (Figs. 22.1 and 22.2). (Source: www.DALEC.net Web-based simulation tool for coupled lighting and thermal simulations)

The percentage of the electricity used for lighting in European buildings is 50 % in offices, 20–30 % in hospitals, 15 % in factories, 10–15 % in schools, and 10 % in residential buildings. At the moment, fluorescent lighting dominates in offices, but 50 % of the connected power is the antiquated T12 (38 mm diameter) or the T8 (26 mm) lamp. Replacing these lamps by the T5-type (16 mm; together with high-frequency ballasts), and meanwhile also by installing up to date and high-performance LED-solutions, allows us to decrease the energy consumption significantly.

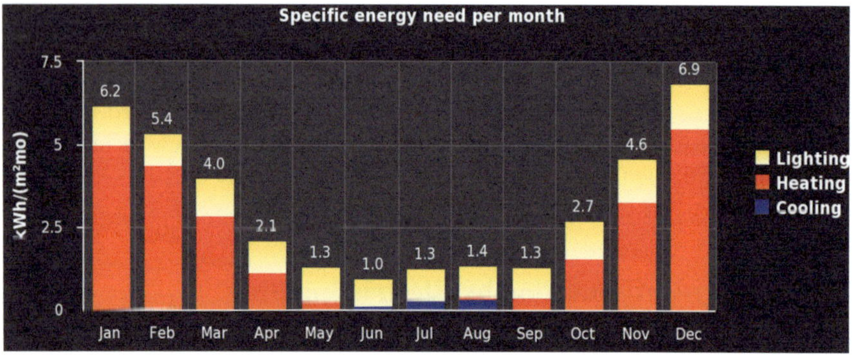

Fig. 22.1 Monthly energy consumption of a standard office room in Innsbruck, Austria

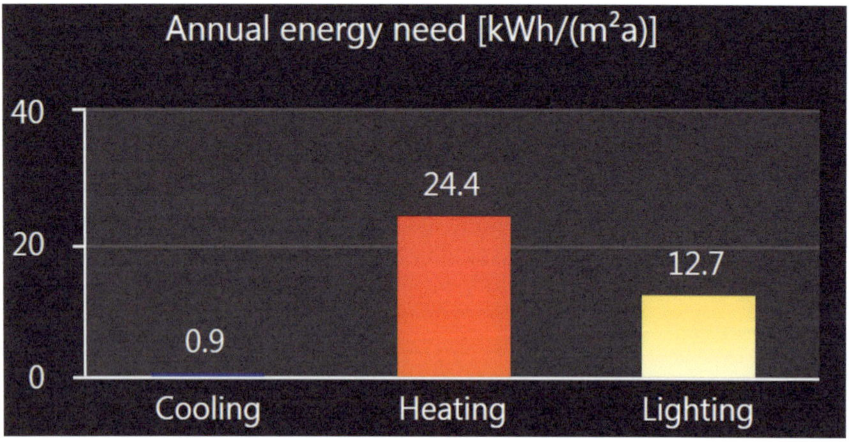

Fig. 22.2 Yearly energy consumption of a standard office room in Innsbruck, Austria

In domestic lighting, the dominant light source (>80 % connected power) is the inefficient GLS bulb (incandescent lamp, 12 lm/W), consuming about 30 % of the electric lighting energy for total residential lighting. Replacing them by CFLs, IRC tungsten halogen lamps, and by LEDs, could save most of the energy used for residential lighting. The development of high reflective (high specular or diffuse) surfaces for lighting purposes, of complex surface calculation methods and of new manufacturing technologies (e.g., injection-molded plastics with Al-coating) has improved the efficiency (light output ratio) of luminaires (reaching 80 % or more); the emerging LED-technique will also follow this trend.

Eighty to 90 % of the running lighting installations are more than 20 years old, i.e., these installations use antiquated and inefficient lighting equipment. Replacing them with energy-efficient components (lamps, controls, ballasts, and luminaires) is potentially a huge energy savings, and, in parallel, the lighting quality could be improved. If we realize the technical potential for energy savings, we can stop the yearly growth of electric power consumption for lighting and freeze the total energy consumption on today's level. (Source: Halonen et al. (2009))

22.2 Policies and Standards

To promote the improvement of the energy performance of buildings within the community, the European Parliament has enacted several directives. For example, on April 22, 2009, the European parliament voted that all new buildings from 2019 on must fulfill the standard for zero-energy buildings. Zero-energy buildings are those that produce as much energy as they consume and represent the cutting edge of energy efficiency in buildings. Whereas in common buildings, the electric energy consumption for lighting may be comparably small, for such low or zero-energy buildings, the demand for electric energy for artificial lighting becomes a decisive matter.

Therefore, the European Parliament has established several energy-labelling and eco-design directives (e.g., Directive 2009/125/EC) that require minimal performances and lead to the outphasing of different antiquated lamp types (e.g., incandescent lamps and mercury lamps).

Procedures to evaluate the fulfillment of energy and quality requirements of artificial lighting are specified in the European standards:

- European standard EN15193 "Energy performance of buildings—Energy requirements for lighting." The standard defines very rough procedures for the estimation of energy requirements of lighting in buildings, and provides guidance on the establishment of national limits for lighting energy based on reference schemes.
- European standard EN12464 "Light and Lighting—Lighting of Work Places." The standard gives quality requirements to be fulfilled by lighting installations.

22.3 Energy Performance Factors for Lighting Installations

Converting electrical power into lighting resp. visibility of the environment takes place in three steps, the lamp, the luminaire, and the room. Furthermore, the operating times and dimming states are decisive for the energy consumption, and so control aspects are very important.

For economic and ecological evaluation of different lighting solutions, a life cycle analysis has to be carried out; this means that costs, energy, and consumed resources for the entire lifetime (from installation until disposal) have to be considered. The awareness of practitioners and end-users that the operating costs, especially the energy costs of a lighting installation during the life cycle, are very often the biggest part of the whole costs, has to be increased.

Modern lighting techniques offer a great variety of energy-conscious solutions, and there is a huge potential to save energy for lighting. In lighting, we can identify three groups of trades, which transform electrical energy into visibility, starting with the power connection (grid) and ending with the luminances of the room surfaces (visual environment): the lamp (light source, including controls and ballasts), the luminaire, and the room. The lamp transforms electric power into light flux, the luminaire distributes the light in the room according to the visual requirements, and the room transforms this light into visible luminances by the surface reflections, thus creating the visual environment.

The energetical performance of these various transformations are characterized by three factors:

- Lamp efficacy (in lm/W, including operating device)
- Luminaire light output ratio (LOR, in %)
- Room utilization factor (RUF, in %)

The product of these factors gives the ultimate (total) utilization of the electric light installation.

The energy consumption of the installation is further defined by the operating times, i.e., the need for artificial lighting should be minimized by intelligent architecture and daylight harvesting. To avoid needless operating of the artificial light, proper controls (occupancy, daylight dependence, etc.) have to be installed. A key point in energy-efficient lighting design is the choice of efficient lamps, which produce the proper spectrum (color temperature CCT and color rendering CRI), and offer the required operating features.

Besides the use of high-efficient lamps, the application of high-quality luminaires with efficient lighting concepts and controls is important for the visual and ecologic quality of all the lighting. A luminaire is a device to operate the lamp and to control the outcoming light flux. It is a complete lighting unit, which comprises the light source, including electric operating devices (transformer, ballast, ignitor), together with the parts that position and protect the lamp(s) (casing, holder, wiring) and connect the lamp(s) to the power supply, and, finally, the parts that distribute the light (optics). The luminaire's function (if not a pure decorative fitment) is to direct light to the appropriate locations, creating the required visual environment, without causing glare or discomfort.

Different lamp technologies require different luminaire construction principles and features. For example, a metal halide lamp HCI 150 W (extremely high power density, very small, luminance 20 Mio cd/m^2, bulb temperature ca. 600 °C), compared with a T5 fluorescent lamp HO35 W (diameter 16 mm, 1.5 m long, surface temperature 35 °C, luminance 20,000 cd/m^2) requires completely different luminaire types (mechanical and thermal construction, wiring and electronics, and optics) (Fig. 22.3).

The efficiency of a luminaire is characterized by the light output ratio (LOR), i.e., the ratio

$$\eta_{\mathrm{Luminaire}} = \frac{\phi_{\mathrm{Luminaire}}}{\phi_{\mathrm{Lamp}}} = \mathrm{LOR}$$

The efficiency (LOR) of a luminaire depends mainly on the lamp with electronic operating device and on the optical components. The usage of highly reflective (specular or diffuse) surfaces, of complex calculation methods, and of new manufacturing technologies (e.g., injection-molded plastics with Al-coating) has improved the efficiency of luminaires now reaching 80 % and even more.

Light-emitting diodes (LEDs) have undoubtedly been the most revolutionary innovation in lamp technology for decades. Powered by several watts, emitting white light with an efficacy of more than 120 lm/W (and envisaging 200 lm/W), and with a color-rendering index (CRI) greater than 90, they meanwhile outperform all traditional lamps.

The benefits of LEDs are a long lifespan—up to 60,000 h, color mixing possibility (flexible color temperature CCT), UV-less and IR-less ("cold," i.e., no infrared) spectrum, design flexibility, and brilliant light due to its small size (i.e., high luminance), easy control and dimming with almost zero response times (digital light source), safety due to low-voltage operation, ruggedness, and a high efficacy (lm/W) compared to conventional lamps.

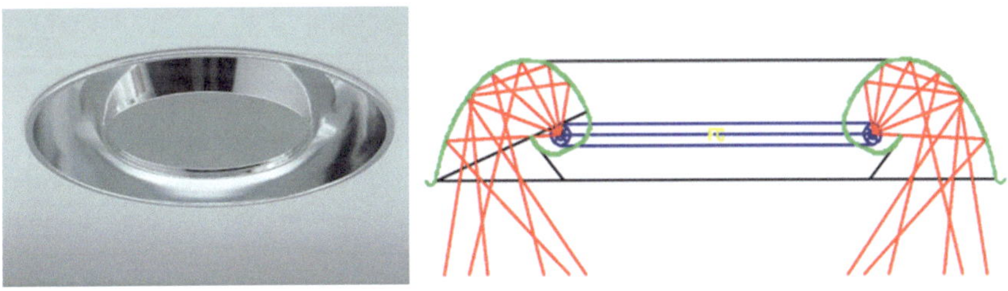

Fig. 22.3 Circular fluorescent fitting with secondary radiation technique and high-quality shielding

Due to the low prices and high lumen output, fluorescent tubes with an efficacy of approximately 100 lm/W and life spans of approximately 20,000 h, are the most economical and commonly used lamps at the moment; today, more than 55 % of artificial lighting is generated by this type of lamp. LEDs are expensive and offer a much lower light flux per piece, i.e., one needs LED arrays respectively LED engines to reach the same light flux output. But this economic gap between conventional light sources, especially fluorescent tubes, and LEDs is decreasing very quickly, and for some installations it is already now more economic to use LEDs than conventional lamps. For home lighting, meanwhile, LEDs are also an economical alternative, compared to the most frequently used lamp types, the GLS-bulb and the tungsten halogen lamp, with very low efficacy (<20 lm/W) and life time (<4000 h).

The other obstacles that have slowed down the mainstream applications of LEDs are missing industrial standards (holders, controls and ballasts, platines, etc.); the required special electronic equipment (drivers, controls); short innovation cycles of LEDs; and special optics that are rather different from conventional metal fabrication. But these restraints will be eliminated more and more.

The efficacy and intensity of LED radiation greatly depends on its temperature, since they are much more sensitive to heat conditions than conventional lamps. It is therefore essential to care for an optimal heat transport to keep the temperature of the solid as low as possible, i.e., a special thermal management is required.

LEDs of one series have a wide spreading of their radiation features (production tolerances), hence they are classified in so-called "binnings," which means that they are graded in different classes regarding luminous flux, dominated wave length, and voltage. For applications with high demands on color stability, it is necessary to compensate and control these production and operating tolerances by micro controllers to reach predefined color features (spectras).

All these features and requirements create high demands on the development of an LED-luminaire.

Following the actual LED performance forecast, white LED lighting will soon (respectively has already) outperform(ed) all traditional lamps (superior lifespan, decreasing prices, increasing efficacy), which opens the way for LEDs to be the light source of the future with a broad field of applications. LEDs will become a competitive lamp also for economical lighting, but will not replace all fluorescent and HID lamps. Due to its spectrum and high luminance, it is the perfect lamp for replacing the GLS bulb and tungsten halogen lamps. One of the challenges will be the management of the maintenance and life cycles of LED-luminaires to enable a sustainable LED luminaire design, thus putting their ecologic potential into practice.

The room utilization factor (RUF), which is the last factor in the efficiency chain, depends on, among others, the room surface reflectances: e.g., for a room with a mean reflectance of $\rho_{mean} = 20$ %, the additional indirect lighting (by multiple room surface reflections) is only 8 % of the direct lighting, whereas for a mean reflectance of $\rho_{mean} = 70$ %, it is 70 %.

22.4 Maintenance and Life Cycle

All system components are aging, and must be replaced at certain times (before dropping out). Lamp performance decreases over time before failure (Fig. 22.4), and dirt accumulation on luminaires and room surfaces decrease the utilization factors.

The lack of maintenance has a negative effect on visual perception, human performance, safety and security, and it wastes energy. Both effects—aging and dirt depreciation—can reduce the entire efficiency of a lighting installation by 50 % or even more, depending on the application and equipment used (Fig. 22.5).

For economic and ecological evaluation of different lighting solutions, a life cycle analysis has to be done; this means that all resources consumed (from the cradle to the grave), including initial (installation) and future costs and wastes (such as maintenance and energy during operation), must be considered over the lifetime of the whole lighting installation (Fig. 22.6).

Fig. 22.4 LED downlight

Fig. 22.5 Inefficient lighting solution (*left*), lack of control (*right*)

Fig. 22.6 Lamp light flux depreciation during lifetime (principal sketch)

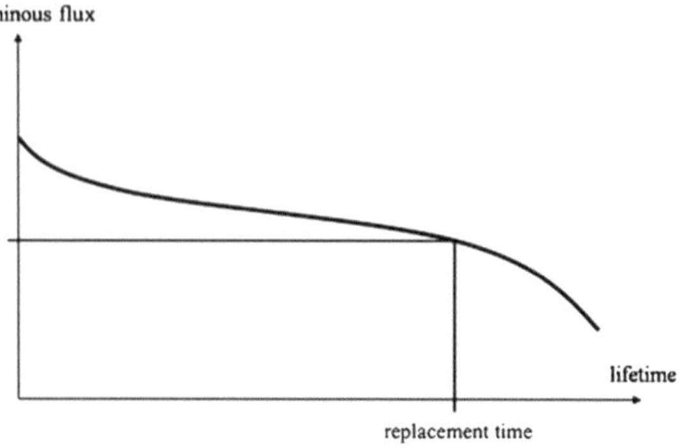

Initial costs are, for example, costs for the lighting equipment, wiring and control devices, and the labor for the installation of the system. Future costs may include relamping, cleaning, energy, and replacement of other parts (reflectors, lenses, louvers, ballasts, etc.) or any other costs that will be incurred.

The following measures should be defined by a regular maintenance schedule:

- Cleaning of luminaires, daylighting devices and rooms (dirt depreciation)
- Relamping (usually before burnout)
- Replacement of other parts
- Renovation respectively retrofitting of antiquated systems and components.

22.5 Comparison of Technologies

22.5.1 Lamps

It is not widely known that the operation costs, especially the energy costs of a lighting installation during the whole life cycle, are usually the largest part of the total costs, and that proper maintenance plans can save a lot of energy during the operating phase of the installation. Due to this lack of awareness in common practice, life cycle costs (LCC) and maintenance plans are very seldom put into practice. The calculations in Fig. 22.7 show that the management of LCC in the planning procedure can change the evaluation of different lighting solutions significantly, adding weight to the energy aspects, and thus influencing the client's final decision regarding more energy-efficient lighting solutions.

Figure 22.7 shows the effect of burning hours for the costs, comparing different lamp typologies. Included are both initial and energy costs. When using the three- and five-year payback period, the total costs with CFLs are from €4 to €4.50 and with an incandescent lamp, €9.50, with 1000 burning hours. Since the energy costs of incandescent lamps are so much higher, the price of the CFL could be €21, and then the total costs would equal those of 1000 burning hours.

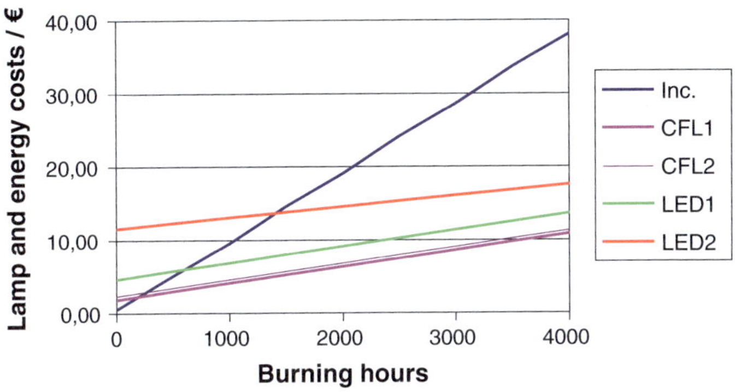

Fig. 22.7 Effect of burning hours on the total costs

A comparison of the total costs of ownership per different lamp typologies (Fig. 22.8) shows that can be determined a halving of the overall costs per year between the standard incandescent lamp and the halogen lamp, and also between the halogen lamp and the compact fluorescent lamp, even though the one-off purchase costs are the same. If compact fluorescent lamps are then compared with LED lamps, the overall costs are reduced by about 50 %, thanks to the actual lifespan of LEDs. OLED lamps are in the beginning stages of development and still have high initial costs. Costs are estimated in relation to a period of 1000 h for a lamp with a luminous flux of 1000 lm. Calculated on the overall lamp-life, these costs can be put into perspective.

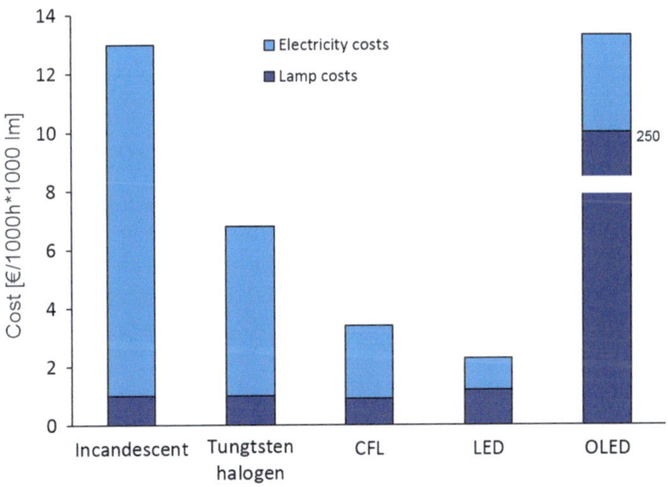

Fig. 22.8 Comparison of the total costs of ownership of incandescent, tungsten halogen, CFL, LED and OLED lamps

22.5.1.1 LED Replacement Lamps

Light-emitting diodes (LEDs) have undoubtedly been the most revolutionary innovation in lamp technology for decades. Powered by several watts, emitting application-tailored white light with an efficacy of more than 120 lm/W (and envisaging 200 lm/W), they meanwhile outperform all traditional general lighting lamps.

In 2013, replacement lamp applications were responsible for most of the LED lighting market, both domestically and globally, with omnidirectional A-type lamps, directional parabolic aluminized reflector (PAR), and multifaceted reflector (MR) lamps comprising the majority of the replacement lamp market. While LED A-type and reflector lamps represent the majority of LED installations in the U.S., they still represent less than 2 % of the total installed stock.

As their quality improves and prices continue to drop, LED lamps will penetrate the general lighting market at a faster pace. The increasing adoption of LED lamps, combined with their extended lifespans, will have a significant impact on regional lighting markets.

22.5.2 Luminaires

Achieving widespread adoption of Solid State Lamps (LED and OLED) and luminaires will depend on the ability to satisfy both performance and economic requirements simultaneously. Demonstrating better efficiency than conventional sources is not enough; the products must also be cost competitive. In this section we give an overview of LED and OLED luminaires.

22.5.3 LED Luminaires

Luminaires are defined as fully integrated lighting products designed to replace an entire fixture (not just the lamp). An example of an LED luminaire would be a fully integrated $2' \times 2'$ troffer replacement. In many instances, integrated LED luminaires can perform outstandingly compared with replacement lamps, since any design constraints imposed by an existing form factor or available space are usually less severe.

Although the installed stock of LED luminaires jumped from an estimated 6,500,000 units in 2012 to nearly 20,000,000 units in 2013, they still represent only about one percent of all luminaires installed in U.S. commercial and industrial applications. Assuming the continued performance improvement and decrease in price, LED luminaires could reach close to 15 % of the installed stock by 2020.

22.5.4 OLED Luminaires

Although some early proponents of OLED lighting envisaged large luminous areas, such as OLED wallpaper or OLED curtains, OLEDs are now mostly being used in modular form, as arrays of small panels of areas of 100 cm^2 or less. These panels can be configured either in two- or three-dimensional forms, offering light sculptures as a new form of architectural lighting. Figure 22.9 shows two examples of OLED luminaires, the Acuity Trilia (left) and Lumen Being (right). In the Lumen Being luminaire, the relative intensity of the individual panels can be varied and controlled by gestures or personal devices, such as smart phones.

Currently, OLEDs can be difficult to be used as the primary source of lighting in a room due to their limited light output and high cost. Many proponents recommend their use in wall sconces and task lights—for example, in desk lamps or under-cabinet lighting, in conjunction with ambient lighting. The low brightness of OLEDs allows them to be placed close to the task surface without being uncomfortable to the user, which improves light utilization. Methods of shaping the OLED light distribution may be required for efficient light utilization at greater distances.

The retail prices of luminaires are even higher than for the panels. Decorative luminaires, such as the K-Blade desk light from Riva 1920, which uses a Lumiblade panel from Philips, and the Bonzai from Blackbody, are priced in the range of \$3000/klm–\$5000/klm. More functional luminaires for commercial applications are now priced at around \$1500/klm (Fig. 22.10).

Fig. 22.9 OLED panel-based luminaires configured as 2-D (*left*) and 3-D (*right*) light sculptures (*source* Acuity Brands)

Fig. 22.10 K-blade and bonsai table lamps. *Sources* Riva, Black-Body

22.6 Daylighting Utilization

One of the most effective measures for saving electrical energy for lighting is utilizing daylight, i.e., choosing a clever architectural concept for the building in the early design stage and using intelligent façades and daylighting constructions.

This is the best exploitation of renewable energy for buildings—for example, it should be possible to light office buildings with daylight for more than 80 % of the working times (see example Fig. 22.11). In general, the levels and hours of daylight availability must be determined to define the operating times for the artificial light.

22.7 Lighting Design

When evaluating economics and energy aspects of a lighting solution, the goal of the system—to light a space—and the special requirements should never be forgotten. Adequate horizontal and vertical illuminance and luminance levels (regarding room utilization, visual tasks, etc.), proper balanced luminances in the field of view, control of direct and reflected glare, and adequate color-rendering are minimal conditions for good lighting.

In this context, lighting design plays a key role in order to simultaneously achieve certain appearances to fulfill the fundamental physiological and psychological visual

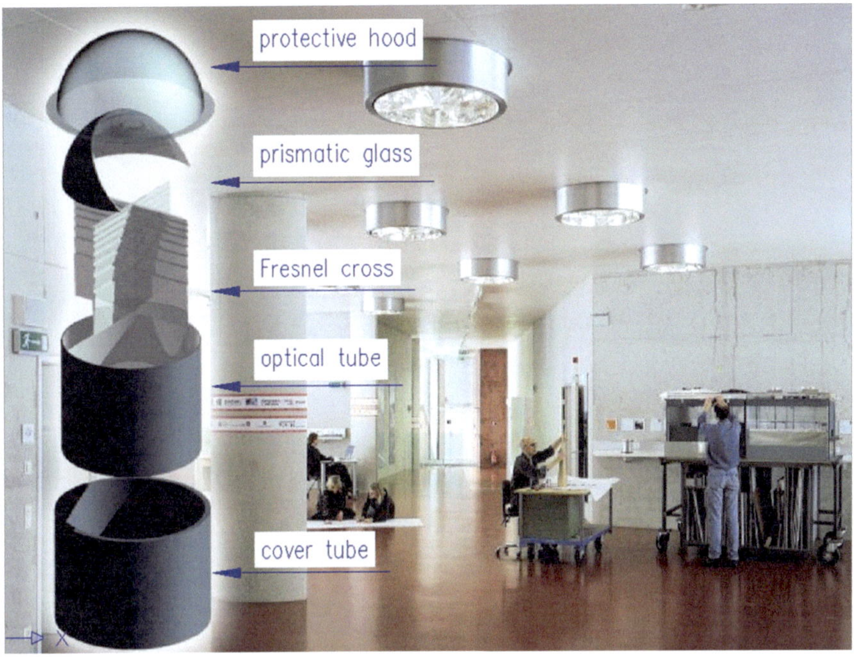

Fig. 22.11 Daylight tube lighting a hall (Bartenbach)

requirements, and to ultimately put the whole concept into effect in an energy-efficient strategy, using resources carefully. For example, the comparison of two antithetic examples for shop lighting is shown in the pictures below (Fig. 22.12): on the left, a lot of glary light sources (no shielding) together with specular surfaces (floor, ceiling, and racks of wares) give a glittery appearance, whereas on the right side, the light sources and luminaires are hidden, and the ware is in the enlightened focus.

Lighting design is more than the planning of stipulated light intensities and luminance levels given by normative guidelines. Lighting design means the creation of an appearance (e.g., of a room), which complies with not only the technical requirements but also with the visual and non-visual (biological) as well as the emotional and aesthetic requirements of the users.

Visual perception is first of all a mental procedure, and not only a pure sensation (like a thermal sensation, which causes hot or cold feelings). Seeing means receiving information about surroundings, about the distances, surfaces, textures, about what happens all round, etc., and all this information arouses emotions. Our perception is very selective, prejudiced by our personal experience, and is also influenced by our actual mental state, history, and expectations. The perception of the visual environment cannot be measured quantitatively, and therefore cannot be mathematically planned or converted. Visualizations by computer simulations (renderings) or scaled models are just aids; ultimately, the true effects can only be experienced in the real situation.

Fig. 22.12 Comparison of shop lighting (at *left*, the light points are the focus; at *right*, the wares)

From an architectural point of view, lighting is a means to express and underline the desired character of the building respectively room, which may be defined by an overall design style of the architect, by the special room using, etc. From a functional point of view, we have to be careful that basic visual requirements are fulfilled for an appropriate visual perception, dependent on the application. For example, in administrative buildings (offices, etc.), functionality is the key element (to satisfy ergonomic, safety, and communication requirements), whereas in residence and tourism, aspects such as comfort, esthetics, value, and social status are in the foreground.

Different places need different ways of designing light. At any rate, it is possible to identify three main typologies of environments, each one characterized by different hierarchies of objectives, with a specific technical, functional, or aesthetic priority:

1. Environments designed for work and service to the public, i.e., places where functionality is the key element guiding the work of the designer, and the main aspects to satisfy are the rules of the ergonomics, safety and communication.
2. Environments designed for exhibitions and sale, i.e., places where the most important need is the image, whether close to the truth or far from the reality, virtual, and fascinating.
3. Environments designed for residence and tourism, i.e., places where light should satisfy the need for comfort, relaxation, esthetic values, status symbols, etc. (Fig. 22.12).

22.8 Conclusions

Lighting causes a significant amount of the energy demand of a building, especially in temperate climates. Replacing antiquated lighting installations by modern lightings, especially with LED, will save a large amount of this energy, and the lighting quality could be improved significantly in parallel: LEDs as a digital light source are predestined to fulfill these special visual, biological, and emotional requirements.

To design energy-efficient lighting solutions, the designer should perform life cycle cost evaluations and economic evaluations (including payback criteria desired by the building owner).

The following rules should be kept in mind to reach or supersede these goals:

- Intelligent architecture and façade construction (use of daylight)
- Efficient lighting concepts (proper illuminance levels, bright room surfaces, use of high-quality components such as luminaires, lamps, operating devices)
- Proper controls (on/off, daylight, occupancy), integrated with sun shading and daylighting systems respectively building management systems
- Good maintenance.

Reference

Halonen L et al (2009) Guidebook on energy efficient electric lighting. IEA, Annex 45

Part IV
The Microclimatic Environment

Chapter 23
Tools and Strategies for Microclimatic Analysis of the Built Environment

Olatz Irulegi

Abstract There are several important challenges facing the construction sector and among them, achieving low energy buildings is the first step towards the attainment of "nearly Net Zero Energy Buildings." For this purpose it is necessary that architects and designers use adequate design strategies especially in early stages of the project. Without a correct interpretation of climatic, geographic and location parameters, meeting the goals in a project a posteriori would be very difficult. For decades, climogramas (or bioclimatic charts) and sun charts have been widely used for the microclimatic analysis of the built environment. However, the interpretation of results, the definition of strategies and design-integration is still the most difficult issue. For this purpose, this chapter aims to provide valuable design strategies and architectural solutions for the case of temperate climates. Furthermore, 21 cities in 10 countries will be studied in more detail.

23.1 Introduction

One of the most relevant issues that the construction sector is currently facing is the achievement of "very low energy buildings." For this purpose, it is necessary for architects and designers to use adequate design strategies, especially in the early stages, either for new construction or for refurbishment. Without a correct interpretation of the climatic, geographic, and localization factors, it would be very difficult to meet the energy savings and efficiency targets.

Energy simulation software tools, such as EnergyPlus, ESPr, TRNSYS, etc., are powerful software programs oriented to predict the energy performance of buildings (heating and cooling demand, artificial lighting needs, etc.). Thanks to them,

O. Irulegi (✉)
Architecture Faculty, University of the Basque Country, Plaza Oñate 2,
20018 San Sebastian, Spain
e-mail: o.irulegi@ehu.eus

S.-N. Boemi et al. (eds.), *Energy Performance of Buildings*,
DOI 10.1007/978-3-319-20831-2_23

it is possible to evaluate and implement some strategies to improve the energy performance of buildings before their construction. However, the use of these tools requires specific knowledge, the modeling process is not immediate, and the modification of parameters (especially those related to geometry) is complex and presents difficulties for their users (Attia et al. 2013). In addition, the amount of information demanded for modeling requires a very advanced stage of design. For all these reasons, these types of software are oriented to final design phases, when the decisions that most affect the energy performance of buildings have been already made.

The main goal of this chapter is to highlight the importance of using highly efficient design strategies in the preliminary design stages. The orientation, amount, and size of windows, the use and design of shading elements, and thermal mass are some of the parameters that most affect the energy performance of buildings. Furthermore, those simple, yet highly efficient solutions, should be decided on during the inception phases (Hausladen et al. 2006).

The strategies should be the result of a detailed analysis of the exterior microclimate, use of the building's interior, and definition of comfort conditions. The higher the comfort, the lower the heating and cooling demand and the lower the energy bills. The goal is not only about energy but also about the well-being of their occupants.

Nowadays, there are some useful and user-friendly information technology programs and methodologies (Lin and Jason Gerber 2014; Asadi et al. 2014; Nembrini et al. 2014) oriented to defining a set of design strategies based on climate data and comfort models, such as ESSAT-EM (Santos et al. 2014), Climate Consultant 5.5, Weather Tool™:: 2011, and Solar Tool™:: 2011.

This chapter tries to study, in detail, the case of temperate climates by analyzing the cases of 21 cities in 10 countries with Mediterranean climates. It is important to stress that despite the fact that all these cities are framed by the Mediterranean climate, it is essential to avoid generalities and conduct detailed studies. With this aim, a list of the most relevant strategies for each city is provided at the end of the chapter. Among them, sun shading of windows, natural ventilation, fan-forced ventilation cooling, internal heat gains, and passive solar direct heat gain, etc., are listed and their suitability quantified.

23.2 Köppen-Geiger Climate Classification

Identifying general climate characteristics of a certain place should be one of the first steps in the design of "low energy buildings." The climate and sub-climate classification allows a first approximation in the microclimate study (Olgyay 2006). However, a building in Venice (Italy) should not look like a building in Tel Aviv (Israel). It is evident from a theoretical point of view, but it is still difficult to be integrated in the professional practice of architecture. Low energy building cannot be achieved by adding hi-tech gadgets and sophisticated systems once the building has been completely designed.

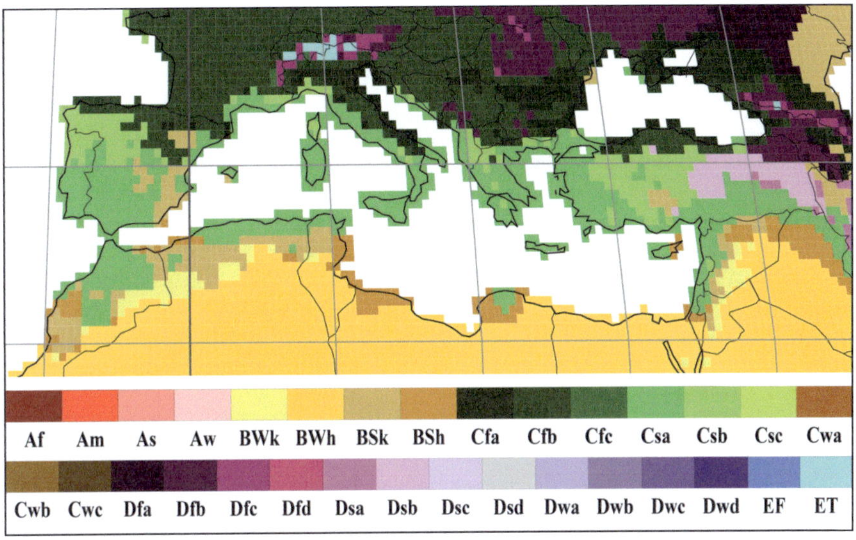

Fig. 23.1 Köppen-Geiger classification for the Mediterranean area. Observed data for 1976–2000 (Rubel 2010)

There are no universal solutions for every building and use—thus, low energy buildings do not follow a determined "style." The design should respond to the microclimatic characteristics of the location and its particular energy potential. For this reason, it would not be correct to define a list of ideal solutions for buildings located in temperate climates; every case is different and should be considered in its complexity and particularity.

Historically, Greeks were the first to develop climate zoning in global terms. Tropical, temperate and polar climates were defined in the early nineteenth century, yet it was not until 1884 when W. Köppen presented his well-known climate classification in which up to five other different climates were identified. The first global climate classification map was published in 1900, and its last review was conducted in 1961 by R. Geiger.

In Kottek et al. (2006) presented an updated and digitized version of the Köppen-Geiger Map. An excerpt for the Mediterranean region is shown in Fig. 23.1. New maps were also published, based on recent climate data recorded by the Climatic Research Unit (CRU) of the University of East Anglia and Global Precipitation Climate Center of the Meteorological Service in Germany. Furthermore, Rubel and Kottek have published maps of future climate scenarios where the impact of global warming is represented.[1] One of the newest visualization options of the digitized maps is that they can be read by Google Earth. Thanks to this, it is possible to identify the climate zone for a specific location by introducing the geographical coordinates.

[1]http://koeppen-geiger.vu-wien.ac.at.

Table 23.1 Köppen-Geiger climate classification for the Mediterranean climate

Climate	Description	Cities
Csa	Warm temperate, summer dry, hot summer	Nicosia (CYP), Barcelona (ESP), Malaga (ESP), Marseille (FRA), Athens (GRC), Thessalonika (GRC), Tel Aviv (ISR), Cagliari (ITA), Lisbon (PRT), Istanbul (TUR), Izmir (TUR)
Csb	Warm temperate, summer dry, warm summer	Nice (FRA), Genoa (ITA), Naples (ITA), Porto (PRT), Ankara (TUR)
Cfa	Warm temperate, fully humid, hot summer	Venice (ITA), Belgrade (SRB)
Cfb	Warm temperate, fully humid, warm summer	Gerona (ESP), Ljubljana (SVN)
BSk	Arid, steppe, cold arid	Alicante (ESP)

The Mediterranean climate is characterized as having a temperate climate, but it is necessary to highlight the importance of its sub-climates. There are main five sub-climates: Csa, Csb, Cfa, Cfb, BSk (Table 23.1). Barcelona (Spain), for example, is characterized by a warm, temperate climate with dry, hot summers that correspond to Csa. On the other hand, Ljubljana (Slovenia) is characterized by a warm, temperate, and very humid climate with warm summers, which corresponds to Cfb. Buildings located in both cities should, therefore, respond differently to their microclimatic conditions.

23.3 Orientation Analysis

Apart from exterior climate factors, one of the parameters that most affects the energy performance of a building is its "orientation." The analysis of the orientation allows minimizing overheating problems and maximizing energy gains in under-heated periods. Depending on the climate and localization characteristics of a given city, shading needs and solar exposure vary significantly. Furthermore, it is important to remember that incident radiation is directly related to the geographical emplacement and that it varies during the year.

The Weather Tool™ 2011 software calculates the incident radiation on a vertical surface by reading climate data[2] and proposes a best orientation. Best orientation for a vertical surface is when there is the most solar radiation during the underheated period and the least during the overheated period. Figure 23.2 shows the calculation for Istanbul and Athens; the red line indicates average daily incident radiation in summer (July, August, September), and the blue line indicates average daily incident radiation in winter (January, February, March)[3].

[2]The results have been obtained with Weather Tool™ 2011 (observed data for 1961–1990 obtained from Meteornorm 7).

[3]Note that the values in each diagram vary. The scale of values is different in Athens and in Istanbul.

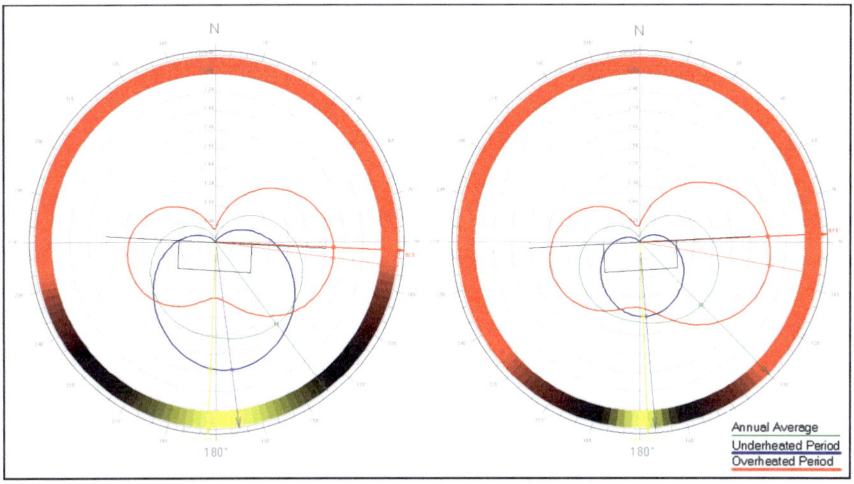

Fig. 23.2 Istanbul's and Athen's optimum orientation calculation

The graphics represent incident radiation values displayed in concentric circles. In Athens, the daily average radiation is 2.69 kWh/m^2 in summer and 1.19 kWh/m^2 in winter. In Istanbul, the daily average radiation is 1.19 kWh/m^2 in summer and 1.08 kWh/m^2 in winter. Therefore in Athens, sun shading is an important passive strategy. In addition, window openings should be minimized in overexposed orientation, e.g., east and west.

In Table 23.2, a summary of the calculation of incident radiation for 21 cities with Mediterranean climate is shown. It shows that incident radiation in the summer varies from 0.81 kWh/m^2 in Nicosia to 1.76 kWh/m^2 in Alicante. In the winter, the values vary from 1.19 kWh/m^2 in Istanbul to 2.89 kWh/m^2 in Gerona. Therefore, the shading need and potential of direct solar heat gain vary significantly within the Mediterranean cities.

Furthermore, the table shows the calculated best orientation for the 21 study cases. In general, the best orientation is towards the south, with light rotation to the east or west (±10º).

23.4 Passive Design Strategies for Mediterranean Climate

Climograms are psychometric charts (with a double entry of values: temperature and humidity) that help to define a number of bioclimatic strategies to counteract discomfort conditions. Climograms are tools to be used in the preliminary design stages, i.e., when the orientation, size, and position of windows, materials, and volume are not yet determined.

Table 23.2 Best orientations for each city

City	Best orientation[a]	Radiation in winter[b]	Radiation in summer[b]
CYP_Nicosia	180.0	2.25	0.81
ESP_Alicante	195.0	2.89	1.76
ESP_Barcelona	186.0	2.65	1.50
ESP_Gerona	182.5	2.89	1.16
ESP_Malaga	190.5	2.79	1.40
FRA_Marseille	182.5	2.47	1.39
FRA_Nice	187.5	2.35	1.74
GRC_Athens	177.5	2.69	1.19
GRC_Thessalonika	180.0	1.99	1.17
ISR_Tel Aviv	192.0	2.35	1.40
ITA_Cagliari	192.5	2.27	1.35
ITA_Genoa	187.5	1.90	1.45
ITA_Naples	190.0	2.06	1.56
ITA_Venice	177.5	1.65	1.19
PRT_Lisbon	187.5	2.35	1.58
PRT_Porto	187.5	2.09	1.59
SRB_Belgrade	180.0	1.63	1.08
SVN_Ljubljana	175.0	1.50	1.00
TUR_Ankara	175.0	1.42	0.93
TUR_Istanbul	177.5	1.19	1.08
TUR_Izmir	182.5	2.38	1.33

Daily average radiation on a vertical surface for the summer and winter periods
[a]Best orientation for a vertical surface when there is the most solar radiation during the underheated period and the least during the overheated period. Weather Tool^{TM} 2011. Clockwise from north
[b]Daily average radiation on a vertical surface (kWh/m^2)

Since B. Givoni and V. Olgyay presented their well-known climograms, some computer applications have been developed. Among them, Climate Consultant, developed by the University of California, Los Angeles, is perhaps one of the most powerful. The software is free for downloading: http://www.energy-design-tools.aud.ucla.edu. One of the greatest advantages of this application is that it uses EPW climate data (Energy Plus Weather Data), and it allows selecting various comfort models (among them ASHRAE 55-2004 PMV,[4] where clothing and activity levels can be defined).

In the Climate Consultant Climogram, a comfort area is plotted in the middle of the graph according to the selected comfort model. Every line in the climogram

[4]ASHRAE Handbook of Fundamentals Comfort Model up through 2005. For people dressed in normal winter clothes, effective temperatures of 20–23.3 °C measured at 50 % relative humidity, which means the temperatures decrease slightly as humidity rises. The upper humidity limit is 17.8 °C) Wet Bulb and a lower Dew Point of 2.2 °C. If people are dressed in light-weight summer clothes, then this comfort zone shifts 2.8 °C warmer.

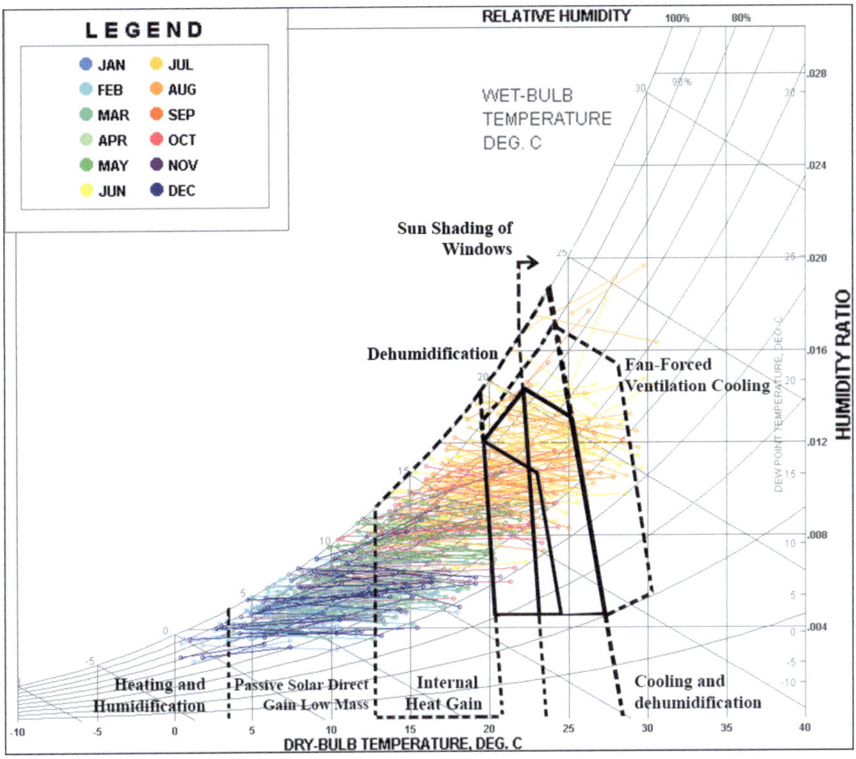

Fig. 23.3 Barcelona's Psychometric Chart. The diagram shows monthly maximum and minimum average values (observed data for 1961–1990 obtained from Meteornorm 7 data)

shows the daily temperature difference, i.e., minimum and maximum hourly values for each day, with a line drawn between them (Fig. 23.3). Climate Consultant uses an algorithm to propose design strategies to counteract discomfort by eliminating the greatest number of heating and cooling hours.

23.5 Passive Design Strategies for Mediterranean Climate

- **Sun shading**

Sun shading is particularly effective in outdoor spaces to control radiant temperatures, and on windows to help prevent indoor dry bulb temperatures from climbing above ambient temperatures.

- **Direct solar gain**

This strategy is very much a function of the building design, but if the building has the right amount of sun-facing glass, then passive solar heating can raise internal temperatures.

- **Internal heat gain**

This strategy represents a rough estimate of the amount of heat that is added to a building by internal loads such as lights, people, and equipment.

- **High thermal mass**

In summer, in hot dry climates, using high thermal mass on the interior is a good cooling design strategy. In winter there is also some positive warming effect of high mass buildings, provided that daytime outdoor temperatures get into the comfort zone.

- **Thermal mass with night flushing**

In summer, in hot dry climates using high thermal mass on the interior is a good cooling design strategy, especially when either natural ventilation or a whole-house fan is used to bring in a lot of cool nighttime air, and then the building is closed up during the heat of the day.

- **Two-stage evaporative cooling**

Evaporative cooling takes place when water is changed from liquid water to gas (taking on the latent heat of fusion); thus the air becomes cooler but more humid. The first stage uses evaporation to cool the outside of a heat exchanger, through which incoming air is drawn into the second stage where it is cooled by direct evaporation.

- **Fan-forced ventilation cooling**

Fan-forced air motion is one of the few ways to produce a cooling effect on the human body. It does this by increasing the rate of perspiration evaporation and giving the psychological sense of cooling (note that ventilation does not actually reduce the dry bulb temperature).

23.6 Climograms—Case Study of Barcelona

In this section, the case of Barcelona is studied in detail in order to show how to define passive design strategies in the early design stages.

According to the Köppen-Geiger classification, Barcelona presents a warm temperate climate with hot, dry summers (Csa, observed period: 1976–2000). In summer, the temperature frequently reaches 30 °C at noon and goes down to 17 °C at night. In winter, the temperature is quite mild and reaches around 17–18 °C at noon and 10 °C at night. For example, shading elements should be correctly designed to avoid overheating problems in summer, but should also allow direct solar gains in colder months. A set of the best strategies for the different seasons is described in Fig. 23.3.

- **Warm months**

During June, July, August, and September, it is common to reach comfort conditions in 50 % of the time (1308 h) without the help of any strategy. Due to this, the efforts should be focused on not disturbing this comfort and solving the discomfort during the rest of the day. At noon, it is common to exceed the upper temperature limit and reach 30 °C. It must be stressed that the thermal lag between diurnal and nocturnal temperature is not very pronounced. In July, for instance, the average minimum temperature is 17 and 29.5 °C maximum. Because of this, in order to reduce overheating problems due to direct solar gains, the building should be correctly oriented, and shading elements should be strategically located in the façade. The use of fan-forced ventilation is also an adequate strategy to improve comfort; it is possible to achieve comfort in an additional 550 h. The use of conventional air conditioning systems could be reduced to occasional moments or unusually hot days (only 99 h). Finally, in order to reduce discomfort at night, the use of internal heat gain is an optimal solution; thanks to it, an extra 869 h in comfort conditions are achievable.

- **Cold months**

In January, February, March, April, November, and December, it is common not to reach comfort conditions either during the day or the night (only 23 h). The average daily temperature fluctuates around 10 °C. Therefore, the building and window openings should be oriented facing the south in order to optimize direct solar gains. Thanks to this strategy, comfort is achievable in an additional 1062 h. The use of internal heat gains is another good strategy that adds an extra 1228 h in comfort. Thanks to these strategies it is possible to postpone and/or reduce the functioning of conventional heating systems.

- **Intermediate months**

In May and June it is common to reach comfort conditions at noon most of the time. For the rest of the day, the use of internal heat gains and solar direct gain add 877 h and 369 h of comfort, and would be the most reasonable passive strategies.

- **Barcelona's "Sun Chart"**

As mentioned, sun control is one of the most important and simplest strategies for improving comfort in Barcelona. The Sun Chart for Barcelona (June 21–December 21) is shown in Fig. 23.4. In the chart, the temperature for every 15-min interval is represented. Red dots indicate temperatures above 27 °C, yellow dots indicate comfort conditions (when sun shading is provided), and blue dots represent temperatures below 21 °C. The chart shows that comfort conditions are achievable with the help of shading devices in 984 h and that in 387 h, overheating problems can occur. Among all the orientations, the western orientation is the most penalized one; thus, in June, July and August—apart from exterior high temperatures—this orientation receives direct solar radiation. Consequently, the size and number of windows in a west-facing façade should be minimized, and vertical shading elements or awnings disposed of in order to avoid overheating problems.

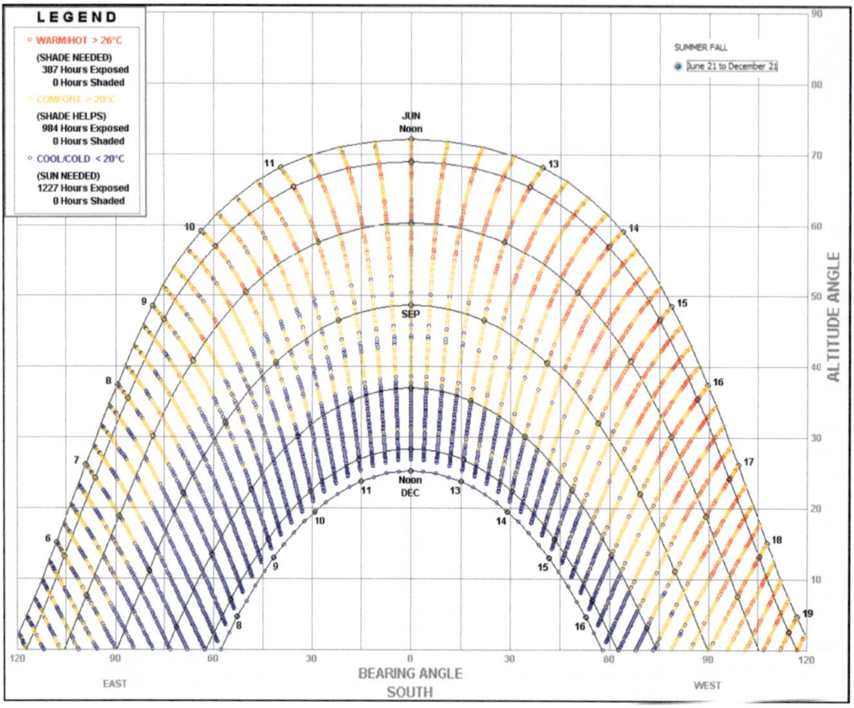

Fig. 23.4 Barcelona's Sun Chart Diagram. Comfort analysis for every 15 min during the summer and fall months (June 21–December 21). The *yellow dots* indicate comfort conditions, *blue dots* indicate under-heated conditions and *red dots* overheated

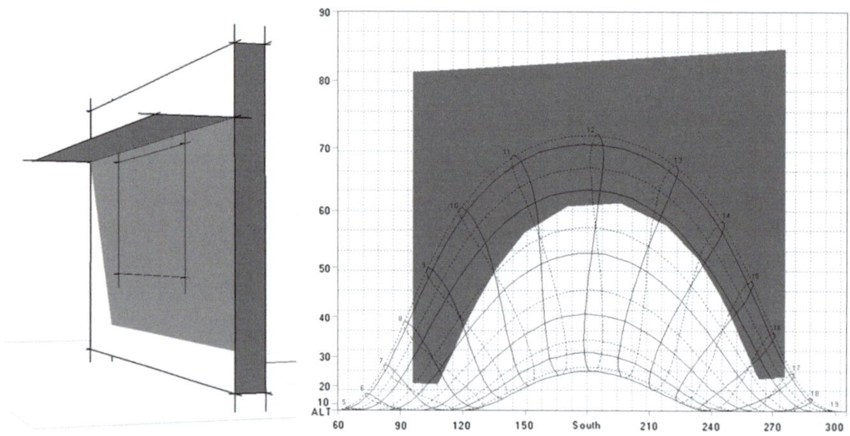

Fig. 23.5 Barcelona's Sun Chart Diagram and shading element optimization

Once the shading need is determined, the software Solar Tool^{TM}:: 2011 allows designing, evaluating, and optimizing the dimensions of a given shading device that best fits the local climate and orientation. Figure 23.5 shows the results of the analysis of a window (1.20 m high and 2.40 m wide) with the best orientation for Barcelona (+186°). The solution is quite simple: with a 0.70 m horizontal element, the window will be protected in the summer and will receive solar radiation during the coldest months. Note that the horizontal element should be wider than the window, as shown in the chart.

23.7 Summary of Design Strategies for Mediterranean Cities

A list of the best strategies for Mediterranean Climate is presented below. Each strategy has been evaluated and its suitability calculated using Climate Consultant 5.5 software. The results are in order, by season and city. The aim of this study is to provide general design ideas to architects and engineers who plan to construct low energy buildings in the Mediterranean climate.

- **Passive strategies for the summer period**

The following chart (Fig. 23.6) shows the suitability of various passive strategies for the hottest months in 21 study-cities (from July 1 to September 30). It must be emphasized that in most of the cases, comfort conditions are achievable without

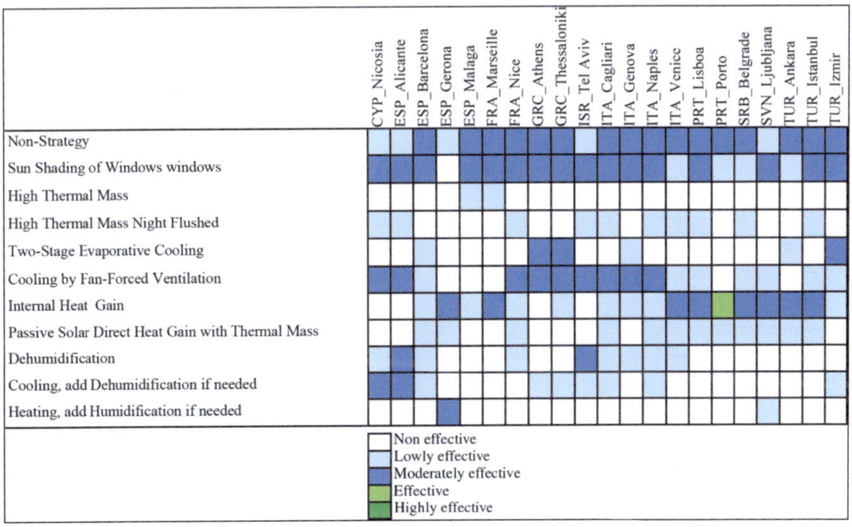

Fig. 23.6 Passive strategies for the July 1–September 30 summer period

the use of any strategy: in 1055 h in Barcelona, 1030 h in Lisbon, and 1002 h in Istanbul. Therefore, a correct design and orientation are key factors for reducing the energy demand. Note that the use of air conditioning systems is not necessary all the time. In addition, it is correct to think that in many cases an incorrect design is the main cause for cooling needs and use of air conditioning in a Mediterranean climate. By reducing the size and number of windows in eastern and western façades, cooling loads could be considerably minimized. When sun shading elements are provided, comfort conditions could be achievable in additional hours: 824 h in Athens, 850 in Tel Aviv and 815 in Nicosia.

Fan-forced ventilation is another good option for cities such as Alicante, Tel Aviv, and Naples, where comfort conditions are achievable in an additional 929 h, 1011 h, and 743 h, respectively. In some cases, such as in Porto and Belgrade using internal heat gains is recommended where comfort is achievable in an extra 1132 and 906 h, respectively.

- **Passive strategies for the spring period**

During the spring (Fig. 23.7), the use of internal heat gains is one of the most effective strategies, particularly in Nice, Genoa, Porto, and Lisbon, with an additional 1261, 1218, 1232, and 1261 h of comfort respectively. In most cases, comfort conditions are achievable without the application of a strategy. The solar direct heat gains accompanied by thermal mass of the building is another good solution, and comfort is achievable in an extra 615 h in Alicante, 537 in Tel Aviv, and 468 in Izmir. The use of conventional heating systems is unavoidable with low night-time temperatures in all the cities. On the other hand, in some other cities– such as Nicosia, Athens, and Cagliari—sun shading starts to be necessary because the temperature starts to exceed 21 °C.

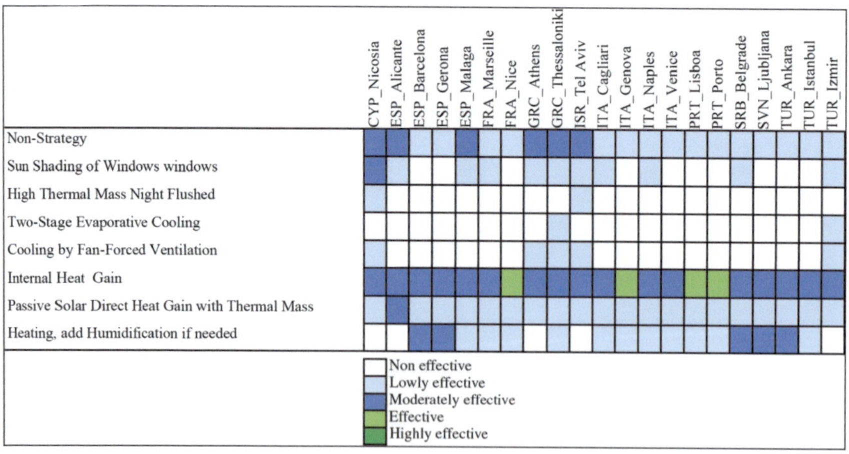

Fig. 23.7 Passive strategies for the April 1–June 30 spring period

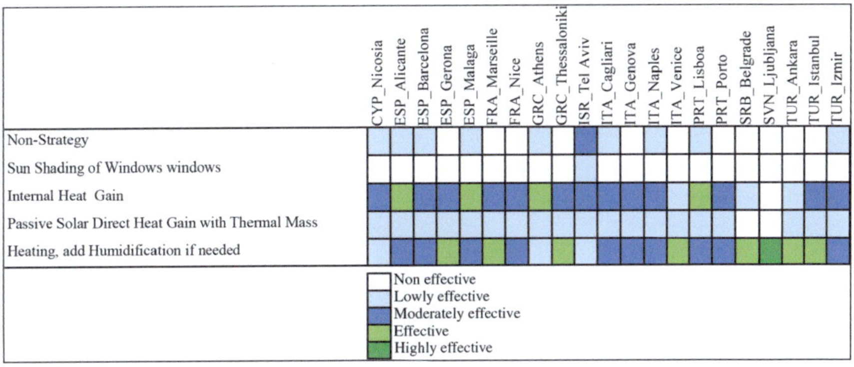

Fig. 23.8 Passive strategies for the October 1st-December 31st autumn period

- **Passive strategies for the autumn**

During the autumn (see Fig. 23.8), the use of internal heat gains is one of the most effective strategies—particularly in Athens, Malaga, Alicante, and Lisbon, where comfort is achievable in additional 1198, 1156, 1112 and 1223 h, respectively. In most of the cases, comfort conditions are achievable with the use of no strategy. The use of solar direct heat gains, accompanied by thermal mass in the building, is another good solution for Gerona (426 h) and Marseille (358 h). The use of conventional heating systems (1908 h) is unavoidable due to low nighttime temperatures in some cities, such as Ljubljana.

23.8 Passive Strategies for Winter

In winter (Fig. 23.9), the use of conventional heating systems is necessary because of low temperatures. At noon, internal heat gains and direct solar gains will help reduce the use of them in cities such as Tel Aviv (an extra 1128 and 273 h, respectively), Cagliari (an extra 624 and 150 h, respectively) and Izmir (an extra 568 and 460 h, respectively).

23.9 Conclusion

There are several important challenges facing the construction sector; among them are achieving low energy buildings, which is the first step towards achieving nearly Net Zero Energy Buildings. For this purpose it is necessary that architects and designers use adequate design strategies, especially in their early design stages, in order to facilitate their decision-making.

To help achieve the main goal of this book, a detailed analysis of 21 cities in 10 countries with Mediterranean climate was conducted. The results show that an

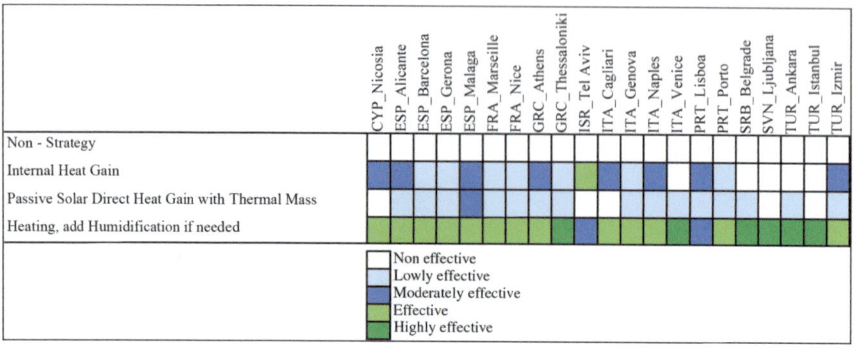

Fig. 23.9 Passive strategies for the January 1–March 30 winter period

incorrect design is, in most cases, the main cause of an unnecessary need for air conditioning in a Mediterranean climate. By minimizing the number and size of windows in eastern and western façades, most overheating problems can be avoided. When a building is correctly oriented (facing south), it is easier to design shading elements that provide shade when necessary and permit solar exposure in under-heated months. In any case, it is necessary to conduct a detailed study for each city and use of the building. On the other hand, fan-forced ventilation is another good solution.

For under-heated months, the use of internal heat gains is a good solution to minimize and postpone the use of conventional heating systems.

References

Attia S, Hamdy M, O'Brien W, Carlucci S (2013) Assesing gaps and needs for integrating building performance optimization tools in net zero energy buildings. Energy Build 60:110–124

Asadi S, Shams Amiri S, Mottahedi M (2014) On the development of multi-linear regression analysis to assess energy consumption in the early stages of building design. Energy Build 85:246–255

Hausladen G, de Saldanha M, Liedl P (2006) Climate skin. Building-skin concepts that can do more with less energy. Birkäuser, Basel

Kottek M, Grieser J, Beck B, Rudolf B and Rubel B (2006) World map of the Köppen-Geiger climate classification updated. Meteorol Z 15:259–263. doi: 10.1127/0941-2948/2006/0130

Lin S, Jason Gerber D (2014) Evolutionary energy performance feedback for design: multidisciplinary design optimization and performance boundaries for design decision support. Energy Build 84:426–441

Nembrini J, Samberger S, Labelle G (2014) Parametric scripting for early design performance simulation. Energy Build 68:786–798

Olgyay V (2006) Arquitectura y clima: manual de diseño bioclimático para arquitectos y urbanistas. Gustavo Gili, Barcelona

Rubel F, Kottek M, (2010) Observed and projected climate shifts 1901–2100 depicted by world maps of the Köppen-Geiger climate classification. Meteorol Z 19:135–141. doi:10.1127/0941-2948/2010/0430

Santos P, Martins R, Gervásio H, Simões da Silva L (2014) Assessment of building operational energy at early stages of design—a monthly quasi-steady-state approach. Energy Build 79:58–73

Chapter 24
Microclimatic Improvement

Francesco Spanedda

Abstract Microclimatic improvement can be defined as the mitigation of microclimatic conditions on a local scale, and it produces, together with the quality of the building envelope, comfortable interior spaces, sheltered from the inclemencies of climate. From this point of view, the mild Mediterranean climate offers very interesting features to explore. Its warm winters and mild, dry summers allow for designing open spaces as buffer zones, close to the building envelope. This favorable climatic condition is not exclusive to southern Europe and North Africa, but extends to parts of California, Chile, South Africa, and Australia. The open areas designed to mitigate climate also blur the boundaries between inside and outside. Since they fall inside the comfort zone most of the year, they can be considered an extension of the living rooms, suggesting a rich interplay between interior and exterior spaces. This chapter introduces some basic notions about microclimate, and illustrates some relevant examples to investigate the fundamental design topics related to microclimatic improvement. From Moorish architecture until the present, the idea of mitigating climate through the design of open spaces inside or around the building has produced a vast array of configurations that feature both proven efficiency and an imposing spatial quality. All these architectural spaces can be considered efforts to tame the energy fluxes flowing through the site and make them part of a spatial concept, improving and enriching the quality of space. The result is an exploration of the complex relationship between inside and outside, a study on the continuity of artifice from the enclosed volume to the outer space, and a different point of view of one of the fundamental questions of architecture: the act of enclosing.

F. Spanedda (✉)
Dipartimento di Architettura, Design e Urbanistica (DADU), Università degli Studi di Sassari, Palazzo del Pou Salit, Piazza Duomo 6, 07041 Alghero (SS), Italy
e-mail: francesco.spanedda@uniss.it

S.-N. Boemi et al. (eds.), *Energy Performance of Buildings*,
DOI 10.1007/978-3-319-20831-2_24

24.1 Introduction

One of the most important goals for architects is building comfortable spaces for everyday life. This is traditionally achieved by subtracting a small portion of space from the harshness of the climate, and building a physical separation from the outside world. The dialectic between interior and exterior space could be considered, to some extent, as the stuff architecture is made of.[1] The act of enclosing is no doubt one of the primary gestures in designing a place.[2]

As Reyner Banham pointed out in his seminal book *The Architecture of the Well-tempered Environment*, there are several models to control energy fluxes between inside and outside. Banham defines a "conservative mode," mostly associated with a massive structure with small openings that withstands energy fluxes and usually acts as thermal storage; a "selective mode," where the building envelope filters the desired flows of air and energy, and expels the unwanted ones, by means of windows, shades, and ventilation openings; and a "regenerative mode," where energy—mostly coming from fossil fuels at the time—is expended in order to maintain a temperature difference between inside and outside. The first two modes are, to some extent, related to climatic conditions. According to Banham, the conservative mode is mostly developed in the Mediterranean dry climate and deeply permeated the European building culture. The selective mode predominates in humid climates, marking buildings with a complex combination of balconies, parasol roofs, floor-to-ceiling windows, and jalousies. The regenerative modes is more associated with the availability of power, and with cultural attitudes (Banham 1984). These models are often combined, and the further development of the climate-conscious design mixes them in ever-new hybrid forms.[3]

The harsher the climate, the tighter the enclosure, and the more controlled and reduced the flow of energy through the outer shell: this seemingly obvious idea grounds several techniques used today in sustainable buildings, which show a coexistence between conservative and regenerative modes, like the well-known "Passivhaus Standard." On the other hand, the opposite is also true: if the climate is mild, the designer can let the energy fluxes through the building envelope, working on the relationship between inside and outside in a more detailed, complex, and intriguing way. The inner space opens up towards the exterior in a blend between the conservative and selective modes. This obviously deals with

---

[1]For more informations about this topic see, among the others: Olgyay and Olgyay (1963), Colombo et al. (1994).

[2]There is a huge bibliography on this topic. Among the most relevant writings, see Norberg-Schulz (1976), and the related Heidegger's analysis of Georg Trakl's *Ein Winterabend* poem (1915). See also Vv.Aa. "Recinti (Fences)" *Rassegna* n.1, 1979.

[3]Between 1980 and 1996, Dean Hawkes updated this classification, describing three fundamental models: the Selective Mode, more oriented toward an interaction with the regional microclimate, the Exclusive Mode, informing buildings whose internal environment is totally sealed from external influences, and the Pragmatic Mode, related to buildings whose environmental control is more a matter of chance than of design. See Hawkes (1980, 1996).

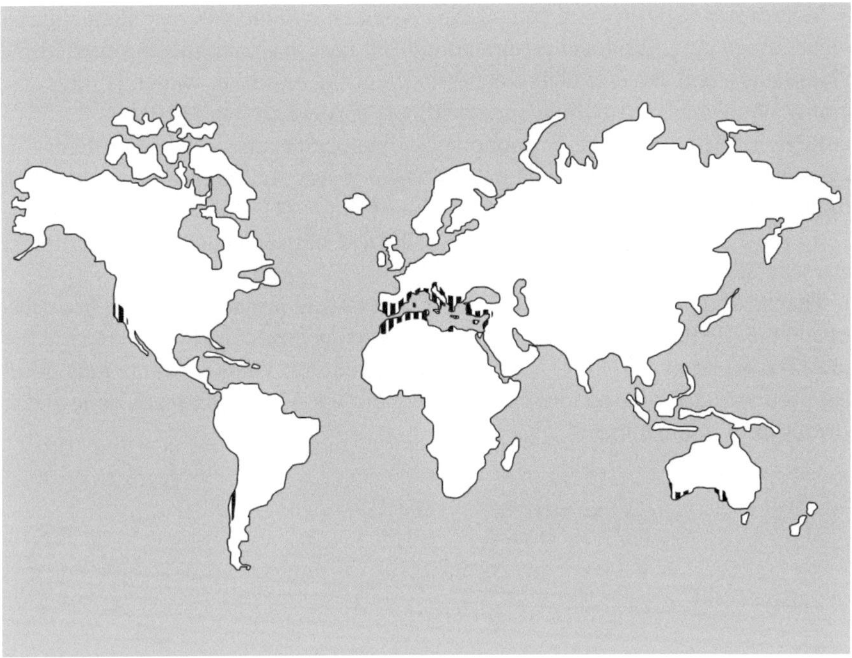

Fig. 24.1 Geographical extension of Mediterranean climate (*data source* Wikipedia)

the uncertainty and continuous changes in climatic conditions, hence the need to design the outer spaces in order to gain a better control of the local microclimate.

This shift in the focus of design, from the simple enclosure to the open space around a building can be fruitful in both climatic and spatial issues. It broadens the concept of building envelope, extending it from the limited thickness of the external walls to a more complex domain where the walls and their adjacent open space become a single artifact that controls the outside climate. Moreover, while the mass of walls becomes a kind of negative space inaccessible to the users' experience, the extended envelope can be considered an amplification of the space of the building.

24.2 Defining Microclimate

24.2.1 Properties of Mediterranean Climate

The Mediterranean climate is a variety of the subtropical climate distinctive to the Mediterranean Basin and to several regions of California, Mexico, Western and Southern Australia, South Africa, Central Asia, and Chile (Fig. 24.1). It features warm to hot, dry summers, and mild to cool, wet winters.

Subtropical high pressure cells during summer, and the polar jet stream during winter, avoid any significant precipitation from four to six months in summertime, while rains—and snow at higher elevations—occur mostly in winter, autumn, and spring. Winters are mostly mild, and summers very warm (Fig. 24.2).

Regional features, such as topography, dampening effects of large bodies of water, protection from seabound or landbound winds, as well as elevation, deeply influence temperatures and comfort on a meso-scale. Thus, great attention must be paid to understanding how the interplay of these elements alters the macro-scale characteristics.

Thanks to its mild weather conditions, a prominent urban civilization flourished around the Mediterranean Sea; its settlements today cover about 40 % of its coastline (Travers et al. 2010). During that prolonged era, various populations developed settlement forms, and building typologies suited to this climatic zone and its various local microclimates.

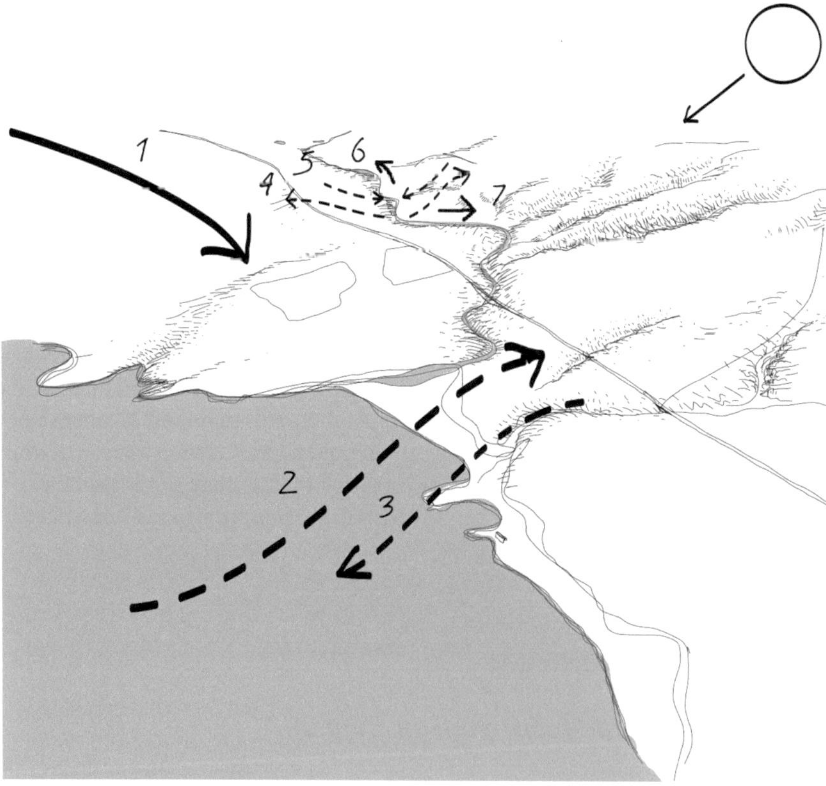

Fig. 24.2 Interaction of topography, wind, and water bodies along the coastline. Prevailing winds (*1*), sea breezes during the day (*2*) and night (*3*), day breezes (*4*), and night breezes (*5*) across river valleys, *upward airflow* along the valley at noon and during the afternoon (*6*) and at night (*7*)

24.2.2 Meso-Scale Conditions

The first step in architectural design is the decision about settlement. Vitruvius reported traditional ways to consider the interaction of topography, winds, and water to decide if a site is suitable or not (Pollio 1914).

24.2.2.1 Topography

Mountains and hills stop airflow, heavily affecting their force and direction. When they are located close to the sea, their presence stops the humidity carried by the landward winds, causing dry zones on the land-facing side, and humid zones on the seaward side.

24.2.2.2 Wind

Wind consists of the flow generated by masses of hot and cold air. In urban environments, uneven surfaces and tall buildings reduce and disrupt the airflow, while the canyon effect between buildings increases its speed. Wind provides cooling or warming, depending on the temperature of upstream areas, and helps in heat transfer, thus playing an important role in defining a microclimate. Wind could either improve comfort or become an unpleasant factor, as a consequence of its interaction with other microclimatic factors, such as bodies of water, humidity, and topography.

24.2.2.3 Bodies of Water

Water's great heat storage capacity absorbs a considerable amount of the incident solar radiation, without changing its temperature very much. Bodies of water thus have a powerful flattering effect on temperature fluctuation. The Mediterranean climate itself is influenced by the constant temperature of the sea, cooling down sea breezes in summer and warming them up in winter, thereby providing a mild inland climate. Water surfaces can also lower the air temperature by absorbing heat during the evaporation process.

24.2.2.4 Vegetation

The role of the plant cover is manifold. By grouping and scattering, plants filter and absorb solar radiation, provide shade, and influence wind direction and speed. Moreover, they provide additional cooling through evapotranspiration: they take water from the ground—through their roots—and let it evaporate through their leaves, subtracting heat to the ambient air (Georgi and Dimitriou 2010; Shashua-Bar and Hoffman 2004). Some species, like figs, contribute to achieving a more comfortable microclimate by increasing the relative humidity (Georgi and Dimitriou 2010).

24.2.2.5 Artificial Elements

The artificial three-dimensional layer made up by urban fabric is called "urban canopy". Human activities, energy leaks, less reflective and more heat-absorbing surfaces, and the absence of a plant layer, lead to higher air temperatures in dense urban areas than those found in the surrounding rural areas. This phenomenon is known as an "urban heat island" (UHI) (Gaitani et al. 2007; Santamouris et al. 2001; Grosso 1998) and is the result of the physical man-made alterations of the urban surface. The thermal balance of the urban region strongly affects the intensity of the heat island. Heat islands can negatively affect the urban environment, especially in summer. Besides general discomfort, and the increase in energy consumption and peak energy demand in the warmer season, they can cause severe health problems (respiratory strain, heat cramps and exhaustion, heat strokes, and heat-related mortality), as happened as a result of the heat waves in Europe in the summer of 2003.

24.2.3 Main Physical Parameters on the Local Scale

A closer look at the methodologies commonly used to measure the comfort level of outdoor spaces can provide a better understanding of the parameters affecting the sensation of comfort. There is a widespread acceptance in defining the main influencing factors,[4] which range from air temperature and wind speed to metabolic energy and clothing. Many variables are independent from the built environment, such as clothing. Some of them descend from global climatic components, e.g., solar radiation. A few are associated to physical activity, and can be regulated through design-related decisions, such as the definition of a functional program. But many parameters can be controlled by means of physical or spatial design that could also influence the way global climate affects a particular place (Fig. 24.3).

Delving more into detail, the physical properties affecting microclimate that can be mostly controlled by design are (Fig. 24.4):

Albedo influences the total solar radiation absorbed by people by two components: the albedo of the ground, and the albedo of objects like walls and other reflecting surfaces. The lighter the ground, the more energy is reflected in the indoor space. It is worth noting that a reflecting ground surface reduces the storage of energy that will later be released in the coolest hours. Albedo influences

[4]The most widely used are the TS_Givoni Method and the Confa Method (Gaitani et al. 2007).

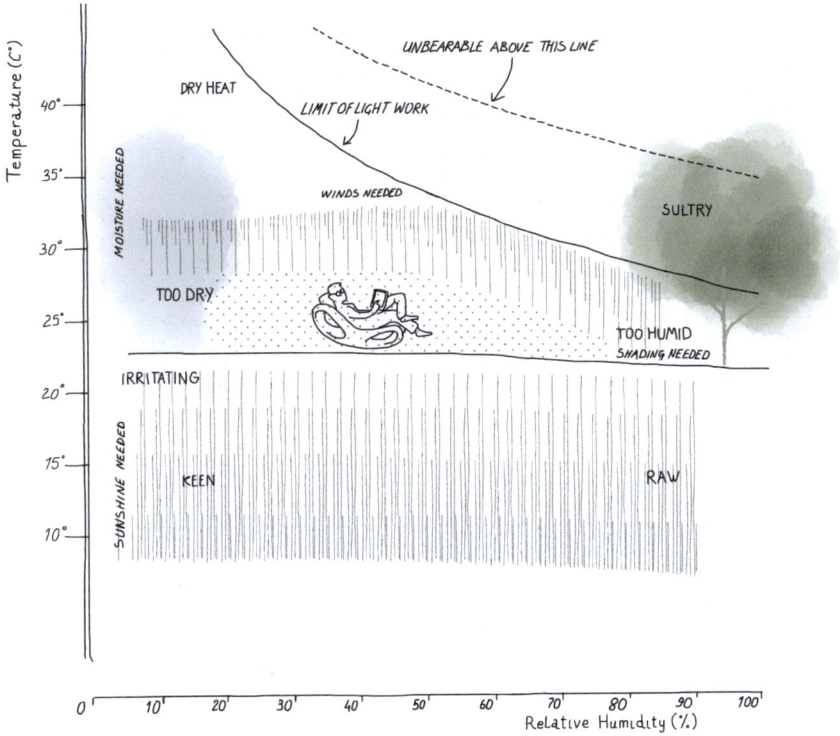

Fig. 24.3 Climate comfort diagram, with the action needed when the parameters fall out of the comfort zone (*data source* Olgyay and Olgyay 1963, Fig. 46)

other factors, too, such as the solar radiation reflected from the ground surfaces. The choice of materials influences the albedo and the emissivity as well, which consequently influence the terrestrial radiation and are ultimately summed up in the total solar radiation absorbed by people in the environment.

Sky obstructions consistently affect diffuse solar radiation. Similarly, the solar transmittance of shading and vegetation acts on the direct solar radiation. Both direct and indirect solar radiation concur to define the total solar radiation absorbed. These parameters represent two important, separate issues in the design of outdoor spaces: the openness to the sky, and the effectiveness of the shading (Figs. 24.5, 24.6).

Wind speed mainly affects the convective heat flow. It can be reduced or increased by careful design, choosing an adequate location related to topography, planning winding or straight streets, narrowing or expanding the street sections, and distributing greenery in order to break wind and lead airflow towards certain points.

Air temperature can be influenced by design, too, subtracting heat from air by using evaporation, designing fountains or water basins, or evapotranspiration,

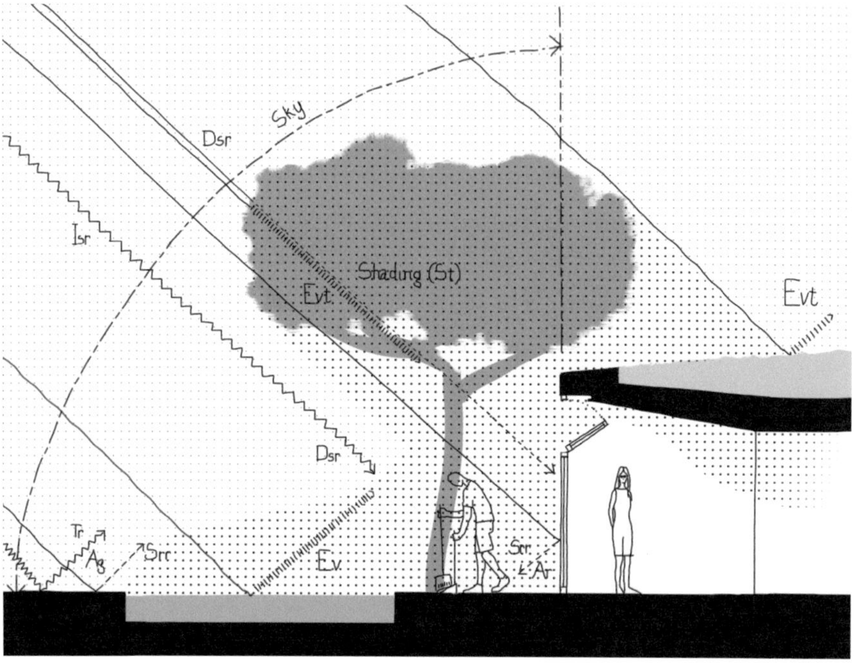

Fig. 24.4 Scheme showing the different parameters and their interaction with spatial design. Main factors are albedo of the ground (*Ag*) and of the reflecting surfaces (*Ar*), solar radiation reflected from the ground surfaces (*Srr*), terrestrial radiation (*Tr*), openness to the sky (*Sky*), attenuation of solar transmission (*St*) through shading or plants, direct (*Dsr*) and indirect (*Isr*) solar radiation, and the subtraction of heat by evapotranspiration (*Evt*)

placing shrubs, and greenery. The large storage capacity of bodies of water could contribute to further stabilizing temperatures, absorbing a great amount of the solar radiation.

24.3 Mediterranean Settlement and Microclimate

Mediterranean settlements always dealt with climate. Although the Roman architect Marcus Vitruvius Pollio is famous for his "firmness, commodity, and delight," his *Ten Books on Architecture* (Pollio 1914) clearly show the relationship between site, climate, settlement, and building. He states that "If our designs for private houses are to be correct, we must at the outset take note of the countries and climates in which they are built (…) as the position of the heaven with regard to a given tract on the earth leads naturally to different characteristics (…) it is obvious that designs for houses ought similarly to conform to the nature of the country and to diversities of climate.[5]" Therefore he gives very detailed advice about the

[5]Book VI, Chap. I.

Fig. 24.5 F. Spanedda, with M. Campus, P. Addis, R. Senes. Consultants: M.G. Marras (landscape), F. Bua (archeology), G. Onni (tourism). Favero and Milan Ingegneria, Refurbishment of Batteria Antinave Carlo Faldi (competition), Sardinia, 2010: shading concept as evocation of camouflage

choice of the building site,[6] the layout of streets with respect to the prevailing winds, and the need to provide different orientation for the different rooms.[7]

As cultures followed each other on either side of the Mediterranean basin, a fruitful exchange of experiences in building and settlement affirmed a few widespread typologies strongly related to climate control and microclimatic improvement, whose permanence through the centuries and civilizations is noteworthy.

On the urban scale, spatial patterns made out a combination of squares, and narrow, deep streets, can be found in the whole Mediterranean area (Grosso 1998; Ragette 2003). They allow for an effective shadowing, both at street level and of the adjoining buildings, building pools of fresh air that can be dragged into the indoor spaces. Although some authors attribute this type of street to Muslim cultures,[8] it is worth noting that Leon Battista Alberti in his *De Re Aedificatoria* prescribed narrow winding streets in order to cool the site, providing shade, and slowing down winds (Alberti 1541). The same pattern, according to Alberti, will confuse and slow down enemies eventually breaking into the town. Temporary textile devices often shade the narrow, deep streets during the warmest period (García-Pulido 2012).

On the building scale, the most diffuse and studied spatial matrix is the ubiquitous courtyard, which is the core of rural houses and palaces as well. As often happens with spatial structures, their validity was claimed by quite different figures, representing opposing ideological positions. The Italian rationalist architect Carlo Enrico Rava saw in the Arab patio house a very modern and efficient way to deal with the North African climate. Rava believes that this typology can be traced back to the Roman house, as patent confirmation of the long-lasting supremacy of Roman culture (Rava 1931). The Egyptian architect Hassan Fathy,

[6]Book I, Chaps. IV, VI.

[7]Book VI, Chap. IV.

[8]For a discussion of the layout and section of Arab streets, see Ragette (2003).

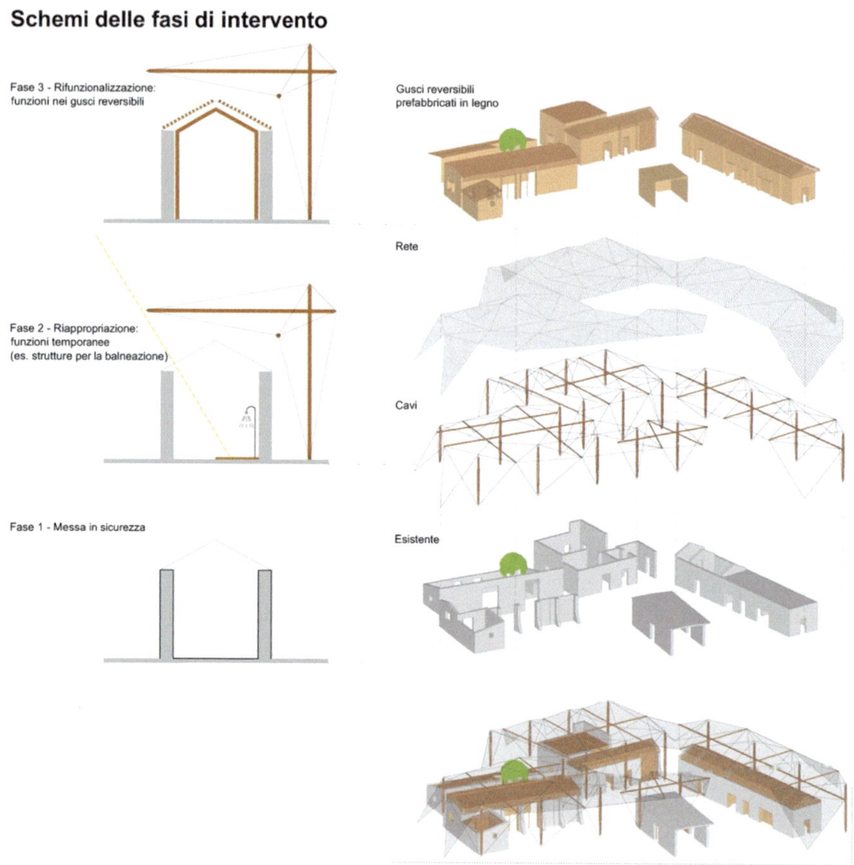

Fig. 24.6 F. Spanedda, with M. Campus, P. Addis, R. Senes. Consultants: M.G. Marras (landscape), F. Bua (archeology), G. Onni (tourism). Favero and Milan Ingegneria, Refurbishment of Batteria Antinave Carlo Faldi (competition), Sardinia, 2010: microclimatic improvement as drive for the social reclamation of the site

in his anti-colonialist review of Arab building traditions, considered the courtyard to be an incomparable space commonly used by all the populations facing the Mediterranean. In his opinion, this place has great value for the Arabs, somewhat reflecting the order of the universe in its quietness (Fathy 2010).

Both the urban structure and the organization of the house clearly show the relationship with the energy flows provided by the climate and the development of a spatial arrangement that could effectively improve comfort during the hot summer days. Better microclimatic conditions are produced by basically controlling winds and airflow and taking advantage of solar obstruction and shading provided by high walls. This spatial configuration made out of walls and shadows provides a transition from the public realm to the private sphere. It organizes a sequence of spatial structures that share features with both inner and outer spaces, suitable to fulfilling many functions other than just climate control.

24.4 Strategies in Microclimatic Improvement

24.4.1 Building as Modifier of Microclimate (North America 1910–1948)

After Banham's seminal book, modern architects areoften accused of neglecting climatic issues in favor of a machine-age esthetic. This is true to a certain extent, but it is worth remembering that widespread beliefs about the availability of incredible amounts of energy and the excitement about technological progress surely played a great role in setting the agenda of architects at that time.

While in Banham's book, Le Corbusier plays the role of the villain, subjugating climate-responsiveness and common sense to his formalistic resolution, Frank Lloyd Wright and the Modernist architects who moved to California and were inspired by him, play a positive role combining sensibility to climatic issues with spatial innovation. For the sake of precision, it should be noted that these American architects were able to design their open structures with the huge glass surfaces needed to connect them with the landscape, because they could count on heat provided by fossil fuels.

Banham himself cites the introduction to Wright's *Wasmuth Portfolio*, where the American master celebrates the opportunities offered by hot water heating. According to Wright, heating provides the possibility to restructure the building, leaving behind the old "compact box" concept needed to reduce heat losses, and providing light and ventilation on several sides, through rows of windows opening to the outer air, the garden, and the landscape (Banham 1984) (Fig. 24.7).

Wright's attention to the relationship with climate is well known. His houses subtly use microclimatic elements as a help in establishing internal comfort. Further in the *Wasmuth Portfolio*, Wright explains how prairie houses are ventilated in the hot season through openings in the long horizontal glass bands under the eaves, letting in the fresh air from the garden. Once warmed, the air will be extracted from the chimney, which plays an important role in summer as well as in winter. In the Robie House (1910), the northern shadowed courtyard has the function of storing cold air to cool down the rooms (Banham 1984). The second Jacobs House, better known as "Solar Hemicycle" (1948), was built on a site chosen by Wright himself, after a careful survey of the client's property. The building is bent along a 120° arc whose convexity, protected by a berm, points northwards, while the southern façade opens with a long, curved, two-story glass wall on a slightly sunken circular garden. A narrow strip of the garden, close to the house, is partially protected by the eaves of a slightly slanted roof. This layout was mainly intended to protect both the rooms and the open spaces from the cold northern winds, and to affect positively the direction of the southern winds, blowing them

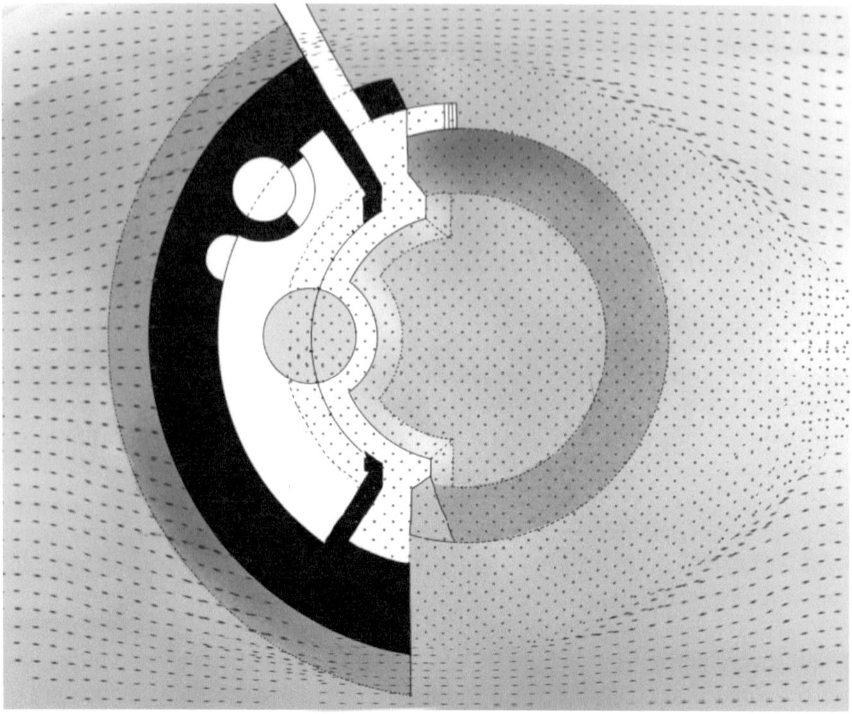

Fig. 24.7 Frank Lloyd Wright, Solar Hemicycle (second Jacobs House), Middleton, Wisconsin, 1948. Plan, showing the circular sunken garden

upwards along the roof, to shelter the sunken open space in front of the house. Sheltered from the cooling effect of the wind, the sun-flooded garden acts as a pool of warm air. The original design featured a pond for fish, that could help lower the air temperature in the garden through evaporation. Actually, the whole garden acts as an extension of the living room, and provides a temperate outdoor space directly connected to the living rooms through the glazed wall, equipped with fixed glass panes and huge glass doors.[9] Wright claimed that his client could stand in front of the southern glass wall and light his pipe in windy weather, and Mr. Jacobs reportedly confirmed that the architect was right.

Both in the Robie and in the second Jacobs House, Wright uses an effective and fascinating strategy to improve microclimate: the building itself is used as a shading and wind-breaking device to temper outdoor spaces, which in turn raise the comfort level of the rooms. In other words, instead of landscaping the outer spaces

[9]The southern doors' height was actually reduced in the building phase from about 4 m (13 ft, 6 in.), as shown in the working plans, to about 2.7 m (9 ft), as previously stated in the preliminary drawings. The two doors at the end of each wall were cut to about 2 m, so the children could easily open and close them (Sprague 2013).

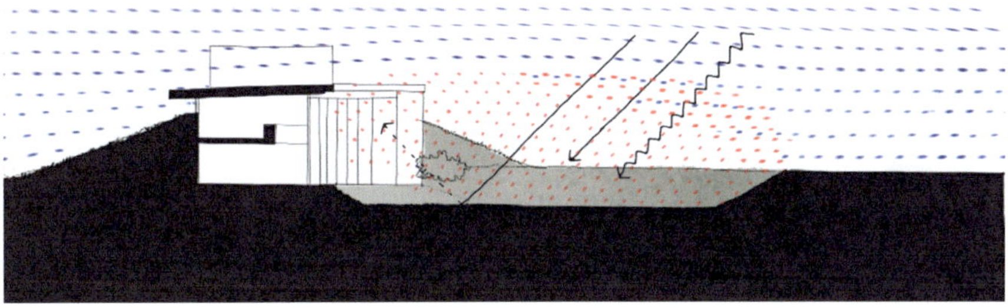

Fig. 24.8 Frank Lloyd Wright, Solar Hemicycle (second Jacobs House), Middleton, Wisconsin, 1948. Section showing the relationship between the berm, the building, the sunken garden, the wind, and the solar radiation

to control the climate around the house and provide comfort inside, the building layout is used to improve the comfort of outdoor space that subsequently contributes to making the indoors comfortable (Fig. 24.8).

24.4.2 Sequences of Dampening Spaces (Andalusia 9th–14th Centuries)

The Moorish architecture in southern Spain shows several remarkable examples of microclimatic control achieved by carefully designed spatial configurations. Its spatial richness and complex interlocking of indoor and outdoor spaces is the result of a long evolution spanning centuries, and common to the different cultures that flourished in that country. In their many details, these buildings bear signs of a fruitful cultural exchange between Christian, Muslim, and Jewish cultures, sometimes reflected by friendly relationships among Spanish kings, caliphs, and their consultants and treasurers (Ruggles 2004).

Many famous buildings of the time, such as the palatial complex of Alhambra and Generalife in Granada, or the Alcazar in Seville, are skillfully designed with consistent attention to microclimate, from the general layout right down to tiny details. Most of their spaces and building elements operate on the various flows of energy in order to provide comfort, as a pragmatic and experience-driven synthesis of several ways to control climatic conditions. More than many others, the medina of Alhambra and the Generalife take advantage of mesoclimate, thanks to an imposing collocation on top of the Sabika Hill, exposed to the flow of breezes and surrounded by dense greenery and by the Darro River.

The first important element is the orientation of courtyards. Many contemporary buildings show a widespread preference for elongated patios along a north-south axis, as can be clearly seen in the Alhambra plan. This kind of orientation exposes the southern inner façade to the rays of the lower winter sun, in order to warm the living spaces behind it. A careful proportioning of the patio, both in plan and elevation, prevents solar penetration during the warmest season. The height of the surrounding buildings influences both the exposure to the sky vault, resulting in control of solar gain, and possibly the strength of the stack effect, related to ventilation. Several other examples show two different courtyards with complementary functions: a larger one surrounded by low façades to provide exposure to sunlight, and a smaller one, carved into a taller part of the building and connected to the previous, to draw out the hot air by temperature difference (Attia 2006). Length, width, and height, the three elemental dimensions of the courtyard, affect many parameters related to microclimate: sky obstruction, direct solar gain, shadows, and ventilation.

Moreover, the denser night air is stored at the lower level of the patio, cooling the walls and allowing for night ventilation. Patios are usually adorned with greenery and water that raise the level of humidity and lower the air temperature through evapotranspiration.

Plants are also used to increase the sky obstruction and provide additional shading when needed. The use of potted trees and shrubs allows for flexibility in the disposition of plants, and is well documented both in Mediterranean and in the Hispanic buildings in the southwestern United States (Poster 1993).

In addition, water pools can reflect sunlight into the shaded windows, providing a soft light without exposing the openings to direct solar gain. Further indirect daylight comes from the reflection of light on the walls, plastered with white limestone, and in some cases from the pavement.[10]

However, the quality and function of patios are not the only important feature of these buildings. The spatial progression between outside and inside is also very important. It was carefully structured through the arrangement of linked spaces with various characters and microclimatic performances. Trees, bushes, climbing plants, and pergolas and porticoes filter the strong light and provide a smooth transition between outer brightness and inner penumbra. Porticoes usually play the role of thermal shock absorber between patios and the inner rooms, whose shape is often influenced by a role in establishing the inner comfort. One of the most important examples, both from climatic and spatial points of view, is the "*qubba,*" a kind of domed tower placed in the northern wing of the patio (like in the Patio de los Arrayanes in the Alhambra). Hot air accumulates in the upper part of the qubba because of temperature differences and stratification, and flows out its upper windows. The resulting suction produces an airflow from the cooler air within the patio and the warmer air in the upper part of the tower that washes out the warm air from the other rooms as well (García-Pulido 2012). Therefore, we could consider the patio to be tightly integrated with the inner rooms, part of a single microclimatic concept (Fig. 24.9).

The effectiveness of this spatial arrangement in taming the energy fluxes of the harsh summer climate has been proved by measurements of temperature and humidity. In the Patio de los Arrayanes in Alhambra, a temperature difference between the patios and outer uncontrolled areas of about 4–8 °C has been

[10]It is not always possible to determine how the patios were originally paved.

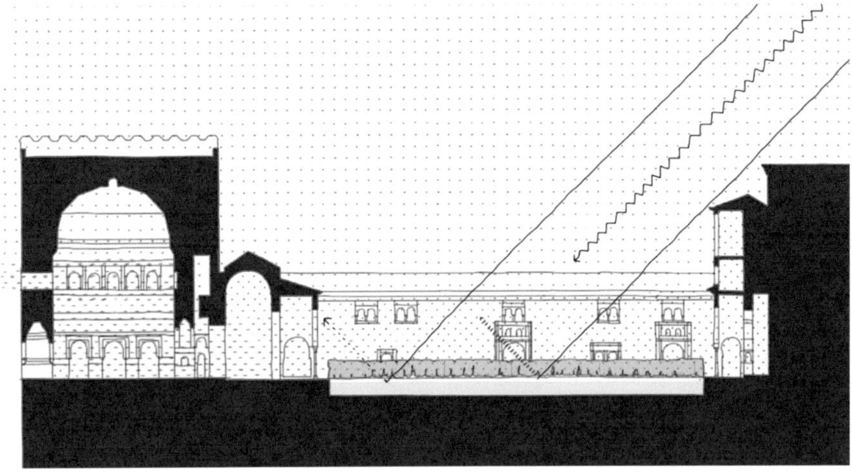

Fig. 24.9 Sequence of spaces: patio-portico-qubba. Alhambra, Patio des Arrayanes, fourteenth century

measured in the warm season (Attia 2006), with humidity, another important factor of well-being in hot dry climates, rising from 11 to 22 %.

24.4.3 Collaboration Between Construction and Microclimate (Corse 2011)

The overall design of Perraudin's Wine Museum and ampelographic garden in Patrimonio (2011) aims to avoid overheating in summer without using air conditioning and limiting the utilization of insulating material. The program is arranged into different buildings, distributed along the south-facing slope. The spaces between the different units provide a natural cross-ventilation through the site, lowering the outside temperature and cooling the surface of the walls. Servant spaces are located on the northern side of the buildings, partially buried underground, to act as buffer spaces (Figs. 24.10 and 24.11).

The roof insulation is made of wood fibers, covered with a layer of earth to improve its thermal inertia. Very thick stone walls act as a dampening system, losing heat at night by the airflow and the thermal exchange with the black body of the sky. Moreover, their porous limestone collects moisture from the sea breezes during the night, cooling the exposed walls. During the day, the sun hits the wall surfaces, dissipating the heat through evaporation of the stored humidity. The overall light color of stone, echoed by the pavement, reflects the solar radiation and helps keep the buildings cool during the daytime.

The water and pergolas that shade the space between the buildings further enhance the interplay between solar radiation, breeze, and moisture. Water takes

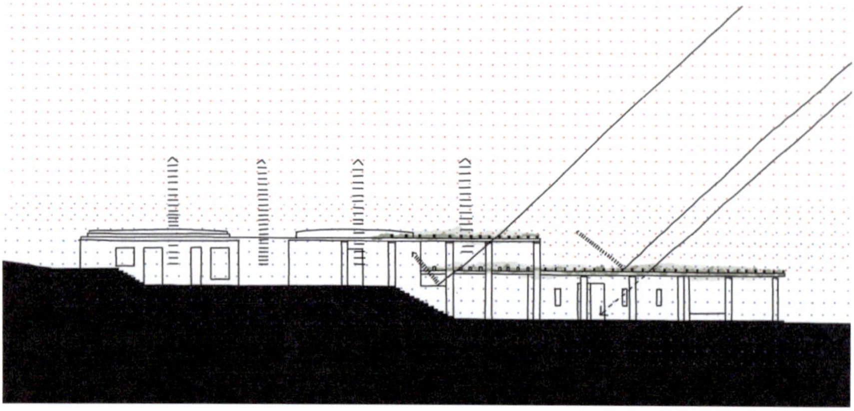

Fig. 24.10 Perraudin Architecture, Musée du Vin et Jardin Ampélographique, Patrimonio, Corsica, 2011. Schematic section, showing the contribution to evapotranspiration of the greenery, the fountain, and the humidity of the walls

Fig. 24.11 G. Perraudin, Musée du Vin, Patrimonio, Corsica, 2011. View of the outer space and of the massive walls enclosing the buildings

the form of fountains, pools and channels, that flow under the windows, cooling the air, increasing humidity during the warm days, and its murmur emphasizes the sensation of a comfortable space. Trellises and pergolas are not only shading devices, but display the different vine varieties that build the ampelographic collection of the museum.

The complex achieves an impressive coherence in using the local microclimate as a resource to provide comfort in both inner and outer spaces. Although such thick walls would usually mark a clear distinction between inside and outside, their monolithic appearance and the constructive and visual continuity between the wooden beams of the roofs and the pergolas, the play of light and and the burble of water, blur the distinction between open and closed spaces. This simplicity and proficiency, together with the use of few building elements for multiple purposes, make this building an excellent example of Perraudin's design approach, which he himself defined as "contemporary vernacular."

24.4.4 Social Spaces and Evapotranspiration (Castile 2004)

The "eco-boulevard" in Vallecas was the winner in a competition of the Municipality of Madrid, with the main goals of providing outdoor comfort and generating public space and social activity in a suburban development area. The place has a hot, dry climate in summer, and all the typical traits of an urban heat island. The proposal features the insertion of three drum-like "air trees," the densification of the existing tree lines, and a reorganization of the vehicular traffic. Other simple actions, like perforation, filling, and painting, alter the surface of the existing pavement in order to reconfigure the urban environment (Fig. 24.12).

The development of a pleasant outdoor experience is crucial for achieving both goals. A comfortable outdoor space could easily generate social activity. Trees are fundamental for improving the comfort experience in public outdoor spaces in the Mediterranean climate (Gaitani et al. 2007), but they need a long time to grow, whereas the social situation of the area called for an emergency solution.

Ecosistema urbano therefore chose to design three "artificial trees" that in essence act as microclimatic machines to chill the hot summer air by evapotranspiration. They use a system already in common use in the greenhouse industry, able to cool air up to 10°C, depending on humidity conditions and temperature. The system is triggered by sensors that detect temperatures above 27 °C in the immediate surroundings. Their steel trusses, covered with climbing plants, can be built quickly and, as the actual trees will grow, they can eventually be relocated wherever the reactivation of urban activities is needed. The open circular space at the base of the "air trees" allows the development of a very flexible program and will still remain after their removal as round clearings within the tree lines (Fig. 24.13).

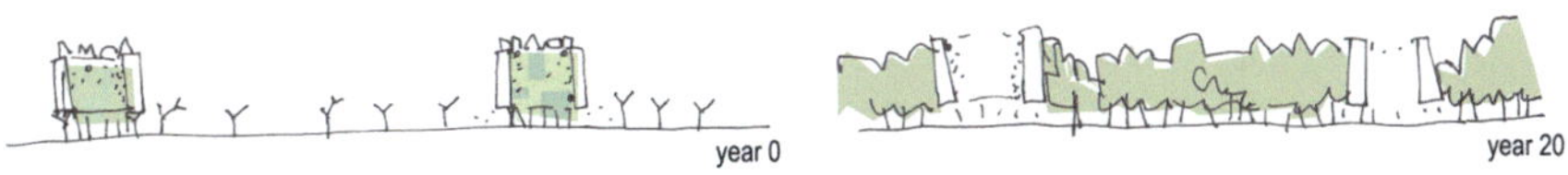

Fig. 24.12 Ecosistema urbano, Eco-boulevard in Vallecas, Madrid, 2004. Planned evolution: in about 20 years, the vegetal trees will grow and the "artificial trees" will be dismantled, leaving room in clearings that can still be used for communal activities

Fig. 24.13 Ecosistema urbano, Eco boulevard in Vallecas, Madrid, 2004. View of the "air tree"

Each "air tree" is topped with photovoltaic panels that supply the energy required to keep the system working. The whole concept uses microclimatic improvement as a way to regenerate an urban area, designing intermediate spaces as a framework for social life (Fig. 24.14).

24.4.5 Blurring and Dematerialization (Southern France 1961 and 2003–2007)

The previous examples show the progression from a combination of sequences of spaces clearly defined by walls or rows of columns, towards a sequence of interlocking spaces whose boundaries are much more blurred. But in some cases, the blurring between inside and outside becomes more radical, to the point that they could be defined by filigree surfaces or simply by different flows. Describing a few of these designs could bring some further suggestion, even if their pertinence to proper microclimatic control and sustainability is somewhat incidental.

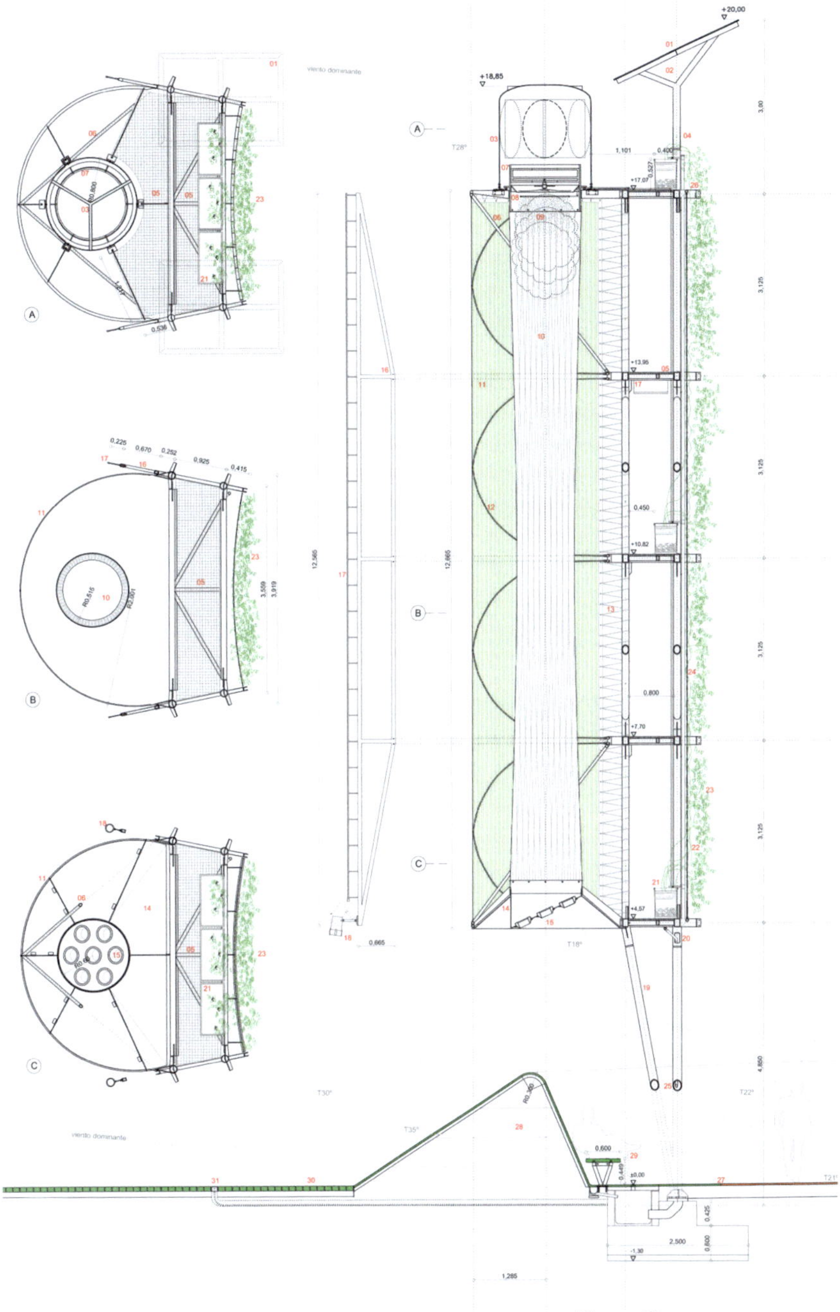

Fig. 24.14 Ecosistema urbano, Eco-boulevard in Vallecas, Madrid, 2004. Construction of the "air tree," temporary structure that improves comfort by means of evapotranspiration

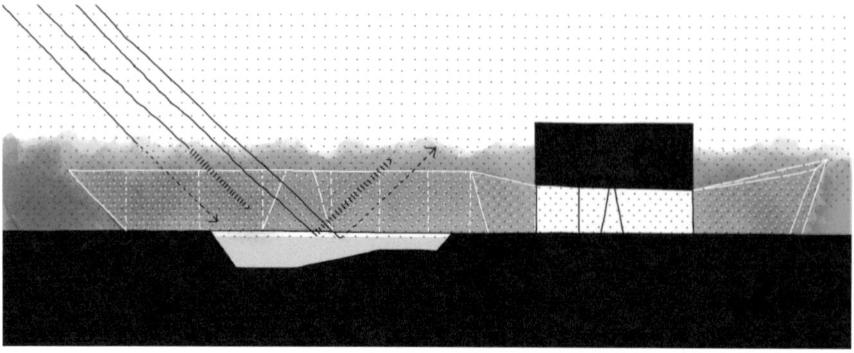

Fig. 24.15 R&Sie(n), Spidernetwood, Nîmes, 2007. Schematic section of the meshwork and of the labyrinth carved into the garden

The first two cases are the "Shearing" and "Spidernetwood" houses designed by R&Sie(n) in southern France, in Montpellier in 2003 and Nîmes in 2007, respectively. The Shearinghouse is a solid entirely covered with green nets commonly used to protect the young trees from rabbits. This folded envelope appears as a bump in the green landscape and deploys a "strategy of confusion"[11] that blurs the building against the listed neighboring fortification, known as "Château de Sommières." In fact, the house was able to be built because the shading system disguised it as a temporary artwork, more than as a permanent building.

The canopy is wider than just the house, sheltering various outer spaces and filtering the light and air that reach the rooms. Despite this, the undeniable shading and cooling effect of the canopy is mentioned in the designers' documents simply as a "subsidiary consequence".

The "Spidernetwood" house is far more rigorous (Fig. 24.15). The house is actually a long labyrinth made of white textile material. The outdoor part is modeled with nets, wires, and poles that contain the free-growing wood all around the house. The plants grow until they become a kind of porous wall pushing and distorting the meshwork.[12] The indoor part is actually a two-story house, where the inner walls and ceiling were clothed with textile white material. There is actually no distinction between inside and outside, except for a huge sliding glass door that seamlessly connects the two. The porous vegetable wall that wraps both the outdoor and indoor parts of the labyrinth, and the water pool close to the glass door, should effectively lower temperatures through evaporation and evapotranspiration. The plants, growing all around up to five meters, should provide shading and protect the whole building from solar radiation, while the white sheets reflect light,

[11]Quote from R&Sie(n), in http://www.new-territories.com/roche%20barak.htm, seen 22/03/15.

[12]The project is presented in a novel by the cyberpunk author Bruce Sterling, which figures out how this strange combination of artifice and nature will look in 2030. See http://www.new-territories.com/roche%20barak.htm, seen 22/03/15.

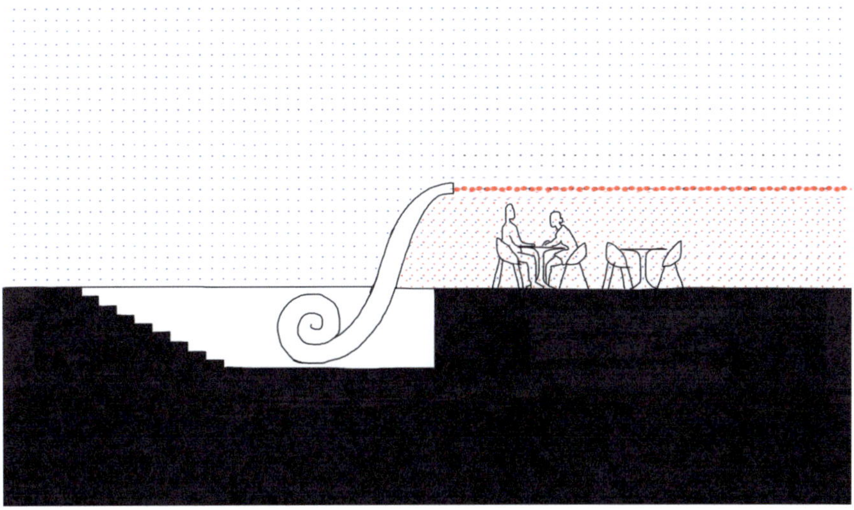

Fig. 24.16 Schematic section of Yves Klein's blown-air roof

softening the shade. However, even in this case, the microclimatic qualities of the design are cited as a secondary feature in the description of the building.

If R&Sie(n) dematerialize the building, deliberately confusing the difference between inside and outside, the French artist Yves Klein attempted to build spaces only by means of introducing further energy flows, in an attempt to control microclimate without establishing any physical boundary. In the 1960s he worked on the concepts of spaces only enclosed by layers of air at different temperatures. He produced some sketches and drawings of such spaces at various scales, and few installations built with the sponsorship of Gaz de France (Fig. 24.16).

One of the sketches shows a pergola made of hot air, blown like a horizontal plan over some chairs and dining tables. Some perspective and section drawings show another "air roof" that shelters "air beds," "fire fountains," and other concepts. Klein built indeed a "fire wall," as an installation in the Museum Haus Lange, Krefeld, Germany, in 1961. Klein asked Philip Johnson to help him obtain a grant to bring his research on dematerialized architecture further, but without success; instead, he found help in Claude Parent, one of the most important avant-garde French architects. At the time, little attention was paid to sustainability and energy consumption, and more emphasis was put on the modernist feeling of these architectures made of pure energy.[13]

However, Klein's architectural concepts are the most radical form of space arrangement based on the manipulation of temperature and air—so radical that

[13]Claude Parent later designed two nuclear plants, attracted by the speed and modernity of atomic energy, admittedly overlooking the problem of nuclear waste. See http://032c.com/2011/the-supermodernist-architect-claude-parent/, seen on 22/03/15.

walls, roofs, and fences appear like an obsolete means of defining the difference between inside and outside. According to Klein, places can be defined simply by creating, controlling, and distorting fluxes of energy, in an attempt to transform microclimate improvement in the ultimate way of building.

24.5 Conclusions

The case studies described in this chapter show the effectiveness of designing open spaces around buildings in order to mitigate unfavorable climate conditions. They highlight the advantages of an integrated design of spatial articulation and energy flows. In the examples shown above, the places' character emerges from the accurate blending of geometry, materials, and all the forms of energy that spread through the site. Building and climate get together in more or less evident ways to define places suitable for comfortable living—an attitude which, combined with today's evolution in building technology and better knowledge of climatic issues, can still produce new, unexpected spaces.

There are many practical descriptions of the various parameters affecting microclimate, and knowledge in this field has greatly increased in recent years—to the extent that computer simulations of thermal behavior in open spaces are now possible. All the basic elements of architectural design concur to microclimatic improvement within a meaningful configuration of space. Basic design decisions that take care of microclimatic issues dramatically improve the quality of architecture with very limited economic means.

The choice of the site, when such a choice exists, could be advantageous for the main climatic features on a meso-scale, exploring topography, winds, and perhaps water bodies. The shape of the building mass itself can be structured so as to provide pleasant outdoor spaces that could improve thermal comfort in the rooms. Materials, greenery, and water could positively influence the climatic design of both outdoor and indoor spaces.

The Mediterranean tradition offers an interesting catalogue of effective typologies that could be creatively reworked with contemporary means and materials so they can fit new lifestyles, but this tradition could help us just for a limited extent, and not simply for the obvious reason, i.e., that lifestyles, materials, labor, and other factors in our society change constantly.

The Mediterranean climate is changing, too (Travers et al. 2010). Model projections foresee increasing temperatures in summer and autumn and decreasing precipitation. The once-moderate climate is becoming warmer and dryer. Vegetal species could change, or shift their locations. This progressive modification could result in water shortages, extended desertification, and floods. The latter could play an important role for settlements, considering the amount of Mediterranean coastline used for human activities.

Although precise projections about climate change are still being developed and debated, the trend undeniably goes in this direction; there is, therefore, a need to critically study tradition, considering it more as a lesson in designing and building processes than as a repository of trusted typologies. Not only is contemporary society changing, but nature around it is changing as well. Including microclimate within the design process can help find spatial structures capable of adapting themselves to this climatic evolution, and possibly contributing to slowing it down with more sustainable architecture.

References

Alberti LB (1541) De re aedificatoria libri decem. Argentorati, Strasbourg, Liber IV, Caput V, p 55

Attia S (2006) The role of landscape design in improving the microclimate in traditional courtyard buildings in hot arid climates. 23rd international conference on passive and low energy architecture-PLEA 2006, pp 23–24

Banham R (1984) Architecture of the well-tempered environment. University of Chicago Press, Chicago, pp 23–27

Colombo R, Landabaso A, Sevilla A (1994) Design handbook—passive solar architecture for mediterranean area. Joint Research Centre Commission of the European Community

Fathy H (2010) Architecture for the poor: an experiment in rural Egypt. University of Chicago Press, Chicago, p 57

Gaitani N, Mihalakakou G, Santamouris M (2007) On the use of bioclimatic architecture principles in order to improve thermal comfort conditions in outdoor spaces. Build Environ 42:317–324

García-Pulido LJ (2012) Bioclimatic devices of Nasrid domestic buildings

Georgi JN, Dimitriou D (2010) The contribution of urban green spaces to the improvement of environment in cities: case study of Chania, Greece building and environment. Build Environ 45:1401–1414

Grosso M (1998) Urban form and renewable energy potential. Renew Energy 15:331

Hawkes D (1980) Building shape and energy use. In: Hawkes D, Owers J (eds) The architecture of energy. Longmans, London

Hawkes D (1996) The environmental tradition: studies in the architecture of environment. E. & F.N. Spon, London

Norberg-Schulz C (1976) The phenomenon of place. In: The urban design reader. Routledge, London, p 127

Olgyay V, Olgyay A (1963) Design with climate: bioclimatic approach to architectural regionalism. Princeton University Press (Some chapters based on cooperative research with Aladar Olgyay)

Pollio MV (1914) The ten books on architecture. Harvard University Press, Cambridge (De Architectura translated by MH Morgan)

Poster C (1993) Sombra, Patios, y Macetas modernism, regionalism, and the elements of southwestern architecture. J Southwest 35:497

Ragette F (2003) Traditional domestic architecture of the Arab region. Edition Axel Menges, pp 50–53

Rava CE (1931) Di un'architettura coloniale moderna—Parte seconda. Domus n. 42, June, p 36

Ruggles DF (2004) The Alcazar of Seville and Mudejar Architecture, Gesta. The University of Chicago Press, Chicago, pp 87–98 (on behalf of the International Center of Medieval Art, 43)

Santamouris M, Papanikolaou N, Livada I, Koronakis I, Georgakis C, Argiriou A, Asimakopoulos D (2001) On the impact of urban climate on the energy consumption of buildings. Sol Energy 70:201–216

Shashua-Bar L, Hoffman ME (2004) Quantitative evaluation of passive cooling of the UCL microclimate in hot regions in summer, case study: urban streets and courtyards with trees". Build Environ 39:1087–1099

Sprague P (2013) Jacobs, Herbert and Katherine, Second House United States Department of the Interior. National Park Service—National Register of Historic Places Registration Form, p 19

Travers A, Elrick C, Kay R (2010) Position paper: climate change in coastal zones of the mediterranean priority actions programme, Split, p 1

Chapter 25
Modelling and Bioclimatic Interventions in Outdoor Spaces

Stamatis Zoras and Argyro Dimoudi

Abstract These days, urban design of open spaces is strongly related to bioclimatic techniques and practices. In this chapter we present the procedure of such bioclimatic studies by the use of simulation tools. Outdoor spaces that are characterized by decreased human thermal comfort conditions during the summer, especially in areas under temperate climatic conditions, justify a bioclimatic intervention. Modelling has contributed to the understanding of both the limitations and benefits of specific interventions that should be made in the open space in order to succeed at the predefined thermal related improvement. Also, it shows how these interventions affect buildings' operations.

25.1 Introduction

It is widely accepted that dense urban developments, in conjunction with the use of inappropriate external materials, the increased human-related thermal energy emission, and the lack of green spaces, increase environmental temperature leading to significant environmental impacts and increased energy consumption (Santamouris et al. 2012a, b; Fintikakis et al. 2011). Open spaces within urban developments are complex, due to thermal energy exchange between structures, shadowing, and wind flow complication in comparison to general flow. Cooling materials (Kolokotsa et al. 2013) and other factors (water surfaces, green roofs) are used which may mitigate the urban heat island effect (Santamouris 2014a; Mastrapostoli et al. 2014; Tang et al. 2014; Tsilini et al. 2014; Santamouris 2013;

S. Zoras (✉) · A. Dimoudi
Democritus University of Thrace, Vas. Sofias 12, 67100 Xanthi, Greece
e-mail: szoras@env.duth.gr; stamatis.zoras@gmail.com

A. Dimoudi
e-mail: adimoudi@env.duth.gr

S.-N. Boemi et al. (eds.), *Energy Performance of Buildings*,
DOI 10.1007/978-3-319-20831-2_25

Piselo et al. 2013; Georgi and Dimitriou 2010; Gartland 2008; Gaitani et al. 2007; Akbari et al. 2001). The main problems that result from poor thermal conditions include decreased human thermal comfort, decreased air quality, increased illness due to heat, and increased energy and water use (Stone 2005; Baik et al. 2001).

Experimental measurements within urban developments must be carried out in order to identify the thermal situation. In the Mediterranean area, there are intense thermal phenomena mainly during the summer (Livada et al. 2002); these are observed in open urban areas throughout countries with temperate climates. Surface temperatures in relation to microclimatic conditions (wind, temperature, radiation) must be analyzed in order to better select the rehabilitation strategy of open developments.

Simulation tools must be used (Stavrakakis et al. 2011) in order to depict the current situation in the opencast area, usually during the warmest day of the hot periods. Material identification and construction configuration must also be taken into account in the simulation process. A new configuration of materials and bioclimatic techniques is then proposed and simulated in order to show its influence on the thermal urban environment. The goal of this procedure is to realize the microclimatic conditions' improvement due to the rehabilitating bioclimatic techniques and practices. The selection of the measures to be proposed depends on the targets that will be defined in an improved thermal environment (e.g., thermal comfort) (Santamouris 2014b; Santamouris et al. 2011; Gulyas et al. 2006).

Due to the complicated urban environment in terms of materials, reflection, emission, wind flows around buildings, altitude differences, etc., the simulation tool must be selected very carefully. It must be able to simulate three-dimensional flows with solar radiation taken into account. This inevitably leads to general codes of computational fluid dynamics tools (e.g., PHOENICS software), but with increased demand of computational resources. Other tools may be useful for the assessment of individual parameters, as surface materials and trees may influence thermal comfort (Matzarakis et al. 2006), but this would not assess the wind flow effect in geometrical detail.

Here we demonstrate the procedure of two bioclimatic studies of open urban spaces in two northern Greek cities—Ptolemaida and Serres. Details of other bioclimatic studies can be found elsewhere (Gaitani et al. 2014, 2011; Santamouris et al. 2013; Skoulika et al. 2014). Experimental measurements, simulation tool verification, and the simulation-based assessment of the proposed architectural reformation of the two cases under temperate climatic conditions are presented.

The effectiveness of the "measurement-simulation-bioclimatic proposal-simulation" procedure depends on the following aspects:

- Knowledge of the area and accurate experimentation (materials, experimental instrumentation)
- Verified simulation against experimental measurement (simulation of the present situation)
- Qualified collaboration between the architectural bioclimatic design and simulation (viable architectural proposals that would improve thermal conditions e.g., green roofs)

- Definition of the thermal related targets (thermal comfort, cooling hours, improved ventilation, surface temperatures, environmental temperature).

The objective of urban and built environmental analysis and simulation lies between the maximal external human comfort and minimal building operational costs. This perspective has increased the corporate customers' concerns at temperate climates regarding the impacts of outdoor versus indoor microclimate conditions on their staff productivity and safety, while those involved in the design and construction stage are concerned with how the operative and passive systems would fulfill customers' requirements.

On the other hand, the ease of use of environmental analysis software packages is questionable, as are the work time required during the simulation of large-scale open spaces and 3-D building geometry parameters. In particular, when the energy design involves the larger scale urban canopy, then the application involves more than a unique simulation package.

Practically, questions have been raised in the description of the needs of modelling during a preconstruction stage or bioclimatic rehabilitation of outdoor urban areas; these are answered throughout this chapter:

- Why should the designer wait until the construction is completed before he knows if his design meets the initial requirements?
- How can operational performance be assessed through energy simulation?
- What is the thermal contribution to an outdoor area design of each implemented bioclimatic technique?
- Is there a wintertime thermal profit for temperate climates?
- What are the benefits of an open-space bioclimatic design for a building's operational energy consumption?
- What are the new future directions in the simulation of environmental flows in the external and indoor environments?

25.2 The Optimum Modelling Tool

A wide range of computer-based physical modelling tools assisting the bioclimatic design for teams to understand the interaction between natural forces and buildings is necessary; these enable them to reformat the external spaces that are responsive to environment. The approximation of such networks is usually achieved by parametric and sensitivity analysis procedures, an approach that results in the assessment of the occupant's interaction with urban climatic conditions.

Engineering is dominated by the physics of natural forces: gravity, temperature, humidity, light, sound, and vibration, and the forces of climate, weather, and microclimate. It has a tendency to resist and exclude natural forces, creating an isolated envelope where energy-intensive machinery is required to maintain the urban environment. The building structure and external façades should be used not to resist, but to filter, channel, and deflect critical natural forces, to create an urban environment geared towards the human's function, and to provide thermal comfort acceptance.

The functions of microclimate and structural skins then merge, allowing the building to become an integrated mechanism that combines comfort and structural efficient operation. Specifically, the building systems and components respond to changing internal demands and adapt to an ambient variation in order to achieve optimum environmental quality for the building's users. Active and passive elements, such as structure and skin, computer control and simple response take advantage of ambient energy sources to achieve maximum energy efficiency, acting as moderators between humans and the environment. The achievement of the optimum modelling tools combination in terms of response to environmental forces and engineering objectives is considered successful only after everything affecting the process's aspects are taken into account.

25.3 Using Computational Fluid Dynamics in the Bioclimatic Design of Open Spaces in Two Greek Cities

The procedure of two bioclimatic studies of an open urban space and a dense urban development that have been carried out in two Greek cities—Ptolemaida and Serres—are discussed here. The thermal targets were associated with the general characteristics of temperate climates in terms of the maximum summer air and surface temperatures at noon on the hottest day in relation to human thermal comfort improvement. The employment of computational fluid dynamics has contributed to the understanding of what interventions should take place at the urban populated routes in order to succeed in creating an improved human environment. The proposed rehabilitation explains what the interventions would contribute to the improvement of the local environment under temperate conditions.

25.3.1 Bioclimatic Thermal Problem

To examine the weather conditions in the Greek cities of Serres and Ptolemaida, meteorological data from periods of at least two years were used from one urban/suburban meteorological station and one rural station; hourly data were available from both stations. The meteorological parameter that was examined was the air temperature (°C).

Compared with the surrounding suburban and rural environment, urban climate varies in terms of solar radiation, characteristics of rainfall, and air temperature. According to Oke (1973) almost every urban center in the world is 1–4 °C warmer than neighboring non-urban rural areas that enforces urban heat island effects. Also, Gilbert (1991) states that the air temperature on sunny days can be from 2.0 to 6.0 °C higher in urban compared to suburban locations.

The comparison of climate data of the two cities during the two-year period has shown a significant temperature difference between rural areas and cities, amounting to more than 3.0 °C, with a mean difference of more than 2.0 °C. Thus, the urban centers act as urban heat islands, and hence open bioclimatic urban upgrading would be suggested.

25.3.2 Bioclimatic Interventions

Seventy percent of Greek power generation takes place in northern Greece. Ptolemaida City is surrounded by power generation activities (lignite mining and combustion processes). In the commercial and social center of the city is an open marketplace that is heavily populated during the week. The urban summer time microclimate in this area is mainly affected by asphalt all over the surface of the open market (see Fig. 25.1). The southern part of the market, along Pontou Street, is bordered by a multipurpose building with vegetation and sidewalks. The rehabilitation strategy for the area targets conserving human activities and improving human thermal conditions in the open public market during intense summer heatwaves. The bioclimatic interventions can be divided into two main directions: green roofs in the open marketplace, and cool surface materials (bricks, asphalt) in the surrounding streets.

The dense residential development of Serres is heavily populated at all times. The urban summer-time microclimate in this area is mainly affected by the presence of conventional bricks and asphalt over all the linking routes, and the pavements are covered by Greek flagstones. The proposed bioclimatic directions are characterized by increasing water surfaces, vegetation, and green, and by installing cool asphalt and flagstones. Table 25.1 describes the thermal properties of the current materials and bioclimatic proposed ones.

25.3.3 Model Verification

Advanced computational fluid models (CFD) models can calculate microclimatic parameters with a high degree of accuracy at every grid point of the meshed space. However, the more complicated the geometry of the urban open space, the more resources for input data and calculation are needed. The CFD model should be able to simulate, in both steady state and transient mode, the microclimatic parameters, heat transfer, and fluid flows, together with long and short radiation distributions in time and space. This is the ideal approach of the three-dimensional urban

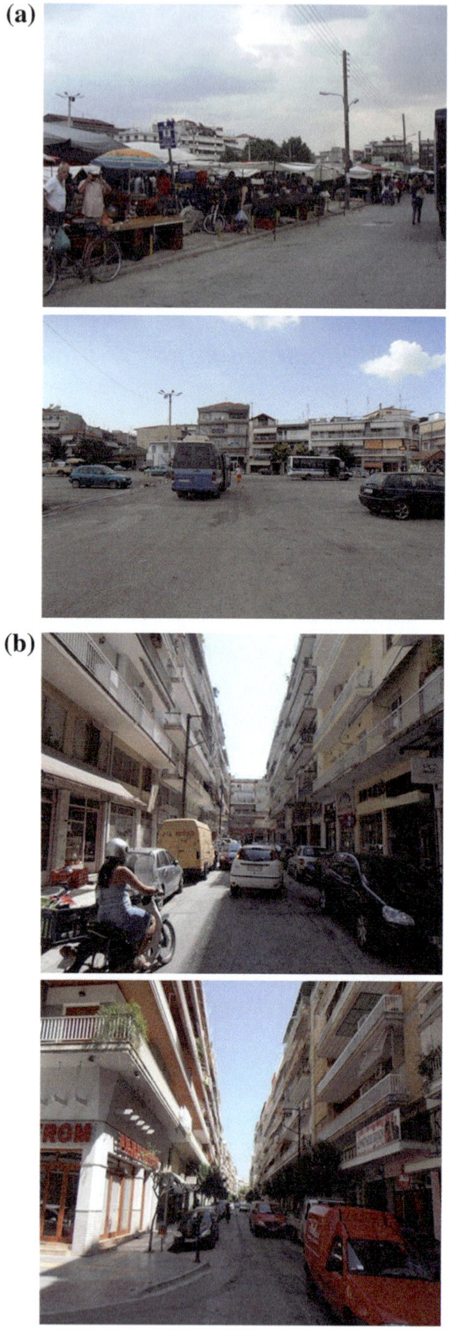

Fig. 25.1 **a** Present situation of the rehabilitated areas: Ptolemaida's open marketplace. **b** Present situation of the rehabilitated areas: Serres's urban residential development

Table 25.1 Thermal and optical properties of materials defined in the CFD model

	Reflection coefficient	Emission coefficient
Conventional flooring and view materials		
Street asphalt	0.10[a]	0.85–0.93 (0.89)[b]
Light-colored covering roofing/roofs (sheathing with pavement flagstones)	0.35[a]	0.90[c]
Light-colored coating	0.60[a]	
Medium-colored coating (beige, gray)	0.40[a]	
Gray color		0.87[c]
Dark-colored coating	0.20[a]	
Conventional structural material		0.80[a]
Cool materials coverings/coatings[d]		
Asphalt Ecorivestimento grigio photocatalitic concrete based mortar (speciment 1)—fotofluid	0.37	0.89
Sidewalk blocks [block CE light gray (N° 5) or CE beige (N° 6)]	0.67	0.89
Pavement flagstones [white flagstone (N° 12)]	0.68	0.92

[a]GTCI (2010), [b]Incropera (1990), [c]Santamouris (2006), [d]ABOLIN

environment. Real experimental data of the thermal conditions in the areas of interest and at surrounding locations are used to obtain the analytical verification of the CFD model, and consequently the accurate simulation of both the current situation and proposed interventions.

For the efficient simulation of the thermal energy condition in the areas of interest, the detailed three-dimensional tool ANSYS CFX 13 has been used. ANSYS CFX is an advanced general code computational fluid dynamics model that solves Navier Stokes differential equations and turbulence by the finite elements technique in the 3-D space. It is a commercial software package that handles very detailed three-dimensional geometry with the ability to solve heat transfer and fluid flow phenomena.

In order to prove the ANSYS CFD model's validity for the open area simulation, it was verified against experimental data that have been carried out during October 2011. From the period of the experiment on the warmest day, it was selected in terms of the completeness of the microclimatic data (i.e., air temperature, air velocity, surface temperature). The data used for validation of the model is the measurement data within the study area, i.e., the air temperature, the temperature of the material surfaces of streets, sidewalks, and façades 1.8 m high as well as the wind speed at the same height.

It was then compared to the simulation results against the measured values of the surface and ambient temperatures and wind speed. Climate data from that period were obtained from the urban/suburban meteorological stations and climatic data were simulated in the intervention area and in places where measurements were taken. Since, as mentioned previously, the surface temperature

Table 25.2 **a** Comparison of experimental measurements and simulation results in Ptolemaida. **b** Comparison of experimental measurements and simulation results in Serres

Location	Air temperature (°C) at 1.8 m height/simulation	Air velocity (m/sec) at 1.8 m height/simulation	Surface temperature (°C)/simulation
a			
Asphalt	36.0/36.6	1.1/1.2	48.2/48.0
Pavement	35.9/36.2	1.1/1.2	47.0/47.3
Building surface	35.8/35.2	1.1/1.2	37.4/37.8
b			
Asphalt	35.586/34.916	0.619/0.709	35.3/35.0
Pavement	37.390/38.116	1.158/1.350	42.2/43.0
Building surface	36.228/36.852	0.608/0.638	35.9/36.2

measurements were taken at midday, the comparisons were done for the same periods. Thermal and optical input properties of the present conventional materials are given in Table 25.1.

The concept of model validity was that if meteorological input from meteorological stations are applied in ANSYS CFD, then this could efficiently calculate the thermal behavior of within the urban complexes. Therefore, the accuracy of the model that was developed for the studied areas was quite good. The above comparison substantiates the high reliability of the model for the assessment of both the current situation and bioclimatic upgrade in the study area. Table 25.2 shows the comparison between experimental data and simulation results at the two locations.

25.3.4 Comparison of the Mean Maximum Air Temperature

These simulations were carried out for the current situation and for the proposed rehabilitated configuration; then the results were compared in order to obtain the microclimatic improvement. In order to clarify matters, the input data of meteorological measurements were different between the verification of the model and the simulation procedure, due to the lack of air and surface temperatures and wind velocity during the selected warmest day, which was July 16, 2011, for Ptolemaida; at Serres it was July 19, 2011—with all data available on the same day. The thermal and optical input properties of the proposed cool materials and the conventional ones are described in Table 25.1.

Simulation of the average maximum summer temperature has been carried out in the open areas at noon on the warmest day. Meteorological data from these days at noon were used in the simulation, in steady state mode, of the present and rehabilitated situation. In each case the same meteorological data from stations were

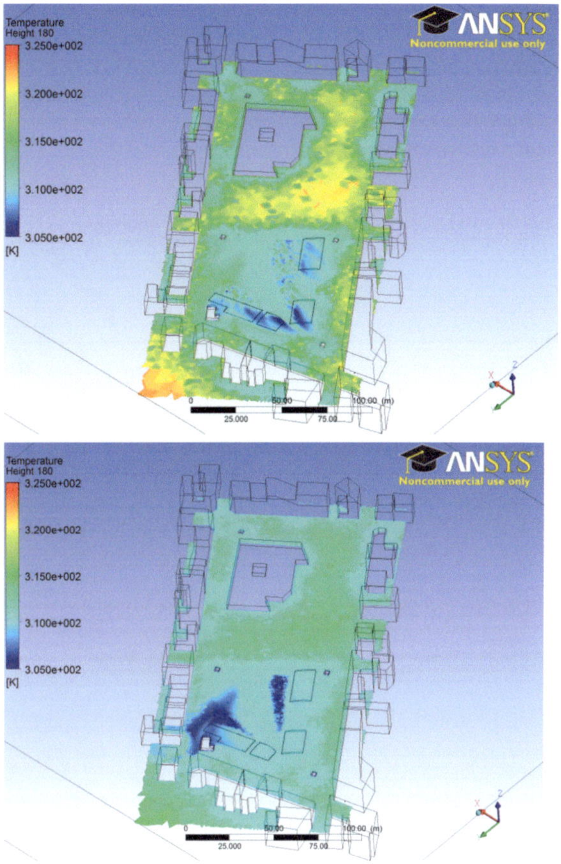

Fig. 25.2 Present mean air temperature at 1.8 m high, calculated from at least 500 grid points at each street in the Ptolemaida open marketplace

applied as input in ANSYS CFD with the respective materials and interventions of each configuration. This way, modeling predicted the thermal situation in the urban complex with input general meteorological data of the city.

Figures 25.2 and 25.3 depict the air temperature field at each street and the open market at a height of 1.8 m at noon on the warmest day before as well as after rehabilitation. The simulated air temperature for each individual space (streets and open market) was obtained from at least 500 grid points of the mesh, at a height of 1.8 m. The calculated total air temperature in Ptolemaida before and after was 37.61 and 35.42 °C, respectively. Therefore, the air temperature improvement—if bioclimatic measures were taken—would be 2.19 °C. In Serres, the respective temperatures were calculated at 37.9 and 36.4 °C, representing an improvement of 1.5 °C.

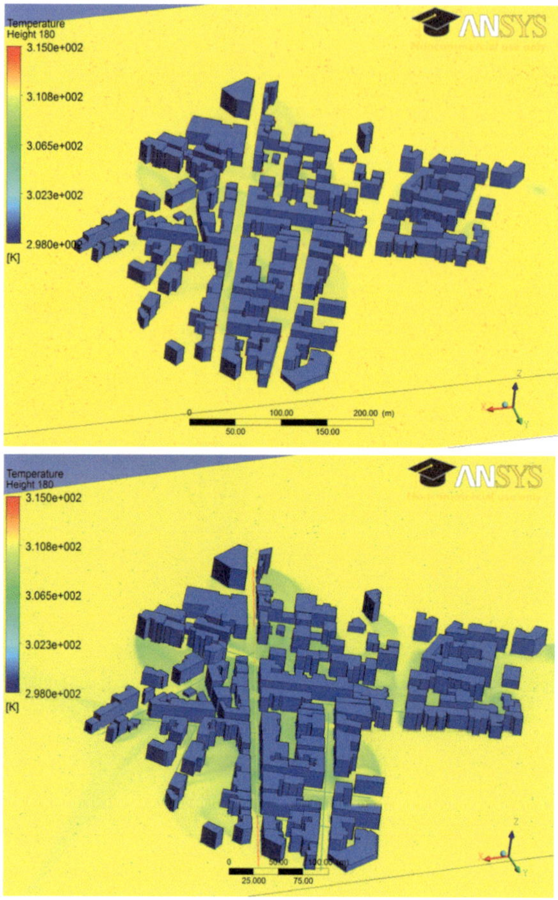

Fig. 25.3 Mean air temperature after rehabilitation at 1.8 m high, calculated from at least 500 grid points at each street in a Serres residential complex

25.3.5 Comparison of the Mean Surface Temperatures

In order to improve the thermal microclimate in the area of the open market, new, cool materials must be used that may reduce the surface temperatures of buildings, streets, and sidewalks. The proposed materials have a relatively high reflectivity of solar radiation and increased emission rate. Green areas in the open square area or where people are activated would also contribute to the reduction of material thermal storage.

CFD simulations have been carried out for this case and for the proposed bioclimatic one during the warmest day for the two cases (Figs. 25.4 and 25.5). It was

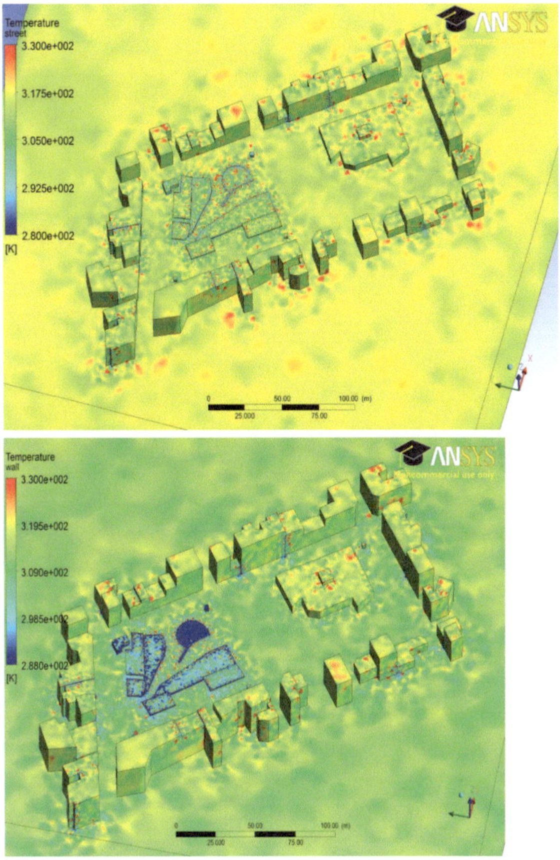

Fig. 25.4 Mean surface temperatures case in Ptolemaida's open marketplace

assumed that the mean surface temperature reaches its peak during the warmest day, and then it was calculated for all the individual roads and the marketplace separately. Therefore, it was assumed that the calculated surface temperatures are the maximums for each individual road or open space. In Ptolemaida's open market, the mean surface temperature at noon on the warmest day reached 43.73 °C in contrast with the proposed bioclimatic configuration (Fig. 25.4), which reached 36.62 °C. The total predicted temperature difference was 7.11 °C. In the residential area of Serres (Fig. 25.5), the respective mean surface temperatures were calculated at 40.3 °C before, and at 33.8 °C after the rehabilitation, resulting in an improvement of 6.5 °C. Significant material surface temperature reduction was predicted due to shadowing from vegetation, water surfaces, green roofs, cool asphalt, and cool flagstones on pavements and sidewalks.

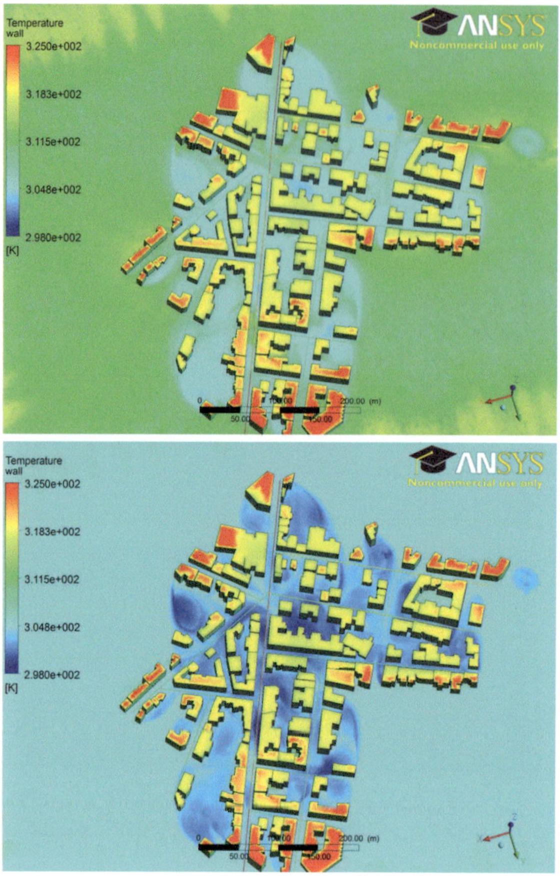

Fig. 25.5 Mean surface temperatures of the predicted bioclimatic case in the Serres residential complex

25.4 Urban Microclimatic Improvement Effects on Building Blocks' Energy Consumption by the Use of Energy Simulation

The energy benefits of the buildings' operation, due to the latter bioclimatic reformation, can be studied by whole building thermal simulation models. The effect of the proposed bioclimatic changes to microclimatic conditions and then to the buildings' energy performances, are demonstrated here by the application of a whole building energy simulation software: TAS-EDSL. TAS simulation estimated the cooling loads of buildings and the reduction of these loads due to microclimatic improvement. Note that the previous bioclimatic work was focused on the cooling loads during the summer and as a result, the energy performances for the summertime are presented here.

The city of Serres was selected for these purposes, due to quite a large number of buildings in comparison to Ptolemaida's case. Serres is located in northern

Fig. 25.6 TAS images of the urban complex buildings

Greece, has an altitude of 50 m, is rather flat, and belongs to climate zone C (according to GTCI)—which means that buildings have rather high energy consumption for heating during the winter. The topography and structure of the city have the effect of creating a major phenomenon of UHI during the summer as shown before, resulting in a significant difference between the temperature of the city center and the suburban areas.

The region has an area of approximately 160,000 m^2 and consists of a few open areas and narrow streets with tall buildings (4–7 floors) (Fig. 25.6). This study examined the energy performance of 207 buildings with a total floor area of 269,657,17 m^2. Their use was mixed (residences, offices, shops on the street level, schools, and one church). The buildings' ages varied from 4 to 50 years, with the majority of them in the 15–35 year category—which is why the buildings were categorized into four groups, depending on the construction materials and thermal insulation:

1. Old buildings (thin walls, no insulation)
2. Newer buildings (no insulation)
3. Newer buildings (with some insulation)
4. New buildings (according to national regulations)

A small number, with a percentage of 8.3 % of the whole building surfaces, belongs in the last category. The mathematical model that depicted a typical or standard summer day corresponded to the average of thermal conditions. This was created for the estimation of the degree-hours (above a base temperature), in which air conditioning was necessary for acceptable thermal comfort.

The meteorological data of the area were gathered from the Greek National Meteorological Service and experimental measurements were taken during the summer of 2011. The required cooling load to achieve thermal comfort depends on the temperature variation during a summer's day; on a typical hot day, it was assumed that the base temperature for air conditioning would be 26 °C, according to national guidelines.

Thermal simulation took into account shadow effects between buildings and the specified internal conditions of all zonal spaces during the buildings' operation in the summer. Results have shown that the cooling load for a large urban area, such as that at Serres which was composed of building blocks, would be reduced by 6.88 % for a single hot day due to external air temperature improvement of 1 °C. The reduction was more significant on the lower floors, in old buildings, and in north-facing zones of the building blocks.

The most significant load improvement, thanks to the reduction of external air temperature, was observed in the north-facing buildings (7.54 %). This was because north-facing rooms were mostly affected by convection rather than solar radiation during the day. The floors that benefitted most from the external air temperature reduction were the lower ones, and the middle and top floors kept on having high temperatures in relation to the lower ones due to undisturbed solar radiation and no interventions outside building surfaces. It is important to note that bioclimatic interventions were proposed only for ground surfaces.

It is vital here to point out that the calculated cooling load per m^2 in older buildings was lower than in newer buildings, because older buildings' heights were significantly lower than the new ones. The higher buildings were warmer—especially the top floors. The newer buildings were the higher ones, and thus the majority of the insulated surfaces were distributed quite high and far from the ground. This was the reason that the external air temperature improvement did not have any significant effect on new constructions.

25.5 The Optimum Modeling Scheme in Bioclimatic Design

The future of CFD codes' employment in large, open urban spaces in combination with energy losses and human comfort software does not rely only on the improvement of computer systems. A cultural shift of the design teams towards the introduction of software packages in their business is quite sufficient for future developments. The kick-off meeting of a construction project between the architect, the engineers, the simulation expert, and the customer is the most important element in the integration of relevant software into the design stage. Simulation is at its most advantageous when used in the initial design phase. Its use at a later stage becomes increasingly difficult, due to narrower margins in alternative solutions. If the team is coordinated from the beginning, according to the following diagram, then it is possible to get the optimum design very quickly.

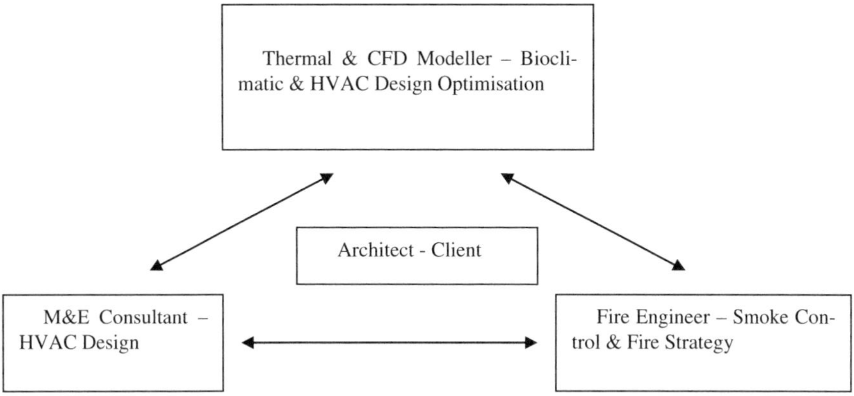

Energy design simulation is in reality more complex, based on the cooperation time spent between the individual parties of a project. However, the cost of this collaborative scheme increases when simulation holds up the desirable result due to delays in data collection and introduction. A promising future prospect in the simulation cooperation would be based on the Internet, with access in acceptable common libraries, e.g., materials, meteorological data time series, etc. In addition, the compatibility between geometric creation models and thermal simulation tools should also be improved in the use of the same three-dimensional model.

In general, it is recognized that environmental simulation is at its most useful when it is used as a combination of thermal and CFD simulation. The former gives the statistical optimum approximation of the energy structural components and is being used in the definition of the inputs in consequent CFD analysis. Up until now, there has been no autonomous model that combines the full capabilities of the two simulation methods in one design environment. However, future efforts should be oriented towards this perspective.

25.6 Conclusions

If a proposed rehabilitation would take place in the future, each thermal index or parameter would be improved. In the previous two studies, it was possible to note why the surface temperatures were significantly reduced in contrast to ambient temperatures. Generally, it is more difficult to reduce air temperature by increasing surface reflectivity that results in a reduction of surface temperatures. This is because air fluid gathers heat from the surface, mostly by convection and not by radiation.

Therefore, increasing plantation is more effective in the improvement of ambient temperatures. Moreover, relatively low air temperature improvement may lead to significant thermal comfort improvement, because thermal comfort is less dependent on air temperature, but mostly correlated to radiant temperature. The wind field of velocities would be approximately the same for the cases before and after rehabilitation and, therefore, it would not significantly influence the human thermal comfort conditions.

Reflection coefficients of cooled materials were higher than the conventional materials but emission coefficients were approximately the same. Therefore the significance of roofs, green areas, and trees within the open urban complex in relation to the improvement of air temperature was rather obvious, which was basically the reason for a better microclimate improvement in Ptolemaida's open market than in the dense residential area in Serres.

Relatively high thermal comfort improvement will succeed more easily if surface thermal exchange is manipulated by any of the above-mentioned ways. The exact influence of a bioclimatic intervention to microclimatic parameters must be studied in relation to urban complexity and climatic zone characteristics. Nevertheless, during the cold period, cool materials' thermal contribution would force air temperatures to lower levels. A simulation procedure should be carried out for a year in order to assess the total balance of cooling effects in the area of interest and, subsequently, the buildings' operation.

References

ABOLIN, Technical material brochures & Certificates

Akbari H, Pomerantz M, Taha H (2001) Cool surfaces and shade trees to reduce energy use and improve air quality in urban areas. Sol Energy 70(3):295–310

ANSYS-CFD-SOFTWARE http://www.ansys.com/Products/Simulation+Technology/Fluid+Dynamics

Baik J, Kim Y, Chun H (2001) Dry and moist convection forced by an urban heat island. J Appl Meteorol 40:1462–1475

Center for Renewable Energy Sources and Saving. www.cres.gr

Fintikakis N, Gaitani N, Santamouris M, Assimakopoulos M, Assimakopoulos DN, Fintikaki M, Albanis G, Papadimitriou K, Chryssochoides E, Katopodi K, Doumas P (2011) Bioclimatic design of open public spaces in the historic centre of Tirana, Albania. Sustain Cities Soc 1:54–62

Gaitani N, Mihalakakou G, Santamouris M (2007) On the use of bioclimatic architecture principles in order to improve thermal comfort conditions in outdoor spaces. Build Environ 42(1):317–324

Gaitani N, Spanou A, Saliaria M, Synnefa A, Vassilakopoulou K, Papadopoulou K, Pavloua K, Santamouris M, Papaioannou M, Lagoudaki A (2011) Improving the microclimate in urban areas: a case study in the centre of Athens. Build Serv Eng Res Technol 32(1):53–71

Gaitani N, Santamouris M, Cartalis C, Pappas I, Xyrafi F, Mastrapostoli E, Karahaliou P, Efthymiou C (2014) Microclimatic analysis as a prerequisite for sustainable urbanisation: application for an urban regeneration project for a medium size city in the greater urban agglomeration of Athens, Greece. Sustain Cities Soc 13:230–236

Gartland L (2008) Heat islands: understanding and mitigating heat in urban areas. Earthscan, Sterling, Virginia

Georgi JN, Dimitriou D (2010) The contribution of urban green spaces to the improvement of environment in cities: Case study of Chania, Greece. Build Environ 45(6):1401–1414

Gilbert OL (1991) The ecology of urban habitats. Chapman & Hall, London

GTCI Greek Technical Chartered Institution 20701-1 (2010). Analytical national standards of parameters of estimating building energy performance energy class certificate production

Gulyas A, Unger J, Matzarakis A (2006) Assessment of the microclimatic and human comfort conditions in a complex urban environment: modelling and measurements. Build Environ 41:1713–1722

Incropera De Witt (1990) Fundamentals of heat and mass transfer. Willey, London

Kolokotsa DD, Santamouris M, Akbari H (2013) Advances in the development of cool materials for the built environment, 385 p

Livada I, Santamouris M, Niachou K, Papanikolaou N, Mihalakakou G (2002) Determination of places in the great Athens area where the heat island effect is observed. Theoret Appl Climatol 71:219–230

Mastrapostoli E, Karlessi T, Pantazaras A, Kolokotsa D, Gobakis K, Santamouris M (2014) On the cooling potential of cool roofs in cold climates: use of cool fluorocarbon coatings to enhance the optical properties and the energy performance of industrial buildings. Energy Build 69:417–425

Matzarakis et al (2006) Modelling the thermal bioclimate in urban areas with the RayMan Model. PLEA II:449–453

Oke TP (1973) City size and the urban heat island. Atmos Environ Oxf 14:769–779

PHOENICS software tool: http://www.cham.co.uk/

Pisello AL, Santamouris M, Cotana F (2013) Active cool roof effect: impact of cool roofs on cooling system efficiency. Adv Build Energy Res 7(2):209–221

Santamouris M (ed) (2006) Environmental design of urban buildings. Earthscan, London

Santamouris M (2013) Using cool pavements as a mitigation strategy to fight urban heat island—a review of the actual developments. Renew Sustain Energy Rev 26:224–240

Santamouris M (2014a) Cooling the cities—a review of reflective and green roof mitigation technologies to fight heat island and improve comfort in urban environments. Sol Energy 103:682–703

Santamouris M (2014b) On the energy impact of urban heat island and global warming on buildings. Energy Build 82:100–113

Santamouris M, Synnefa A, Karlessi T (2011) Using advanced cool materials in the urban built environment to mitigate heat islands and improve thermal comfort conditions. Sol Energy 85(12):3085–3102

Santamouris M, Gaitani N, Spanou A, Saliari M, Giannopoulou K, Vasilakopoulou K, Kardomateas T (2012a) Using cool paving materials to improve microclimate of urban areas Design realization and results of the flisvos project. Build Environ 53:128–136

Santamouris M, Xirafi F, Gaitani N, Spanou A, Saliari M, Vassilakopoulou K (2012b) Improving the microclimate in a dense urban area using experimental and theoretical techniques—the case of Marousi, Athens. Int J Vent 11(1):1–16

Skoulika F, Santamouris M, Kolokotsa D, Boemi N (2014) On the thermal characteristics and the mitigation potential of a medium size urban park in Athens, Greece. Landscape Urban Plann 123:73–86

Stavrakakis GM, Tzanaki E, Genetzaki VI, Anagnostakis G, Galetakis G, Grigorakis E (2011) A computational methodology for effective bioclimatic-design applications in the urban environment. Sustain Cities Soc 1:54–62

Stone B Jr (2005) Urban heat and air pollution: an emerging role for planners in the climate change debate. J Am Plann Assoc 71(1):13–25

Tang L, Nikolopoulou M, Zhang N (2014) Bioclimatic design of historic villages in central-western regions of China. Energy Build 70:271–278

TAS-EDSL http://www.edsl.net

Tsilini V, Papantoniou S, Kolokotsa DD, Maria EA (2014) Urban gardens as a solution to energy poverty and urban heat island. Sustain Cities Soc 14:323–333

Index

S.-N. Boemi et al. (eds.), *Energy Performance of Buildings*,
DOI 10.1007/978-3-319-20831-2

MIX
Papier aus verantwortungsvollen Quellen
Paper from responsible sources
FSC® C105338

If you have any concerns about our products,
you can contact us on
ProductSafety@springernature.com

In case Publisher is established outside the EU,
the EU authorized representative is:
Springer Nature Customer Service Center GmbH
Europaplatz 3, 69115 Heidelberg, Germany

Printed by Libri Plureos GmbH
in Hamburg, Germany